Victorian Popularizers of Science

Victorian Popularizers of Science

Designing Nature for New Audiences

BERNARD LIGHTMAN

The University of Chicago Press

CHICAGO AND LONDON

BERNARD LIGHTMAN is professor of humanities at York
University, Toronto. He is author of *The Origins of Agnosticism*,
editor of *Victorian Science in Context* and the journal *Isis*, and
coeditor of *Figuring It Out* and *Science in the Marketplace*.

The University of Chicago Press, Chicago 60637
The University of Chicago Press, Ltd., London
© 2007 by The University of Chicago
All rights reserved. Published 2007
Printed in the United States of America

16 15 14 13 12 11 10 09 08 07   1 2 3 4 5

ISBN-13: 978-0-226-48118-0 (cloth)
ISBN-10: 0-226-48118-2 (cloth)

Library of Congress Cataloging-in-Publication Data

Lightman, Bernard V., 1950–
Victorian popularizers of science : designing nature for new audiences /
Bernard Lightman.
p. cm.
Includes bibliographical references and index.
ISBN-13: 978-0-226-48118-0 (cloth : alk. paper)
ISBN-10: 0-22648118-2 (cloth : alk. paper)   1. Science—Great
Britain—History—19th Century.   2. Technical writing—Great
Britain—History—19th Century.   3. Great Britain—Social conditions—
19th century.   I. Title.
Q127.G4L54   2007
509.41'09034—dc22
2007015179

∞ The paper used in this publication meets the
minimum requirements of the American National
Standard for Information Sciences—Permanence of Paper
for Printed Library Materials, ANSI Z39.48-1992.

Contents

Preface

IN 1875 an article appeared in the *Saturday Review* harshly condemning those "scatter-brained auditors" who frequented the Royal Institution to hear lectures about science. The anonymous author divided such audiences into two types. One group dabbled in science "out of shifting caprice, or in deference to the dictates of fashion." Another group attended lectures "with the best and steadiest intentions," but they were "incapacitated by lack of general education from grasping any special subject." The author's low opinion of the audience for science lectures was matched by an equally dismal evaluation of what they heard. "It is to accommodate such feeble votaries," the *Saturday Review* critic declared, "that 'popular science' has been invented." Although the author acknowledged that there were benefits to "real science from being fashionable and popular," nevertheless its hard facts could not be expected to "retain the favour of the multitude." "Real science" therefore used a "counterfeit" to "stand for it upon public platforms" in order to secure the "patronage of the vulgar." Popular lecturers tended to "garnish the information" they conveyed to their audience with "rhetorical flourishes" and to stress peculiar, sensational, and strange phenomena rather than those that were intrinsically important. The popular lecturer had a "strong propensity for paradox" and sought to "surprise and astonish his auditors." In order to illustrate the most abstruse points of the lecture, the speaker referred to "familiar objects and circumstances" that seemed to illuminate the subject but in the end merely muddled it.[1]

1. "Sensational Science" 1875, 321

In the opinion of this author, it was really a "hopeless task" to attempt to make "sound disquisitions on any scientific subject thoroughly intelligible to those who have not undergone special training, unless they have received a thorough general education." Conveying scientific information to an uninformed reading public was doomed to failure. It created two unwanted creatures, the dilettante and the incompetent lecturer. As to the first, the *Saturday Review* critic denied that "sensational science" advanced the culture of the age. A literary dilettante was tolerable, but a "*dilettante* in science proud to make small-talk out of Huxley's or Tyndall's lectures, inflated with fallacies of his or her own extraction . . . is a social pest." Worse still had been the creation of the incompetent lecturer who merely imitated the dramatic lecturing styles of Thomas Henry Huxley or John Tyndall without possessing their expertise. The existence of incompetent lecturers was a "collateral evil entailed by the really competent teachers catering for the frivolous and illiterate." Under the present system "anybody of sufficient social standing to draw an audience considers himself entitled to hold forth at penny readings, Mechanics' Institutes, and so forth, on scientific subjects perfunctorily crammed up for the occasion." There was only one "right way" to instill "a healthy enthusiasm for science into all classes of society," the critic insisted. The representatives of science had to "exhibit themselves in their true character, and to abandon all the meretricious allurements of thaumaturgy and 'sensation.'" They had to make the public understand that the "genuine pursuit of scientific truth is work and not play."[2] If the critic from the *Saturday Review* had had his way, only the men of science would be permitted to communicate knowledge to general audiences, and they would make no concessions to the public's rudimentary level of understanding. But who, exactly, were the purveyors of "sensational science"? The critic never mentions any by name. Was there an army of them or only a handful? Were they really as pernicious to the cause of scientific truth as this journalist claimed?

This book examines those popularizers who offered "sensational science" to the British public in the second half of the nineteenth century. The emphasis is on those who were not practitioners of science. Many of them were professional writers and journalists. The overwhelming majority of them were members of the educated middle class. There were a significant number of them. I deal with over thirty of these figures. I could easily have included

2. Ibid., 322

many more, but I have chosen to limit myself to the most prolific, the most influential, and the most interesting among them. They assumed the role of interpreters of science for the growing mass reading audience in this period. I do not attempt to analyze this audience in great detail. I am more concerned with how the popularizers conceived of their audience and how this conception affected the way they wrote and lectured. They saw themselves as providing both entertainment and instruction to their readers. Cognizant that they were operating in a market environment, they considered the use of "thaumaturgy" as a necessity for those who wished to be a commercial success. Some were extremely successful, producing best sellers that were as widely read as the *Origin of Species* or the other key scientific texts of the day. For many of these popularizers, nature was full of meaning, charged with religious significance. They looked back to the natural theology tradition and in their writings offered new audiences a vivid glimpse of the design they perceived in nature. Since they were influential, and since their interpretation of the larger meaning of scientific ideas was often at odds with the agenda of elite scientists, they cannot be ignored. Any attempt to investigate how the British understood science in the second half of the century must take them into account. By focusing on Britain I can examine the development of science writing and lecturing for a broad audience in a country that was among the first to experience a communications revolution.

This book is divided into eight chapters. I begin with a chapter that sets the scene. Here I explore the transformation of the publishing scene, and how it intersected with the changing world of science. I consider why the sciences were thought to be especially important and exciting in the second half of the century—why this period is sometimes referred to as the age of the worship of science. I discuss the traditions of science popularization that were important from the late eighteenth through the first half of the century. The varied approaches adopted by scholars to study popularizers of Victorian science are also analyzed.

In the next two chapters I examine two distinct groups of popularizers active in the second half of the century. Chapter 2 centers on the large number of Anglican parsons who wrote about science for the general reader, including Ebenezer Brewer, Charles Alexander Johns, Charles Kingsley, Thomas William Webb, Francis Orpen Morris, George Henslow, and William Houghton. This was one of the groups that would-be professionalizers like Thomas Henry Huxley were trying to push out of science. Through their work the Church of England maintained an active presence in the

British scientific world, and they kept the relevance of religious themes to contemporary science before the minds of the public, even after the appearance of Charles Darwin's *Origin of Species*. Chapter 3 turns to the role of women as scientific authors. Another group that Huxley and his allies wanted to exclude, women were faced with the daunting task of presenting themselves as authoritative authors while lacking the status of the practitioner or the clergyman. I discuss their appeal to an older tradition of women's writing in science and how they refashioned it for a new age. I also investigate why each of them decided to become popularizers, their working relationship with publishers, how they defined their audience, their literary experiments, the centrality of aesthetic, moral, and religious themes in their books, and their seemingly deferential attitude toward practitioners.

The next two chapters deal with how popularizers used visual and literary techniques to attract their audiences. Chapter 4 focuses on two very important popularizers active from the middle of the century until the 1880s, John George Wood and John Henry Pepper. Both understood that they needed to be entertaining as well as instructive. Living in a time when Victorian audiences had been dazzled by exhibitions such as the Crystal Palace, gigantic panoramas, and spectacular theater performances, both appealed to the eyes of their audience in their lectures and writings. Chapter 5 treats the literary parallel to the spectacle: the evolutionary epic. It centers on David Page, Arabella Buckley, Edward Clodd, and Grant Allen, authors who adapted the evolutionary epic as a vehicle for communicating contemporary scientific ideas to a general reading audience. Based on the template provided by Robert Chambers in his *Vestiges,* combined with the theories of Darwin and Herbert Spencer, these popularizers contributed to the formulation of a literary format that is still used today.

The sixth chapter is devoted to discussing the important role of periodicals for popularizers of science, using Richard Proctor's editorship of the journal *Knowledge* (founded in 1881) as the key example. Proctor's journal was designed to compete with *Nature* for control of the science periodical market for the popular audience and to question the role and dominance of would-be professional scientists. Proctor's format for *Knowledge* drew on a "republican," rather than an "elite," image of the scientific community. The seventh chapter treats T. H. Huxley and Robert Ball as practitioners who became active popularizers of science only after the early 1870s. A discussion of Huxley and Ball provides us with an opportunity to compare their writings for a general audience with those who were not practitioners of science. It also demonstrates that Huxley and Ball were indebted to the popularizers who preceded them. Often viewed as the foremost "popular-izer of science" of his time, Huxley's initial attitude toward writing for a popular audience was quite ambivalent. This chapter analyses the shift in his views during the late 1860s and concentrates on the three projects he became involved in during the 1870s. Ball, who was far less aggressive in pushing the naturalist agenda than Huxley, was more successful as a popu-larizer. He was willing to give more time and energy than even Huxley to his activities as scientific author and lecturer. As a result, he made a small fortune. In the final chapter I turn to important popularizers of science in the final decades of the century, including Agnes Giberne, Eliza Brightwen, Henry Hutchinson, Alice Bodington, and Agnes Clerke. The first three are representatives of earlier traditions of popularization, while the last two presented a new justification for science writing in an age of specialization.

Examining the evolution and impact of popularizers of science in the sec-ond half of the century leads to a number of rich areas of research. It takes us to the world of Victorian authors, publishers, and audiences. It brings us to the fascinating territory of Victorian visual culture and the display of nature in museums and exhibitions. It transports us to the literary dimensions of science. We find ourselves exploring the terrain first discovered by scholars interested in recuperating the role of women in science. We also have to cover the ground traversed by those who have researched the relationship between science and religion. It raises interesting questions about how we map the process of professionalization and the contest for cultural author-ity. In sum, the issue of popularization stands at the intersection of many of the most intriguing spaces of Victorian science. A study of the important popularizers of science shows how they transformed many of these spaces while subverting the agenda of the would-be scientific professional.

Acknowledgments

THIS BOOK has taken over fifteen years to reach fruition. Parts of it have already appeared in some of my previously published journal articles and book chapters. I am grateful to the publishers who have allowed me to reproduce some of this material in this book. Components of the section on T. W. Webb in chapter 1 were drawn from *The Stargazer of Hardwicke: The Life and Work of Thomas William Webb*, edited by Janet and Mark Robinson (Gracewing, 2006). Segments of the discussion on J. H. Pepper in chapter 4 are taken from *Science in the Marketplace*, edited by Bernard Lightman and Aileen Fyfe (University of Chicago Press, 2007). Several paragraphs from my article "The Visual Theology of Victorian Popularizers of Science," *Isis* 91 (2000), appear in chapters 4, 6, and 8. Portions of chapter 6 were published in *Culture and Science in the Nineteenth-Century Media*, edited by Louise Henson, Geoffrey Cantor, and others (Ashgate, 2004). Finally, bits and pieces of chapter 8 come from "Constructing Victorian Heavens: Agnes Clerke and the 'New Astronomy,'" in *Natural Eloquence: Women Reinscribe Science*, edited by Ann Shteir and Barbara Gates (University of Wisconsin Press, 1997).

Many archives and libraries have provided me with access to material in their possession and have allowed me to quote from it. Quotations from archival materials are reproduced by courtesy of the British Library, the Cambridge County Record Office, the Syndics of the Cambridge University Library (for permission to quote from the Darwin correspondence), the Particle Physics and Astronomy Research Council and Syndics of the Cambridge University Library (the Royal Greenwich Observatory Archives),

xiii

the British Library and the Royal Literary Fund (the Royal Literary Fund archives), the Dittrick Medical History Center at Case Western Reserve University, the John G. Wolbach Library and Information Resource Center at the Harvard-Smithsonian Center for Astrophysics located in Cambridge (Massachusetts), the College Archives at the Imperial College London (Huxley correspondence), the Trustees of the National Library of Scotland, the Lambeth Palace Library, the Leeds University Library (Brotherton Collection, Clodd Correspondence), the Mary Lea Shane Archives of the Lick Observatory at the University of California, Santa Cruz, the Trustees of the Natural History Museum (London), the Pennsylvania State University Libraries (Mortlake Collection, Rare Books and Manuscripts, Special Collections Library), the Archives of the Reading University Library, the Royal Astronomical Society, the Royal Institution of Great Britain, the Director of Culture of the Sheffield City Council and the Sheffield Archives, the Master and Fellows of St John's College (Cambridge), the University College of London Special Collections, the University Librarian and Director of the Johns Rylands University Library at the University of Manchester, the University of Exeter, the University of London Library Special Collections, the Robinson Library Special Collections of the University of Newcastle upon Tyne and by permission of the Special Collections Librarian of the Robinson Library at the University of Newcastle upon Tyne, the Trustees of the Trevelyan Family Papers of the Robinson Library at the University of Newcastle upon Tyne, the University of Wisconsin–Madison Archives, the Governors of Dunford House and with acknowledgement to the West Sussex Record Office and the County Archivist, and the Special Collections Department of the Vancouver Public Library.

While I worked on this project I was fortunate to receive the support of granting agencies, York University staff, research assistants, colleagues, and my family. I received several grants from the Social Sciences and Humanities Research Council of Canada. A number of York University Faculty of Arts Research Grants provided funding for research trips to England while the Faculty of Arts Fellowship awarded to me for the academic year 2001–2 gave me the time to do a significant portion of the research. The staff working at the Scott Library's interlibrary loan office, especially Gladys Fung, cheerfully obtained countless books from near and afar. Tafila Gordon-Smith, from the McMaster University Library's interlending and document supply office, facilitated my use of numerous reels of microfilm from their British Publishers' Archives collection. Countless archivists have helped me locate elusive letters and documents over the years. I am especially indebted to

Anne Barrett of the Imperial College, Michael Bott of the University of Reading, Peter Salt of the Jarrold and Sons Ltd., and Brenda Weeden of the University of Westminster. I was blessed with a series of hardworking research assistants who tracked down obscure sources for me, including Erin McLaughlin-Jenkins, Liza Piper, Katrina Sark, Sharrona Pearl, Aileen Fyfe, Wesley Ferris, Andrea Koritko, Jessica Poole, and Kady Shear. Steve Bunn took on the onerous task of organizing the illustrations and permissions.

Colleagues in the field have not only stimulated me, they have also located important information for me. I learned a great deal from discussions with James Elwick, Gowan Dawson, and Paul Brinkman. Martin Fichman helped me work out the relationship between Arabella Buckley and Alfred Russel Wallace. Alan Rauch, Michael Collie, and Leslie Howsam provided indispensable help in determining print runs for a number of key books. I learned a lot about Robert Ball in the course of long e-mail exchanges with Mark Butterworth. I am grateful to Richard Bellon, Michael Collie, Adrian Desmond, Richard England, Paul Fayter, Iris Frye, Jim Paradis, Anne Secord, Ann Shteir, Jennifer Tucker, and Paul White for looking over drafts of chapters and sharing their stimulating responses with me. Jim Secord and Aileen Fyfe generously agreed to read over the entire manuscript and gave me excellent advice that led me to make significant changes to the structure of the book. I am also indebted to the two anonymous referees who evaluated the manuscript on behalf of the University of Chicago Press. Due to the long gestation of this book I have worked with four different editors at the University of Chicago Press. The late Susan Abrams provided inspiration at the early stages of the project; Catherine Rice handled the evaluation of the manuscript and offered good advice as I revised the work in response to referees' reports; and Alan Thomas and Karen Darling carefully supervised the last stage of the publication process. I have discussed some aspect of my research on popularizers of science with nearly all of my colleagues in the Science and Technology Studies program at York University. I feel privileged to be able to work alongside such a congenial and dynamic group of scholars. Early in my career I was guided by three generous mentors: Sydney Eisen, William Johnson, and Frank Turner. Without their encouragement I would never have endured an eight-year search for a permanent position in academe. As usual, my family provided me with the stability and love that I require in order to do any scholarly work at all. My father and my in-laws cheered me on from the sidelines. I have been inspired by the stellar academic achievements of my son, Matthew, with whom I share a deep interest in science. I have marveled at the growing artistic talents of my daughter, Ilana, and treasure the times we have spent together listening to, and playing, rock, folk, and blues music. Finally, Merle, my wife of over thirty years, and the love of my life, keeps me feeling young, and gives me the strength to take on new scholarly challenges.

Victorian Popularizers of Science

——— ✳ ———

CHAPTER ONE

Historians, Popularizers, and the Victorian Scene

SIGNS of a remarkable change in attitude toward science were evident to many mid-nineteenth-century British observers. They had only to contemplate the popularity of the Great Exhibition of the Works of Industry of All Nations, held in London in 1851. The Exhibition was housed in an immense glass and iron building of unique architectural design, earning it the nickname "Crystal Palace." Never before had an industrial exhibition drawn such huge crowds. The aspiring young biologist Thomas Henry (T. H.) Huxley wrote to his future wife in 1851 that visitors to the Crystal Palace approached it with awe and reverence, as if they were on a sacred pilgrimage to a holy shrine. "The great Temple of England at present," Huxley told her, "is the Crystal Palace—58,000 people worship there every day. They come up to it as the Jews came to Jerusalem at the time of the Jubilee."[1] Contemporary observers could also consider the natural history crazes of the 1850s as another indicator of a growing interest in science. Just two years after the Exhibition opened, the naturalist and popularizer Philip Henry Gosse predicted in his *Naturalist's Rambles on the Devonshire Coast* (1853) that the marine aquarium would soon be found in many Victorian parlors. Almost overnight the aquarium became a national craze, and members of the British middle class traveled to the coast to comb the beaches for specimens. At the same time, fern collecting became a widespread fad.[2]

1. Huxley to Heathorn, September 23, 1851, letter no. 165–66, Imperial College, Huxley Collection, Huxley/Heathorn Correspondence.

2. D. Allen 1976, 136–37.

The Victorian fascination with aquaria and ferns was followed by an intense curiosity in dinosaurs. When the Crystal Palace Company relocated the Great Exhibition as a permanent site in Sydenham, just south of London, a new exhibition of extinct reptiles and mammals became one of the most popular attractions. The public was treated to the first life-sized restorations of the *Ichthyosaurus*, the *Plesiosaurus*, pterodactyls, the *Megalosaurus*, and the *Iguanodon*, built by the natural history illustrator Benjamin Waterhouse Hawkins in collaboration with the anatomist Richard Owen. Opening on June 10, 1854, over a million people a year for the next fifty years saw these full-scale models. Dinosaurs became part of the popular imagination. They appeared regularly in the pages of *Punch* magazine and fought in Jules Verne's *Journey to the Centre of the Earth* (1864).[3] In addition to the aquaria, fern, and dinosaur crazes, in the early 1860s the British public became interested in gorillas, in part due to curiosity about the relationship between humans and apes in light of Darwin's theory of evolution and because of the activities of the explorer Paul du Chaillu. In 1861 du Chaillu toured England with his collection of decapitated ape heads. His extremely popular book, *Explorations and Adventures in Equatorial Africa*, appeared in the same year. It contained lurid tales of du Chaillu's experiences hunting aggressive gorillas. Patronized by Richard Owen, du Chaillu's gorillas became entangled in the celebrated debate between Owen and Huxley on the anatomy of human and ape brains. The public's curiosity about apes and evolution was piqued when the controversy was lampooned in *Punch* and immortalized in Kingsley's *Water Babies* (1862).[4] Why was there such a sudden upsurge of interest in all things scientific? As historian David Allen has pointed out, one reason was the development of a public for science largely composed of a new generation of middle-class consumers. In contrast to the 1840s, a period of economic depression in Britain, the following decade was one of prosperity. During the 1850s and 60s Britain was the leading commercial, industrial, and imperial power in the world. More people had more money to spend on leisure activities. Science became fashionable and respectable within a broader spectrum of the populace, not just within the circles of the well to do.[5]

3. J. Secord 2004e, 138; Cadbury 2001, 289–90, 293, 298–99. British interest in dinosaurs was occasionally bolstered during the rest of the century by discoveries of fossils in the United States starting in the late 1870s.

4. Hodgson 1999, 231; Raby 1997, 178–95; Rupke 1994, 314–22; Dawson 2007, 26–81.

5. D. Allen 1976, 137.

Science continued to captivate the Victorians right up to the end of the century. They came into contact with science through witnessing the spread of dazzling new technologies, through encountering exotic animals and plants, and through experiencing heated controversies about the validity of novel theories. With its promise of continued progress through technological advancement, science sparked utopian visions of the future and seemed to offer the means by which British imperial aims could be accomplished. Cable telegraphy, for example, was one of the exciting, new technologies based on the electrical research of scientists such as William Thomson. Introduced in 1851 when the first successful undersea cable was laid across the English Channel, by the 1870s cables spanned the globe and transformed the nature of communications. From the beginning the worldwide cable industry was dominated by British capital and by British engineering expertise. Through its global cable system Britain could exercise direct control over its vast empire.[6] Britain was a global center for the trade of specimens as well. Flora and fauna from across the empire were brought into London, some to be examined by British scientists, others to be displayed to the public in the London Zoo, the British Museum, or Kew Gardens. The Victorians were also excited by new scientific discoveries, such as Darwin's theory of natural selection, and became engrossed in debates over their validity. These controversies involved scientists like Huxley, John Tyndall, and Thomson, who were larger than life public figures. New scientific instruments seemed to open up hitherto unexplored worlds of nature. William Huggins's application of the spectroscope to astronomy, for example, which made it possible to ascertain the composition of distant heavenly bodies, in combination with more sophisticated photographic techniques, profoundly modified astronomy. As one contemporary observed, the adoption of the spectroscope and the camera had widened the borders of astronomy, "inviting it to undertake novel tasks, endowing it with previously undreamt-of powers. Realms of knowledge deemed inaccessible to human faculties have, as if at the touch of a magician's wand, been thrown open."[7] Scientific knowledge seemed to offer the magical password—the "open sesame"—that unlocked the doors to exhilarating new worlds in the second half of the century.

But the fascination with science operated at an even deeper level. For some, it provided the basis for making sense of themselves and their place

6. Hunt 1997, 316.
7. Clerke 1898a, 33.

in the universe, either in conjunction with revised Christian notions or completely on its own terms. At the beginning of the nineteenth century the dominant, though by no means universally accepted, worldview was legitimized by Christian modes of thought. It was tied to the old order and reflected its pastoral, agrarian, and aristocratic nature. By the end of the century British society had been profoundly transformed, and the worldview of the old order no longer seemed relevant to many members of the intellectual elite, the middle classes, and the working classes. To those who were dissatisfied with traditional beliefs, scientific modes of thought seemed to offer the glue holding together a new worldview far more relevant for living in an urban, industrialized, and middle-class society.[8]

In his "On the Advisableness of Improving Natural Knowledge" (1866), T. H. Huxley argued that the progress of scientific knowledge had resulted in the invention of great ships, railways, telegraphs, factories, and printing presses, "without which the whole fabric of modern English society would collapse into a mass of stagnant and starving pauperism." But Huxley pointed out that science was much more than "a sort of comfort-grinding machine." Only the "ideas" of science, he maintained, could "still spiritual cravings." Huxley believed that scientific progress had laid "the foundations of a new morality."[9] In his review of the progress of science from 1836 to 1886, the popularizer Grant Allen asserted that as a result of the penetration of evolutionism into "all the studies which bear upon the divisions of human life," the "entire view of man and his nature has been reversed." Allen, like Huxley, believed that scientific developments dictated the adoption of a more secular perspective. Science, Allen declared, offered a coherent worldview through its synthesis of "all our concepts of the whole consistent system of nature, and [it] sets before our eyes the comprehensive and glorious idea of cosmos which is one and the same throughout, in sun and star and world and atom, in light and heat and life and mechanism, in herb and tree and man and animal, in body, soul, and spirit, mind and matter."[10] The sciences therefore assumed tremendous significance in the second half of the nineteenth century as every theory, and every new discovery, seemed to contain huge implications for all facets of human life. Interpreting, and arguing over, the social, political, and religious meaning of scientific ideas became the focus of intellectual activity.

8. R. Young 1985, 240.   10. G. Allen 1887, 875, 884.
9. T. Huxley 1895a, 6, 10–11.

The vital importance of science did not end there for many Victorians. In addition to supplying the glue for a new worldview, it was also touted by its defenders as providing the best method for determining all truth. In her autobiography, Beatrice Webb, a disciple of Herbert Spencer who later married the leading Fabian socialist, recalled the "cult of science" that had inspired many mid-Victorians. One of its key tenets was "the current belief in the scientific method . . . by means of which alone all mundane problems were to be solved."[11] In order to be deemed intellectually legitimate, ideas and theories had to be determined through scientific method. Increasingly in the second half of the century, truths based on an appeal to sacred texts, to religious authorities, to an inner conscience (placed in humans by God), or to intuition of any kind were no longer seen as possessing intellectual credibility. Therefore those who could claim to speak on behalf of science gained immense cultural authority and intellectual prestige. They could assert that they spoke truthfully, and they could argue that they understood the broader significance of scientific ideas. Since the modern worldview was held together by scientific ideas, they essentially maintained that they could pronounce with authority on all issues. They could present themselves as the genuine arbiters of truth to the British public. But a modern, professionalized body of scientists was still in the making in the second half of the nineteenth century, which left a number of key questions to be answered. What, exactly, was proper scientific method? For that matter, what was science? Which groups could participate in the debates on these questions? The stakes were therefore quite high in the fight to be recognized as an intellectual who spoke on behalf of science.

HISTORIANS AND THE TOPOGRAPHY
OF BRITISH SCIENCE

How have historians dealt with this complex period in the development of science? Mapping out the topography of nineteenth-century British science has proven to be a challenging task. The map has already been redrafted several times over the course of the last forty-five years and remains a work in progress. Although the "externalism/internalism" debate from the 1960s to the 1980s led to a rejection of approaches drawn solely from intellectual history, even the more contextualist works that began to appear up to the end of the 1980s tended to focus on the transformation of the scientific elite

11. B. Webb [1950], 112.

as Britain evolved into a modern, industrialized nation.[12] In the map drawn by such influential scholars as Frank Turner and Robert Young, the gentlemen of science, predominantly Oxbridge-educated Anglicans, controlled scientific sites during the first half of the century, and provided British society with a vision of culture and social order based on natural theology. In the second half of the century, Young and Turner argued, the towering peaks of scientific naturalism dominated the topography of British science.[13] Middle-class Young Turks of science like Thomas Henry Huxley and John Tyndall, who came from outside the Oxbridge environment, began at the middle of the century to vie with the gentlemen of science for the leadership of the British scientific world. At the same time, they engaged in a debate with the Anglican clergy over who would provide the best leadership for modern British society.

Referred to as "scientific naturalists" or "evolutionary naturalists" by scholars, these men put forward new interpretations of nature, society, and humanity derived from the theories, methods, and categories of empirical science. Scientific naturalists were naturalistic in the sense that they ruled out recourse to causes not present in empirically observed nature, and they were scientific in that they interpreted nature in accordance with three major midcentury scientific theories, the atomic theory of matter, the conservation of energy, and evolution.[14] This group of elite scientists gained coherence when some of its most active members formed the X Club. Starting in 1864, George Busk, Edward Frankland, Thomas Hirst, Joseph Dalton Hooker, Thomas Huxley, John Lubbock, Herbert Spencer, William Spottiswoode, and John Tyndall met for dinner on a monthly basis to strategize on how to accomplish their objectives.[15] The activities and ideas of the evolutionary

---

12. Lightman 1997c.

13. R. Young 1985; Turner 1993. Although published in 1993, Turner's book contained essays that first appeared much earlier. At first glance, Turner and Young seemed to be offering quite different accounts of the development of nineteenth-century British science. While Young emphasized the continuity between the Anglican natural theology of the first half of the century and the scientific naturalism that succeeded it, Turner stressed the conflict accompanying the transformation of the scientific elite. But both see the dominance of scientific naturalism as the hallmark of the second half of the century (Lightman 1997c, 5-7).

14. Turner 1974, 9-35; 1993. Besides Huxley and Tyndall, Turner includes a long list of Victorian scientists and intellectuals under the rubric of "scientific naturalists." On the list are Herbert Spencer, W. K. Clifford, Francis Galton, Frederic Harrison, John Morley, G. H. Lewes, Edward Tylor, John Lubbock, E. Ray Lankester, Henry Maudsley, Leslie Stephen, Grant Allen, and Edward Clodd (Turner 1974, 9).

15. Barton 1990, 1998a; Jensen 1970; MacLeod 1970.

naturalists, including Charles Darwin, have provided historians with a focus for their map of the scientific landscape. Historians therefore concentrated on themes such as the Darwinian revolution, the secularization of science, and the professionalization of science. Even those scholars who wanted to explore new territory, whether it was in the area of science and literature or science and gender, tended to concentrate on the scientific elite.[16]

The map of nineteenth-century British science that prominently featured scientific naturalism began to be redrafted at the end of the 1980s as a result of studies of groups who were not a part of the scientific elite. Adrian Desmond's *Politics of Evolution* (1989), a history of science "from below," was among the first to look beyond the establishment science offered by both the scientific gentry and Oxbridge clergy and the middle-class scientific naturalists who challenged their authority. Exploring the world of radical, lower-class evolutionists that existed in the secular anatomy schools and Nonconformist colleges of London in the 1830s, Desmond demonstrated that a thriving scientific culture existed outside, and in opposition to, the elite establishment. Since Desmond's groundbreaking book, the field of the history of nineteenth-century British science has gradually shifted its focus as scholars painstakingly drew a revised map that incorporated the presence of new groups while scaling down the exaggerated features of scientific naturalism.

As a result of Crosbie Smith's *Science of Energy* (1998), the dominance of scientific naturalism within the scientific elite has been reassessed. Effective opposition to scientific naturalism came from a group of scientists who from the 1850s to the 1870s constructed the science of energy. Like the scientific naturalists, they too had a reform program for the whole range of physical and even life sciences. Bearing the impress of Scottish presbyterianism, representing whig and progressive values, and linked to the industrialists of northern Britain, energy physics was founded by a "North British" group composed of Glasgow professor of natural philosophy William Thomson, Scottish natural philosophers James Clerk Maxwell and Peter Guthrie Tait, and the engineers Fleeming Jenkin and Macquorn Rankine. These men found the perceived anti-Christian materialism of the metropolitan scientific naturalists quite distasteful, and they were prepared to enter into an alliance with Cambridge Anglicans to undermine the authority of Huxley and his

---

16. Gillian Beer, for example, in her *Darwin's Plots* (1983), stuck, for the most part, with the shared discourse between Darwinian scientists and novelists, while in her *Sexual Science* (1989), Cynthia Russett was interested mainly in the gendered theories of the men of science.

allies. They promoted a natural philosophy in harmony with, though not subservient to, Christian belief.[17]

Questions about the dominance of scientific naturalism within the intellectual, and not just the scientific, elite have also been raised. T. H. Green, F. H. Bradley, Edward Caird, Henry Jones, John Watson, William Wallace, J. S. Mackenzie, David George Ritchie, and Bernard Bosanquet formed an idealist school that dominated British philosophy in the last three decades of the nineteenth century. Drawing on evolutionary theory to construct a distinctive social philosophy, British idealists nevertheless embedded it in a quasi-Christian metaphysical system.[18] Scholars have also reminded us not to underestimate the power of members of the intellectual elite with strong ties to the old Anglican-aristocratic establishment. Their influence did not disappear just because Huxley and his allies gained control over many of the important institutions of science. Even in the last decade of the century, Arthur Balfour, a Tory aristocrat, one of the most important political figures of the day, and a future prime minister, could write a book, *The Foundations of Belief* (1895), in which he denied that Huxley and his fellow scientific naturalists spoke with the authority of science behind them.[19]

Our map of nineteenth-century British science has also been reconfigured to allow groups from outside the intellectual elite to emerge from under the shadows cast by the mountainous elevations formerly occupied by scientific naturalism. Instead of directing our attention to the gendered theories of male practitioners, scholars have demonstrated how women actively participated in scientific activities despite the efforts of Huxley and others to exclude them from the scientific community.[20] Desmond's contention that members of the British working class could embrace a scientific culture in marked contrast to the one offered to them by the scientific elite has been extended to the second half of the century. Working-class intellectuals and their readers reinterpreted Darwinism to reflect their own social aspirations.[21] We have recently been reminded that despite the sustained campaign launched by scientific naturalists to discredit spiritualism and

17. C. Smith 1998.

18. Otter 1996.

19. Lightman 1997b. Both Opitz and Schaffer have pointed to the continuing importance of the aristocracy within the intellectual and scientific elite (Opitz 2004a; Schaffer 1998c).

20. Gates 1998; P. Gould 1997, 1998; Le-May Sheffield 2001; Ogilvie 2000; E. Richards 1997; Richmond 1997; Shteir 1996.

21. McLaughlin-Jenkins 2001a, 2001b, 2005; Paylor 2004.

associated movements of thought, occultism was attractive to members of the educated middle class at the end of the century.[22] In sum, the power of scientific naturalism, both inside and outside intellectual circles, seems to have been overestimated by the older scholarship. Scientific naturalists and the Anglican clergy were not the only players in the contest for cultural authority. North British physicists, neo-Hegelians, socialists, secularists, women, spiritualists, and occultists drew on the credibility of scientific ideas to join the contest. All of them are now visible alongside the scientific naturalists on our map of the terrain. My purpose in this book is to complicate the historian's mapmaking task even further by finding a place for popularizers in the topography of Victorian science.

WHAT'S IN A NAME?

But locating popularizers on the map drawn by historians has not been an easy task. I have even had difficulty deciding on an appropriate term to label this important group of writers and lecturers. I have opted to use the term "popularizers." Using this designation is problematic. The modern meaning of the terms "popularizer of science" or "popular science" contain such negative connotations that any use of them to discuss nineteenth-century figures introduces an ahistorical distortion that seems to justify their dismissal as unimportant. The late Stephen Jay Gould, one of the most eminent popularizers of science of the twentieth century, objected to the tendency in America to equate "popular writing with pap and distortion," or with "grandstanding." He pointed out that in France this genre of writing was named *vulgarization,* which had entirely positive implications.[23] It would seem to be advisable for historians dealing with the nineteenth century to avoid the use of terms that are loaded down with so much baggage, much of it collected after 1900.

Some historians of science have experimented with the use of different terms to avoid prejudicing their examination of scientific books and lectures for the layperson. Katharine Pandora has argued that the term "vernacular" can be helpful "in providing a broader sense of the history of 'everyday scientific knowledge' than is currently signified by the terms 'popular culture' and 'popularization.'" She discusses how a vernacular sphere of discourse, more accessible than professional discourse, constitutes "a kind of 'intellectual

22. Owen 2004.
23. S. Gould 1991, 11.

commons' where social and theoretical comment can circulate without regard for scientific propriety."[24] Pandora's term "vernacular," which may be more appropriate for the early twentieth-century American context that she has analyzed, assumes the existence of two quite separate realms of discourse that did not exist in Britain in the second half of the century. James Secord has proposed the term "commercial science" in his study of Robert Chambers and his readers.[25] He rejects the use of "popular science" since the contemporary meaning of the term was established, in part, as a result of the debates he investigates in his *Victorian Sensation*. The pejorative use of the term "popular," Secord argues, "as it stabilized in the late nineteenth and twentieth centuries was designed to render readers as invisible members of a mass audience."[26] Secord's "commercial science" is not broad enough to include some of the figures in this book, who became popularizers out of a sense of religious calling, and it is too broad since it includes those who made money out of instrument making, museum curating, and showmanship. However, it has the virtue of reminding us that popularizers were involved in a form of activity involving pay.

Since the alternatives are unsatisfactory, I will retain the label "popularizer of science," although it remains a problematic term. I will use it to place the questions of authorship, authority, and audience front and center. What did it mean to be a popularizer of science in the Victorian period, and what motivated an individual to become one? Who had the authority to write about scientific subjects and to communicate their larger philosophical or metaphysical significance to a general reading audience? Did popularizers always defer to the authority of practitioners? How did popularizers address their audiences? What new literary genres did they develop in order to reach more literate and more sophisticated audiences? I will refer to lectures and books as being popular—popular in the sense of being highly successful or because they were intended for a mass audience—since the medium is popular, but I will refrain as much as possible from referring to "popular science."[27]

24. Pandora 2001, 491–92.

25. James Secord prefers the term "commercial science" because it underlines how science became a part of the commercial culture of exhibition in the early nineteenth century. He argues that the revolution in communication transformed opportunities for making money from the display of knowledge in the prose of journalism, in lectures, in panoramas, and in museums (J. Secord 2000, 437).

26. Ibid., 524–25.

27. I am indebted to Aileen Fyfe and Jim Secord for helping me think through these issues, though they may not agree with all of my conclusions.

The historian of science cannot dispense with these terms and their cognates altogether. These words and phrases were used by the historical actors under study and they will turn up in quotes from their writings and lectures. Scholars have traced the introduction and history of the terms "popular," "popularizer," "popularize," and "popularization," both in general and as applied to science in particular, in an attempt to understand what they meant at specific times and in particular places. The *Oxford English Dictionary* cites the earliest uses of a number of these terms, "popularize" in 1593 and "popularization" in 1797. These terms and their cognates were not being used in reference to science until the early nineteenth century, beginning likely in the 1830s.[28] The label "popular science" already had wide currency by the mid-nineteenth century. The term could be found in public addresses, book titles, journal headings, and lecture programs.[29] Morag Shiach argues that the meaning of the term "popular" has been constantly redefined and fought over, that it is ideologically loaded. By the mid-nineteenth century, she asserts, the term is increasingly applied to aspects of cultural forms appealing to, or favored by, people generally. But she finds that already the use of the term by some writers positioned within the dominant culture carried negative connotations of crudity, ignorance, and tyranny.[30] Raymond Williams, however, sees the negative sense of the term as being accepted somewhat later. He has asserted that "popularize" was a political term until the nineteenth century, meaning belonging to the people, but then it took on the meaning of presenting knowledge in generally accessible ways. He states that nineteenth-century uses were still "mainly favourable," and that it was not until the following century that a strong sense of "simplification" became predominant.[31]

We must be careful in how we understand each use of the phrase "popular science" by nineteenth-century figures, as during that period there were many interpretations of its meaning and purpose. For Royal Institution managers, it meant attracting large audiences and financial prosperity. For

28. The first entry where the meaning of popularize is used to refer to technical subjects (under "2c") lists a quote from J. S. Mill in 1833. The phrase "popularizer of science" first appeared in 1848 (Simpson and Weiner 1989, 126). But the *OED* cannot be regarded as a neutral record of linguistic usage. Like previous dictionaries, it is a product of a particular historical and geographical context, in this case the British context between 1880 and 1920, and it reflects dominant cultural discourses by excluding some meanings. It can, however, provide clues as to the ways in which the dominant culture constructed the "popular."

29. Hinton 1979, 6.

30. Shiach 1989, 19, 21–22, 29, 32.

31. R. Williams 1984, 237.

Tyndall, popular lectures enabled him to earn a living, provided an entrée to elite London society, offered a basis for cultural authority, and constituted a hindrance to scientific research. To the physicist Peter Guthrie Tait, in a piece that criticized Tyndall's activities, popular lecturing was a potentially dangerous enterprise that could only be trusted to select individuals.[32] Even Huxley, the great champion of professionalizing science, protested against the tendency to cast aspersions on the abilities of those practitioners who were also excellent communicators. The "popularization of science," he pointed out, "whether by lecture or essay, has its drawbacks. Success in this department has its perils for those who succeed." Those who failed took their revenge by "ignoring all the rest of a man's work and glibly labeling him a mere popularizer." Huxley rejected the pejorative use of the term "popularizer" as implied by some of his contemporaries.[33] The phrase "popularizer of science" meant different things to different actors, and could depend on whether they saw it in a positive or negative light.

Just as the meaning of "popular" and "popularize" was in flux during the nineteenth century, so also was the understanding of what constituted "science." This brings us to the thorny question of the professionalization of science. For those like T. H. Huxley, professionalization was all about redefining the meaning of science. Huxley wanted science to be associated with expertise, laboratory research, and naturalism, and he wanted to break its connection with the Anglican clergy, amateurism, and natural theology. The term "professionalization" can be misleading if we accept Huxley's desires as reality or as completely identical to the twentieth-century definition of the term.[34] Stefan Collini has discussed the diversity of, and limits to, the professionalization of intellectual life in the late nineteenth century. He warns that a use of the term "professionalization" to describe what differentiated intellectual life in 1930 from that of 1850 assumes a uniform and complex process that did not exist. The notion of what constituted the "professional" was not yet fixed in the second half of the nineteenth

32. J. Howard 2004, 755–56.

33. T. Huxley 1897a, vii.

34. Huxley's desire to place laboratory work at the center of science, for example, did not reflect the reality until late in the century. It was still considered to be an innovation for laboratory methods to be extended from chemistry to physics, physiology, geology, and engineering in the late 1860s and early 1870s. The proliferation of physics laboratories in British academic institutions took place between 1865 and 1885 (Gooday 1990, 25–28). Laboratory physiology was not ascendant in the universities, colleges, and hospital schools until after 1870. Huxley's own laboratory at South Kensington was not completed until 1872 (Gooday 1991, 333–34).

century.[35] Desmond, dealing more specifically with this issue in science, has also pointed out that professionalization can no longer been seen as triumphal in its "Whiggish inevitability." Huxley does not fit into the mold of the twentieth-century professional scientist. When it came to forging alliances, the members of the X Club were more concerned with an individual's commitment to naturalistic science rather than their "professional" qualifications.[36] I have therefore avoided using the phrase "professional scientist." There are connotations built into the term that are misleading, and it implies that the process of professionalization was complete in the second half of the century. I have tended to use the term "would-be professionalizer of science," to indicate that some of the characteristics of modern science were desired by nineteenth-century figures but not necessarily realized. I have also made use of the phrase "practitioner of science" to distinguish between those who were engaged in conducting experiments or analyzing the natural world and those popularizers whose activities were mainly focused on writing about nature.

The main reason for fussing about all of these labels is to underscore that what later came to be known as professional and popular science in the twentieth century and thereafter were not yet in existence in the latter half of the nineteenth century. Their predecessors were in a state of fluctuation in this period, and they were being worked out in relationship to each other. The distinction between the popularizer of science and the scientific practitioner should not be seen as being too rigid. After all, some practitioners took on the role of popularizer and wrote and lectured to general audiences. The primary question raised by this book concerns both practitioners and popularizers. What happened to the notion of popularization in the wake of a profound communications revolution that transformed the way science was understood? The new medium of the mass publication press radically altered the possibilities of debate and the parameters of disciplinary authority by changing the dynamics of authorship and audience.

HISTORIOGRAPHICAL APPROACHES TO POPULARIZATION

At least four approaches could be adopted in undertaking a study of the development of science for the general reading audience in the nineteenth

35. Collini 1991, 203, 220.

36. Desmond 2001, 5, 7, 11, 41.

century. The first, and the most pervasive up until the early 1990s, focused on members of the scientific elite, such as Huxley and Tyndall, and examined the beneficial impact that their writings and lectures had on their audiences.[37] Presented as great scientific heroes and educators, Huxley and Tyndall brought the light of knowledge to a grateful public. The older scholarship using this approach was based upon the "positivist diffusion model." According to this model, the scientific elite produced genuine, privileged knowledge. "Popularizers" disseminated simplified accounts of this knowledge to a passive readership. The best "popularizers" were members of the scientific elite. Since the 1980s, scholars have offered telling criticisms of the positivist diffusion model. It grants to scientists the sole possession of genuine scientific knowledge and serves to support their epistemic authority. This model cannot be adopted as a heuristic guide to research because it uncritically assumes the existence of two independent, homogeneous cultures, elite and popular, and forces the latter into a purely passive role. Popular culture can actively produce its own indigenous science, or can transform the products of elite culture in the process of appropriating them, or can substantially affect the nature of elite science as the price of consuming the knowledge it is offered.[38]

More sophisticated work on elite British scientists of the nineteenth century has appeared since the early 1990s that treats the diffusionist notion of "popularization" as an object of historical study while rejecting it as a historiographical principle.[39] A good book remains to be written on the involvement of the scientific elite in writing and lecturing for the public. In-depth studies of the history of the journal *Nature*, the International Scientific Series, and lecturing at the British Association for the Advancement of Science are sorely needed. I will deal with the scientific elite insofar as it helps to locate our popularizers within the scientific community, and I will focus on Huxley and Robert Ball in particular in the seventh chapter.

37. Two important exceptions are Hinton 1979 and Kitteringham 1981, but they both tend to be overly descriptive.

38. Hilgartner 1990; Whitley 1985; Cooter and Pumfrey 1994; Lightman 1997d, 188–90.

39. Topham 2000, 560. Examples of more recent works on elite science include studies by Golinski, Howsam, and Howard. Golinski's examination of the chemist and Royal Institution lecturer Sir Humphry Davy analyzes how elite scientists attempted to create a passive role for their public audience (Golinski 1992). Howsam has deconstructed the strategy adopted by the organizers of the International Scientific Series, which included Huxley, John Tyndall, and Herbert Spencer, to revolutionize the dissemination of science in society (Howsam 2000). In her finely nuanced examination of Tyndall as lecturer at the Royal Institution, Jill Howard has explored the complexities of communicating science to popular audiences (J. Howard 2004).

The three other approaches point us away from a focus on elite science. They have been developed, for the most part, since the early 1990s, and they all share the goal of recovering the agency of groups who participated in the making of science for the general audience in nineteenth-century Britain. The second approach draws our attention to marginalized groups and sciences. Scholars have turned their attention to alternative sciences, such as mesmerism and phrenology, and how various audiences took them up.[40] The critique of science offered by feminist scholars in the 1980s generated an interest in neglected women, which in turn has led to a number of major studies, each dealing with a specific female popularizer, including Mary Somerville, Agnes Clerke, Arabella Buckley, and Margaret Gatty.[41] Roughly half of the figures I will deal with in this book are women, which is an indication of their importance as popularizers of science in this period. Likewise, scholars are paying more attention to the part played by members of the working class in the making of science for general audiences. In addition to pioneering studies of radical evolutionary thought and artisanal culture in the early nineteenth century, we now have more recent work on secularists, socialists, and working-class science in the second half of the century.[42] For the most part, however, I will be sticking to middle-class popularizers.

A third historiographical approach to the subject of science for the public concentrates on Victorian publishing, including authors, readers, publishers, and the periodical press. While historians of science have recently discovered the importance of the popularizer, historians of authorship have paid little attention to writers who devoted a significant amount of their time and energy to producing scientific works.[43] Yet a comparison of the professionalization of scientists with the professionalization of authors writing about science is revealing. Many popularizers stood at the intersection

40. Some examples are Winter 1998; Cooter 1984.

41. Neeley 2001; Brück 2002; Gates 1993; Le-May Sheffield 2001. Barbara Gates and Ann Shteir have taken the lead in writing and editing volumes that investigate the role of women as popularizers more broadly, either as a formidable group within a specific discipline such as botany, or across the entire spectrum of scientific activities (Shteir 1996; Gates 1998; Gates and Shteir 1997).

42. Desmond 1992; A. Secord 1994; Paylor 2004; McLaughlin-Jenkins 2001a, 2001b.

43. Cross (1985) and Bonham-Carter (1978) focus primarily on novelists, playwrights, and essayists. However, their insights into the material conditions of authorship, the situation of female authors, and the evolution of publishing during the century are important for understanding the careers of popularizers of science.

between the world of authorship and the world of science. Recent work on the Victorian reader has been particularly valuable as an access point for examining how science was popularized.[44] I will discuss how popularizers defined their audience and the textual strategies they used to reach their readers. I will also deal with popularizers as readers of the works produced by elite scientists. As mediators between elite scientists and the developing mass readership, the figures in this book were both consumers and producers of knowledge. An approach that examines the role of publishers in the development of science for a popular audience has been closely connected with a focus on readers. A communications revolution took place in the second quarter of the nineteenth century, spurred on by new developments in print technology.[45] The resulting changes in both the reading and publishing of science had a profound effect on the nature of science in the early nineteenth century.[46] Publishers, not authors, were often the driving force behind the production of scientific works for a popular audience.[47] Popularizers could not avoid becoming enmeshed in a network of trade relationships that shaped the practice of authorship.[48] The place of popularizers of science in the world of publishing will be an important theme in this book. The chief source of power and authority for popularizers lay in the institutions of publishing rather than those of science.

Studies of the role of science in the periodical literature represent another strand in this approach.[49] From the point of view of the periodical reader, scientific subjects were "omnipresent," so much so that some scholars have

44. Although highly suggestive for historians, Richard Altick did not focus on science in his *English Common Reader* (1957). In *Victorian Sensation,* James Secord has analyzed the various interpretations of Robert Chambers's *Vestiges of the Natural History of Creation* (1844) held by different communities of readers, each with their own set of reading practices (J. Secord 2000). This kind of in-depth study of the readership of a specific book can only be undertaken for a small number of influential nineteenth-century scientific works. Since my aim is to deal with a large number of science writers and lecturers, I will be forced to limit myself to a discussion of their readership in general terms, and mostly from their own point of view. For a succinct overview of the new history of reading see Topham 2004, as well as Topham 2000.

45. J. Secord 2000.

46. Topham 2000.

47. Fyfe 2004.

48. Topham 2000, 587.

49. The recent interest in science and Victorian periodicals has largely been due to the tremendously productive SciPer project. The SciPer database and the three edited collections on science in the general periodical press, the work of Broks on late Victorian journals, and Barton's and Sheets-Pyenson's studies of commercial popular journals, now give us a solid base for further

argued that general periodicals played a far greater role than books in molding the public understanding of new scientific discoveries, theories, and practices.[50] Many of the popularizers of science to be discussed in this book were contributors to a wide range of periodicals for general readers, and one chapter will be devoted to the establishment of a particular science periodical for the public. However, I am not attempting to give a comprehensive survey of the full range of writing in periodicals. It is an enormous subject deserving of separate studies.

The first three approaches remain, for the most part, in the realm of print culture. The fourth, and final, approach stresses the varying sites in which science for general audiences can be found and takes us into new and vast territory. By comparing the public places to the elite spaces in which knowledge has been produced and disseminated we become more aware of the range of scientific sites. Science has been communicated in different ways in libraries, lecture theaters, salons, nurseries, zoos, observatories, churches, workshops, artists' studios, mechanics' institutes, stock farms, shipyards, game reserves, and in countless other places.[51] By taking us beyond print culture, the emphasis on sites opens up a whole series of exciting possibilities for investigation.[52] Since some popularizers were active lecturers, while others were involved with science museums, I will occasionally wander into the areas of oral culture and the culture of display. In this book, I will be quite eclectic and opportunistic. I will draw on the various approaches to studying science for the general public whenever necessary. They are not mutually exclusive. However, my emphasis will be on the scientific writer and his or her relationship to practitioners of science, to the growing mass reading audience for whom the author interpreted the larger meaning of scientific ideas, and to the publishers who were so essential to their chances for success.

SETTING THE SCENE: UP TO 1840

Scholars adopting the historiographical approaches sketched out above have recently established a clearer picture of how the complex relationship between

projects. See Cantor and Shuttleworth 2004; Cantor et al. 2004; Henson et al. 2004; Broks 1988, 1990, 1993, 1996; Barton 1998b; Sheets-Pyenson 1976, 1985.

50. Cantor et al. 2004, 1–2.

51. Livingstone 2003, 85.

52. Some of the Victorian sites are explored in Lightman and Fyfe 2007.

popularizers, practitioners, reading audiences, and publishers evolved over the course of the nineteenth century. They have tried to place the history of "popular science" publishing into the larger context of developments in publishing in general. The origins of "popular science" are inextricably connected to the transformation of all publishing and reading in the early nineteenth century. William St. Clair has shown that by the 1820s, it was clear to contemporaries that the surge in reading was not "a temporary blip which would soon level off or go into reverse." He has persuasively argued that the romantic period "marked the start of a continuing, self-sustaining expansion, a take-off in the nation's reading" that made Britain into a "reading nation" by the end of the century.[53] The size of the British reading audience grew, as literacy rates changed dramatically over the course of the last sixty years of the century. Whereas the number of literate and illiterate Britons was roughly equal at the end of the 1830s, by the close of the century illiteracy had fallen to 1 percent.[54] The composition of the reading audience, not just the size, changed significantly, as new segments of the population were added. By the 1830s, some publishers were beginning to reach out to a new audience composed of members of the middle class and the wealthier working class.

Publications produced under the designation "popular science" first appeared in the 1820s and 1830s, at the same time that St. Clair detects the take-off point in British reading. Their target audience differed from the numerous publications in the seventeenth and eighteenth centuries attempting to present the findings of the new philosophy to nonmathematical readers. The new publications of the 1820s and 1830s were aimed at audiences defined by the new social and intellectual divisions of the industrial age. Up until the appearance of these cheaper books, most readers had had little access to scientific reading matter. The "popular science" publications of the 1820s and 1830s had their origins in experiments in cheap educational publishing that began in the second decade of the nineteenth century, in the wake of the transformation of the book trade. These ventures were important in serving to develop and define "popular" publishing in general and "popular science" in particular. Cheap children's books published by Longman and Richard Phillips, and William Pinnock's educational catechisms, were followed by the inception of the new weekly periodicals. Relatively inexpensive literary journals such as *Literary Gazette* (1817) and *Literary Chronicle*

53. St. Clair 2004, 13.
54. Vincent 1989, 22.

(1819) were founded, along with the new two- and three-penny weeklies, such as the *Mirror of Literature* (1822), the *Mechanic's Magazine* (1823), the *Lancet* (1823), and the *Chemist* (1824).[55]

The children's books and the weeklies of the 1810s and early 1820s confirmed the existence of a new and profitable market for cheap publications. This led to the production in the 1820s and 1830s of publications by the Society for the Diffusion of Useful Knowledge (SDUK), John Murray, and Longman with the designation "popular science." Although the development of "popular science" was a result of the recognition by publishers that new commercial opportunities existed, it was also driven by fears that cheap publications from radical presses were a threat to the religious and social order. These fears remained throughout the turbulent 1830s, when the debate over political reform raged, and into the 1840s, when Chartism became a potent force. The SDUK was set up in 1826 by a group of liberals led by Henry Brougham. Their goal was to provide publications that were inexpensive enough to be purchased by members of the working classes, especially the artisans, who were expected to take advantage of the recently founded Mechanics' Institutes. SDUK published several cheap nonfiction series in which many new scientific works appeared, including the Library of Useful Knowledge and the Library of Entertaining Knowledge. John Murray, the publisher of the Tory *Quarterly*, produced a series of original nonfiction between 1829 and 1834 titled the Family Library, which included David Brewster's *Life of Isaac Newton* (1833) and his *Letters on Natural Magic* (1832). Murray's aim was to bridge the widening divisions between classes. Longman published the 133-volume Cabinet Cyclopaedia edited by Dionysius Lardner composed largely of introductions to the sciences, including John Herschel's *Preliminary Discourse on the Study of Natural Philosophy* (1832). Since the books in this series could be bought individually for six shillings, middle-class readers and some of the more prosperous artisans could afford to buy them. Other publishers attempted to tap into the same market by producing series of nonfiction works that were relatively cheap.[56] The success of the penny weekly magazines of the early 1830s, such as *Chambers's Edinburgh Journal* and the SDUK's *Penny Magazine*, also demonstrated to the book trade that an immense reading audience did exist if the price was pennies rather than shillings.[57]

55. Topham 2007.   57. Fyfe 2004, 48.
56. Fyfe 2004, 43; J. Secord 2000, 48–50.

The publication of books on science in a series produced in the cheap format during the 1820s and 1830s was only one of several important developments. Influential science books for children appeared on the market, and women writers began to establish reputations as reliable authors of nontechnical works. While the need for simple introductory works for adults with limited education was not recognized until the 1820s and 1830s, science books for children had been an established genre since the late eighteenth century. John Newberry, the publisher, and likely the author, of Tom Telescope's *Newtonian System of Philosophy, adapted to the capacities of young gentlemen and ladies* (1761) created the genre in the 1740s. The genre featured the inclusion of moral and religious lessons drawn from nature, drawing on easily understood examples from everyday life, and the use of a conversational format. Authors writing in this genre included Sarah Trimmer, John Aikin and Anna Barbauld, Jane Marcet, and Jeremiah Joyce. By the 1830s some authors, such as Samuel Clarke in his *Peter Parley's Wonders of Earth, Sea, and Sky* ([1837]), experimented with a third-person narrative and attempted to attract the reader by focusing on wonders and marvels. Clarke's book was part of a widely known series, all appearing under the pseudonym "Peter Parley."[58] "Peter Parley" was originally the nom de plume of the American educational writer Samuel Goodrich. His Parley books sold so well in the United States that British publishers stole both the pseudonym and the idea behind the series, which was to provide a safe entry into the sciences for young children. Informative and entertaining, they linked knowledge of nature with knowledge of God. The British imitations sold well. Clarke's *Wonders* was popular in both Britain and the United States, reaching a "seventeenth edition" thirty years after its first publication.[59] Children's writing was not seen as a high status activity in the early part of the century, and it was often undertaken by women or by men involved in teaching or publishing.[60] The success of the Peter Parley series may have led more popularizers in the middle of the century to consider writing for children.

In addition to writing children's books, female popularizers of the first half of the nineteenth century addressed themselves to an audience of receptive, but uninformed, women. A number of important popularizers were active during this period, including Maria Edgeworth, Sarah Trimmer, Priscilla Wakefield, Jane Marcet, and Margaret Bryan. Even though Marcet's

58. Fyfe 2003, xii–xvii.
59. J. Secord 2003b, vi–ix.
60. Fyfe 2003, xx.

Conversations on Chemistry (1806) became known as the book that sparked Michael Faraday's interest in science, her primary goal was to make knowledge more accessible to women.[61] In order to reach their intended audience of women and children, Marcet and the other female popularizers in the period from 1780 to 1840, adopted the "familiar format," a fictional literary format that used letters, dialogues, and conversations, customarily situated in a domestic setting.[62] Since a mother or mother substitute, such as governess or teacher, often played a central role in the "familiar format," those who adopted it belonged to a "maternal" tradition in science writing.

Marcet recalled that "familiar conversation" had been a "most useful auxiliary source of information" in helping her to understand Humphry Davy's lectures at the Royal Institution. It was therefore appropriate that she record the ideas on chemistry that she had "first derived from conversation" in "the form of dialogue" in which Mrs. Bryan teaches two sisters, Emily, and Caroline.[63] In Wakefield's *Introduction to Botany,* the reader encounters a series of letters from Felicia to her sister Constance, describing what she had learned from her governess during her botany lessons. Letters, conversations, and dialogues fostered drama and debate while encouraging consensus.[64] The "familiar format" also presented women who were knowledgeable about science in positions of authority, though their locus of authority was situated in the home and in their roles as religious and moral educators of the young. Marcet and her fellow popularizers, as well as the female characters they created, taught the reader how to understand the moral and religious lessons to be found in nature. Their work was underpinned by a natural theology that counseled submission to the social status quo. In his *Natural Theology* (1802), the Anglican divine William Paley had argued that evidences of divine design existed in nature. Just as the contrived mechanism of a watch pointed to the existence of an intelligent human agent who created it, the adaptations of living beings to the natural world demonstrated that an omniscient, omnipotent, and benevolent divine being had created the universe. The Paleyian theme was so important to these women of the first half of the century that the "familiar format" was contained within what scholars have referred to as the "narrative of natural theology."[65]

Mary Somerville (1780–1872), the most well known female popularizer of the first half of the century, did not belong to the maternal tradition. In

61. Bahar 2001, 30.
62. Shteir 1996, 83.
63. [Marcet] 1817, 1: vi–vii.

64. Gates 1998, 38.
65. Benjamin 1991, 39; Gates 1998, 39.

many ways, she was atypical for a female popularizer. Whereas Marcet and the other women imagined their audience as being composed of receptive but uninformed women and children, Somerville addressed knowledgeable adults, including men. Since she was aiming at a different audience, Somerville did not use the "familiar format." Already a Whig icon by the mid-1830s, she became a symbol of self-education, liberalism, and women's rights. The bulk of her major works, *Mechanism of the Heavens* (1831), *The Connexion of the Physical Sciences* (1834), and *Physical Geography* (1848), were published before the midcentury mark.[66] Her early work won her recognition as a scientific author of great skill. *Mechanism of the Heavens* established her place among the British scientific elite, winning her the support of such powerful savants as John Herschel and William Whewell. In *Mechanism* she translated, contextualized, elucidated, and interpreted the mathematics of the French astronomer Pierre-Simon Laplace, sanitizing the impious implications of the French Enlightenment in the process. Somerville aimed to demonstrate that the higher analysis of the French could be used to enhance understanding of the manifestations of God's divine goodness and power. Although Brougham had originally invited Somerville to write *Mechanism* for the SDUK, Murray published it. The book was too difficult for SDUK's readership. Somerville's intended audience was the mathematically sophisticated reader, not the beginner.[67]

In the *Connexion of the Physical Sciences,* Somerville addressed a wider public than she had in *Mechanism.* Murray published it in the cheap format he had used for the Family Library.[68] Instead of resorting to the "familiar format," she presented here, and in subsequent works, a synoptic overview of rapidly advancing fields of science.[69] Somerville developed an interpretive framework for viewing the intricate operations of nature from a "cosmic platform" in the *Mechanism of the Heavens.* She applied this same technique in her more accessible *Connexion of the Physical Sciences* and established a new literary form, the extended synthetic literature review. The *Connexion* provided a visual survey or guided tour of the entire physical sciences, emphasizing the links between areas of enquiry. This aspect of her work had endeared her to the polymath and Cambridge master William Whewell, who feared that the sciences were losing their unity, disintegrating like

---

66. J. Secord 2004d, xi–xii.

67. J. Secord 2004a, xxxi–xxxii; Neeley 2001, 73, 77, 228; J. Secord 2004b, xiv.

68. J. Secord 2000, 50.

69. By 1842, Somerville's *Connexion* had reached a sixth edition and 10,500 copies had been printed. By the end of the century a total of 17,500 had been printed (J. Secord 2004c, xi).

"a great empire falling to pieces." He praised Somerville's *Connexion* for "showing how detached branches have, in the history of science, united by the discovery of general principles."[70] But Murray and his son, with whom Somerville published the rest of her works, expressed concern on several occasions that she was in danger of losing the general reader by writing at such a demanding level.[71] Somerville established a tradition of science writing distinct from that of the maternal tradition.

Religious themes were noticeably present in science books for children as well as in works by female popularizers of the maternal tradition in the first four decades of the nineteenth century. Elements of the natural theology tradition appeared in many "popular science" works of this period. Although the influential Bridgewater Treatises were not initially produced in a cheap format, they contained a largely nontechnical compendium of contemporary science. To that extent, Jonathan Topham has asserted, "the treatises arguably represented a nascent publishing form that would later be called 'popular science'—a form that publishers were very soon to find highly remunerative." They were intended from the beginning to be works dealing with "the Power, Wisdom, and Goodness of God, as manifested in the Creation." The eighth earl of Bridgewater had bequeathed £8,000 to securing a person or persons to write a work on natural theology. A Royal Society committee distributed the sum equally among eight authors, and William Pickering published their books from 1833 to 1836. Although the authors represented all shades of religious opinion, including evangelicalism, the High Church, and liberal Anglicanism, and though they were not agreed on a common purpose, the series came to be seen as representing a conservative and religiously safe account of the sciences. None of the authors designed his treatise to be an exposition of the philosophy of the design argument in the manner of Paley's *Natural Theology*.[72]

The success of the Bridgewater Treatises may have encouraged popularizers to incorporate natural theology themes in their works. Like the authors

70. Neeley 2001, 114; J. Secord 2004c, x; [Whewell] 1834, 59-60.

71. Patterson 1969, 331. Claire Brock has persuasively argued that Somerville's popularity was an illusion. She herself was ambivalent about assuming the role of popularizer. She undertook this role largely for financial reasons, and she found it difficult to make her writing more accessible to a popular audience. Later she regretted that she had not devoted herself to mathematical research in order to make an original contribution. Periodical reviews of her work and the political debates surrounding the plan to grant her a pension in 1837 also demonstrate that many contemporaries did not perceive her as a writer who sought to inform and enlighten the public (C. Brock 2006).

72. Topham 1998, 241, 238.

of the Treatises, they did not adopt the demonstrative natural theology of Paley as their model. They presented a "theology of nature." John Brooke first suggested the distinction between natural theology and a theology of nature. Brooke argued that whereas natural theology was grounded in natural reason and offered a demonstration of God's existence and attributes without recourse to revelation, a theology of nature was concerned "only with the status and significance of the *natural world*."[73] More recently, Jonathan Topham and Aileen Fyfe have both found this distinction useful when discussing the expression of theological ideas in scientific works of the first half of the century. Topham has argued that passing references to the evidence of design in early nineteenth-century religious magazines were intended to stop short of any form of inductive inference or philosophical rigor. He asserts that these references should be described as a "discourse of design," since they express a theology of nature based on a prior commitment to the truth of Christian revelation, and that their pervasiveness should not be mistaken for the pervasiveness of natural theology.[74] Fyfe has pointed out that while evangelicals subscribed to a theology of nature, since they based their faith on Revelation, they rejected natural theology when used to ground Christianity on a rational proof of God's existence.[75] The phrase "theology of nature" is useful for describing theologies that incorporate beliefs about the natural world, many of which have little to do with demonstrative natural theology as presented by Paley. Many popularizers of science presented theologies of nature in their books in the early nineteenth century. While they saw nature as a world full of divine purpose, and though they often pointed to instances of design, they did not attempt to present philosophical arguments to prove God's existence.

THE MIDCENTURY SCENE

Theologies of nature informed the work of influential popularizers of science during the 1840s, such as Hugh Miller. A Scottish Free Churchman and editor of the newspaper *The Witness*, Miller has been regarded by scholars as the leading popular expounder of geology in the 1840s and 50s.[76] His first scientific work for a general audience, *The Old Red Sandstone* (1841), was reissued at least twenty-five times.[77] Miller conceived of science as revealing

---

73. Brooke 1974, 8–9.
74. Topham 2003, 38.
75. Fyfe 2004, 7.

76. Oldroyd 1996, 77.
77. Cribb 2004, 1399.

the sublime works of the Creator, and he focused on the aesthetic dimensions of the natural world.[78] Theologies of nature were also expounded in
the works published in the 1840s by religious publishers, especially in the
case of the Religious Tract Society (RTS), an important evangelical publisher. The interdenominational RTS was founded in 1799. Concerned by
the increase in the 1840s of cheap works that seemed to support unbelief,
the RTS began to transform its publishing program by moving beyond religious works to general nonfiction, including science. According to Aileen
Fyfe, during the midcentury the RTS became the most significant player in
cheap religious publishing in Britain. The RTS aimed to publish affordable
books that demonstrated that all forms of knowledge, especially the sciences,
remained part of a Christian framework. One hundred sixpenny volumes
were published from 1845 to 1855 in the RTS's "Monthly Series," including
books on scientific subjects by authors such as Thomas Dick, Thomas Milner, and Anne Pratt. Since the "Monthly Series" was the cheapest and most
widely distributed source of introductory treatises on the sciences available
in the 1840s and 50s, it brought its version of Christian natural knowledge
to readers who could not be reached by other series.[79]

At the same time that the RTS was transforming itself in the 1840s into a
producer of books in areas that had previously been the domain of secular
presses, the Scottish firm W. and R. Chambers made its move into the world
of large-scale, cheap book publishing. Those who were behind the new direction of the RTS were well aware of the fact that they were in competition
with the Chambers brothers. Both were using the same sorts of techniques
to reach a similar audience. With their experience in periodical printing,
the Chambers publishing firm was well positioned to take this new path.[80]
Due to *Chambers's Edinburgh Journal* (f. 1832), the firm had established a
reputation as "publishers for the people." They aimed their publications
at a new "polity of consumers" composed of a middle- and working-class
family readership.[81] When Robert Chambers wrote his anonymously published *Vestiges of the Natural History of Creation* (1844) he drew on all of his
experience as a publisher and journalist for this audience. It allowed him
to present evolutionary theory, hitherto linked to working-class radicalism,
in a favorable, appealing light to his readers. The result was a sensation.
Immensely popular—it reached a sixth edition by 1847—the book sparked
a controversy on the issue of scientific authority. When the men of science

78. Brooke 1996, 176, 185.   80. Ibid., 55, 72.
79. Fyfe 2004, 58–59, 65, 271.   81. J. Secord 2000, 69.

judged that *Vestiges* lacked scientific credibility and mercilessly attacked its evolutionary theory as a hasty generalization, Chambers appealed directly to the public in his *Explanations, A Sequel* (1845). Here he rejected the authority of practitioners and refused to accept their verdict on the scientific value of his book. Practitioners had become too specialized and narrow to appreciate the synthesis of recent ideas that he had provided. By appealing to the nonscientific public for a judgment on *Vestiges,* Chambers challenged the right of the practitioner to restrict the scope of conjecture and upheld the right of the layperson to speculate in matters scientific. Practitioners argued that only those with the proper training and knowledge should be writing books on science for the general public.[82] The issues raised by the *Vestiges* controversy reverberated throughout the rest of the century.

Although the *Vestiges* controversy had led some practitioners and intellectuals to question the credibility of books written by popularizers, the value of such works was presented in a new light in the beginning of the latter half of the century. In the second quarter of the nineteenth century, scientific literature for the general public was valued as a counter to attacks on the Anglican-aristocratic establishment. But during the 1850s and 60s, when prosperity and social stability defused fears about radicalism, such literature came to be valued more for its potential to elevate the level of public culture, to inspire the British public with a love of science, and to offer accurate information. In an article in the *Quarterly Review* for 1849, Whitwell Elwin, who later assumed the editorship of the *Quarterly Review* from 1853 to 1860, defended the value of good "popular science." Elwin established a distinguished pedigree for this genre of scientific writing. The publication of Isaac Newton's *Principia* had given an "impulse to popular science" in England. Since then it had been "the humble attendant on mathematical philosophy, like the squire on the knight in the days of chivalry," but he was critical of some of the more recent attempts to bring science to the public. Although they had "probably diffused" some love of science, the Mechanics' Institutes "have remained insufficient schools for the labouring classes, and done more to justify the fears of opponents than the hopes of friends." Elwin also had little sympathy for the productions of the Society for the Diffusion of Useful Knowledge, very few of which "rose above mediocrity." But he approved of Herschel's *Discourse on Natural Philosophy* and his *Treatise on Astronomy*. Dionysius Lardner's treatises, Mary Somerville's *Connexion of*

82. Yeo 1984, 5–31.

the Physical Sciences, and the Bridgewater Treatises were also singled out for praise.[83]

Seventeen years later, an anonymous reviewer in the *Popular Science Review* asked if the "desire and supply" of popular books "is injurious to the true interests of science? Are we to suppose that, because the masses cannot follow the philosopher into the discussion of propositions which involve a consideration of abstract questions, therefore they cannot be improved by being taught the interesting facts and the grand general principles of science?" The reviewer answered, "we think not," and declared, "a little learning is <u>not</u> a dangerous thing." "Popular scientific works" were useful as they excited a "taste for scientific pursuits" and in some cases inspired individuals to attain the "rank of the philosopher." Despite some misconceptions that arose in such literature, "there is a good grain of truth; and who will contend against its beneficial influence?" The reviewer offered a passionate argument against excluding "masses from the prosecution of Nature's grandest problems" on the "pitiful" pretense that the scientist possessed perfect knowledge and that unless an individual could reach the same level of understanding they should be allowed to perish in ignorance.[84]

Although there was agreement that scientific literature for the public had the potential to have a beneficial impact, an ongoing debate arose over who should write it. Who could legitimately assume the role of popularizer of science? Who had the authority to write about the natural world and to communicate the meaning of contemporary scientific theories to a general reading audience? As one critic pointed out in the *Atlantic Monthly* in 1860, producing this kind of literature was extremely demanding. Pointing to errors in an article on "Meteorology" in the *Atlantic Monthly* for July, the reviewer asserted that any attempt at *"popularizing science"* without "misleading and confounding the general reader is a task which claims the largest and most exact knowledge, and the greatest perspicuity of statement, no less than a flowing style and felicitous illustration. It is a task in which true success, though apparently frequent, is in reality extremely rare."[85] Some, like the *Saturday Review* critic who condemned "sensational science," believed that only practitioners like Huxley and Tyndall had the ability to lecture to, and write good science books for, a general reading audience.

83. [Elwin] 1849, 307, 316–17.

84. "Reviews. Popular Zoology" 1866, 214.

85. "July Reviewed by September" 1860, 383.

Others were willing to view nonpractitioners as allies in instilling a love of science among members of the public. In 1867, the astronomer Warren de la Rue, then secretary of the Royal Astronomical Society, wrote to Mary Ward, a popularizer, about her astronomical article in the *Irish Times*. "Permit me to say," he declared, "that such articles do much to encourage the cultivation of science by amateurs—and that England (in which word I of course include Ireland) has much to be proud of the work done by non-professional cultivators of science."[86] Although de la Rue may have intended to be supportive of Ward's efforts, his phrase, "non-professional cultivators of science," reminded her that she was subordinate to the "professional" and that her role was limited to garnering support from amateurs. Both de la Rue and the *Saturday Review* critic would have maintained that only the practitioner could truly speak on behalf of science. Their expertise gave them the authority to produce knowledge and to be the final arbiter as to its cultural meaning. The second half of the century was, after all, a period when cultural authority was particularly important to the men of science. They were especially keen on pushing clergymen and women out of science, since both were seen to be strong supporters of the Church and as barriers to the establishment of a self-defining community of experts. Huxley and his allies were not about to cede their authority to a cadre of writers and journalists, many of whom who were sympathetic toward Christianity. When they engaged in popularizing activities their aim was to enroll the support of the public for their own secular objectives, such as securing state funding for science, increasing the presence of science in educational institutions, or banishing religious considerations from the search for knowledge.

Others were not so sure that a practitioner was necessarily the best candidate to communicate with a popular audience. John George Wood, a prolific writer and speaker, agreed that lecturers had to be selected with care, and insisted that the primary consideration should be whether or not they knew how to put themselves "in the mental condition of their hearers." This meant that "the most learned men are not necessarily the best teachers," as they tended to make the mistake of "assuming that their hearers are already somewhat versed in the subject, and in consequence are unintelligible just where they wish to be especially lucid."[87] W. T. Stead, editor and proponent of the "new journalism," agreed with Wood. He recommended that newspaper editors "never employ an expert to write a

---

86. As quoted in Harry 1984a, 194.   87. Wood 1887, 395.

popular article on his own subject." It was far better, he insisted, to employ "someone who knows nothing about it to tap the expert's brains, and write the article, sending the proof to the expert to correct." The expert always forgot that "he is not writing for experts but for the public, and will assume that they need not be told things which, although familiar to him as ABC, are nevertheless totally unknown to the general reader."[88]

In many ways, the views of editors like Stead, and the influential Victorian publishers, were more important than those of the practitioners. They were the ones who decided who did the writing and which books and articles were published. Publishers such as W. and R. Chambers, Charles Knight, and George Routledge, who desired to reach the masses, and who reckoned that cheap prices and nontechnical language were crucial, believed that well-known men of science were not necessarily the most appropriate writers.[89] From the point of view of publishers, the best popularizers of science were those with sound literary skills and those who could meet tight deadlines, and not necessarily those who had expertise in science. Ministers and teachers, who had experience in making themselves understood to audiences with little or no prior scientific knowledge, were valued by publishers such as Chambers and the Religious Tract Society for their ability to take specialist knowledge and translate it into understandable language.[90]

THE REVOLUTION IN PUBLISHING AND THE NEW MARKET

Publishers played an important role in the explosion of books on science in the middle of the nineteenth century. The publisher John Churchill was vital to the success of the *Vestiges of the Natural History of Creation*. Even though Robert Chambers was himself a successful publisher, he needed Churchill's knowledge of the London scene and his experience as a specialist in medical and scientific publishing. Churchill was personally involved in decisions involving the nature of the text and was responsible for actions that were essential for the reputation of the book and its message.[91] Publishers could have a tremendous impact on the market and the reading habits of Victorians through the creation of series with a scientific theme. In 1857 George Routledge began to plan out a series of shilling handbooks on natural

88. Stead 1906, 297.
89. Fyfe 2005a, 203.
90. Fyfe 2005b, 118–20.
91. J. Secord 2000, 151.

history. He recruited John George Wood to undertake at least one of the volumes, and his *Common Objects of the Seashore* appeared later that year. It was the first book in the series, and it met with immediate success. Routledge was barely able to keep pace with the demand from eager readers. This boosted the sales of the entire "Common Objects" series and led Routledge to undertake a series of projects with Wood.[92] A number of publishers subsequently set up their own scientific series.

The most important series from the second half of the century can be divided into middlebrow general, general and working-class, and natural history categories. The "International Scientific Series" (begun in 1872 by the publisher H. S. King), the "Nature Series" (1873, Macmillan), the "Modern Science" series (1891, Kegan Paul), the "Romance of Science Series" (1889, Society for the Promotion of Christian Knowledge), and the "Contemporary Science Series" (1889, Walter Scott Publishing Company) were among those in the first category. Cheaply priced series for working-class readers included Macmillan's "Science Primers" (1872), the "Manuals of Elementary Science" (1873, Society for the Promotion of Christian Knowledge), "Chambers Elementary Science Manuals" (1875, W. and R. Chambers), and "Simple Lessons for Home Use" (1877, Stanford). Natural history series were also popular, including Routledge's "Common Objects" series, the "Series of Natural History for Beginners" (1866, L. Reeve and Company), the "Natural History Rambles" series (1879, Society for the Promotion of Christian Knowledge), Longman's "Fur, Feather, and Fin Series" (1893), and Allen's "Naturalist's Library" (1894).[93] Although many British publishers produced scientific books for the general reader in the latter half of the nineteenth century, the leaders were Macmillan, H. S. King, Longman, John Murray, Edward Stanford, Cassell, and George Routledge.[94] The men who ran these firms were the major power brokers in the world of scientific publications. They, and their editorial assistants, decided who wrote science books for the public, not the practitioners.

Publishers in this period were the beneficiaries of a two-part industrial revolution in the production and selling of books. This revolution allowed publishers eventually to reach a mass reading audience in the second half of the century that included more and more members of the working class. The first phase took place from 1830 to 1850, and has been referred to by Simon Eliot as the "distribution revolution." The introduction and develop-

---

92. Mumby 1934, 75–79.
93. Ring 1988, 69–90.

94. P. Gould 1998, 149–50.

ment of the Foudrinier machine, steam-driven presses, case binding, as well as the reduction of the "taxes on knowledge" and the development of the railway system, characterized this revolution.[95] James Secord has labeled this same process the communication revolution and asserts that it represented the "greatest transformation in human communication since the Renaissance" that led to "opening the floodgates to a vastly increased reading public."[96] The 1840s was the key decade in the first phase of this publishing revolution. Many steam-printing technologies developed in the early nineteenth century had been utilized by newspaper and penny periodical publishers (who saw the advantages of faster production), but not book publishers, who relied on hand-press technology. In the 1840s some book publishers decided to attempt to reach the audience of the penny periodicals and began to adopt steam-print technologies. The second phase, labeled the "mass-production revolution" by Eliot, began in the 1870s and had its major effects near the end of the nineteenth century. It was characterized by technical processes such as rotary printing, hot-metal typesetting, use of lithographic and photographic techniques, and the displacement of steam power by electricity. The features of this phase included the rise of the public library, the sixpence paperback edition, the professional literary agent, the rise of professional associations, the development of the royalty system, and the mass circulation daily newspaper.[97]

By adopting a two-phase model for the industrial book revolution of the nineteenth century, and rejecting the notion of a continuous process, it becomes clearer why there were surges in production between 1830 to 1855, and 1875 to 1914. The production of titles per annum increased dramatically in the 1840s and early 50s, reached a plateau from 1858 to 1872, and then accelerated again from the late 1870s until 1913. The plateau was the result of the interval between the first and second phases of the industrial revolution in book production. As book production increased over the course of the century, the cover price of the average book decreased. In his discussion of the overall structure of book pricing, Simon Eliot has pointed to three groups: a low price group (1d. to 3s. 6d.), a mid-price group (3s. 7d. to 10s.),

95. Intended to constrain the radical publishers, the "taxes on knowledge" applied to the entire publishing trade. There were taxes on paper (except Bibles), on political content, and on advertisements. Before they were reduced these taxes made cheap publishing for the masses legally risky and economically difficult (Fyfe 2004, 45–46).

96. J. Secord 2000, 2.

97. Fyfe 2004, 55; Eliot 1994, 107.

and a high price group (10s. 1d. and more). In the early 1850s the cover price of the average book fell radically. By 1855 the price structure characteristic of the early nineteenth century had been turned upside down. Low price books now accounted for the largest percentage share, mid-priced books were second, and high price books dropped to the lowest percentage ever recorded.[98]

Although this pattern altered slightly in the 1860s, it was reestablished and developed from 1875 to 1905. Eliot asserts that the steady decline of the high price group was the most consistent feature in this period.[99] Even those books sold in the mid-price group were aimed at a wider audience than the upper middle class. As a result of the increase in production and the fall in prices, commercial book publishing became a mass-market business in the Victorian period.[100] Despite the increase in output and the decrease in prices, the combination of population growth and improving literacy rates (adult literacy had increased to about 60 percent by midcentury) enlarged the potential market for books to the extent that production did not catch up with demand until the 1880s.[101] To one contemporary in the 1880s, the positivist Frederic Harrison, it seemed as if he were watching a "remorseless cataract of daily literature," so enormous that it was difficult to decide what to read. Harrison wondered how much of this "harmless, entertaining, and even gently instructive" literature deserved to be read instead of the great books of the world. These thoughts led him to "almost reckon the printing press as amongst the scourges of mankind."[102]

The publishing revolution had its impact on the production of science books. During the 1840s and 50s there were four times the number of titles on the sciences published annually than at the start of the century.[103] Before 1850, there were only a handful of scientific best sellers. In his lists of nineteenth-century English best sellers in the scientific category, Richard Altick lists George Combe's *The Constitution of Man* (1828), which sold 11,000 copies in eight years, and Robert Chambers's *Vestiges of the Natural History of*

---

98. Eliot 1995, 39; 1994, 60, 63, 76, 106–7.

99. Eliot 1994, 76. The three-decker novel continued to sell for 31s. 6d. until the end of the century. But it was not intended solely for the wealthiest, as it could be obtained through one of the circulating libraries. The strong library market may have been one of the important reasons why the high price group declined at a slower pace. James Secord has argued that the price of the three-decker remained high due to the influence of Charles Mudie, who established the fastest growing circulating library in 1843 (J. Secord 2000, 140).

100. Weedon 2003, 1.

101. Ibid., 157; Fyfe 2005a, 200.

102. Harrison 1886, 4–5, 10, 16.

103. Fyfe 2005a, 201.

Creation (1844), which sold 24,000 copies in sixteen years.[104] William Astore has argued persuasively for including Thomas Dick's *Christian Philosopher* (1823) to the list of best sellers, since it sold 22,000 copies in thirty years, as well as his *Solar System* (1846), a two-part Religious Tract Society publication, which sold 30,510 and 26,890 copies.[105] After 1850, the list of best sellers increased significantly. Larger print runs and cheaper prices meant larger sales.

I will pay close attention to print runs and book sales by popularizers in Britain, as they are a powerful indicator of the impact that they had on Victorian readers. Where figures are not available from publisher's archives, I can roughly estimate from the number of editions, if they are known.[106] The average print run of an edition from 1856 to 1896 was 1,000 copies.[107] To help gauge the success of popularizers of science of the latter half of the century in terms of sales of their books in Britain, it is important to distinguish between steady sellers and best sellers. Whereas steady sellers continue to sell, perhaps only modestly, for years and decades after they first appear, a best seller describes a book that sells a large number of copies in a short period of time.[108] I will provide figures that allow an evaluation of books by popularizers as both steady sellers and best sellers. Selling 39,000 copies by 1890, Robert Chambers's *Vestiges of the Natural History of Creation* (1844) was an extraordinarily successful scientific steady seller.[109] The same can be said about Darwin's *Origin of Species* (1859), as by 1899 it reached the

---

104. Altick 1986, 240; 1969, 205. Altick also includes William Buckland's Bridgewater Treatise, *Geology and Mineralogy Considered with Reference to Natural Theology* (1836), on his best-sellers list (Altick 1986, 241).

105. Astore 2001, 136, 142.

106. In future references to the number of editions, or copies printed, of books, if a reference is not given to a publisher's archive or another source, the figures are taken from either the National Union Catalogue or the British Library Catalogue.

107. Weedon 2003, 49. However it can be risky to estimate copies sold from number of editions without knowing the exact size of each print run (Altick 1986, 235). Fortunately, I have been able to locate accurate figures in most cases for books that were very, or extraordinarily, successful. For many of those books that were only moderately successful I have had to rely on determining the number of editions and multiplying by one thousand.

108. I am indebted to Aileen Fyfe for this neat distinction. In her article on Paley's *Natural Theology*, Fyfe has discussed the difference between best sellers and classics. A classic is a work that continues to be read several generations after it was written. These works "speak to later generations because there was something timeless and unchanging about them." But Fyfe argues that historians should not focus on classics alone, as this would exclude consideration of works that historical actors may have considered as classics but which did not continue to sell in the long run. A best seller does not necessarily become a classic (Fyfe 2002, 731–35).

109. J. Secord 2000, 131.

56,000 mark.[110] However, figures that allow us to consider a book's status as a best seller are more important. They are better indicators of the immediate impact of a book. I will therefore provide, where possible, sales within ten years.[111] Some benchmarks may be useful. Books that had a print run of up to 10,000 copies, I would place in the category of moderately successful; those with print runs between 10,000 and 20,000 can be described as being very successful; and those above 20,000 can be referred to as being extraordinarily successful. Since Darwin's *Origin of Species* sold close to 10,000 copies within that period, it can only be described as being at the very bottom of the very successful category. However Chamber's *Vestiges* reached 21,250 within a decade and should therefore be categorized as being at the lower end of the extraordinarily successful best-seller group.[112] In addition to the *Vestiges,* at least half a dozen books by popularizers surpassed the *Origin.* Of course, the impact of the *Origin* cannot be underestimated, but, as the print-run figures suggest, the post-1859 era cannot be seen solely in terms of Darwin's enduring theory of evolution.

POPULARIZERS: THEIR CAREERS AND THEIR NARRATIVES

In addition to making the publication of more scientific best sellers possible, the revolution in publishing ensured that more literary work was available and that writers were more in demand. By the 1860s literary journalism was becoming secure enough as a profession to attract a steady flow of talent from the universities.[113] In 1871, about 2,500 people classified themselves as writers

---

110. Freeman 1965, 45.

111. A decade is long enough to give an indication of subsequent editions, but not so long as to give the protoclassic an advantage over the books that sell hugely and quickly.

112. The *Origin*'s sales got off to a rather slow start. Murray's first edition consisted of 1,250 copies. Runs of 3,000 (2nd edition, 1860), 2,000 (3rd edition, 1861), 1,500 (4th edition, 1866), and 2,000 copies (5th edition, 1869) followed. So by 1869, ten years after it had first appeared, Murray had printed only 9,750 copies of the *Origin.* But Darwin's *Origin of Species* continued to sell long after 1869. At the time of Darwin's death in 1882, it had reached the 24,000 mark (Desmond and Moore 1991, 477–78; Freeman 1965, 44–49; for figures on Chambers see J. Secord 2000, 131). Darwin's *Descent of Man* (1871, Murray) sold better than the *Origin* in the first decade, at 14,000 copies, but not as well as the *Origin* by the end of the century, at 35,000 copies (Freeman 1965, 52). It should be kept in mind that at times I will be comparing works from different genres of science books for popular audiences. Science books that were intended to appeal to general readers as well as serving as a school textbook, for example, had the potential to sell far more copies than Darwin's *Origin of Species.*

113. Gross 1969, 25.

in the census, five times the number at the start of the century. The increased opportunities due to the revolution in publishing rendered the career of writer more appealing and viable, even in the area of science writing.[114] It was possible to eke out a living by science writing in the mid-nineteenth century. William Martin and Thomas Milner were earning between £150 and £250 a year on the average. This was enough to subsist on but not to accumulate substantial savings. As a result, unexpected events, such as illness, publishers' bankruptcies, or a depression in the book trade, could have serious financial consequences. It also meant that it was impossible for them to retire. In the terms of their day, both Martin and Milner were considered to be professional writers on the sciences.[115]

Successful popularizers had to keep in mind that they were writing for an audience that expected to be entertained, as well as instructed. Clever popularizers of science paid attention to what was successful in other publishing genres. Subject statistics reveal a shift over the course of the nineteenth century from the publishing of religious materials toward more secular subjects, especially literature and, within literature, prose fiction. Slow at first, and almost imperceptible for the first half of the century, this shift gathered speed in the 1850s and 60s.[116] The popularity of prose fiction may have led popularizers to adopt a style of writing with a strong emphasis on entertaining storytelling. Greg Myers, Barbara Gates, and Ann Shteir have explored the narrative structure of the work of popularizers writing for a general readership. In his *Writing Biology,* a study of literary form and the social construction of scientific knowledge, Myers develops a typology that he applies strictly to the twentieth century. He asserts that all twentieth-century science writing falls into one of two categories, the "narrative of nature," or the "narrative of science." The "narrative of science" describes any scientific publication written to meet the standards of a discipline and designed to establish the credibility of a scientist within the scientific community. It emphasizes the work of the author or authors, and stresses the importance of the results for other scientists in the field. The scientific work itself involves experiment, perhaps in combination with model building. The argument is embodied in a present tense narrative describing a parallel series of simultaneous events all supporting the claims of the author(s).

Myers contrasts the "narrative of science," favored by professional scientists when they are writing for one another, to the "narrative of nature," which structures popular accounts of nature that are diverting, full of

114. Fyfe 2005a, 199.
115. Ibid., 205, 214, 216, 223.

116. Eliot 1994, 58; 1997, 1998.

anecdotes, and nontheoretical. The features of the "narrative of nature" allow the popular science writer to make science accessible to a popular audience. Here the plant, the animal, and the fascinations of the natural world, not the activity of the scientist, are the focus. Here the unmediated encounter with nature is detailed, rather than the expertise of the observer. Here the observer seeks the singular, not the typical, and his or her response to what seems remarkable or strange is deemed worthy of being noted. Animals are treated as individuals, sometimes even anthropomorphized. The chronological, past tense narrative allows an exciting story to be told to the reader as they follow the activities of plants or animals. The purpose of the story is to make the reader feel as though they are walking beside the naturalist on an adventurous quest for knowledge. Rather than repressing, or denying, the narrativity in science writing, the popularizer using this narrative structure openly acknowledges the storytelling that went into their "narrative of nature."[117]

In the introduction to their collection of essays *Natural Eloquence: Women Reinscribe Science,* Gates and Shteir argue that Myers's distinctions can usefully be applied to the Victorian era, not just the twentieth century. Gates has subsequently applied these distinctions with success to an examination of female popularizers in her *Kindred Nature.* Gates and Shteir claim that many mid- to late-century women utilized the "narrative of natural history," while practitioners adopted the "narrative of science."[118] Gates and Shteir's "narrative of natural history" is virtually indistinguishable from the "narrative of nature" observed by Myers in the works of twentieth-century popular science writers. Elsewhere I have argued that popularizers of the second half of the century, a significant number of who wrote fictional, as well as scientific works, drew on the fictional framework of natural history to create the "narrative of natural history."[119]

Popularizers of science also looked to other forms of mass entertainment, in addition to literature, for inspiration on how to amuse their readers. They would have noticed the London crowds attracted to panoramas and spectacles, and the success of heavily illustrated magazines. The pictorial character of popular culture was difficult to ignore. Panoramas were a popular form of entertainment in London in the first half of the century. Offering immense images of cities, historical events, or military battles, the panorama offered

---

117. Myers 1990, 142–89.   119. Lightman 1999.
118. Gates and Shteir 1997, 11.

the viewer a bird's-eye view of the scene.[120] The development and refinement of new reproductive media, including wood engraving, lithography, and photography, made diverse imagery widely available and affordable for the first time. The result was the rise of the illustrated weeklies from 1830 to 1860, such as the *Penny Magazine,* the *Illustrated London News,* and the *London Journal,* and the proliferation of illustrations in books.[121] After the invention of photography in the 1830s by Louis-Jacques-Mandé Daguerre and William Henry Fox Talbot, photographs became increasingly ubiquitous, especially after the 1850s.[122] A stream of spectacular new visual images bombarded Victorian audiences.

To some, scientific instruction could be enhanced by catering to the craving for the visual and the spectacular. The Crystal Palace's series of life-sized reconstructions of dinosaurs, opened in 1854, was designed as both a visual treat and as a factory for improving public taste.[123] Artisan instrument makers, eager to appear before the public as men of science, displayed electrical science in spectacles at such London venues as the Adelaide Gallery during the 1830s and 1840s.[124] In general, practitioners did not draw on the visual culture of the second half of the century in their lectures and publications, although Darwin seems to be an important exception.[125] Popularizers responded to the pictorial turn in Victorian culture by increasing the use of illustrations and vivid literary images in their books, and, in the case of those who were active speakers, by incorporating spectacles into their lectures.[126] They also explained the scientific principles behind the new optical gadgets that were invented during the nineteenth century, such as the camera,

120. Altick 1978.

121. Anderson 1991.

122. See Tucker 2005 for a study of how photography was used in science from 1839 to the end of the century. The introduction of photographic prints into books was not feasible until technical difficulties were overcome. But the appearance of the *carte de visite* in the 1850s led to the widespread cultural exposure of photography (J. Smith 2006, 31).

123. J. Secord 2004e.

124. Morus 1998.

125. Jonathan Smith argues that Darwin's account of aesthetics and the illustrations in his books were "carefully aligned with this vibrant visual culture rather than the high art of the Royal Academy and National Gallery." Smith has demonstrated that Darwin consciously grappled with the visual problem of illustrating his theory of natural selection in his books. Not only was the process-oriented theory almost impossible to illustrate directly, conceptions of species' fixity were entrenched in existing visual conventions of the natural sciences (J. Smith 2006, 32, 1).

126. Lightman 2000.

camera lucida, stereoscope, kaleidoscope, phenakistoscope, and spectroscope.

In his *Victorian Sensation*, James Secord has demonstrated how significant Chambers's *Vestiges* was in relation to the development of a scientific culture and the first industrial society. It was Chambers, and not Huxley, or any of the other scientific naturalists, whose experience in publishing had led him to understand how the communications revolution had transformed "the relations between knowledge, the market, and the reader."[127] A host of popularizers followed in his footsteps in the second half of the nineteenth century, taking advantage of the space he had opened up. Together with their publishers and readers, Chambers and his successors played an integral role in defining the meaning of science in modernity.

127. J. Secord 2000, 506.

—— ✳ ——

CHAPTER TWO

Anglican Theologies of Nature in a Post-Darwinian Era

IN HIS lecture "How to Study Natural History" (1846), the Reverend Charles Kingsley called attention to a startling change in the availability of natural history books since he had been a child. Now it was possible to buy these books "at a marvellous cheapness, which puts them within the reach of every one, and of an excellence which twenty years ago was impossible. Any working man in this town might now, especially in a class, consult scientific books, for which I, as a lad, twenty years ago, was sighing in vain." In fact, Kingsley stressed, twenty years ago the richest nobleman could not have purchased these books for the simple reason that "they did not exist."[1] A little less than fifty years later, another Anglican clergyman made a similar observation. In the preface to his *Extinct Monsters* (1892), the Reverend Henry N. Hutchinson drew attention to the explosion of popular books on natural history in the second half of the nineteenth century. He exclaimed that "never before was there such a profusion of books describing the various forms of life inhabiting the different countries of the globe, or the rivers, lakes, and seas that diversify its scenery. Popular writers have done good service in making the way plain for those who wish to acquaint themselves with the structures, habits, and histories of living animals."[2] A significant number of the authors of affordable science books for the public from the middle to the end of the nineteenth century were Anglican clergymen like Kingsley and Hutchinson.

1. Kingsley 1890, 309.   2. Hutchinson 1892, ix.

The clergyman-naturalist and clergyman-academic had played an important role in British science since the seventeenth century. John Ray, Joseph Priestly, John Stevens Henslow, Adam Sedgwick, William Buckland, and William Whewell all belonged to this distinguished group. Before the nineteenth century such men were often contributing members of the Royal Society and were honored for their work. During the 1830s clergymen played a key role in the founding of the British Association for the Advancement of Science (BAAS). For these men natural science and Christian theology were complementary, as were their clerical and scientific callings.[3]

It may seem somewhat surprising that Church of England parsons were so well represented within the ranks of popularizers of science in the second half of the nineteenth century. Theories in geology, biology, and physiological psychology in this period were taking on a naturalistic direction that made the attempt to reconcile science with revelation and theology more difficult. Studies on the professionalization of British science in the nineteenth century have shown that the numbers of clergyman-naturalists in positions of importance in scientific societies and institutions declined during this period. In his classic article "The Victorian Conflict between Science and Religion: A Professional Dimension," Frank Turner argued that from the 1840s onward "the size, character, structure, ideology, and leadership of the Victorian scientific world underwent considerable transformation and eventually emerged possessing most of the characteristics associated with a modern scientific community." Physics and chemistry faculties expanded and membership in major scientific societies increased. However, professors and scientific society members were, more and more, practicing men of science rather than wealthy amateurs or aristocrats. The "young guard" of science in the 1850s, men such as Huxley, and Tyndall, publicly championed the professionalization of science. By the 1870s they dominated the editorships, professorships, and offices in the major societies. Committed to a naturalistic approach to science, the "young guard" believed that science should be pursued without regard for religious dogma, natural theology, or the opinion of religious authorities. As a result, Huxley and professionally minded scientists worked to eliminate from their ranks the clergymen-scientists who saw the study of nature as a handmaiden to natural theology or as subordinate to theology and religious authority. Participation by Anglican clergy in scientific societies such as

3. Turner 1993, 183.

the British Association dropped significantly in the second half of the century.[4]

Huxley and his allies may have intended to drive Anglican clergymen out of the institutions and societies that they controlled, but their power did not extend to the periodical press and the great publishing houses.[5] The growth of a reading public eager to learn about natural history created an opportunity for members of the Anglican clergy to pursue their scientific interests as popularizers. They had the option to work with one of the religious publishers, such as the Society for the Promotion of Christian Knowledge (SPCK), which was firmly linked to the Church of England, or the interdenominational Religious Tract Society. By the middle of the nineteenth century both were publishing books on science. But with the exception of Charles Alexander Johns and George Henslow, they rarely worked with religious publishers. Nevertheless, they were able to draw on their authority as clergymen to speak in public about their views on science. Although they did not collaborate as a close-knit group akin to the scientific young guard's X Club, and though they were not leading a Church backed campaign to discredit scientific naturalism, through their work the religious ideals of the Anglican Church were present in efforts to convey the larger meaning of contemporary science to the British reading public. Kingsley, Francis Orpen Morris, Thomas William Webb, George Henslow, and William Houghton were all prolific writers, but they all remained in their clerical posts. Johns, an ordained priest, worked as a schoolmaster for most of his life. Only Ebenezer Brewer decided to pursue a career as a writer instead of assuming an official role within the Church.

Huxley and his naturalist friends not only failed to eliminate the power of the clergyman-naturalist tradition, they also were unable to destroy the appeal of themes drawn from a theology of nature. Despite their use of

---

4. Ibid., 179–87. While men of science were becoming professionalized, so too were the clergymen. The marks of professionalization within the Anglican Church included the rise of theological seminaries in cathedral closes, the establishment of church congresses on ecclesiastical topics (starting in the 1860s), the rise in the number of churches, awareness of social problems, and the specialization of Church literature and periodicals. Combined with the reform of the universities, the removal of religious tests, and new opportunities for employment of scientifically trained persons in government, in school boards, in the civic universities, and in industry, meant that the Church and ecclesiastical patronage were no longer routes to a scientific career. Turner asserts that by the third quarter of the century it was becoming increasingly clear "that to be a scientist was one vocation and to be a clergyman was another" (Ibid., 189).

5. Lightman 2004f, 199–237.

Darwin's theory of natural selection to undermine the argument that God's handiwork was evident in the many intricate adaptations of structure to function in living beings, the discourse of design by no means disappeared in the second half of the century. Although it seemed self-evident to agnostically inclined Darwinians that natural theology and natural selection were incompatible, others seized upon the traces of design embedded in Darwin's theory. As Richard England has asserted, "in the decades after the publication of the *Origin*, British men of science, clergy and philosophers turned their attention to just what adjustments evolutionary theory demanded for teleological arguments for the existence of God." These authors revised natural theology in light of evolutionary theory.[6] Brewer, Johns, Kingsley, Webb, Morris, Henslow, and Houghton all shared a determination to provide a religious framework for science, even after the publication of Darwin's *Origin of Species* in 1859. In this sense they presented a unified Christian agenda that differed significantly from that of the would-be professionalizers of science. They had allies among the Nonconformists, though popularizers like Philip Henry Gosse and Hugh Macmillan professed a form of natural theology with distinct differences.[7] But they adopted a variety of strategies to keep a Christian agenda for science before the eyes of the public. Although they spoke out of the clergyman-naturalist tradition, they modified it in order to offer a religious interpretation of nature in new forms of writing that were attractive to a diverse reading audience.

This chapter will explore the writings of Anglican clergymen, treating them as an important group within the ranks of Victorian popularizers.

6. England 2003, v–viii.

7. Gosse has recently been the subject of renewed interest. Ann Thwaite's first-rate biography explores his activities as scientific author in detail (Thwaite 2002). Amy King has stressed the parallels between Gosse's observational style and literary realism (King 2005). Jonathan Smith has pointed out that Gosse's natural theology, in keeping with the evangelicalism of the Plymouth Brethren, was distinctive in its use of natural objects as symbols or types of truths. He echoes John Brooke's emphasis on how different versions of natural theology reflected various religious cultures (J. Smith 2001, 251–55). No doubt a more finely nuanced analysis than I have offered would do the same for differing religious cultures within Anglicanism. However, in examining the theme of natural theology my main focus has been on what unites the Anglican figures in this chapter. While Gosse has received a lot of attention from historians, partly due to his part in the controversies over evolutionary theory, other Nonconformists who took on the role of popularizer of science have been neglected. The Scotsman Hugh Macmillan, a minister in the Free Church, wrote seven books on science for popular audiences. Some of his books sold extremely well. *Bible Teachings in Nature* (1867), for example, reached at least nine editions, the last being published in 1893. More research on Nonconformist figures like Macmillan, comparing them to their Anglican brethren, would certainly illuminate another important facet of scientific writing for the public.

Anglican clergymen established themselves as authorities in a wide range of scientific areas, from astronomy to botany, entomology, geology, and ornithology. Regardless of the scientific topic of their books, these men shared the goal of providing a religious framework for science in opposition to the secularizing agenda of scientific naturalists. To them, nature was filled with religious significance. However, their diverse responses to Darwin's *Origin of Species* reveals how a wide array of strategies could be used to offer a discourse of design in a work aimed at a popular audience. Morris, who I will deal with first, chose to oppose evolutionary theory directly. I will then turn to Johns, whose strategy was to ignore Darwin altogether, and then to Webb and Brewer, who engaged Darwin indirectly. Then I will discuss Kingsley as one who attempted to co-opt Darwin into a new vision for natural theology. Finally, I will examine Houghton and Henslow, clerical authors from a later generation whose work appeared in the 1870s, 80s, and 90s. I do not mean to use the term "post-Darwinian era" to suggest that a revolution occurred in 1859 that led to the end of natural theology. I use it as an indication that Darwin's explosive book altered the intellectual scene to the extent that some Anglican popularizers of science felt compelled to think carefully about how to handle religious themes in their books. But others, like Johns, and even Webb and Brewer to some extent, did not perceive Darwin's impact to be so great that they had to respond directly to the challenge.

OPPOSING DARWIN: MORRIS AND THE "HANDIWORKS OF CREATION"

Particularly prolific, with over twenty natural history books to his credit, Francis Orpen Morris (1810–1893) did not start off as a writer (Fig. 2.1). When he was a student at Worcester College at Oxford his interest in natural history had been encouraged by John Shute Duncan, keeper of the Ashmolean Museum. However, after graduating in 1833 he served in a series of clerical posts.[8] It was not until he arrived at Nafferton in 1845 to take up the position of vicar that he began to undertake serious literary work. Here he began his *History of British Birds* (1851–57), which made his reputation as a scientific author while stimulating an interest in ornithology. It was the result of Morris's collaboration with Benjamin Fawcett, a bookseller and printer from the nearby town of Driffield who had made his name by

8. K. Smith 2004, 1427–29.

FIGURE 2.1 Francis Orpen Morris. (Rev. M. C. F. Morris, *Francis Orpen Morris: A Memoir* [London: John C. Nimmo, 1897], frontispiece.)

publishing illustrated children's books. Fawcett provided the illustrations and Morris contributed the text. The work was brought out in monthly parts, each part dealing with four species of birds and priced at one shilling. The price was kept low by Fawcett, who handled some of the work connected with the production of the book. The first part appeared on June 1, 1850, published by Groombridge and Sons, whose cautious first run of 1,000 copies did not meet demand. Sales increased and reprints of different numbers had to be done at Driffield. The low price, the vivid colored illustrations, and the popular writing style all contributed to the success of this work, which took seven years to complete. Fawcett eventually gave up his retail business and concentrated on book production and invented a new process for fine printing in colors.[9]

9. M. Morris 1897, 67–71.

While at Nafferton, Morris also completed his *History of British Butterflies* (1853), a sumptuously illustrated book with seventy-one colored plates, one for each species of butterfly covered in the book. Again, Morris worked with Fawcett, who engraved the illustrations, and published with Groombridge. A moderate success, sales of the *History of British Butterflies* required three printings after ten years, and eight editions by 1895.[10] The book appeared in parts, the first in the beginning of 1852. Morris told his readers where each butterfly could be located in Britain and the rest of the world, its size, colors, and food. The first-person anecdotes of other naturalists, recounting how they captured a specific specimen, were included to enliven the narrative. Morris also had more to say about religion and natural history than in the earlier book. In the introduction, Morris affirmed that God had implanted in every human mind "an instinctive general love of nature" or in "the works of God." A feeling of admiration moved young and old alike when they looked upon the "handiworks of Creation," whether it was "a rich sunset, a storm, the sea, a tree, a mountain, a river, a rainbow, a flower." Through his use of visual images and lush verbal descriptions Morris attempted to convey to his readers a sense of the beauty of nature so that he could elicit the same reaction. The colors of the orange-tip butterfly were a reminder of the endless varieties in natural objects that would never have occurred to the most fertile human imagination. "There is but ONE of whom we can say with truth, 'His work is perfect,'" Morris declared. Watching the red admiral butterfly in his "quiet retirement in the country," Morris gave thanks for the opportunity to "enjoy in tranquility the 'Thousand and one' beautiful sights in which the Benign Creator displays such infinite wisdom of Almighty skill."[11] Although politically a staunch Tory, Morris opposed partisanship in the Church and did not like to be thought of as belonging to a particular party. He once asserted, "for myself, as far as mere names go, I am not, never was, and never will be either a Low or a High Churchman; unless, indeed, I am both."[12]

After the publication of the *Origin,* Morris became a vigorous opponent of evolutionary theory. The depth of Morris's hostility is evident in his acrid exchange of correspondence with T. H. Huxley. Writing on September 16, 1869, Morris asked about Huxley's statement at the British Association meeting at Exeter that "all my objections to Mr. Darwin's theories had been already answered. I shall feel much obliged if you will tell me where I can

10. K. Smith 2004, 1428.   12. M. Morris 1897, 118.
11. F. Morris 1853, iii, 31, 84.

find these answers." Huxley's response, condescending to say the least, was sent on September 30. He wrote to Morris that all of the answers to his objections to Darwin's theories could be found in "five or six years' serious and practical study of physical and biological science" accompanied by instruction in the "principles and practices of inductive logic" and "a return to the 'Origin of Species.'" As a final dig, Huxley strongly recommended that Morris study the *Origin of Species* "with the same earnest desire to grasp their real meaning as, I doubt not, animates you when you read your Bible." Huxley denied that Morris had the training or expertise necessary to understand Darwin's work. Moreover, he implied that Morris's commitment to finding scientific truth was in question.

In his response of October 8, Morris sarcastically thanked Huxley for the suggestion of this course of study, and in return he recommended that Huxley enter, "as soon as possible," one of the "ancient colleges, or, better still, at one of the small new Halls at Oxford," and "I really have no doubt but that after your five or six years study therein, (such as I have long since gone through myself), you will, when you shall have passed your 'Little Go,' have learnt sufficient of Logic . . . to be able to understand and explain the meaning of the terms 'Petitio principii'; 'ignoratio elenchi'; 'undistributed middle' etc.—the want of which knowledge is the 'weak point of the disciplines of Darwin, and of their oracle himself.'" After implying that Darwin's and Huxley's grasp of logic was not even up to the standard of an Oxford undergraduate, Morris asserted that he was quite familiar with the *Origin*. "I have long since extracted whatever meaning there is to be found in Mr. Darwin's book on 'Origin of Species,'" Morris declared, "and have elsewhere given it the credit it deserves as a valuable collection of interesting facts."[13]

By "elsewhere," Morris was likely referring to the papers he read at the British Association meetings at Norwich (1868) and Exeter (1869), which were reproduced in his *Difficulties of Darwinism* (1869, Longman). Here Morris depicted Darwinism as a "flimsy fancy" and published his exchange with Huxley at the end of the book. Expressing a sense of "foreboding" at the news that Huxley had been appointed president of the BAAS for the year 1870, Morris asserted his right, as a life member of the British Association since 1844, to speak on behalf of a large number of members that "Section D should no longer be left in the hands of a small busy-body

13. Imperial College, Huxley Collection, XXXIII, 87–89.

clique who have banded themselves together to cry down every attempt to disabuse the public mind of the pernicious principles to which the doctrines in question necessarily tend." Whereas Darwin's supporters had claimed, when attacked, that only they could understand the meaning of the *Origin of Species*, Morris insisted, "every ordinary person of fair average ability is perfectly competent to approach the discussion." While the Darwinians attempted to cloud the issues by using "high-flown words which must in most cases leave all ordinary readers just where they were," Morris claimed that his book was written for the common reader. Morris maintained that species had remained basically the same throughout time and that any powers acquired in the course of life were not handed down to offspring. If the strongest creatures were ordained to prevail by a law of nature, Morris asked, why did the great dinosaurs vanish from the face of the earth and the rabbit and mouse flourish to this day? Morris then presented a whole series of specific questions concerning natural selection, asking why his opponents had been unable to give "a definite and intelligible answer" to any of them. He asked, "By what act of Natural Selection was the pouch of the camel formed," and "How are the electric organs in fishes accounted for by Natural Selection?" He moved on to questions about moral issues, asking, for example, how Darwin's theory explained the origins of the sense of ethical responsibility in humans. Morris then turned to the matter of domestication by humans, and argued that Darwin had never proved his point. The changes made by humans through breeding were artificial, not permanent.[14] Morris's *Difficulties of Darwinism*, together with his *All the Articles of the Darwin Faith* (1875, W. Poole) and *The Demands of Darwinism on Credulity* (1890, Partridge), were, as James Moore has put it, part of his crusade "to eradicate Darwinism from English intellectual life."[15]

Even Morris's less polemical works after 1859 were designed to undermine the credibility of evolutionary theory. His *Records of Animal Sagacity and Character* (1861, Longman) presented a series of anecdotes intended to provide "abundant evidence" of "the mental capacities of animals furnished by their actions." In the tradition of natural history, the anecdotes were taken from books, journals, secondhand from friends, and from his own experiences. Morris started with anecdotes on dogs by various authorities, including Sir Walter Scott and Plutarch, affirming their ability to think, to

14. F. Morris 1869, iv–v, 1–2, 11, 13–15, 35, 54.

15. Moore 1979, 196–97.

be wise, and to be faithful. After dealing with dogs in almost a third of the book, subsequent chapters focused on elephants, horses, cats, storks, geese, and others. In the preface, Morris argued that there was nothing irrational in entertaining the probability "of a future resurrection or restoration of the animal creation." He presented a series of quotes in support of this notion from men usually considered to have possessed great "mental powers," such as Bishop Butler, Tertullian, and the Reverend John Wesley. Finally, he maintained that no passage from the Bible contradicted the idea of animal immortality, while some appeared to sanction it. By endowing animals with a soul, Morris automatically granted one to humanity, thereby refuting Darwinian theory and its implied materialism. Morris later became a committed antivivisectionist, and in 1866 he was involved in founding an association for the protection of birds. Priced at five shillings, *Records of Animal Sagacity* was not a huge success in terms of sales. Of the close to one thousand copies printed in 1861, Longman had sold about 350 a year later. By 1876 almost 300 copies remained unsold.[16] Nevertheless, Morris produced a similar book in 1870 that focused exclusively on dogs, titled *Dogs and Their Doings* (S. W. Partridge). Whereas *Records* contained no visual images, this time he included over twenty-five full-page illustrations, and the book was priced at over seven shillings. Again, anecdotes from books, journals, and correspondents were presented as providing "facts" that proved that dogs displayed humanlike characteristics, such as bravery, fidelity, and wisdom. In one case, Morris claimed that a dog discerned that it was the Sabbath.[17] Morris, then, was one of a small group of popularizers who opposed Darwinism publicly, though he was the only Anglican clergyman. He was joined by Philip Henry Gosse, author of a series of popular works on marine life and a member of the Plymouth Brethren, and by Frank Buckland, son of the Oxford geologist William Buckland and author of the widely read *Curiosities of Natural History* (1857).

IGNORING DARWIN: CHARLES ALEXANDER JOHNS

Whereas Morris confronted Darwin head on, Charles Alexander Johns (1811–1874) virtually ignored him. Johns was the prolific author of more than fifteen science books published over a thirty-year span, the majority of them

16. F. Morris 1861, vii, xviii, xxvi; M. Morris 1897, 143; *Archives of the House of Longman, 1794–1914* 1978, B12, 15; B16, 334.
17. F. Morris 1872, 52, 61, 81, 135.

with the SPCK, from the early 1840s to the early 1870s. Some of his works, well-illustrated field guides for the amateur, were published right up to the middle of the twentieth century. *Flowers of the Field* (1853, SPCK) reached four editions by 1860, and thirteen editions by 1878. The thirty-fifth and final edition was published in 1949. *British Birds in Their Haunts* (1862, SPCK) went through twenty-five editions, the last appearing in 1948. Some of Johns's works were very cheap SPCK works, and they were likely to have had print runs in excess of 1,000 copies per edition.[18] The SPCK was firmly linked with the Church of England and focused on publishing materials for members of the Church.[19] His intended audience may have been an Anglican one. Although ordained as deacon (1841) and priest (1848) after graduating BA from Trinity College, London, in 1841, Johns spent most of his life teaching, writing, and studying natural history. He became headmaster of Helston Grammar School from 1843 to 1847. Later he established Winton House in 1863, a private school for boys in Winchester. In 1870 Johns founded and presided over the Winchester Literary and Scientific Society.[20]

Johns's early works in the 1840s explored the religious themes in the study of botany. In *Flora Sacra; or, The Knowledge of the Works of Nature Conducive to the Knowledge of the God of Nature* (1840, J. Parker), which sold for six shillings, Johns focused on a select number of diverse plants. In each section he juxtaposed a series of quotes drawn from the Bible, poetry, and prose, accompanied by an illustration, to lead the reader to think about the plant's religious significance. Johns drew from poets such as Cowper, Wordsworth, Coleridge, and Keble, and from prose writers such as Jeremy Taylor and Linnaeus. With its high literary approach, *Flora Sacra* does not seem to have been aimed at a popular audience. However, in his *Botanical Rambles* (1846, SPCK) Johns geared his presentation to the curious botanical beginner. He assumed a first-person narrative style, addressing the reader directly throughout the book. By adopting this literary technique, a reworking of the familiar format, Johns aimed to draw the reader into a participatory experience of nature. He begins by pretending that the reader is an acquaintance that had recently seen him return from a botanical ramble. "You asked me a few days ago," Johns wrote, "of what use were all the dried plants which I

18. I am indebted to Aileen Fyfe for pointing this out.

19. Fyfe 2004, 29.

20. Lightman 2004e, 1082–83; Boulger and Hudson 2004, 223.

FIGURE 2.2 An illustration of the heath. Pastoral scenes such as this enabled the reader to visualize the setting for the next ramble. (Rev. C. A. Johns, *Botanical Rambles* [London: SPCK, 1846(?)], 113.)

was so carefully fastening to paper." Instead of answering the question, the narrator had promised to take the reader on his next excursion. The reader ends up accompanying Johns on a series of rambles. The book is divided into eight parts, dealing with generalized natural settings that would be familiar to any reader: the meadow, the cornfield, the hedge bank, the wood, the heath, the mountain, the bog, and the seashore. Johns included a large number of illustrations in his *Botanical Rambles*. In addition to woodcuts of the plants discussed, full-page illustrations helped Johns to bring the reader with him to each new setting (Fig. 2.2).[21]

Before they set off for the meadow, Johns warned the reader that hard work was involved in studying botany. Not only did it require the development of good observational skills, it also demanded an appreciation for the divine design to be found in nature. While on a ramble in the cornfield, Johns criticized those who studied botany merely to "learn from it some new and useful property of a certain herb." When told that an examination of a leaf

21. Johns [1846], 3.

might provide "a glimpse of the creative and protecting wisdom God," the utilitarian botanist left "the study to others." Those who pursued scientific knowledge only "if they can thereby heap up another bushel in their barns" adopted an attitude of "thoughtless selfishness." By contrast, Johns insisted that even the "meanest and most insignificant part of the creation, is worthy of being looked into, because it is His work; and when we admire His consummate wisdom and goodness in causing the earth to bring forth grass for the service of man, we should not be influenced wholly or principally by selfish motives, but join to our gratitude for benefits conferred on ourselves an acknowledgement of His universal benevolence." Each botanical ramble was designed to illustrate another aspect of divine wisdom and goodness in nature. While traipsing through the hedge bank, Johns points to the nettle and its sting. Although "despicable" in appearance it "is furnished with an apparatus which may with truth be called wonderful. So minute, yet so exquisitely contrived! So simple, yet so perfect. A wise Creator must indeed have been engaged here!" Trips to the heath reveal how lichen and mosses, the "humble plants," also "serve as instruments in the hand of Providence for converting rocky districts into fertile pastures and woodlands." During the visit to the heath, Johns explained to his companion why the botanist felt such great delight in finding a new plant. "It *must*," Johns asserted, "however, afford him a fresh instance of 'the wisdom of God in Creation:' there must be something about it different from any other plant that he has seen. Possibly he may discover organs hitherto unnoticed, proving the skilful design of Him, who called it into being."[22]

During the 1850s Johns produced a series of works all written for beginners, but priced at different levels in order to reach different audiences. His *Flowers of the Field* was priced well above the others and targeted better-off segments of the middle classes. The fourth edition (1860) was sold by the SPCK for six shillings and eight pence. More or less a field guide of flowering plants indigenous to Britain, Johns listed them by classes broken down further by orders based on the Natural System of Jussieu and De Candolle. Religious themes were absent from this work. For each plant Johns included a brief description of the leaves and flowers, where the plant was commonly found in England, its various uses, and, quite often, an illustration. The price reflected the large number of illustrations and the book's massive length. But Johns presented *Flowers of the Field* as an introductory handbook. The

22. Ibid., 6, 26–27, 85, 156, 180–81.

opening section contained a description of the organs of plants and the terminology used to refer to them. Johns explained, "before a novice can commence the study of any science he must make himself acquainted with the terms employed by writers on that science." Nevertheless, he maintained that in a popular description of the plants growing wild in a single country it was not necessary to burden the reader with too many scientific terms. "The Author, therefore," Johns wrote, "has endeavoured to keep technical terms as much as possible out of sight." He also told the reader that it was not necessary to discuss the internal structure of plants or the functions of their various organs in a work "which professes merely to teach the unscientific how to find out the names of the flowers they may happen to fall in with in the course of their country rambles." Although he acknowledged that the knowledge of plants he offered was "not Botany; nevertheless it is a step towards Botany."[23]

In 1853, the same year as the *Flowers of the Field* was published, Johns wrote another book that also brought readers to the threshold of true botany. Johns wrote his *First Steps to Botany* with the lower classes in mind. He chose the National Society for Promoting the Education of the Poor as the publisher, which priced the book cheaply at one shilling and four pence per dozen. In this book Johns covered the elementary facts of botany, including the seed, the stem, the leaf, the flower, modes of inflorescence, and the fruit. No illustrations were included. Although the book was only thirty-two pages in length, Johns turned frequently to religious themes. In the section on leaves, Johns could not resist commenting on "the wisdom of Him" that created vegetables that absorbed carbonic acid exhaled by animals and produced oxygen. In his discussion of fruit, Johns acknowledged that "we cannot help admiring the goodness of God in thus marking so plainly this large tribe of plants (for it is a very large one), which men may use either as good or medicine."[24]

The year following the appearance of *Flowers of the Field* and *First Steps to Botany*, Johns produced yet another book, titled *Birds' Nests*. Published by the SPCK, at four shillings and eight pence, it was priced between the other two, indicative of how publishers targeted different audiences with different products. The book is notable in two respects. First, it was well illustrated, containing full-page color illustrations of eggs as well as black-and-white woodcuts of nests. Second, Johns was experimenting with a novel

23. Johns 1860, i–ii.   24. Johns 1853, 18, 29.

narrative technique that alternated between a fictional and a factual format, and that featured the father as mentor figure common to earlier science books that adopted the familiar format. The book was aimed at an audience of young boys. The first chapter told the story of a boy named Henry Miller who encounters three young trespassers attempting to take birds' eggs from a tree. After catching them in the act, Henry's father delivers a stern lecture to them on the biblical injunction against causing suffering to one of God's creatures. Henry is interested in what the boys were after, and with his father's permission he climbs the tree and finds a nuthatch nest containing several eggs. His father looks up the nuthatch in Yarrell's *British Birds* and conveys information to his son, and the reader, on its nest and eggs. This dramatic opening is followed by a nonfictional chapter where Johns addresses the reader directly, shifting from third- to first-person narrative, and offers hints on how the sport of "birds'-nesting" can be conducted humanely and for amusement. Whereas the "mere robber of nests" gains no knowledge, the "considerate and humane collector" leaves the nest undisturbed, observing and taking notes on the eggs and the habits of the birds. Johns then offers a classification system for eggs, based on size, shape, and color, though he realizes that some might consider this approach unphilosophical. "At the risk," Johns declared, "however, of incurring the censure of scientific men (if any such should condescend to read what I have written,) I intend to describe, in separate chapters, those eggs which are most alike, classifying them according to their colour and markings." The third chapter returns to Henry, who has become the embodiment of Johns's humane collector, followed by a chapter that begins to describe "Eggs White, Without Spots," the first grouping according to the system of classification laid down previously. In addition to discussing the number of eggs, color, size, shape, and the young, Johns provides information on where the nests are built and their composition.[25]

During the rest of the book Johns continued to alternate formats, and the insights they provide reinforce each other. In one of the descriptive chapters, Johns discussed how though the cuckoo put its eggs in the nests of other birds, the young later joined their own species. To Johns this was a telling illustration of divine wisdom. "Here, at least," Johns proclaimed, "the infidel must acknowledge the teaching of HIM who is invisible." In the next chapter, the narrator describes Mr. Miller as one who was acquainted

25. Johns [1854], 22, 25.

with the names and properties of British plants, insects, and birds and who was "in the habit of looking out for fresh instances of the 'Wisdom of God in Creation.'" Now, the narrator says approvingly, Henry was "following his example," though he was too young to understand "the hard words which it is necessary to use in scientific books." Just as his father had been able to "foster the taste for natural history" without the use of these books, Johns encouraged his young readers to study nature through his unconventionally double-formatted book. Johns used the fictional format to dramatic and didactic advantage at the end of the book. Henry and his father learn that one of the boys who had tried to steal eggs at the beginning of the book had now been caught killing a pheasant and was to be put in prison.[26]

Johns never publicly attacked Darwinism in his works appearing after the publication of the *Origin*. He virtually ignored evolutionary theory. Two of his books, published by the SPCK in 1859 and shortly thereafter, are written in the same style as his earlier works. In *Picture Books for Children: Animals* (1859) Johns again adopted a modified version of the familiar format. The narrator spoke directly to the reader as they examined a series of animals, such as the redbreast, the white bear, the fox, the mole, the weasel, the lizard, the pheasant, the swallow, the tiger, and others. Each chapter was formatted to be easily digested by young readers, ran about four pages on average, and contained a full-page black-and-white illustration. In the chapter on "The Red-Breast," the narrator learns from a friend, Mary Miller, whose tame robins provide a focus for discussion. In another chapter the narrator converses with a kangaroo that explains why he is designed so well. But in most chapters the narrator provided the reader with information about the animal and sometimes relayed an anecdote. As in the chapter on the kangaroo, religious themes arose on occasion. How does the swallow know when to migrate and how to find its way across the sea to return to its summer home? The narrator asserted "it can only be that GOD who gave them their lives, teaches them the seasons and guides them in the way." Although the lot of the flying fish seems to be an unhappy one—hunted by great fishes in the sea and by birds in the air—the narrator assured the young readers that "the great and good GOD who has fitted it so wonderfully for living in two elements, has no doubt made its life as happy as that of the lark soaring in the clouds, or of the minnow sorting in the brook."[27]

Priced at one shilling, *Sea-Weeds* (1860, SPCK), another small book but written for a slightly older audience, began with an imaginary trip to the

26. Ibid., 140, 142.   27. Johns [1859], 39, 51.

sea. Johns asked his readers to suppose that they have just disembarked from a train at a station a few miles from the shore. As in *Botanical Rambles,* Johns's literary technique draws the reader in as they explore the natural world together. When they come closer to the water the stunted vegetation leads them to doubt that they will find any life at all, but upon reaching the sea they are surprised to find strange and unfamiliar plants. This is Johns's cue to provide the reader with information on the three primary groups of seaweeds and on how to collect, preserve, and study them. Although there are no references to a divine being, Johns does emphasize the beauty of seaweeds. Not just poets, but even the unimaginative botanist "insist on the resemblance between sea-weeds and flowers." Twelve full-page color illustrations are included to demonstrate vividly the aesthetic attraction of studying seaweeds.[28]

Published later in the 1860s by the SPCK, *British Birds in Their Haunts* (1862) and *The Forest Trees of Britain* ([1869]) dealt more extensively with religious themes. In the preface, Johns explicitly denied that his *British Birds* was a substitute for Yarrell's more comprehensive *History of British Birds* (1843). Rather, Johns's book was intended to provide "the lover of nature with a pleasant companion in his country walks, and the young Ornithologist with a Manual which will supply his present need and prepare him for the study of more important works." A catalogue of British birds with brief descriptions ranging from one to five pages, each section contained a basic description in italics followed by information on geographical distribution, number of eggs laid, suitability for eating, literary references, anecdotes by other naturalists, and accounts of Johns's own encounters with the bird. Like some of his other books, *British Birds* was well illustrated, containing close to 190 black-and-white woodcuts. This pushed the price up to twelve shillings.

Two themes, drawn from a theology of nature, run throughout *British Birds.* First, Johns often pointed to the useful role that birds play as scavengers. He introduced this theme almost immediately in his discussion of the vulture. Due to the rapid decomposition of carrion in warmer countries, "the providence of God" stocked them "with carnivorous animals of great voracity" like the vulture, which was always ready to undertake its "useful and wholesome" work of scavenger. Although "stigmatized by gamekeepers with an evil name," the kestrel was an invaluable ally to the farmer as it

28. Johns [1860], 15.

ate destructive beetles and caterpillars, while the "little appreciated" barn owl rendered good service to the agriculturist by its consumption of rats and mice. Although some birds, like the lesser whitethroat, ate a portion of the fruit grown on trees, they also fed on insects in the spring and "may be looked on as instruments employed by the providence of GOD to protect from injury the trees which are destined to supply them with support when insect food becomes scarce." Second, Johns pointed to instances of the adaptation of means to an end. In his discussion of the common crossbill, Johns rejected Georges-Louis Buffon's assertion that the beak of this bird was a useless deformity. Yarrell, a less dogmatic but more trustworthy authority, asserted that the beak, tongue, and muscles in the crossbill were beautiful examples of the adaptation of means to an end. Johns agreed with Yarrell, describing in detail the crossbill's beak and how it was "exquisitely adapted to its work" of splitting, opening, and securing the contents of a fir cone.[29]

The Forest Trees of Britain handled religious themes in a similar manner. It did not match the twenty-five editions of *British Birds*, but its run of eight editions by the end of the century must be considered a moderate success. In his introduction, Johns constructed a religious frame for the entire book. He stated that the object of the book was to supply "the lover of nature" with such information on trees either "natives of Great Britain, or naturalized in it, as will impart additional interest to his wanderings in the country." Johns warned the reader not to expect the announcement of any new botanical discoveries or suggestions of new methods of planting. If the reader was interested in "exploring the wonders of nature as displayed in the more stately vegetable productions of his native country," Johns hoped to provide a stimulus "to fresh research." Then Johns assured the reader that "even his own slender amount of scientific attainments, can crowd the hedges and byways with countless miracles, which for the untrained eye have no being." Johns saw his task as disciplining the senses of the reader so that they could perceive traces of divine wonder in the natural world. Although he vowed to avoid scientific nomenclature as much as possible, he insisted that it was necessary for the reader to learn the terms associated with the anatomical structure of a tree. After discussing cells, woody fiber, pith, bark, and leaves, Johns returned to the religious significance of botany. The process whereby leaves take in carbonic acid gas and produce the oxygen required

29. Johns 1862, viii, 1, 24, 53, 103, 133, 226, 231.

by animals constituted an "arrangement of the all-wise Creator of the universe."[30]

The body of the two-volume work contained chapters on sixty trees and was organized according to the conventions of natural history. Each chapter included a physical description of the tree, its flower, nuts, and leaves; a discussion of whether it was native or of foreign origin; where it was to be found in Britain; its place in history; and its uses. The narrative was spiced up with anecdotes, with generous quotes from poetry by Chaucer, Shakespeare, Wordsworth, Goldsmith, and Milton, and with numerous black-and-white woodcuts. Darwin and the would-be professionalizer of botany, John Lindley, were among those presented as acknowledged scientific authorities. But their discoveries were pressed into the service of Johns's vision of a designed nature watched over by a caring God. The winged seeds of the sycamore suggested to Johns "pleasing and instructive reflections on the wise superintending Providence of the Almighty." After a lengthy explanation of the coolness of the beech grove in the summer, Johns recognized that the discussion might appear to the reader to be a "long and uncalled-for digression." Yet he was "unwilling to pass by any opportunity of drawing the attention of my readers to those instances of design on the part of our Heavenly Father, which, though mainly instrumental to the production of other effects, are greatly conducive to the comfort and enjoyment of mankind."[31] Those who read *Forest Trees of Britain,* and Johns's other post-1859 works, would have been forgiven had they concluded that either that Johns has not bothered to read Darwin's *Origin* or that he had read it and dismissed it as unimportant.

ENGAGING DARWIN INDIRECTLY (1): WEBB'S CHRISTIAN HUMILITY

The Rev. Thomas W. Webb was not one to seek out controversy. In 1883 he confessed to Arthur Ranyard, his friend and fellow astronomer, that it was not to his taste to sign his name to a review in *Nature.* "I love to be more quiet," he wrote, "and I would not give it any time when it might be more likely to lead to what I so cordially dislike—paper-skirmishing."[32] Nevertheless, Webb's emphasis on the relative ignorance of the astronomer

---

30. Johns [1869], 1: ix.

31. Ibid., 1: 99, 327.

32. Webb to Ranyard, February 2, 1883, Royal Astronomical Society, Archives, 296.

FIGURE 2.3 T. W. Webb, the father of amateur astronomy. (Rev. T. W. Webb, *Celestial Objects for Common Telescopes,* 2 vols. [London: Longmans, Green, 1904], frontispiece.)

and the glory of the divine heavens was actually quite provocative and represented an indirect response to evolutionary naturalism. In the preface to his *Celestial Objects for Common Telescopes* (1859, Longman), he informed his readers that "a personal examination" of the wonders of the heavens would lead to "the most impressive thoughts of the littleness of man, and of the unspeakable greatness and glory of the CREATOR."[33] Webb's book was designed to be a testimony to the wisdom and goodness of God. In his *Celestial Objects,* Webb presented a subtle theology of nature worthy of the Anglican clergyman-naturalist tradition.

An Oxford BA in 1829, and MA in 1832, Thomas William Webb (1806–1885) assumed a series of clerical positions after he was ordained in 1829 (Fig. 2.3). In 1856, he accepted the living of Hardwicke, a parish near the

33. T. Webb 1868, x.

Black Mountain district of Wales, on the western border of Herefordshire. Here Webb studied the heavens and wrote his *Celestial Objects*. He was also a prolific contributor to a wide range of periodicals that addressed a number of different audiences. He would have reached working-class readers in his essays for the *English Mechanic* and middle-class readers in his articles for the *Intellectual Observer* and the *Popular Science Review*. Starting in the early 1870s Webb also became a frequent contributor to *Nature,* the weekly established in 1869 as a forum for would-be professional scientists to communicate with the lay public. Webb was involved in several scientific societies. He was elected to the Royal Astronomical Society in 1852, served on the British Association's moon committee, and was active in the Selenographical Society.[34]

Historians have accorded Webb a prominent place in the development of amateur astronomy. "Modern amateur astronomy," Allan Chapman once asserted, "as a pursuit for serious observers whose principal motivation was pleasure, fascination, or the glory of God, as opposed to fundamental research, began in a Herefordshire vicarage in the 1850s."[35] To Richard Baum, Webb's *Celestial Objects* "did more to create observers and popularize astronomy than any other single volume."[36] *Celestial Objects* was a moderately successful seller. For the first edition of 1859, Longman published 1,000 copies, priced at seven shillings, of which a little over a half sold in the first year. The rest of the first edition was purchased by 1865. The second edition (1868) also had a print run of 1,000. So ten years after it first appeared, 2,000 copies of *Celestial Objects* had been printed. By the end of the century, Longman had sold approximately 5,500 copies total.[37] *Celestial Objects* did not have broad appeal, but since it was "designed for so limited a class as that comprised by working astronomers," as the reviewer in the *Observatory* pointed out, it was considered to be a success within the astronomical community.[38]

In the introduction to *Celestial Objects,* Webb declared that one of the goals of the book was to "furnish the possessors of ordinary telescopes with plain directions for their use, and a list of objects for their advantageous employment." By telling the serious amateur what to look for and how to

---

34. Baum 2004b, 2127–28; Robinson 2004, 858–59.

35. Chapman 1998, 225.

36. Baum 2004b, 2127.

37. *Archives of the House of Longman, 1794–1914* 1978, A5, 599; A10, 399; A13, 435–36.

38. "Webb's *Celestial Objects*" 1882, 11.

look for it, Webb was filling a gap in the astronomical literature of the day. Although materials existed to guide the amateur, "some of them are difficult of access," Webb asserted, "some, not easy of interpretation, some, fragmentary and incomplete." For the more advanced observer, he recommended Admiral W. H. Smyth's *Cycle of Celestial Objects* (1844). Webb actually imitated the structure in Smyth's *Cycle,* starting off with telescopes, observing practices, and helpful hints, before moving on to a detailed description of planetary and stellar objects the amateur could observe in an affordable telescope of low power.[39] In part 1, "The Instrument and the Observer," Webb discussed what he meant by a common telescope. "By 'common telescopes,'" Webb wrote, "are here intended such as are most frequently met with in private hands; achromatics of various lengths up to 5 or 5 1/2 feet, with apertures up to 3 to 4 inches; or reflectors of somewhat larger diameter, but, in consequence of the loss of light in reflection, not greater brightness." Webb went on to give suggestions on how to distinguish functioning from faulty instruments, how to set up, operate, and care for the telescope, and how to record observations. In part 2, "The Solar System," Webb discussed the sun, Mercury, Venus, the moon, Mars, Jupiter, Saturn, Uranus and Neptune, and comets and meteors. In each case, Webb drew attention to the distinctive features of each celestial object, and to the history of the observations and theories of famous astronomers from the past. Part 3, "The Starry Heavens," focused on double stars, clusters, and nebulae, and included detailed descriptive catalogues.[40]

For Webb, the primary purpose of studying the heavens was to bring the astronomer closer to God. His work as a popularizer of science was almost an extension of his clerical duties. In his introduction to his *Celestial Objects,* he insisted that "to do justice to this noble science" of astronomy meant appreciating firsthand "the magnificent testimony which it bears to the eternal Power and Godhead of Him" who made the heavens. But Webb's commitment to a subtle theology of nature is evident in his *Celestial Objects.* Comments on the religious meaning of astronomy were restricted for the most part to sections that provided an interpretative framework for reading the book, such as the introduction and the introductory portions of each major part. In the opening of part 3, "The Starry Heavens," Webb asserted that the instances of order and beauty in the solar system are sufficient to bring home to the astronomer the presence of the divine. "If the Solar System

39. Chapman 1998, 225.   40. T. Webb 1868, vii–viii, 1.

had comprised in itself the whole material creation," Webb stated, "it would alone have abundantly sufficed to declare the glory of GOD, and in our brief review of its greatness and its wonder we have seen enough to awaken the most impressive thoughts of His power and wisdom." But our solar system is "but as a single drop in the ocean," and Webb promised that the third section of his book will deal with thousands of systems that contain "more amazing regions, and fresh scenes will open upon us of inexpressible and awful grandeur." Yet, all of them are "bound together by the same universal law which keeps the pebble in its place upon the surface of the earth, and guides the falling drop of the shower, or the midst of the cataract." The catalogues of double stars, clusters, and nebulae were the keys to viewing "this great display of the glory of the Creator."[41]

In his later works Webb avoided a direct attack on Darwinism while continuing to deal with religious themes as he had in *Celestial Objects*. Faithful to his Christian beliefs throughout the fifty-three years of his ministry, and uninfluenced by controversy or reform movements that affected some groups within the Church of England, it is no surprise that Webb did not alter the way he presented astronomical information in a religious framework.[42] In the opening pages of *Optics without Mathematics* (1883, SPCK), Webb reminded his readers that light was one of the "very best and chiefest gifts of Him who is pleased to describe Himself under the name of Light." Throughout the book Webb stressed the wonderful properties of light, and then linked them in the conclusion to God's "wonderful works." The purpose of *Optics,* he declared, had been to teach the reader enough about light to awaken "a greater interest than we had before, in wonders that lie on every side of us, but pass unnoticed because we see them every day."[43] In *The Sun* (1885, Longman), Webb's goal was to overwhelm the reader with the magnitude of the heavens. The distance between the earth and the sun was so great that even if an individual took an express train traveling 60 miles an hour they would die before the journey of 175 years had ended. The sun itself was so large that "109 of our Earths touching one another would hardly cross the Sun from side to side" and an express train would take more than seven years to make a round trip. Webb concluded the book by asserting that the "magnificent Sun," a "mighty exhibition of Creative Power and Wisdom, is but one among the countless host of heaven;—is no other than a star."[44]

41. Ibid., 168, 170.
42. Robinson 2006, 72.

43. T. Webb [1883], 4, 121.
44. T. Webb 1885b, 26, 31, 78.

Webb's reflections on the greatness of the Creator were often coupled with discussions of the need for humility, which can be read as a response to the arrogance of scientific naturalists. This theme runs throughout his work, from the appearance of *Celestial Objects* in 1859 to his last essays in *Nature* shortly before his death. In the introduction to *Celestial Objects*, Webb pointed to the value of astronomy as a leisure activity that also led to "the most impressive thoughts of the littleness of man, and of the unspeakable greatness and glory of the CREATOR."[45] After the publication of the *Origin*, Webb began to make the theme of humility more central to his work. In his *The Earth a Globe* ([1865], Thomas Hailing), Webb emphasized the extent of human ignorance in the face of the vastness of the universe. The "great multitude of the stars stand merely as silent witnesses of their Creator's power, and refuse to answer the enquiries of man," he declared.[46] In a number of his articles in the *Intellectual Observer* in the latter half of the 1860s, Webb picked up on the same theme. An examination of the disagreement between astronomers as to the form of the shadows that Saturn and its ring mutually cast upon one another led Webb to note how ironic it was that "the mystery of the subject has increased under closer, more powerful, and more extended scrutiny." To those disappointed that no progress had been made in this area of astronomical study, Webb reminded them that research into the heavens brought its own compensation, an appreciation of God's great works.[47] The impossibility of determining whether or not a lunar atmosphere existed was the cue for Webb to make a plea for more research, which he "commended to those who love to trace the footsteps of the Maker of all things in the manifold exercise of His creative power."[48]

Due to his friendship with Norman Lockyer, editor of *Nature*, Webb was able to place a series of short articles on astronomy in this important journal in the 1870s and early 1880s that explored the theme of human ignorance in the face of the divine wonders of the heavens.[49] Although the unprecedented multiplication of telescopes in his time meant that Jupiter had never been "subjected to such an extended scrutiny as the present," Webb impressed upon the reader the disagreement between observers. As a result of the differences in the instruments used, in the sharpness of the eyes of astronomers, and in their experience, agreement on the appearance of Jupiter was hard to obtain.[50] Astronomers fared no better when it came

---

45. T. Webb 1868, x.
46. T. Webb [1865], 23.
47. T. Webb 1866, 201.

48. T. Webb 1867, 222.
49. Lockyer and Lockyer 1928, 22, 26.
50. T. Webb 1871, 430.

to the other planets. The best observers could not reach agreement as to the features of Mars.[51] Despite advances in optical power, little progress had been made in recent years in penetrating Saturn's mystery. "What material progress have we to boast of?" Webb asked. "What further light have the same instruments, or others of greater power, thrown on the minute subdivisions of the rings, or the abnormal and inexplicable outlines of the shadow of the globe?" Since Saturn was unique—astronomers could find no analogy in human experience—they were confronted by their "entire ignorance of the real nature of our subject."[52]

Many other celestial objects were equally elusive. In "The Theory of Sunspots," Webb again stressed the ignorance of astronomers, apparent in their inability to reach agreement as to the true nature of sunspots. The observer of the solar disk "knows absolutely nothing as to what he is looking upon," Webb declared. The best astronomers offered no help. "Shall we listen to Wilson," Webb asked, or "Herschel, or Kirchhoff, or Nasmyth, or Secchi, or Faye, or Zöllner, or Langley? More or less, they all disagree." In light of the protracted discussion over this issue, observers could "hardly bring to our telescope an unbiased eye or an impartial judgment." Solar phenomena lent themselves to "very dissimilar and even opposite interpretations." If the telescope disappoints us, Webb asserted, then perhaps the spectroscope could resolve the problem. Here, too, Webb found only equivocal evidence that was sometimes very perplexing.[53] Unsurprisingly, Webb found astronomers to be even more unenlightened as to the nature of celestial objects outside the solar system. In his article on "The Great Nebula in Andromeda," this mysterious object became a symbol of human ignorance. Webb speculated that astronomers had neglected the Andromeda Nebula, despite its "enormous magnitude," because it has "hitherto resisted all inquiry." After outlining the history of the nebula, he went on to discuss the disparity between observers using telescopes of the greatest power. Discrepancies were "illustrative of the uncertainty that hangs about such observations," and he concluded that the "telescope has comparatively failed." Even the spectroscope offered indecisive evidence as to the physical constitution of the nebula. As the largest body in the universe, Webb described it as "the greatest display as to magnitude of its incomprehensive Creator." He ended the article with a profound admission of the limits of astronomy. "And with these inquiries as to a mystery never in all probability

51. T. Webb 1880, 213.
52. T. Webb 1885a, 485.

53. T. Webb 1884, 59.

to be penetrated by man," Webb concluded, "our imperfect remarks shall close."[54]

Although Webb may have preferred to avoid controversy, his emphasis on the relative ignorance of the astronomer and the glory of the divine heavens was actually quite provocative, particularly when it appeared in the pages of a journal such as *Nature*. Professionalizing scientists like Huxley had hoped *Nature* would become the chief organ for disseminating a secular vision of nature to a popular audience.[55] Webb's insistence on the ignorance of astronomers and the corresponding majesty of divine creation flew in the face of the positivistic inclinations of evolutionary naturalists. Huxley and his allies were interested in stressing the boundless capacity of knowledge to increase and in convincing the common reader that scientists had already discovered enough knowledge to qualify them, and not the Anglican clergy, as the proper cultural authorities for the modern, industrialized age.

ENGAGING DARWIN INDIRECTLY (2): REVISING NATURAL THEOLOGY

In 1874, in the preface to the thirty-second edition of his *A Guide to the Scientific Knowledge of Things Familiar* (1847), the Reverend Ebenezer Cobham Brewer boasted that his book had attained "almost unparalleled success." Brewer claimed that a staggering 113,000 copies had been printed since the book had first been published. The immense sales of the book, he claimed, were "incontrovertible proof of its acceptability" to the scientific reader. He had gone to great lengths to ensure that the book contained accurate information. "The most approved modern authors have been consulted," he declared, "and each addition has been submitted to the revision of gentlemen of acknowledged reputation for scientific attainments." The object of the book was to answer about 2,000 of the most common questions asked about natural phenomena "in language so simple that a child may understand it, yet not so foolish as to offend the scientific." Question number twenty-two read as follows: "What should a FEARFUL person do in order to be most SECURE in a storm?" According to Brewer, the scientific reader should not have been offended by the answer: "Draw his bedstead

---

54. T. Webb 1882, 341-45.

55. Barton has shown that Huxley and his fellow scientific naturalists did not control *Nature*. Lockyer's handling of controversies in the early 1870s alienated Hooker, Tyndall, and Huxley, who ceased to be regular contributors (Barton 2004, 228).

FIGURE 2.4 Brewer at work at his desk at the age of eighty-two. (*The House of Jarrolds, 1823-1923: A Brief History of One Hundred Years* [Norwich: Jarrold Publishing, 1924], opposite p. 41.)

into the middle of his room, commit himself to the care of God, and go to bed; remembering that our Lord has said, 'The very hairs of your head are all numbered.'"[56] Brewer was convinced that the impressive sales of his book showed that readers accepted religious answers to scientific questions. Many of the answers in Brewer's *Guide* drew on a discourse of design. After the appearance of the *Origin*, Brewer wrote a book titled *Theology in Science* (1860). Although Darwin's name is never mentioned, Brewer intended in this work to demonstrate to his readers that evolutionary theory had not demolished the natural theology tradition.

After a distinguished undergraduate career at Trinity Hall, Cambridge, Ebenezer Cobham Brewer (1810-1897) devoted himself to his writing (Fig. 2.4). Brewer took his degree in the civil law (first class) in 1835, and later obtained the degrees of LL.B. in 1839 and LL.D. in 1844, though he ultimately decided against a legal career. He was ordained Deacon in 1834, and priest in 1836. But he never seems to have held ecclesiastical preferment. Although headmaster of King's College School, Norwich, shortly after the first edition of his *A Guide to the Scientific Knowledge of Things Familiar* first appeared, Brewer dedicated himself to writing books on a wide range of

56. Brewer 1874, v–vi, 22.

topics, including the literary, social, and political history of Europe, book-keeping, dictionaries of phrases and fables, school textbooks, and scientific works aimed at a popular audience. During the 1860s, Brewer wrote a series of introductory books on astronomy, chemistry, and science in general, which were published by Cassell and Company. He also worked closely with Jarrold Publishing in Norwich, having become a friend of the Jarrold family when one of the children was a pupil at his school.[57]

Brewer's *Guide to the Scientific Knowledge of Things Familiar* was among his most popular works. After an initial print run in 1847 of 2,000 copies priced at three shillings and sixpence, Jarrold and Sons could only keep up with the demand over the next few years by increasing the number of copies in each edition. Jarrold printed 3,000 copies of the second edition in 1848. It was followed by a third (January 1849) of 5,000, a fourth (July 1849) of 7,000, and a fifth (1850) of 8,000. From the fifth edition until the twelfth (1858), the print run held steady at 8,000 copies per edition, and then from the thirteenth edition (1859) until the thirty-second (1874) it was reduced to 4,000 copies. When Brewer asserted that 113,000 copies of the book had been bought by 1874 he had actually underestimated the sales. The Publication Register for Jarrold and Sons shows that over 160,000 copies had been printed by that date.[58] The print runs for Brewer's *Guide* are among the highest of any scientific book published in the second half of the nineteenth century. Within the first ten years of publication, Brewer's *Guide* reached 75,000 copies printed.[59] By 1892, when the book had reached its forty-fourth edition, the number had risen to 195,000 copies printed.[60] Brewer made a small fortune as a result of the extraordinary sales. His contract, dated November 5, 1847, stipulated that he would receive £20 for the first edition of the *Guide,* and a little over twelve pounds for every 1,000 copies sold from then on.[61] Jarrold paid him £207 for the first four editions, and then, when the print run jumped to 8,000 copies, Brewer began to earn £100 per edition.[62]

57. Lightman 2004a, 266–67.

58. Archives of Jarrold and Sons Ltd., Publication Register, 1848 to-, Brewer's Guide to Science.

59. Archives of Jarrold and Sons Ltd., Royalty Ledgers, Publication Register, 1848 to-, 7.

60. Archives of Jarrold and Sons Ltd., Royalty Ledgers [1876–92], Dr. Brewer's, 1, 7, 17, 35 46, 59, 6, 35, 50, 71, 102, 126, 13, 36, 60, 188, 108.

61. Archives of Jarrold and Sons Ltd., Brewer's Agreements and Letters.

62. Archives of Jarrold and Sons Ltd., Ledger, 13.

Advertised as a "book for the fireside and the school room" in the *Publishers' Circular*, Brewer's *Guide* sold well, in part, because of his focus on the "knowledge of things familiar."[63] "No science is more generally interesting," he declared in the preface to the thirty-second edition, "than that which explains the common phenomena of life. We see that salt and snow are both white, a rose red, leaves green, and the violet a deep purple, but how few persons ever ask the reason why."[64] Brewer's *Guide* was the type of book favored by those involved in the influential movement for teaching the science of common things in the 1850s. Leading proponents of English popular education, such as James Phillips Kay-Shuttleworth, endorsed an emphasis on the knowledge of common things in schoolbooks. This approach to science was seen as being particularly suitable for working-class children.[65]

Brewer's *Guide* may have also appealed to Victorian audiences due to its incorporation of religious themes in both its structure and content. The book was divided into two parts. The first part on heat included sections on various sources of heat, such as the sun, electricity, chemical action, and mechanical action, as well as questions on the effects and communication of heat. Part 2, on air, dealt with carbonic acid gas, carbureted hydrogen gas, phosphuretted hydrogen gas, wind, the barometer, snow, hail, rain, water, ice, light, and sound. The format was quite simple. First, a question was put to the reader, such as "Q. What is HEAT?" A short, often one sentence, answer followed, "A. That which produces the sensation of warmth."[66] The question and answer format would have been familiar to audiences through their reading of religious catechisms. Brewer was not the first to write a scientific catechism. In the second decade of the nineteenth century, William Pinnock, publisher and educational writer, produced a series of eighty-three educational works with the title "Catechisms" collected into the *Juvenile Encyclopaedia* (c. 1828). The Catechisms were among the first of the cheap educational books and may have even played a major role in establishing the genre.[67] Alan Rauch has argued that the scientific catechism represented an "acceptable transition from scriptural to secular material." Although seemingly secular in content, Brewer's *Guide* was still connected to a recognizable theological tradition.[68] Brewer's use of the question and answer format allowed him to place the presentation of scientific fact in a religious frame.

63. "Just Published" 1847, 432.
64. Brewer 1874, v.
65. Layton 1973, 95, 111–12.

66. Brewer 1874, 1.
67. Kinraide 2004, 1602.
68. Rauch 2001, 52.

Brewer also seductively waited until about a third of the way into the book before including more questions and answers that were shaped by a religious perspective. For example, after Brewer established that air was a bad conductor, he queried: "Q. Show the wisdom of God in making AIR a BAD conductor." The correct response was, "If air were a good conductor (like iron and stone) heat would be drawn so rapidly from our body, that we should be chilled to death." Brewer raised a series of similar questions, challenging the reader to link established scientific facts to divine wisdom or goodness:

Q. Show how the goodness of God is manifested, even in the clothing of BIRDS and BEASTS.
Q. Show the WISDOM of GOD in making the EARTH a BAD conductor.
Q. Show the WISDOM of GOD in making grass, the leaves of trees, and ALL VEGETABLES, excellent radiators of heat.
Q. Show the goodness and wisdom of GOD in this constant tendency of air to equilibrium.

Brewer included many more questions of this type in the last two-thirds of the book, leading the reader by the hand to discover the numerous signs of design scattered throughout nature. Brewer never developed a step-by-step argument for design in his *Guide*. The accumulated force of his many examples is intended to overwhelm the reader.[69]

Unlike Johns or Webb, Brewer systematically restated natural theology after the publication of Darwin's *Origin of Species* in his *Theology in Science; or, The Testimony of Science to the Wisdom and Goodness of God* (1860, Jarrold). A line on the title page indicated the intended audience: "a book especially suitable for Sundays both in schools and private families." For this audience Brewer adopted the "simplest language" and wrote short paragraphs so that they could be "committed to memory" as in a catechism. Priced at three shillings and sixpence, Jarrold and Sons sold 4,100 copies a decade after its appearance, and over 7,000 copies in seven editions by 1892.[70] In the preface, Brewer spelled out the main object of the book, "to point out indications of Divine wisdom and goodness in the phenomena that science has unveiled." Brewer claimed that *Theology in Science* was "wholly unique," largely owing to its scope. Divided into five sections, the book covered geology, physical geography, ethnology, philology, and astronomy (focusing

69. Brewer 1874, 183–84, 189, 215, 318.

70. Archives of Jarrold and Sons Ltd., Ledger, 55–56, 58–59, 82; Royalty Ledgers, 1, 103, 109.

on the extraterrestrial life issue). Brewer claimed that dealing with these subjects allowed him to tackle the most important questions of the day: "Does Geology contradict Scripture? Was the penalty, imposed upon Man for disobedience, extended to dumb animals? Did the deluge deposit the fossils? Can all mankind be of one blood? Are the languages of the world so diverse, because they were confounded at the building of Babel?" However, Brewer was careful to distinguish his approach to natural theology from Paley's. Whereas Brewer's goal was to "show how science is the handmaid of Religion, and confirms what Scripture has revealed," Paley's "admirable work, entitled 'Natural Theology,' takes a different line of argument." Paley's focus was on "the adaptation of certain *organs* and *functions* to the work they have to perform." Although he never mentioned evolutionary theory explicitly, Brewer seems to have recognized that Darwin's theory undermined the scientific credibility of a natural theology based solely on adaptation. His strategy was to outflank Darwinism, by constructing a much broader version of natural theology (Fig. 2.5). As a second safeguard, Brewer stressed that as a "handmaid of Religion," science had no validity outside a religious framework.[71]

In the first section of the book, titled "The World before Man," Brewer reviewed recent geological discoveries and accepted the fact that extinct plants and animals existed before the first human was created. He acknowledged that it was difficult to reconcile biblical passages from Genesis with the new time scale demanded by geology. Nevertheless, he offered a number of different interpretations that, though not problem free, were all "better than rejecting the great facts of geology, or supposing that science is antagonistic to revelation." Then Brewer pointed out that revelation and geology agreed on a number of crucial points, including that the universe had a beginning and was created by a divine power; that since its creation the earth had been changed by the agency of fire and water; that the work of creation was progressive; that the earth was round; that humans were the last of created animals; and that the surface of the globe was ravaged by a flood after the creation of humanity. Brewer then listed a series of proofs of the wisdom and goodness of God as revealed by geology. The manner in which soils are formed by the disintegration of rocks allows the world to be a garden. The geological activity that brought coal, rock salt, marble, chalk, and other valuable minerals to the surface for the benefit of humanity was

71. Brewer 1870, v–vi.

THE ROCKS AND ANTEDILUVIAN ANIMALS.

FIGURE 2.5 Brewer's illustration of earth's savage past that emphasizes the brutal struggle for existence. Using this illustration as the frontispiece of his book on the testimony of science to the wisdom and goodness of God implied that Darwin's theory could be contained within the larger framework of a revised natural theology. (Rev. Dr. Brewer. *Theology in Science; or, The Testimony of Science to the Wisdom and Goodness of God* [London: Jarrold and Sons, n.d.], frontispiece.)

another proof of divine benevolence. Acting as "safety-valves of the earth," volcanoes were evidence of divine wisdom and goodness, as without them continents would be torn asunder.

Part 2, "The World as It Now Is," dealt with the shape, weight, and dimensions of the earth. Again, Brewer found fertile ground for indications of divine wisdom and goodness, this time in the shape and density of the earth and in the physical differences between the old and new worlds. More signs flowed from a consideration of the ocean and the atmosphere. Often, he invited the reader to imagine what the world would be like if nature were organized differently. If there were more water, the sea would inundate our shores, destroy our ports, and devastate our fields. If there were less water, "it would change our coastline, dry up our bays and harbours, render our ports useless, and frustrate much of the skill and labours of man." Scientific study of the atmosphere revealed that the proportion of oxygen within the air was "wisely adjusted to our requirements." If, on the one hand, nitrogen were increased in quantity and oxygen diminished, fires would lose their strength, plants would wither, and living beings would perform their functions with difficulty and pain. If, on the other hand, the amounts of nitrogen were diminished and oxygen increased, the effects would be no less disastrous. The least spark would set combustible substances on fire and they would be consumed in the blink of an eye.[72]

In part 3, "Man in the World," Brewer discussed the wisdom and goodness of God as demonstrated in the physical diversity of the human race. Those races living in regions of the earth exposed regularly to the sun were "distinguished by high projecting cheek-bones, and eyes deep-set in the head," an arrangement that protected their eyes from the glare of the sun. The following section, "Man Dispersed over the World," dealt more with the biblical notion of "the oneness of the human race" rather than with natural theology themes. Brewer believed that the affinity of all the languages best known by philologists provided "an incontrovertible argument" for monogenesis. Finally, the very short fifth section on "The Plurality of Worlds" presented both positions on the vexed question of the existence of extraterrestrial life. Although he refused to take sides in the debate, he showed how both the pluralists and their opponents could be in line with the principles of natural theology. Strikingly, throughout the entire book, Brewer managed to avoid broaching the issue of Darwinian theory while constructing a revised natural theology framework for science.[73]

CO-OPTING DARWIN: KINGSLEY AND THE FUTURE OF NATURAL THEOLOGY

In his *Glaucus; or, The Wonders of the Shore* (1855, Macmillan), Charles Kingsley (1819–1875), Christian socialist, novelist, and liberal Anglican cleric, recommended a series of natural history books to his readers (Fig. 2.6). In his opinion, "all Mr. Johns's books are good (as they are bound to be, considering his most accurate and varied knowledge), especially his 'Flowers of the Field,' the best cheap introduction to systematic botany which has yet appeared." Although "self-trained in a remote and narrow field of observation," Kingsley reckoned that Johns had "developed himself into one of our most acute and persevering botanists, and has added many a new treasure to

72. Ibid., 100–109, 201, 207.   73. Ibid., 306, 325.

FIGURE 2.6 Charles Kingsley, novelist, Christian socialist, liberal Anglican, and popularizer of science. (Frances Kingsley, *Charles Kingsley: His Letters and Memories of His Life* [London: Paul Trench, 1885], frontispiece.)

the Flora of these isles." Then Kingsley revealed that "one person, at least, owes him a deep debt of gratitude for first lessons in scientific accuracy and patience,—lessons taught, not dully and dryly at the book and desk, but livingly and genially, in adventurous rambles over the bleak cliffs and ferny woods of the wild Atlantic shore."[74] Kingsley, of course, was referring to himself. He met Johns when he was a pupil at Helston Grammar School, where they became close friends despite the eight-year age difference. Johns cultivated Kingsley's passion for natural history by taking him on floral collecting expeditions in the Plymouth area.[75]

Although Johns had encouraged his interest in natural history while he was a student at Helston, Kingsley did not begin writing or lecturing on scientific topics until the middle of the 1840s. During his lifetime, Kingsley produced a series of books for the reading public, including *Glaucus, The*

74. C. Kingsley 1908, 309.

75. Even after Kingsley left school, the two remained friendly. Kingsley's son Grenville was a student at the school Johns had started at Winton House in Winchester. See Colloms 1975, 36–37; Martin 1959, 29–30; W. Brock 1996, 25.

Water-Babies (1863), *Madame How and Lady Why* (1870, Bell and Daldy), and *Scientific Lectures and Essays* (1880, Macmillan). Writing on science was only one activity among many for Kingsley, and not even the most prominent among his various literary projects. Beginning in 1836 Kingsley studied at King's College London. He entered Magdalene College, Cambridge, in 1838, was ordained in 1842, and then became curate of Eversley Church in Hampshire. His experiences working with the underprivileged parishioners led to a passionate sympathy for the poor and his involvement with Christian socialism. He became a leading figure within liberal Anglicanism, closely associated with Frederick Maurice, A. P. Stanley, and Thomas Hughes. He was appointed Reguis professor of Modern History at Cambridge in 1860, canon of Chester in 1869, and canon of Westminster in 1873.[76]

Kingsley's first foray into popularizing science was in 1846 when he delivered a lecture at Reading on "How to Study Natural History." Like Johns, Kingsley adopted an imaginative approach to natural history. As if to stress the point that the wonder of nature lay literally at everyone's feet, Kingsley showed the audience a pebble that he had picked up on the street as he came to the lecture. "It shall be my only object tonight," Kingsley told them. If only we listen to the pebble patiently, Kingsley insisted, it would tell "a tale wilder and grander than any which I could have dreamed for myself; [it] will shame the meanness of my imagination, by the awful magnificence of God's facts." Then the pebble tells its story, speaking through Kingsley's lips. Eons ago the pebble was a living sponge in the depths of a great chalk ocean. It eventually became a stone buried in chalk-mud for ages; then it dropped from the face of a chalk cliff far away and was carried by water until it became a pebble on the beach while Reading was a sandbank in a shallow sea. Kingsley stressed that the pebble's amazing story could only be heard by those who studied natural history, and he began to list the positive effects of such study on the logical faculty and the imagination. As to the latter, Kingsley pointed out that the human imagination would find "inexhaustible wonders, and fancy a fairy-land" not just in a pebble, but also in "the tiniest piece of mould or a decayed fruit, the tiniest animalcule." Kingsley also maintained that the study of natural history had a beneficial religious effect. "I have found the average of scientific men," he declared, "not less, but more, godly and righteous men than the average of their neighbours." Finally, Kingsley drew attention to the practical utility of natural history. The

76. Endersby 2004, 1138–40.

knowledge gained from natural history had enabled the English race to "replenish the earth and subdue it."[77]

Kingsley's love of natural history did not result in a written work until the mid-1850s when he spent the winter of 1853–54 at Torquay with his ailing wife. Here Kingsley became enthralled by marine biology. He spent hours on the shore collecting specimens, kept a daily journal, and sent off sea beasts, shells, and seaweeds to Philip Henry Gosse in London, with whom he had started to correspond. The upshot was his book *Glaucus* (1855), based on an article "The Wonders of the Shore" published in the *North British Review*.[78] In *Glaucus,* Kingsley addressed the Victorian father on summer holiday, appealing to the notion that even leisure time should be spent in the pursuit of something profitable. Instead of reading another silly novel or engaging in "ineffectual attempts to catch a mackerel," why not "try to discover a few of the Wonders of the Shore" that are there "around you at every step, stranger than ever opium-eater dreamed"? Natural history was not only a fascinating field of study, it had "become nowadays an honourable one," pursued by dukes and princes. "Nay," Kingsley declared, "the study is now more than honourable, it is (what to many readers will be a far higher recommendation) even fashionable." Natural history books were "finding their way more and more into drawing-rooms and schoolrooms" and knowledge of the subject was no longer "considered superfluous." Kingsley insisted that the study of natural history led to happiness, as an understanding of nature lifted the individual outside his petty world to a realm of higher meaning. "Everywhere he sees significances," Kingsley remarked, "harmonies, laws, chains of cause and effect endlessly inter-linked, which draw him out of the narrow sphere of self-interest and self-pleasing, into a pure and wholesome region of solemn joy and wonder."[79]

In Kingsley's opinion, fathers had a responsibility to teach natural history to their children. This "age offers no more wholesome training, both moral and intellectual, than that which is given by instilling into the young an early taste for outdoor physical science." The training provided by natural history was especially important for "middle class young men," the "frightful majority" of which were "growing up effeminate, empty of all knowledge but what tends directly to the making of a fortune." To undermine the attitude that natural history was a pursuit "fitted only for effeminate or pedantic men," Kingsley argued that the qualifications required for the

77. C. Kingsley 1890, 296, 298–99, 304, 308.   79. C. Kingsley 1908, 217–18, 220, 224.
78. F. Kingsley 1877, I, 404–5, 412.

perfect naturalist were akin to those needed for the ideal "knight-errant of the middle-ages." The naturalist must be strong in body, ready to face the elements in order to work in the field; he must be gentle, courteous, and capable of being friendly with the poor, the ignorant, and the savage, so he can collect valuable local information; and he must be brave, enterprising, and devoted to the pursuit of knowledge without reward. Kingsley's perfect natural historian seems to have been a close relative to his vision of a muscular Christian. In his sermons and stories Kingsley emphasized the need for action and physical prowess. Even in his description of the natural historian, Kingsley included a Christian dimension. For most of all, the naturalist needed to be reverent, "wondering at the commonest, but not surprised by the most strange; . . . able to see grandeur in the minutest objects, beauty in the most ungainly," and capable of estimating things spiritually, "by the amount of Divine thought revealed to him therein." He must believe that "every pebble holds a treasure, every bud a revelation."[80] *Glaucus* was widely praised by the press, lay and scientific, when it was first published.[81] Initially, it sold well, reaching four editions within four years of its appearance. Afterward, as more and more popularizers began to write about the seaside, it sold more slowly, and reached a fifth edition in 1873. By 1900 it had gone through ten printings.

After the publication of the *Origin*, Kingsley, like Brewer, also attempted to revamp natural theology, but he engaged evolutionary theory head on in the process. In pursuing this strategy he diverged significantly from his former teacher, Johns. Kingsley had been an early convert to Darwinism, responding enthusiastically to a prepublication copy of the *Origin*.[82] In the second edition of the *Origin*, Darwin tried to defuse religious hostility by drawing attention to Kingsley's positive response, referring to him as "a celebrated author and divine."[83] In the early 1860s, Kingsley established friendships with some of Darwin's leading supporters. He met Charles Lyell and Joseph Hooker in 1860. It was Lyell who seconded the motion to elect him a Fellow of the Geological Society in 1863.[84] Kingsley consoled Huxley on the death of his son Noel in 1860, initiating an intimate correspondence between them. Kingsley also wrote to Darwin frequently, keeping him updated on the progress that his theory was making within Anglicanism. On January 31, 1862, he wrote to Darwin about a trip to Lord Ashburton's, where

80. Ibid., 238–39, 241–42; Savage 1988, 326.
81. Brock 1996, 31.
82. Moore 1979, 306.

83. Endersby 2004, 1139.
84. F. Kingsley 1877, 2: 119, 153.

Bishop Wilberforce and the Duke of Argyle were present. He claimed that of the six men there, only one (likely Wilberforce) regarded Darwin's theory as absurd, and that this showed "how your views are steadily spreading." He wrote to Darwin again on December 15, 1867, reporting that the number of converts to Darwinism in Cambridge had increased in the past year.[85] Kingsley did more than merely report to Darwin on the growing success of evolutionary theory, he wrote a charming book, *The Water Babies* (1863, Macmillan), that introduced evolution in a nonthreatening way. He wrote to Maurice that in the book he had tried "to make children and grown folks understand that there is a quite miraculous and divine element underlying all physical nature."[86] However, Kingsley was still thinking through the implications of Darwinism for natural theology in the early 1860s, and *Water Babies* was not intended to offer a sophisticated perspective on this issue. The same year that *Water Babies* appeared, Kingsley wrote to Maurice, "I am very busy working out points of Natural Theology, by the strange light of Huxley, Darwin, and Lyell. I think I shall come to something worth having before I have done." At this point he believed that Darwin had destroyed the notion of an "interfering God," leaving a choice between "the absolute empire of accident, and a living, immanent, ever-working God," and he praised Asa Gray as having made the best step forward in reformulating natural theology.[87]

Kingsley's work on a form of natural theology inflected with evolutionism did not bear fruit until the end of the decade. Kingsley quit his Cambridge post in 1869 in order to spend more time on his scientific studies. Within the next few years he produced three major works that articulated a revised natural theology. On May 19, 1869, he wrote to W. B. Carpenter that he was "intending henceforth to devote myself to my first love, physical science, as far as is compatible with my parish duties," and he looked for help to those who, like Carpenter, had "been able to carry on through life a study which in my case has been interrupted for many years."[88] In that year Kingsley's *Madam How and Lady Why* began to appear in serial form in *Good Words for the Young,* subsequently published in 1870 in book form by Bell and Daldy. It reached a third edition by 1878, two more printings a decade after its first appearance, and four more printings by the end of the century. Although it was written for children, *Madam How and Lady Why* contained the key to Kingsley's updated natural theology, the separation

85. Ibid., 2: 135, 249.
86. Ibid., 2: 137.

87. Ibid., 2: 171.
88. Ibid., 2: 294.

between how and why. The structure of Kingsley's book was reminiscent of Johns's *Botanical Rambles*. It was divided into generalized natural settings, including the glen, the coral reef, and field and wild, though it also covered scenes of geological activity such as earthquakes, volcanoes, and glaciers. Strikingly, Kingsley adopted the dialogue format, by then abandoned by most popularizers. The adult narrator teaches a young boy by taking him on a series of imaginative rambles. Greg Myers has argued that Kingsley tried to revive the didactic dialogue as a strategy for resisting the naturalistic science of his time. Through the dialogue Kingsley could "teach Darwinism as a natural theology while avoiding its more unsettling aspects," Myers asserts.[89]

In the first chapter, "The Glen," Kingsley began to teach the reader how to see the design in nature. A glen might look dreary, but if you have trained your eyes properly you will see that it is "beautiful and wonderful,—so beautiful, and so wonderful, and so cunningly devised, that it took thousands of years to make it." The narrator knows this because a fairy named Madam How told him so. Whereas Madam How could be seen at work, there was another fairy "whom we can hardly hope to see," Lady Why. In order to see Lady Why, it was necessary to talk first with Madam How. So the initial sections of Kingsley's book dealt with how Madam How worked, through processes that turned out to be very much like the agents that geologist Lyell had emphasized in his *Principles of Geology* (1830–33). Lyell had argued that the earth's crust had been shaped largely through the action of slow-acting visible causes still at work. Kingsley highlighted these same geological agents in the way he depicted Madam How's modus operandi. In the case of the glen, Madam How had formed it slowly and patiently, using water. The narrator later asserted that Madam How's ways do not change and her laws have never been broken. "As that great philosopher Sir Charles Lyell will tell you," the narrator declared, "when you read his books, Madam How is making and unmaking the surface of the earth now, by exactly the same means as she was making and unmaking ages and ages since." Likening the work of water to Madam How's rain-spade, the narrator goes on to discuss her steam-pump (earthquakes) and her ice-plough (glaciers), as the main tools at her disposal to fashion the earth's crust.[90]

Lady Why is not glimpsed until the end of the book, after Madam How's operations are explored in coral reefs and in the diversity of living things. In the final chapter, titled "Homeward Bound," which helped Kingsley to

89. Myers 1989, 181, 197.   90. C. Kingsley 1870, 2–3, 20, 108, 115.

maintain the illusion that the reader had been taken on a trip, at least in the imagination, he remarked that to learn the laws of Madam How it was only necessary to look at the smallest thing. If you start asking questions about a "pin's head or pebble," you would find that the answer to one question would lead you to another question, "and to answer that you must answer a third, and then a fourth; and so on for ever and ever." There was no way to put a stop to the infinite regress. Once this was recognized, it was possible to catch a fleeting glance of Lady Why. "All things," Kingsley declared, "we shall find, are constituted according to a Divine and Wonderful Order, which links each thing to every other thing; so that we cannot fully comprehend any one thing without comprehending all things: and who can do that, save He who made all things?" The purpose in nature was only revealed once it was understood that Madam How was totally earthbound. Science, Kingsley was implying, yielded knowledge of how nature operated, not why it existed.[91]

In *Town Geology* (1872, Strahan and Company), based on a series of lectures delivered in 1871 in Chester, Kingsley laid out his vision of uniformitarian geology as informed by natural theology. *Town Geology* reached a second edition the same year in which it was published and ran through at least one more printing before the end of the century. In the opening pages he explained that he aimed to teach his readers "the method of geology" rather than its facts, to "furnish the student with a key to all geology, rough indeed and rudimentary, but sure and sound enough, I trust, to help him to unlock most geological problems which he may meet, in any quarter of the globe." Through geology, Kingsley hoped to teach his readers the scientific habit and method of mind. He believed that of all the physical sciences, geology was the "simplest and the easiest" to learn as it appealed the most to "mere common sense," required fewer difficult experiments and expensive apparatus, demanded less previous knowledge of other sciences, and contained little puzzling terminology. Geology was "the poor man's science." Kingsley maintained that learning a branch of science offered several advantages. It provided amusement during leisure hours, put the learner in touch with the important intellectual currents of the time, and granted admission to a brotherhood that "owns no difference of rank, of creed, or of nationality." Expanding on the last advantage, Kingsley pointed to Michael Faraday and Hugh Miller as examples of men of humble origin who "became the companions and friends of the noblest and most learned on earth, looked

91. Ibid., 348.

up to them not as equals merely but as teachers and guides." Kingsley suggested that the true path to political equality for all lay in the cultivation of the scientific frame of mind. He recollected that as a youth he believed that perfect freedom and social reform could be obtained by altering "the arrangements of society and legislation," but later he realized that change at the political level was effective only if society were composed of rational and wise men. From that point on he was determined to train himself to think scientifically and to make it his duty to "train every Englishman over whom I can get influence in the same scientific habit of mind." Kingsley implied that his earlier devotion to Christian Socialism lacked an engagement with scientific principles to be effective as a force for genuine social change.[92]

Kingsley then raised the question: "why should I, as a clergyman, interest myself specially in the spread of Natural Science? Am I not going out of my proper sphere to meddle with secular matters? Am I not, indeed, going into a sphere out of which I had better keep myself, and all over whom I may have influence?" Kingsley acknowledged that in the current climate science was seen as being "antagonistic to religion," and that a clergyman's duty was to "warn the young against it, instead of attracting them towards it." But he denied that he stepped beyond his role as clergyman and into a forbidden secular sphere when he lectured or wrote about science for the public. Nothing was secular, Kingsley maintained, especially "anything which God has made, even to the tiniest of insects, the most insignificant atom of dust." Since the laws of nature were really "the laws of God," the study of nature led to a knowledge of the "works and of the will of God." Having argued that nature revealed the mind and character of its creator, Kingsley asserted his duty, as a clergyman, to meddle in the so-called secular sphere. He asked rhetorically, "can it be a work unfit for, unworthy of, a clergyman—whose duty is to preach Him to all, and in all ways,—to call on men to consider that physical world which, like the spiritual world, consists, holds together, by Him, and lives and moves and has its being in Him?" As an Anglican clergyman in particular, Kingsley believed that he was charged with the responsibility of teaching natural science. He pointed to a hymn on the blessedness of the works of God sung in the Anglican Church service, titled the "Song of the Three Children." "On that one hymn I take my stand," Kingsley announced. "As long as that is sung in an English Church, I have a right to investigate Nature boldly without stint or stay, and to call on all who have the will, to investigate her boldly likewise." Here was

92. C. Kingsley 1890, 4–5, 18–20.

Kingsley's spirited defense of his right, as an Anglican clergyman, to inter-
vene in the debates swirling around Darwin's theory of evolution. Kingsley's
subsequent discussion of gradual geological change through such agents as
water and ice, planned on "Sir Charles Lyell's method," was designed to
demonstrate that "the great book of nature" is "the Word of God revealed in
facts."[93]

Kingsley's "The Natural Theology of the Future" (1871), contained his
most systematic statement of a revised natural theology in light of evolutionary theory.[94] Kingsley began by expanding on one of the central themes
in his *Town Geology:* the role of the Anglican Church in exploring the potential of natural theology. Kingsley presented a historical analysis of the
changing relationship between the Church and natural theology. From the
time of the founding of the Royal Society in the seventeenth century, the Anglican clergy had done more than the clergy of any other denomination to
develop a natural theology that kept "pace with doctrinal or ecclesiastical
theology." The three greatest natural theologians, Kingsley believed, were
Berkeley, Butler, and Paley—all Anglicans. Unfortunately, orthodox thinkers of the last one hundred years had not followed in their footsteps, as
Wesley had turned Anglicanism toward questions of personal religion. The
religious temper of England for the last two or three generations was therefore unfavorable to the development of a scientific natural theology, and
the result was the present divorce between science and Christianity. But
Kingsley maintained that a viable natural theology was still possible for the
Church of England, since its theology was "eminently rational as well as
scriptural." If the Church approached nature "with a cheerful and reverent
spirit, as a noble, healthy, and trustworthy thing," instead of as a fallen
world, then a "scriptural and scientific" natural theology could be envisioned. For just as science required a belief in the permanence of natural
laws, that was "taken for granted" throughout the Bible. The existence of
laws in nature, confirmed both by science and the Bible, guaranteed the
continuing relevance of some form of natural theology.[95]

Having at first addressed his clerical brethren, the next section of the piece
was aimed at the man of science, and it drew on the distinction between
"how" and "why" issues developed in his *Madam How and Lady Why.* To

93. Ibid., 22–23, 27, 33, 151.

94. Kingsley's "The Natural Theology of the Future" was first delivered as a public lecture at Sion
College on January 10, 1871. He used it as the "Preface" to his *Westminster Sermons* (1874), and
then later included it among his *Scientific Lectures and Essays* (1880).

95. C. Kingsley 1874, vi–xiii.

those who argued that scientific research did not reveal marks of divine design, Kingsley restated the basic premise of natural theology. "We can only reassert that we see design everywhere," he wrote, "and that the vast majority of the human race in every age" has seen it. To Kingsley, it was self-evident that "wherever there is arrangement, there must be an arranger; wherever there is adaptation of means to an end, there must be an adapter." Kingsley unabashedly asserted the existence of final causes, even though the modern scientific man was "nervously afraid" of mentioning them. Kingsley argued that the man of science had no business with final causes. "Your duty is to find out the How of things," Kingsley stated, "ours, to find out the Why." Kingsley included himself among the natural theologians, and he maintained that they were legitimately concerned with final causes. Through this division of science and natural theology into two spheres of authority, Kingsley hoped to put an end to the notion of a conflict between science and religion and to define a specific role for natural theologians in the post-Darwinian era.[96]

In the final section of the paper, Kingsley grappled directly with the implications of evolutionary theory for natural theology. He denied that the doctrine of evolution had done away with the theory of creation and with the idea of final causes. "We might accept all that Mr. Darwin, all that Professor Huxley, has so learnedly and so acutely written on physical science, and yet preserve our natural theology on exactly the same basis as that on which Butler and Paley left it," he avowed. "That we should have to develop it, I do not deny." The new theory did not interfere with the central ideas of a designer, contrivance, and adaptation. "If there be evolution," he declared, "there must be an evolver." For those who disagreed, Kingsley recommended that they read Darwin's *Fertilisation of Orchids*, a "most valuable addition to natural theology." Evolution, to Kingsley, was simply the divine means through which God worked. The details of the "how" could be left up to the scientists. The task for natural theologians was to take into account both the struggle for existence and the love and self-sacrifice to be found in nature.[97]

A LATER GENERATION OF ANGLICAN POPULARIZERS:
HOUGHTON AND HENSLOW

Houghton and Henslow began to popularize science well after the publication of Darwin's *Origin*. They were most active during the 1870s and 1880s

96. Ibid., xxi–xxiii.   97. Ibid., xx, xxiv–xxvi.

and belong to a later period in the history of popularizers than Brewer, Webb, Morris, Kingsley, and Johns. Their differing approaches to providing a religious framework for science reflect the same tensions within the earlier group and illustrate how strategies for dealing with evolutionary theory continued to be used later in the century. Whereas Houghton avoided discussions of evolutionary theory, Henslow directly confronted the implications of Darwinism. Versatile and prolific, William Houghton (1828–1895) wrote nine natural history books covering such topics as insects, fish, and the microscope. Houghton was educated at Brasenose College, Oxford University, obtaining his BA in 1850 and his MA in 1853. He was ordained deacon in 1852 and priest in 1853. From 1858 to 1860 he was headmaster of Solihull Grammar School. Fascinated by natural history, he became a Fellow of the Linnean Society in 1859. In 1860 he accepted the position of rector of Preston-on-the-Wild-Moors near Wellington, Shropshire, a post that he held until his death.[98]

Houghton's first two books were aimed at children, though not below the age of nine. In his *Country Walks of a Naturalist with His Children* (1869), priced by Groombridge at three shillings and sixpence, Houghton expressed the hope that his book would be "intelligible to boys and girls of nine or ten years old, with a little explanation from parents or teachers." The book was organized into walks, ten in all, titled by month. Not every month of the year was represented, and some months, such as May, June, and July, were the subject of more than one chapter. The book was illustrated with eight colored plates and numerous wood engravings, some of which were borrowed from Gould's *Birds of Great Britain.* Houghton adopted a version of the familiar format to draw in his young audience. The reader is addressed as if they are part of a group of children accompanying their father on spontaneous excursions into the country. In the first walk, the children come upon a swallow while taking advantage of a pleasant day in April "for a ramble in the fields." This is the father's cue to discuss different kinds of swallows and their migration patterns, interrupted from time to time by questions from the children. Just as he concludes, they see a splash in the nearby river, and a section on the water vole begins. In this first walk Houghton moves through moles, herons, kingfishers, and other creatures that could be seen on a country walk. Through this modification of the traditional dialogue, Houghton is able to provide a more fluid narrative. However, in other walks, the father is asked questions about animals of

98. Boase 1965b, 709.

which he has no firsthand knowledge, and he resorts to quoting natural historians at length. This serves as a reminder to the young reader that even a knowledgeable adult has limited experience that must be supplemented by reading natural history books. Explicit religious themes intrude only at the conclusion of the book. "Let us never forget our great Creator," the father tells the children, "who has made all the beautiful things we see around us." In contemplating "the works of the Almighty," the lesson to be learned is that "as all created things are fulfilling their appointed work, so we too should fulfil ours."[99] *Country Walks* met with moderate success, reaching its fifth edition a decade after it was first published, and at least six editions by the end of the century.

Houghton intended his *Sea-Side Walks of a Naturalist with His Children* (1870) to be a "companion volume to my 'Country Walks,' hoping that it may induce some of the numerous young people visiting the sea-side, to take an interest in the study of Marine Natural History." It too was published by Groombridge and sold for three shillings and sixpence. It was also a moderate success, reaching a fourth edition by 1880. Like the first book, it was illustrated with eight colored plates and many wood engravings, some from Gould's *Birds of Great Britain*. Again, it contained a series of walks—this time twelve—taken by the same family while on holiday at the seaside. The children are constantly discovering new curiosities requiring explanation. They encounter shells, sea plants, birds, fish, and other sea creatures. At the conclusion of the last walk, the religious meaning of their rambles is highlighted. The father expressed the hope that his children would all continue "to use your eyes in the examination of those countless forms of plants and animals which surround us on all sides, whether in the country or at the sea-side." For he believed that finding amusement in nature would lead to something more solemn. The book ends with a poem about the beauty and sublimity of the stupendous sea, reminding us "if overwhelmed by thee / Can we think without emotion / What must thy Creator be?" Both books climax with a recognition that a divine being lies behind the beauty and wonder in nature encountered on all of their rambles.[100]

Houghton's later works are geared more toward adult readers, and in them he struggled to work out a strategy for speaking authoritatively in a post-Darwinian environment while presenting a theology of nature. For Houghton, an investigation of nature was a devotional exercise and a means

---

99. W. Houghton 1869, iii, 1, 107, 153.   100. W. Houghton 1870, [iii], 153–54.

for learning more about the works of the biblical God, rather than an opportunity for proving that God existed or that He was wise, good, and omnipotent. Houghton's later works contained less explicit references to God and they also avoided any discussion of evolutionary theory. In his *The Microscope and Some of the Wonders It Reveals* (1871), which Houghton referred to an as "Elementary Hand-book," he emphasized the beauty and wonder of the microscopic world. Published by Cassell, Petter, and Galpin, it was priced at two shillings and sixpence. In this book, which quickly reached a third edition in 1872, Houghton pointed to the "beautiful iridescent hues" in the wings of some insects and the "great beauty" of insect eggs. Animal skin was composed of a "wonderful structure" while the study of the feet of insects would afford the reader "delight." The microscope offered an incredible display to the eyes of the observer. "One of the most interesting spectacles afforded by the microscope," Houghton declared, "is that which is furnished by the circulation of the blood." The circulation of blood in a young tadpole was "a most astonishing spectacle."[101]

Similarly, *Sketches of British Insects* (1875), a handbook for beginners sold at three shillings and sixpence by Groombridge, appealed to the sense of wonder and beauty in its readers. The metamorphosis of insects, though striking and remarkable, was matched and perhaps surpassed by the "still more wonderful spectacles" in other animals. "It is not easy to imagine," Houghton wrote, "anything more beautiful and delicate than the Lace-wing-fly, with its eyes of burnished gold, its wide gauze-like wings, reflecting varying hues of pink on green, according to the incidence of the angle of light.[102] All of the language of wonder reminiscent of natural theology is in these books but there is no attempt to demonstrate that design in nature proves the existence of a God. To enhance his appeal to the reader's aesthetic sensibilities, Houghton included a large number of visual images in both books. The *Microscope* contained full-page and some small illustrations, none in color. However, *Sketches of British Insects* was illustrated with colored plates as well as wood engravings (Fig. 2.7). Houghton's *British Fresh-Water Fishes* (1879, William Mackenzie) was even more lavishly illustrated. Full-page colored plates of each species were included, as well as a landscape scene in black and white of a freshwater location where the fish could be found (Fig. 2.8). Each section of the book discussed where the fish thrived in Britain, its size, its color, and whether or not it was edible, and if so, how

101. W. Houghton n.d., 5, 56, 58, 74, 81, 92.   102. W. Houghton 1875, 21, 62.

FIGURE 2.7 The vividly colored frontispiece signaled Houghton's intention to engage the aesthetic sensibilities of his readers. (Rev. W. Houghton, *Sketches of British Insects* [London: Groombridge and Sons, 1875], frontispiece.)

it tasted. Houghton believed that the description and the colored drawings would make it possible for "any one to identify any fish that may be met with." Interest in Houghton's book continued until the end of the century, as it reached a second edition in 1895 and a third edition in 1900.[103]

103. W. Houghton [1879], vii.

FIGURE 2.8 Colored plate of the ruff and Miller's thumb. (Rev. W. Houghton, *British Fresh-Water Fishes* [London: William Mackenzie, 1879], opposite p. 5.)

In a delicate balancing act, Houghton avoided "God-talk" as well as any discussion of evolutionary theory in his later books. He also deferred to professional scientists in order to bolster his own authority. *British Fresh-Water Fishes* was dedicated to "one of the most eminent of naturalists and most generous of men," George Busk. Houghton's *Gleanings from the Natural History of the Ancients* (1897) contained a dedication to Sir John Lubbock. Both Busk and Lubbock were chartered members of the X Club. In *The Microscope* Houghton included not only long quotes from the works of naturalists such as Gosse but also from professional scientists like Huxley and Lionel Beale. *British Fresh-Water Fishes* contained Houghton's thanks to Albert Günther, then keeper of the zoology collections at the British Museum, and a note that he had followed Günther's classification system in the book. He also expressed his gratitude to a fellow naturalist, Frank Buckland, for sending him specimens.[104] Anxious to persuade his readers that the contents of *Sketches of British Insects* were accurate, Houghton claimed in the preface to have adopted a classification system that had "the sanction of Entomologists eminent in their respective departments," and he listed the names of over fifteen authorities whose works "have been constantly before me and freely

104. Ibid., vii–viii, xxii.

FIGURE 2.9 George Henslow while at the Bury School, 1847–54. (Suffolk Record Office, Bury St. Edmunds, GD 502/121.)

used." In the body of the book he drew on the natural history tradition of earlier times, including literary references from ancient literature and contemporary poetry.[105]

Like Houghton, the Reverend George Henslow (1835–1925) was also faced with a delicate balancing act in his attempts to be properly deferential to practitioners while upholding his commitment to the religious significance of nature (Fig. 2.9). He was the son of John Stevens Henslow, Darwin's mentor, and he tried to remain on friendly terms with Darwin. But he became more and more uncomfortable with Darwin's theory of natural selection and de-emphasized its importance. Unlike Houghton, Henslow followed Kingsley's path of confronting evolutionary theory directly. Henslow was yet another prolific author, producing over half a dozen books on botany for a popular audience and roughly the same amount of apologetic works that aimed to reconcile science and religion. He was educated at Christ's

105. W. Houghton 1875, v, 37.

College, Cambridge, where he studied in the Natural Science Tripos and in Divinity. In 1858 he received his BA, and in the same year was appointed as curate for the parish of Steyning in Sussex. In 1861 he received his MA. From 1861 to 1865 he was headmaster of Hampton Lucy Grammar School in Warwick, after which he accepted another position as headmaster at the Store Street Grammar School in London, where he stayed until 1872. Henslow juggled his headmasterships and clerical duties. He assumed the position of curate of St. Johns Wood Chapel in 1868, and then became assistant minister of St. James, Marylebone, in 1870. But he also kept a finger in science. He was selected to be an examiner for the Natural Science Tripos at Cambridge in 1867 and in 1874 for botany in the College of Preceptors in 1874. He had also begun to lecture on botany at St. Bartholomew's Medical School in 1866. He was already writing books for the Victorian public in the early 1870s. During the 1880s, Henslow elected to put more of his time into his scientific work by taking on three new positions. In 1880 he was appointed Professor of Botany to the Royal Horticultural Society, a job that he held until his retirement in 1915. In his capacity as professor he lectured to the student gardeners and presented public botanical demonstrations at flower shows. He also held botanical lectureships at two London schools of higher education, Birkbeck College for mechanics, and Queen's College for women. Shortly after taking on these positions he resigned all of his clerical duties.[106]

Henslow saw himself as continuing in his father's footsteps, particularly in his educational activities. In the preface to his *Botany for Children* (1880, E. Stanford) he claimed that his father had been the first to render botany "capable of being taught to children not only with great simplicity, but also on a thoroughly scientific basis." He urged country clergyman to "follow his example." Henslow's *Botany for Children* began with a chapter on the parts of a plant, followed by chapters on the major classes, subclasses, and families. He dealt mainly with common British plants. Insisting that it was not enough to read about flowers, Henslow emphasized the importance of observation. In each chapter he instructed the reader to find the plant under discussion and then to be guided by him in a dissection exercise. Each chapter also contained a description of the plant's leaves, flower, and fruit; the time of year when the flower bloomed and where it was to be found; the way the flowers were grouped together; and how the plant was used by humans. He concluded the book with short chapters on the principles of variation and

106. Elliott 2004, 933–34; "Henslow, Rev. George" 1929, 488.

the principles of classification. Priced at four shillings, *Botany for Children* met with some success, reaching a third edition in 1881. Elementary and descriptive, *Botany for Children* contained no discussions of evolution or the religious issues it raised. In this sense it remained true to his father's memory. John Stevens Henslow died in 1861, limiting his participation in the debates surrounding his former student's theory. Although Henslow senior defended Darwin on occasion, his reaction to the *Origin* was cautious. Meanwhile, his son explored the relationship between evolution and natural theology in his other works. He is among those who de-emphasized the theory of natural selection in order to effect a reconciliation between evolution and Christianity.[107]

In 1865, following his father's death, Henslow began to correspond with Darwin on the advice of his brother-in-law, the botanist Joseph Dalton Hooker. For a time, Darwin became Henslow's mentor, just as Henslow senior had been Darwin's advisor. Henslow asked if he could consult Darwin on "one or two little botanical matters," including the irritability of the stamens in the flower of the *Medicago sativa*. In December he wrote to thank Darwin for his response and informed him that he had been commissioned by the *Popular Science Review* to write a précis of Darwin's paper on climbing plants.[108] In 1866, Darwin, at Henslow's request, loaned him books on hybridism for an article he was preparing for the *Popular Science Review*. Darwin also examined the proofs, and sent some criticisms on June 12, 1866. Although he was happy to help, he preferred that Henslow refrain from making it known that he had read over the proofs. Darwin was not sure he agreed with all of Henslow's conclusions. Henslow thanked Darwin for his "valuable criticism" on June 13 or 14, and a few days later told him that he had altered his paper in light of Darwin's remarks.[109] On March 20, 1868, Henslow initiated a brief correspondence on the relationship between natural theology and evolution. He asked Darwin for "candid criticism" on an article on "Natural Theology" for the *Educational Journal*. He was thinking of expanding it, but he wanted Darwin's criticism first. He affirmed his belief in evolution but also wanted to prove "the argument of Nat Theology to be as sound, if not sounder, on that Hypothesis as upon the old Creative one." Eight days later he thanked Darwin for his candid criticism as they had

---

107. Henslow 1881, viii–ix; Walters and Stow 2001, 170–173; Moore 1979, 221.

108. Burkhardt et al. 2002, 288, 317–18.

109. Burkhardt et al. 2004, 95–96, 99–100, 103, 117–18, 183, 201–2, 204–6, 210.

shown him "where I have not only failed to convey any meaning but I think failed in reasoning also."[110]

Henslow was to have misgivings about Darwin's theory of natural selection, which manifested themselves in both his scientific works and his theological writings. In his *The Theory of Evolution of Living Things* (1873, Cambridge), Henslow attempted to demonstrate that a revised natural theology could accommodate evolutionary theory. In part 1, Henslow presented the evidence for evolution. He asserted that the theory had "not yet received that uniform acceptance to which it is undoubtedly entitled" and he regretted that many theologians still supported the "Creative hypothesis" so obstinately. There were two reasons for the resistance. First, theologians held an erroneous idea of the method of creation derived from a misreading of Genesis. Second, due to their lack of scientific training they could not appreciate the "arguments of the scientific man." Henslow then boldly criticized those clerical brethren who stubbornly refused to accept the truth of evolution and who thereby undermined efforts to rehabilitate natural theology. "In order, therefore, that the proof of the Wisdom and Beneficence of the Almighty as shown in the processes of Evolution may not be considered as based on unsound premises," he insisted, "it will be desirable to point out the untenableness of the present theological position, as well as the grounds upon which Evolution is founded."[111]

Although Henslow was attacking Christian theologians, he maintained that he was a devoted believer. He assumed "God to be the Author of Creation and believes Him to have adopted Evolution as the method by which He chose to bring about the existence of successive orders of beings until Man appeared upon the scene of Life." Henslow maintained that the argument from design was actually strengthened by the theory of evolution. Paley himself had alluded to mechanism as proof of an intelligent maker and added that our idea of the greatness of that intelligence would be enhanced if watches could produce offspring like themselves. The degree of intelligence would appear to be far greater if the maker could have infused into the watch "the law that slight variations should appear in the offspring, which, by accumulating in successive generations, ultimately produced all the varieties not only of watches but also of clocks of every description in the world. Yet this would be exactly analogous to the production of animal

110. Henslow to Darwin, March 20, 1868, and March 28, 1868, Cambridge University Library, Charles Darwin Papers, MS.DAR.166.164, MS.DAR.166.165.

111. Henslow 1873, 1–2.

and vegetable organisms by Evolution." The design perceived in a structure pointed to a divine intelligence "irrespective of the *process* by which that structure was brought into existence." Henslow made it clear that he was defending evolution, not Darwin's theory of natural selection. While Darwin's argument in the *Origin* posited the imperfection of the geological record in order to explain why intermediate forms were absent, Henslow's position was that there were enough transitional forms in the geological record and enough living forms in existence to support the doctrine of evolution.[112]

In the second part of the book Henslow proposed to focus on human evolution, and his growing disagreement with Darwin became even clearer. Here Henslow sided with Alfred Russel Wallace and denied that humans had evolved through exactly the same process as other living things. The gap between humans and animals led him to argue "that Man cannot have been evolved solely by Natural laws, at least such as we are acquainted with in the Evolution of plants and animals." Yet, Henslow acknowledged that evolutionary theory, with its recognition that evil was an integral part of nature, required a readjustment of natural theology. He was critical of Paley and other natural theologians who tended to base their arguments on the analogy between human and divine design, as this implied that contrivances existed only for the benefit of humanity. Design benefiting one organism could be utterly destructive for another organism, as in the case of the ichneumon fly that laid its eggs inside the caterpillar. The world, Henslow believed, was "inideal" or relatively perfect. If everything were perfectly adapted to human wants and desires there would be no progress. The "inideal" condition of things was a result of the evolutionary process and would continue until the end of the world. God had chosen to make struggle for existence a law of nature in order to prepare humanity for heaven. In contrast to previous natural theology, Henslow declared that his new theodicy highlighted "the recognition of physical evils of the world as part, and a very important part, of the scheme of Creation as bearing upon the probationary condition of man."[113]

After breaking with Darwinian theory, Henslow was free to explore the religious significance of botany in his science books for the general reader. His *Plants of the Bible* (1896), written for the Religious Tract Society and priced at one shilling, belonged to a natural history genre that brought contemporary scientific knowledge to bear on the Bible. Johns's *Flora Sacra*

112. Ibid., x–xi, 31, 39. 113. Ibid., 115, 161, 191, 197, 214, 217.

was indebted to that genre. Henslow's book was arranged into chapters on textile materials, herbs, odorous gums, resins and perfumes, fruit and timber trees, and desert trees and plants. Drawing on the tools of both the natural historian and the biblical scholar, Henslow identified the 120 plants mentioned in the Bible and outlined their uses in ancient times, how they came to be in Palestine, where they were cultivated, and how they operated as a religious symbol. In the section on grains, Henslow even pointed to the analogy between a religious symbol and a scientific theory. "The description which our Lord gives of the various conditions in which the grains find themselves when they fall into the hands of the sower," Henslow wrote, "and their resulting efforts to germinate and grow, is as beautiful an illustration of what is called 'natural selection' as can well be found." *Plants of the Bible* was intended to interest religious readers in botany and to attract scientific readers in the study of the Bible.[114]

Another book intended for the classroom, *How to Study Wild Flowers* (1896), contained more discussion of religious themes in botany than the earlier *Botany for Children*. Published by the Religious Tract Society, it sold for two shillings and sixpence and reached a second edition in 1908. In the preface, Henslow declared that the object of the work was to "familiarize the beginner with the majority of the commonest of our wild flowers." But he also hoped that the book would lead the young student to look more deeply into the mysteries of plant life, "and when trying to trace out the effects of causes, and to discover the origin of things, he will be thereby led to see how—judging by analogy from the capacities of man—there must be a Mind somewhere and somehow directing the many forces of nature." Henslow then referred directly to Paley and the watch analogy in *Natural Theology*. "So far from the supposed truth underlying Paley's celebrated argument of the watch being disproved at the present time," Henslow asserted, "scientific knowledge of to-day greatly extends that argument." If the mechanism of a watch reveals the presence of an intelligent maker, how much more would the discovery of its having "self-repairing powers if it went wrong—such as all animals possess—be proof of a far greater power and skill than man possesses?" Having provided in the preface a natural theology framework for the book, Henslow avoids extended discussions of religious themes, although there is an emphasis on the power of adaptation and some discussion of allusions to flowers in the Bible. In the introduction Henslow outlined the parts of the flowering plant, their division into

114. Henslow n.d., 63.

dicotyledons and monocotyledons, and the issue of classification. For the rest of the book he examined, systematically, the common British wildflowers, their uses, and where they could be found. The book also contained twelve color foldout plates.[115]

Part of the prestigious International Scientific Series, Henslow's *The Origin of Floral Structures* (1888, Kegan Paul) spelled out in more detail his critique of Darwinism.[116] In the preface he asserted that "the belief that we must look mainly to the environment as furnishing the influences which induce plants to vary in response to them—whereby adaptive morphological (including anatomical) structures are brought into existence—appears to be reviving." Henslow then recounted the development of his own opinions. He recalled, "having been early and greatly interested in Paley's 'Natural Theology,' and well as the 'Vestiges' when Mr. Darwin's work appeared, the great difficulties I felt in accepting natural selection as any real *origin* of species lay, first, in the seeming impossibility of the histological minutiae of the organs in adaptation having been selected together; and, secondly, in the idea that all those wonderful and 'purposeful' structures which Paley thought could only have been 'designed,' could be the ultimate result of any number of accidental and apparently at first 'purposeless' variations." Since the publication of the *Origin*, Henslow asserted that he began to put more emphasis on the effect of the environment, as did Wallace, Spencer, and even Darwin. Henslow did not reject natural selection completely. To him, the true Darwinist used Jean-Baptiste Lamarck to explain modifications of structure, and then brought in natural selection to account for extinction. Henslow warned that the idea of natural selection was often used to "hide our ignorance of its *concrete* representatives, that is to say, the real causes at work to induce change." In his *Origins of Floral Structures*, Henslow aimed to "refer every part of the structures of flowers to some one or more definite causes arising from the environment taken in its widest sense." He emphasized the role of insects as the cause of changes in flowers. Henslow's break with Darwin was even more evident in his *The Origin of Plant Structures by Self-Adaptation to the Environment* (1895, Kegan Paul). Also part of the

---

115. Henslow 1908, iii–iv, vi.

116. Henslow's work appeared after Huxley and his friends were no longer involved in the series. *The Origin of Floral Structures* was not a huge publishing success in Great Britain. Selling at five shillings, twelve years after it first appeared not much more than 2,000 copies had been sold in Great Britain (*Archives of Kegan Paul, Trench, Trübner, and Henry S. King, 1858–1912* 1973, C1, 212-15). However, as part of the arrangements for the series it was published in the United States by Appleton and Company.

International Scientific Series, it sold for five shillings. Here he pushed the Lamarckian tendencies in his evolutionary theory even further. He declared that the sole causes of structural modifications in organisms were because of the direct action of the environment and the responsive power of proto-plasm.[117] Henslow moved even further away from evolutionary naturalism when he later developed an interest in spiritualism.[118]

The Anglican clergymen who wrote for the mass reading audience during this period were not willing to concede dominion over science to the scientific naturalists. They were joined by Anglican laymen, and by Nonconformist popularizers, in their attempts to retain a place for religion in science. This group of popularizers aimed to frustrate the plans of scientific naturalists to secularize nature. They form the missing link between the natural theologians of the first half of the century and Peter Bowler's group of intellectually conservative scientists, liberal religious thinkers, and popular writers who in the early twentieth century attempted to effect a reconciliation between science and religion. Bowler argues that a "new, nonmaterialistic science had emerged that could serve as the basis for natural theology appropriate for the modern age," based on a notion of evolution as a moral force controlled by minds that acted as divine agents.[119] Brewer and his fellow Anglican clergyman not only helped to keep religious themes alive in the minds of the Victorian public in the post-Darwinian age, they also creatively developed them in new directions and in new ways.

117. Henslow received a royalty of nine and half pence per copy on all copies sold after 1,250 from Kegan Paul, Trench, Trübner, and Co. (*Archives of Kegan Paul, Trench, Trübner, and Henry S. King, 1858–1912* 1973, F, Reel 21, September 21, 1894). But the book did not sell well. Of the initial run of 1,250 copies, a little more than 950 copies sold by 1906 (*Archives of Kegan Paul, Trench, Trübner, and Henry S. King, 1858–1912* 1973, C26, 137-38).

118. Henslow 1888, v-vi, xi, 179, 335; Gershenowitz 1979, 25-30; Elliott 2004, 934.

119. Bowler 2001, 3, 407.

———— ✳ ————

CHAPTER THREE

Redefining the Maternal Tradition

IN THE late 1850s, an anonymous writer for the *Englishwoman's Domestic Magazine* proudly announced that literary women had made "great progress" within the last few years due to the introduction of new printing technologies. The steam press had "effected nearly as great a revolution in letters as the invention of printing itself." This was followed by a huge demand for "fresh wholesome food" to feed "that living monster." To keep up with the demand, "many new pens were dipped in ink, and not a few of these quills were held by female fingers." The author of this article declared that publishers were now just as willing to consider literary work of women as from men. To prove the point, the journalist launched into a discussion of the principal female writers of the period, covering journalists, translators, novelists, historians, travelers, and scientific authors. Jane Marcet, Jane Loudon, and Rosina Zornlin were mentioned by name as important natural history writers. Mary Somerville was singled out for praise as being "a genius of the highest order" who produced "books of which no man need be ashamed, and which few men could produce." There was no doubt in the mind of the anonymous writer that the recent works by literary women were "but the first-fruits of a rich harvest, for, since the rough ploughing of the literary field has produced such results, what may we not expect when the ground shall be carefully cultivated, watered and kept?"[1]

The prediction of a "rich harvest" of literary fruit by women proved to be accurate. Women distinguished themselves as novelists, but the second half

1. "Literary Women of the Nineteenth Century" 1858–59, 341–43.

of the nineteenth century was also a golden age for female popularizers of science.[2] They wrote about virtually every aspect of the natural sciences, though natural history topics tended to dominate. Lydia Becker, Phebe Lankester, Anne Pratt, Elizabeth Twining, and Jane Loudon all explored the world of botany. Arabella Buckley and Alice Bodington wrote primarily on evolutionary biology, while Margaret Gatty was more interested in marine biology. Other women, such as Mary Roberts, Anne Wright, Sarah Bowdich Lee, Annie Carey, Eliza Brightwen, and Elizabeth and Mary Kirby, moved across topics in natural history, from geology, to conchology, ornithology, and entomology. A smaller number of women tackled natural philosophy. Agnes Clerke and Agnes Giberne concentrated on astronomy. Some women were agile and knowledgeable enough to range over both the physical and life sciences. Mary Ward covered astronomy in one book and the use of the microscope to study living things in the other. Rosina Zornlin penned works on electricity, geology, geography, astronomy, and hydrology. Mary Somerville began with astronomy, but also dealt with other physical sciences and the life sciences in her later books. Enough women were involved in writing (though not lecturing[3]) about science in the second half of the century that several of them should be included along with Brewer, Kingsley, and the other Anglican clerics as influential figures in the popularization of science.

There are good reasons to deal separately with female popularizers of science in this period, at least initially. The men of the last chapter drew on their authority as clergymen to write on scientific issues. The women did not have this option, and the historical features of their activities as popularizers are therefore somewhat different. The maternal tradition that had given agency

---

2. I will be dealing both with women who started writing after 1850 and those female popularizers whose careers began in the early nineteenth century but who were still active after the midpoint of the century. The changes in writing style adopted by this latter group are sometimes very revealing.

3. None of the women lectured as extensively as male popularizers like J. G. Wood or J. H. Pepper. Relatively few of them were active at all as science lecturers. Twining's *Short Lectures on Plants for Schools and Adult Classes* (1858, Nutt) were first given at classes for young women at the Working Men's College in London (Twining 1858, x), while Arabella Buckley's *The Fairyland of Science* (1879) was originally delivered in lecture form to an audience of children at St. John's Wood (Buckley 1879a, v). Naturalist, ethnographer and commentator on West African affairs, Mary Kingsley established a reputation for herself on the lecture platform (Early 1997, 215-36). Before she became deeply involved in the feminist movement, Lydia Becker delivered several papers at the British Association for the Advancement of Science (Bernstein 2004, 163-68). Another woman with feminist sympathies, Rosa Grindon (née Elverson) of Manchester lectured at the Manchester Geographical Society, the Chester Society of Natural Science, and the Manchester Working Men's Clubs Association on such topics as the life history of a mountain, Chaucer as field naturalist, and the families of common plants ("Mrs. Leo H. Grindon, L.L.A." 1895, 7-12). I am indebted to David Riley for drawing my attention to Grindon.

to women in the early nineteenth century to write for an audience of women and children had to be redefined in order for female popularizers to communicate with the new audiences formed at the middle of the century. Women did not have to claim to be mother figures in the second half of the century in order to have narrative voice, as they could capitalize on other writing strategies. In addition, unlike their male counterparts, women had to come to terms with the towering figure of Mary Somerville. They also encountered obstacles erected by male practitioners tied directly to their status as women. In this chapter I will examine the general pattern of female authorship in the sciences—a pattern shaped by historical circumstances rather than determined by their biology. However, not every female popularizer conformed to this pattern in all respects. I will begin by discussing the different routes taken by these women to scientific authorship. I will explore their conception of their audience, how they experimented with literary formats in order to reach their readers, and their emphasis on the aesthetic, moral, and divine qualities of the natural world. Finally, I will consider their responses to the views of scientific naturalists on the implications of evolutionary theory for understanding the role of Christianity and of women in Victorian society. Although in public they deferred to the authority of eminent scientific practitioners, their agenda was closer to that of the Anglican clergymen.

WOMEN WRITING SCIENCE: NINETEENTH-CENTURY TRADITIONS

Beginning in 1864, Arabella Buckley (1840-1929) served as Charles Lyell's secretary, managing his extensive correspondence. This placed her at the center of the London scientific scene, where she met some of the most eminent scientific men of the day. It also put her into contact with the world of publishing, as she helped Lyell with his various writing projects. Finding herself without employment after Lyell's death on February 22, 1875, she decided to undertake her own publishing projects and contacted John Murray, Lyell's publisher, to see if he had any work for her. On March 19, 1875, she wrote, "As you have opportunities of hearing of literary work I hope you will excuse my writing to say that I shall be glad to find some, to replace that which I have lost."[4] Murray later asked Buckley if she would be interested in editing a tenth edition of Mary Somerville's *On the Connexion of the Physical Sciences,* originally published in 1834.

---

4. Buckley to Murray, March 19, 1875, Archives of John Murray.

On November 18, 1875, Buckley wrote to Murray that she had "looked carefully through Mrs. Somerville's beautiful work" and wondered if she had the ability to revise and update it. "If it is necessary to recast the work as Mrs. Somerville would have done if alive," Buckley declared, "and to bring it as a philosophical work on all points up to the level of modern theories I have not the wide grasp of knowledge or scientific experience necessary to do this and I should only lessen the value of the work by attempting it." However, Buckley offered to act as editor on the condition that she undertake a more circumscribed role. If the work could be "kept substantially as Mrs. Somerville left it, and my part may consist in comparing it with modern works, correcting obvious errors, cutting out parts that are obsolete, and so making room for the addition of new facts on the authority of competent men, this I think I could do usefully if you can allow me time to get up the subjects thoroughly."[5] Murray accepted Buckley's conditions, and by December 1876, she had completed her work on Somerville's *Connexion,* though she admitted, "It has been such a complicated piece of work."[6] Buckley's edition of 1877 turned out to be the last.

Somerville's sophisticated and broad-ranging writings could not provide a model for female popularizers of the second half of the century to imitate. Somerville dealt with astronomical topics, sound, refraction, light, heat, electricity, and magnetism. Of all the female popularizers active in the middle of the century, Zornlin came the closest to emulating Somerville's synoptic style. Zornlin's range was far more limited, and in her books she could only present overviews of one scientific discipline, such as geology, hydrology, or physical geography. The increasing specialization of the sciences would have made it even more difficult for those in the second half of the century to acquire the expertise necessary to range across all of the physical sciences. Female popularizers of science of the second half of the century may have admired Somerville, but they were unable to imitate her.

If Somerville did not provide a model for emulation, neither did women writers from the late eighteenth and early nineteenth century who established a maternal tradition in science writing geared toward an audience of women and children. By positioning themselves within the maternal tradition, female popularizers of the latter half of the century could obtain some measure of authority as scientific, moral, and religious educators. However, by midcentury the use of the "familiar format" had declined, and questions

5. Buckley to Murray, November 18, 1875, Archives of John Murray.

6. Buckley to Murray, December 26, 1876, Archives of John Murray.

were being raised about Paleyian natural theology.[7] Those who belonged to the maternal tradition had hitherto restricted their audience to women and children. For those who wished to reach the new mass reading audience this limitation was problematic. Moreover, by the middle of the nineteenth century, women who had written in this tradition seemed dated. Gatty told a correspondent that "I *decline* to read Mrs. Marcet altogether." She viewed Marcet as a great bore and wrote, "I believe I hate Mrs. Marcet."[8] Neither Somerville nor the maternal tradition offered women a viable model of scientific authorship during the midcentury period, when the communications revolution and the growth of a reading public offered new opportunities.

The attitude of male practitioners in the second half of the nineteenth century posed another challenge for female scientific popularizers. To those male practitioners bent on professionalizing their discipline, women were considered to be doubly disqualified from full participation in science, including the role of popularizer. Not only were they more easily seduced by the lure of Christianity, they also did not possess the required intellectual power to engage in genuine scientific research. By nature they were religious, emotional, and subjective. Essentialism pervaded the thinking of even the most liberal scientific minds. Darwin's *Descent of Man* (1871) provided an evolutionary rationale for the alleged intellectual inferiority of women.[9] Writing to the geologist Charles Lyell, T. H. Huxley declared, "five sixths of women will stop in the doll stage of evolution, to be the stronghold of parsondom."[10] Huxley made it his special mission to drive women from would-be professional scientific societies and from positions of importance in scientific institutions. Part of his strategy for amalgamating the Ethnological Society and the Anthropological Society, in order to bring anthropology under Darwinian control, involved the reconstitution of the Ethnological Society into a "gentlemen's society." By excluding women from the "Ordinary Meetings" of the Society, where the serious scientific discussion took place, Huxley could upgrade its professional status and remove a major impediment to the union of the two societies.

Other male practitioners, such as John Lindley, Professor of Botany at London University from 1829 to 1860, redefined the intellectual focus of

7. Gates 1998, 44; Gates and Shteir 1997, 11.

8. Gatty to Harvey, May 30, 1864, Sheffield Archives, HAS 48/438.

9. E. Richards 1983, 57–111; 1997, 119–42; Russett 1989.

10. T. H. Huxley to Lyell, March 17, 1860, Huxley Papers, Imperial College, 30.34, as cited in E. Richards 1989, 256.

their disciplines in order to exclude women. Before the middle of the century women had more culturally sanctioned access to botany than to any other science.[11] Botany was widely associated with women and was "gender coded" as feminine. Lindley's rejection of Linnaean botany in favor of a utilitarian botany was also a dismissal of the polite botany previously identified with women. His attempt to envision a new kind of botanist, a new identity for the scientific practitioner, contributed to the creation of a masculine "culture of experts."[12] Excluded from universities until the end of the century, prevented from joining many scientific societies, faced with an intellectual redefinition of science hostile to women as a result of the growing force of professionalization, and portrayed by Darwin as intellectually inferior due to the evolutionary process, women in Britain were confronted by a multitude of obstacles when they attempted to be a part of the scientific world.

However, many women in the second half of the nineteenth century did not allow these obstacles to deter them. Women participated in the scientific enterprise as illustrators, lecturers, invisible assistants, marital collaborators, explorers, and animal protectionists.[13] But most of all, they became popularizers of science, perhaps because it was a more accessible route to science than many of the others. In opportunistic fashion they seized upon elements of previous traditions of female scientific authorship to forge a powerful, new tradition. From Somerville, they took the idea of reaching out to an adult male audience, as well as to women and children. Later in the century they explored the potential of the synoptic review. As they navigated around the barriers erected by the professionalization and masculinization of science, they retained the role of religious and ethical guides established by their predecessors in the maternal tradition. Adopting a religious or moral purpose provided female popularizers, or any women who undertook the role of literary author, with the justification for leaving the private sphere and entering the public world of literary self-display.[14]

SCIENTIFIC AUTHORSHIP AS A VOCATION

Relegated to the home, writing was one of the few respectable outlets for middle-class women in the midcentury period who were ambitious and who desired to earn a living. The widening publishing market opened up

11. Shteir 1997b, 29.

12. Shteir 1996, 156–58; 1997b, 29–38.

13. Shteir 1996; Gates 1998; Le-May Sheffield 2001.

14. Mermin 1993, xiv, 109.

FIGURE 3.1 Mrs. Phebe Lankester. (Mary P. English, *Victorian Values: The Life and Times of Dr. Edwin Lankester* [Bristol: Biopress, 1990], 165. Courtesy Biopress, Ltd. and Ipswich Borough Council Museums and Galleries.)

new possibilities, and from 1830 to 1880 women's impact was greatest in prose fiction, moderate in poetry, and somewhat significant in nonfictional prose.[15] The decision to undertake the role of popularizer of science often came as the result of family circumstances, ill health, or the death of a loved one, sometimes in combination with one another and with financial necessity. Some women undertook science writing, at least in part, out of a sense of religious duty. Phebe Lankester (1825-1900) first entered the world of scientific authorship as wife and helpmate of Edwin Lankester, whom she married in 1845 at the age of twenty (Fig. 3.1). Lankester had studied science and medicine at the University of London and was an active medical reformer and scientific writer. Phebe helped him research and write articles for the *Penny Cyclopaedia* from 1846 to 1847 and for the natural history section of the *English Cyclopaedia* (1854-55). After her last child was born in 1859, she pursued an independent career as a science writer under the name

15. Ibid., xv.

"Mrs. Lankester." Capitalizing on the fern craze in England, she produced *A Plain and Easy Account of the British Ferns* (1860, Hardwicke), a heavily revised edition of Edwin Bosanquet's earlier botanical manual that continued in print up to the 1890s under the title *British Ferns*. The following year she wrote *Wild Flowers Worth Notice* (1861, Hardwicke) and contributed three botanical articles to the *Popular Science Review* from 1861 to 1862. During the 1860s Lankester also began writing the popular portion of J. T. Boswell Syme's *English Botany* (1863–86), to which she eventually contributed four hundred entries over nine years. Thus began a career as a London-based popularizer that lasted from the 1860s to the 1880s. When she was widowed in 1874 at the age of forty-nine, and left in difficult financial circumstances, writing became more of a financial necessity.[16]

Like Lankester, Sarah Bowdich Lee (1791–1856) first became involved in scientific writing projects due to her husband, Thomas Edward Bowdich, a writer in the British African Company. Thomas set his sights on exploration, and as part of his training moved his family to Paris in 1819 to study natural history under Georges Cuvier and Alexander von Humboldt. To finance the proposed expedition, they translated French natural history works, including some of Cuvier's work. The expedition turned into a disaster when Thomas died shortly after they reached Gambia in 1823, stranding a penniless Sarah and her three small children. Returning home, Sarah completed her husband's records of the Madeira-Gambia trip, including some of her own illustrations. She secretly married Robert Lee in 1826, but she wrote travel stories for popular magazines, as she needed the additional income. From 1828 to 1838, she brought out the lavishly illustrated *The Fresh-Water Fishes of Great Britain* in twelve fascicles. Among the subscribers were F. W. Herschel, Sir Humphry Davy, and Roderick Murchison. When the inheritance she received upon the death of her mother was less than she had expected, she turned to new forms of writing, including short stories and natural history for children, novels for young adults, and anecdotal natural history. Over the course of the 1840s and 50s she wrote six natural history books, of which *Anecdotes of the Habits and Instincts of Birds, Reptiles, and Fishes* (1853, Grant and Griffith) was likely the most popular, reaching a second edition in 1861 and a third, and final, edition in 1891.[17] Her *Elements of Natural History* (1844), published by Longman, sold out the initial print

16. Shteir 2004a, 1181–83; 2003, 157–59, 162.

17. Creese 2004b, 243–44; Beaver 1999.

FIGURE 3.2 The Hon. Mary Ward, sitting beside a Smith, Beck, and Beck stereo-
scopic viewer. (David H. Davison, *Impressions of an Irish Countess: The Photographs of
Mary, Countess of Rosse, 1813–1885* [Birr, Ireland: Birr Scientific Heritage Foundation,
1989], 21. Copyright the Irish Picture Library.)

run of 1,000 copies by 1849. A second edition of 1,000 copies was gone by
1855, leading Longman to print another 500 copies.[18]

Like Lankester and Bowdich Lee, Mary Ward (1827–1869) was strongly
influenced by a family member (Fig. 3.2). In contrast to the other women discussed in this chapter, who came from a middle-class background, Ward was
part of an aristocratic family. Her first cousin was the astronomer William

18. *Archives of the House of Longman, 1794–1914* 1978, A4, 265; A5, 290; A5, 354; A6, 301.

Parsons, third Earl of Rosse and builder of the Leviathan of Birr, the world's largest reflecting telescope right up until 1919. Since she lived only fifteen miles from Birr Castle, Ward was able to chronicle the construction of the telescope from the time she was thirteen years old until its completion five years later in February of 1845. She was among the first to make observations. Ward was also indebted to other members of her family for her interest in science. Her father, the Reverend Henry King, a wealthy landowner, and her mother, Harriette Lloyd (whose sister was William Parson's mother), encouraged their children to study natural history. After marrying Henry Ward in 1854, she began to produce a series of books written for a broad audience, while enduring eleven pregnancies over the next fifteen years. In addition to her *A World of Wonders Revealed by the Microscope* (1858, Groombridge), *Telescope Teachings* (1859, Groombridge), and her *Entomology in Sport* (with Lady Jane Mahon, 1859, Paul Jerrard and Son), she also wrote articles for the *Intellectual Observer* on comets, telescopes, and the natterjack toad. An expanded version of *A World of Wonders* was published by Groombridge as *Microscope Teachings* in 1864, and then as *The Microscope*, which was in its third edition within a decade, and a fifth edition by 1880. *Telescope Teachings* was later published under the title *The Telescope*, which reached a fourth edition in 1876. Ward remained in Ireland throughout her life, working out of her home. Ward's promising career as a popularizer of science was cut short in a tragic accident in 1869 when she was thrown from the seat of a steam carriage while visiting Birr Castle.[19]

For other women, illness, rather than the support of family members, was a major factor in how they became involved in scientific writing. Anne Pratt (1806–1893) owed her introduction to science to her delicate health as a child (Fig. 3.3). Daughter of a wholesale grocer from Kent, Pratt's poor health left her incapacitated with a bad knee. Since active outdoor pursuits were out of the question, she became a voracious reader. A family friend offered to teach her about botany, a subject she took up with enthusiasm with the help of her older sister, who did the collecting. In her early thirties she secretly wrote her first book and sent it to Charles Knight, who published it as *The Field, the Garden, and the Woodland* (1838). By 1847 this book had reached a third, and final, edition. Throughout her career she wrote over a dozen books, mostly on botany, with the evangelical Religious Tract Society and with the Society for Promoting Christian Knowledge. Pratt lived in Chatham and Rochester, in Kent, for the first forty years of her life, resided at Brixton for several years,

19. McKenna-Lawlor 1998, 31; Harry 1995, 37, 40; 1984a, 194; 1984b, 472; Kavanagh 1997, 60; Creese 2004e, 2102–3.

FIGURE 3.3 Anne Pratt. (Library, Linnean Society of London. By permission of the Linnean Society of London.)

and then in 1849 went to live in Dover, where she wrote *Flowering Plants and Ferns of Great Britain* (1855, SPCK), initially a five-volume work that was valued by field botanists well into the twentieth century. Her writing may have been an important source of income until she was sixty years old, for after she married John Pearless in 1866 books ceased to flow from her pen.[20]

20. Shteir 1996, 202–8; Graham 1977, 1500–1501; Britten 1894, 205–7; D. Allen 2004, 1629–30.

FIGURE 3.4 Mrs. Anne Wright. (E. H., *A Brief Memorial of Mrs. Wright, Late of Buxton, Norfolk* [London: Jarrold and Sons, 1861], frontispiece.)

    Illness also played a pivotal role in the short career of Anne Wright (d. 1861), author of several books on geology, ornithology, and zoology who began writing rather late in life (Fig. 3.4). Married to John Wright of Buxton, Norfolk, in 1816, and devoutly religious, Anne undertook philanthropic work and wrote a book on scriptural study titled *Passover Feasts, or Old Testament Sacrifices* (1849). But she suffered from frequent illness, and during a particularly long interval of indisposition, she decided to learn about natural history. As her illness lingered, she wrote a series of letters on lower forms of life addressed to a young audience. Friends urged her to have them published, and they appeared anonymously in three successive parts in 1850, titled *The Observing Eye* (Jarrold). *The Observing Eye* passed through five editions by 1859 and 17,600 copies had been printed. By the end of the century the number of copies printed was up to 20,100.[21] This was followed by *The Globe Prepared for Man; A Guide to Geology* (1853, W. J. Adams). In 1853 her husband was involved in the founding of a reformatory school for youthful offenders. Wright taught at the school, and her lectures on birds later appeared in the volume *What Is a Bird?* (1857, Jarrold). Jarrold persuaded her to write a treatise on geology for the young and lower-class readers, and in 1859 it began to appear in small monthly numbers under the title *Our World: Its Rocks and Fossils*. In three weeks 1,500 copies of the first number were sold and by 1868 8,000 copies had been printed by Jarrold.[22]

Like Wright, Margaret Gatty (1809–1873) owed her career as popularizer of science to enforced idleness as a result of illness (Fig. 3.5). After marrying her husband Alfred, a Yorkshire clergyman, in 1839, she gave birth to ten children. Slow to recover from her seventh confinement in 1848, and suffering from a bronchial condition, she recuperated by the sea at Hastings. When Gatty became bored she took the advice of the local doctor to collect seaweeds and read William Harvey's *Phycologia Britannica* to pass the time. She became a life-long marine biology enthusiast. This led her to establish a friendship with Harvey, who was later to become Professor of Botany at Trinity College Dublin in 1857, and to write a number of widely read books on phycology and other zoological topics.[23] Her *British Sea-Weeds* (1863, Bell and Daldy), an introductory book to the topic, established her credentials as a knowledgeable collector, but her series of didactic and scientifically informed short stories, *Parables from Nature* (1855–71, Bell and Daldy), became an international bestseller that made her a household name in Britain. Six editions of the first series were published by 1858. Reaching an eighteenth edition in 1882, it was reissued many times by different publishers right up until 1950. Gatty's *Parables* contained a mixture of science, morality, and religion that was considered to be appropriate Sunday reading for Victorian families.[24] Although working from a remote vicarage in Ecclesfield, well outside of Sheffield, Gatty was able to achieve considerable success as a popularizer.

21. Publications Register 1847–1875, Archives of Jarrold and Sons, 31.

22. E. H. [1861], 3–4, 5–6, 8–10, 16; Publications Register 1848 to-, p. 31, Archives of Jarrold and Sons Ltd.

23. Drain 1994, 6–11; Rauch 2004, 761–64.

24. Flint 1993, 193.

FIGURE 3.5 Portrait of Mrs. Gatty from *Illustrated London News* obituary, which described her as "one of the best authors of wholesome and pleasant reading for young people" ("The Late Mrs. Alfred Gatty" 1873, 370). ("The Late Mrs. Alfred Gatty," *Illustrated London News* 63 [October 18, 1873]: 379.)

Rosina Zornlin (1795–1859), an invalid for much of her life, took great comfort in her writing. With over nine science books to her credit, and a career that spanned the 1830s up to the 1850s, Zornlin was well known to the early Victorian reading audience. Zornlin was the second daughter of John Jacob Zornlin, a successful investment broker in London, and Elizabeth Alsager, whose brother Thomas played a major role in the operation of the *Times* as part owner and financial lead writer. Since her family was wealthy it is unlikely that earnings obtained through her writing represented the main source of her income. Like Somerville and Marcet, Zornlin wrote about the physical sciences, including geology, physical geography, astronomy, electricity, and hydrology. A nondogmatic evangelical Anglican, religious themes played a significant role in her books. Many of her earlier works, such as *What Is a Comet, Papa?* (1835, James Ridgway and Sons), *The Solar Eclipse* (1836, James Ridgway and Sons), and *What Is a Voltaic Battery?* (1842, John Parker), were written for children. Her other works were pitched at a higher level and geared toward adults and to school use. Several of them sold well enough to go into multiple editions, including *Recreations in*

Geology (1839, John Parker) and *The World of Waters* (1843, John Parker), both reaching third editions, in 1852 and 1855 respectively. *Recreations in Physical Geography* (1840, John Parker) hit its third edition within the first ten years of publication, and a fourth, and final, edition in 1851.[25]

While illness played a key role in leading Pratt, Wright, Gatty, and Zornlin to undertake scientific authorship, for others the pivotal event involved the death of a loved one, usually a father or husband. Mary Kirby (1817–1893) began her career as popularizer of science after the sudden death of her father, John Kirby, a devout Dissenter and Leicestershire businessman in 1848. Due to business losses suffered in the late 1840s, John Kirby was unable to leave as much as he had hoped for his daughters. Mary and her unmarried sisters were forced to find work. In her autobiography she recalled that she and her sister Elizabeth "soon began to plot and to plan for book-writing." Their first effort, *A Flora of Leicestershire* (1850, Hamilton, Adams, and Company), came out of Mary's botanical interests. It catalogued over nine hundred flowering plants and ferns from the Leicestershire region, grouping them according to the orders of the natural system while discussing their habitats and location. Mary, listed as the author on the title page, was the driving force behind the project. Elizabeth contributed the descriptive notes. Rather than writing more technical manuals, many of their subsequent works were more geared toward a juvenile audience. They adapted stories from the classics for children, wrote fiction, and churned out a series of books on natural history. They became a writing team who worked from their home. In 1855 they moved to Norwich and hooked up with the publisher Jarrold, who invited them to write what would become *Plants of Land and Water* (1857) for his "Observing Eye" series. Besides working with Jarrold on subsequent projects, their natural history books were published by T. Nelson and Sons and the Religious Tract Society. In 1860 Mary married the Reverend Henry Gregg. Elizabeth lived with the Greggs in Brooksby, Leicester, a small village in a rural parish. The sisters continued to write together until Elizabeth's death in 1873. Over the course of their careers they wrote over ten natural history books, many of them selling well enough to call for more than the first edition. *Stories about Birds of Land and Water* ([1873], Cassell), the last book they wrote together, sold 18,000 copies.[26]

25. J. Secord 2004d, v–xi; Creese 2004f, 2230–31.

26. Kirby 1888, 13, 30, 70; Creese 2004d, 838–40; Shteir 1996, 216–19.

FIGURE 3.6 Portrait of Jane Loudon. (Bea Howe, *Lady with Green Fingers: The Life of Jane Loudon* [London: Country Life Limited, 1961], frontispiece.)

Mary Roberts (1788–1864) found herself in a similar position to Mary Kirby after the death of her father in 1811. Daniel Roberts, a Quaker merchant, cultivated Mary's scientific interests when she was younger. After his death, she began a writing career that spanned the early 1820s to the early 1850s, during which she produced over ten natural history works on conchology, zoology, vegetables, and trees. Several books reached multiple editions, including *Domesticated Animals* (1833, John Parker), which was in its fourth edition by 1837, and its seventh edition by 1854. Devoutly religious, Roberts resigned her Quaker membership in 1826 in order to become a follower of the Millenarian preacher, Edward Irving. Many of her books can be viewed as Christian meditations on natural history themes.[27]

Like Kirby and Roberts, Jane Loudon (1807–1858) turned to writing as a career when, in 1824, the death of her father Thomas Webb, a Birmingham businessman, forced her to earn a living at the age of seventeen (Fig. 3.6). In her first major work, *The Mummy: A Tale of the Twenty-Second Century* (1827), Loudon created a science-fiction novel that contained elements of the

27. Opitz 2004c, 2004b; Shteir 1996, 96–99; Lindsay 1996.

Gothic tradition while presenting speculations about the technology of the future and reflections on the political instability of her time. John Claudius Loudon, a well-known landscape gardener, town planner, and writer, was so taken with the book that he arranged to meet the anonymous author. Although there was a twenty-four year difference between the two, they married on September 14, 1830, several months after their first meeting. Jane moved into her husband's villa residence in Porchester Terrace, Bayswater, a semirural area outside London, where John had set up a magnificent garden. By this time John Loudon's right arm had been amputated, and Jane became his amanuensis, helping him with his gardening and writing projects. Embarrassed that she knew so little about gardening or botany, Jane undertook a private study of the subjects and attended John Lindley's lectures. Her career as popularizer did not begin until the 1840s, when the cost of the illustrations in her husband's *Arboretum et Fruticetum Britannicum* (1838) saddled them with a crippling debt of £10,000. When John died in 1843, he left Jane in severe financial straits. Writing had become a financial necessity for Jane Loudon, and, based in Bayswater, she churned out eight botany and gardening books after 1840. Her best-selling science book was *The Ladies' Companion to the Flower Garden* (1841, W. Smith), which reached five editions in a decade, and a ninth edition by 1879, with over 20,000 copies sold.[28] Loudon was led to become a popularizer both as a result of personal losses and her role as helpmate to her husband.[29]

For Elizabeth Twining (1805–1889), the role of popularizer was undertaken out of a sense of religious duty (Fig. 3.7). Born into a wealthy family— her father was the tea merchant Richard Twining—Twining had no need for whatever she earned from her writing. She wrote five botanical books over the course of nearly thirty years, beginning in 1849 with the first appearance of her ambitious *Illustrations of the Natural Order of Plants* (1849–55, Cundall). Her book *The Plant World* (1866, Nelson), her most popular work, reached its second, and final, edition by 1873. Twining was a member of the Botanical

28. Rauch 1994, ix; Gloag 1970, 61. Some of her other books reached at least a third edition, including *Gardening for Ladies* (1840), the *Entertaining Naturalist* (1843), *British Wild Flowers* (1844), and *The Amateur Gardener's Calendar* (1847, Longman). In the case of the *Amateur Gardener's Calendar*, Longman's first edition of 2,000 copies was later followed by a second edition in 1857 of 1,000 copies. The editions of the *Lady's Country Companion* (1845, Longman) followed a similar pattern, with a print run of 2,000 copies for the first edition, and a smaller run of 500 copies for the second edition in 1860 (*Archives of the House of Longman, 1794–1914* 1978, A5, 265, 283; A6 415, 416).

29. Fussell 1955, 192; Linfield 2004, 1263–66; Shteir 1996, 220–27; Taylor 1951, 17–39.

FIGURE 3.7 Elizabeth Twining. (*Highlights of the Richmond Borough Art Collection* [Riverside, Twickenham, England: Orleans House Gallery, 2002], 31. Reproduced by kind permission of the London Borough of Richmond upon Thames, Orleans House Gallery.)

Society of London, attended meetings of the British Association, and in 1847 contributed a paper on the flora of Britain and Continental Europe. Deeply involved in philanthropic activities, she helped found the Ladies College, Bedford Square, London, which eventually evolved into the Bedford College for Women, a groundbreaking institution of higher female education. She was also involved in restoring almshouses and other charitable social undertakings. She was especially devoted to the Christian Temperance Society, as she believed that "the *chief cause* of all the trouble of the poor—their weak health, their insufficient food and clothing, and their dullness and hardness in seeking after spiritual instruction—is *the drink*" (Twining 1877). Her *Leaves from the Note-Book of Elizabeth Twining* (1877, W. Tweedie) chronicled over sixty-five visitations to the sick and dying, most of them brought low by alcoholism.[30] Twining was a practicing member of the Church of England.

30. Browne 2004, 2047–48.

Women who became popularizers did not find it easy to build success-ful careers as writers. With the possible exception of Gatty's *Parables from Nature,* women did not produce extraordinary successful sellers of the mag-nitude of Brewer's *A Guide to the Scientific Knowledge of Things Familiar.* Many of them realized that they could not base their careers on science writing alone, so they wrote fiction, children's literature, memoirs, and translations. For some women, such as Lydia Becker (1827–1890), known for her role as one of the leading lights of the women's suffrage movement in Britain, science writing was an activity engaged in briefly before going on to another career (Fig. 3.8). Upon the appearance of an obituary in the *Journal of Botany* shortly after Becker's death, the anonymous author remarked that it might surprise some readers to find that Becker "had any claim to notice in these pages. But at one time of her life she paid much attention to Botany, in which indeed she always retained her interest."[31] Becker's *Botany for Novices* (1864, Whittaker) was not a commercial success, and the companion vol-ume, "Stargazing for Novices," was never published. She was most active as a popularizer from 1864 until 1870. However, already in the late 1860s, her energies were directed more and more into the women's suffrage movement. In 1867 she was appointed secretary of the newly formed Manchester Soci-ety, and in 1870 she was selected as editor of the *Women's Suffrage Journal,* a post she held until the end of her life. In 1887 she was elected president of the National Society for Women's Suffrage. Even after 1870, she continued to be involved in scientific activities. Her correspondence with Darwin on botanical matters, begun in 1863, did not come to an end until 1877. She addressed the Economic Science and Statistics section of the British Asso-ciation in 1871, 1872, and 1874 on such issues as political economy and the employment and education of women. She faithfully attended the annual meetings of the British Association for the Advancement of Science from 1864 until her death, though she believed that this association discouraged scientific study by women.[32] Becker was a strong churchwoman who was opposed to the disestablishment of the church.[33]

Unlike Becker, other women decided to stick with science writing al-though they earned barely enough to live on. The records of the Royal Lit-erary Fund (RLF) are filled with accounts of the suffering of women writers who experienced economic hardship. Based on her examination of the

31. "Obituary [Becker]" 1890, 320.

32. Bernstein 2004, 163–68; Blackburn 1971; Parker 2001, 631–32, 637; Shteir 1996, 227–31.

33. "Late Miss Becker" 1890, 5; Hallett 1890, 4.

FIGURE 3.8 The formidable Lydia Becker. (Audrey Kelly, *Lydia Becker and the Cause* [University of Lancaster: Centre for North-West Regional Studies, 1992], 2. Portrait of Lydia Becker by Suysan Dacre. Copyright Manchester Art Gallery.)

female applications to the RLF, Susan Mumm concludes that these women were not earning enough "to preserve a standard of living with which they grew up." Of the women who applied to the RLF, about 13 percent wrote nonfiction works, which includes science as well as travel, history, textbooks, and biography.[34] Jane Loudon was one of those female authors of nonfiction works who applied to the RLF in severe financial distress, not once, but twice. In 1829 she was awarded £25 after reporting that a "severe attack of illness" had completely exhausted her means. Shortly after the death of her husband, Loudon was forced to apply again. She explained in her application of May 1, 1844, that he had died before being able to pay off his debts. Although she had possession of thirteen copyrights that produced about £500 a year, they were being held in trust by Longman until all of her

34. Mumm 1990, 31, 35.

husband's publishing debts, amounting to £3,207, were paid.[35] This time, Loudon received £50.[36] But it was not enough, and Loudon contacted Sir Robert Peel's wife on March 6, 1846, about obtaining a pension for her daughter Agnes, aged thirteen. She recounted how her husband had ruined himself by publishing the *Arboretum Britannicum* at his own expense and how she had maintained herself and her daughter through her own writings. She feared that a "disease of the heart" would "carry me off before I can make any provision for my daughter." On March 24, Robert Peel wrote to Loudon that he was recommending that £100 be granted as pension to her for life "in consideration of the merits and services of your husband the late Mr. Loudon."[37] Loudon was again in financial difficulty only three years later. Writing to Mrs. Gaskell on October 20, 1849, she reported that all of her husband's publishing debts had been paid off the previous Christmas and that she had been hoping to "indulge in a life of idleness for a year or two, but alas! the times are so bad that Mr. Loudon's books have not produced enough for us to live on during the past year." As a result, she was prepared to accept the editorship of the journal *Ladies Own Companion*, "though I am fully aware it will be a most arduous and most laborious undertaking."[38] After editing the journal for only two years, she was suddenly dismissed without explanation, leaving her "depressed in spirits."[39] By 1855 she was depending on the generosity of friends like the Trevelyans.[40] Despite being a prolific popularizer of science, Loudon struggled financially for most of her life.[41]

35. Loudon had decided not to apply to the RLF in the winter after her husband's death in the hopes that a public appeal would bring in a sizable sum. But the appeal raised only £2,000. The income from her husband's copyrights would pay the remainder owed in about two years, she believed, so in the meantime she had to maintain her family through her writing. But due to the "shock of nerves I've sustained by what I have gone through since the loss of my poor husband," Loudon was "incapable of writing" (Cambridge University Library, Royal Literary Fund, File no. 1101, Letter 4).

36. Cambridge University Library, Royal Literary Fund, File no. 648, Letter 2, File no. 1101, Letters 1, 12.

37. British Library, Peel Papers, Add MS 40586, f. 167 and f. 173.

38. University Library of Manchester, Rylands English MS 731/112.

39. University of Newcastle, Robinson Library, Special Collections, Trevelyan Papers, J. W. Loudon to W. C. Trevelyan, May 29, 1850, WCT 175/13.

40. J. W. Loudon to Sir Walter Trevelyan, October 10, 1855, University of Newcastle, Robinson Library, Special Collections, Trevelyan Papers, WCT 175/29-30.

41. Bowdich Lee's attempt to live off of her earnings as a scientific author was also a struggle. She too applied to the Royal Literary Fund (twice) and received a Civil List Pension of £50 per year in 1854 (Beaver 1999).

In comparison to Loudon, Mary Kirby's career as popularizer of science seems to have been less stressful and lucrative enough to keep her and her sister from financial difficulty. Kirby was neither saddled with debt nor was she responsible for raising a young daughter. Whereas Loudon found it a chore to write, especially in later years because of her ill health, Kirby did not find it to be particularly onerous. As a member of a writing team with her sister, the division of labor was the key to success. After an interview with Jarrold on a book for the "Observing Eye" series, likely *Plants of the Land and Water* (1857) Mary and Elizabeth quickly wrote a few chapters. "It was very easy to do," Mary remarked, "for I had the botanical knowledge at my finger ends; and Elizabeth had such fluency in writing that the sentences seemed to flow from her pen as readily as the pearls and diamonds drop from the mouth of the fairy, in the fairy tale." Two years later the team was hard at work for T. Nelson and Sons on *Things in the Forest* (1861). By now, their work was in steady demand. Mary recalled that "earned money seems always the sweetest and best of any; and we were glad to find a ready sale for our manuscripts, and also to put the profits into our pockets." By the 1860s, Mary declared that "our engagements with the publishers were increasing, and we were obliged to devote two hours or more every morning, and a couple of hours in the evening, to pens and paper." Like other female popularizers, the Kirby sisters cultivated a working relationship with a number of different publishers, including Jarrold, Nelson, Cassell, and the Religious Tract Society, ensuring that there was always work available. Rather than limit their writing projects to natural history topics, the sisters successfully adopted a diversification strategy similar to many other women who became popularizers of science. They became absorbed in novel writing in the latter part of their careers, and wrote serial tales in the *Quiver* and *Cassell's Magazine*. But the importance for Mary of working as part of team was evident after Elizabeth's death in 1873. In her autobiography, she admitted that "I was never the same again," and she remembered that her husband "used to observe that now Elizabeth was gone, he looked upon me as only 'half a person.'" Mary undertook no new writing projects after her sister died.[42]

WORKING WITH PUBLISHERS

Shortly after moving to Norwich, Norfolk, in 1855, Elizabeth and Mary Kirby were invited by the publisher Thomas Jarrold to spend an evening at his

---

42. Kirby 1888, 144, 165, 213, 221, 224, 232.

home. Jarrold came from a family of Nonconformists involved in missionary activities and the temperance movement. Jarrold was also interested in science, and by the time the Kirbys were living in Norwich, he had already established a close working relationship with Ebenezer Brewer and other popularizers of science such as Robert James Mann and Anne Wright.[43] During their evening at Jarrold's he broached the subject of the value of natural history books. Mary Kirby recalled, "Mr. Jarrold began to talk about books of Natural History, and how much good they were calculated to do. Bishop Stanley's two volumes on birds, were full of interest, and would lead many on to the study of ornithology."[44] Without the interest and support of publishers like Jarrold, who saw the opportunities for selling books on science to the Victorian reading public, the Kirbys would never have been able to build a career for themselves as popularizers of science.

Making contact with publishers and establishing a good working relationship with them was crucial for women who wished to become popularizers of science. Some women already had connections with a publisher prior to the publication of their first book. As Lyell's secretary, Buckley had had extensive contact with John Murray before she approached him in 1875 with the manuscript for her *Short History of Natural Science* (1876). The families of publisher George Bell and Margaret Gatty had been neighbors when they were younger.[45] Gatty worked closely with Bell and Daldy from 1854 to 1872. They published her *British Sea-Weeds* and numerous editions of the *Parables*. While Gatty and Buckley benefited from their prior relationships with their publishers, the Kirbys were lucky enough to have Jarrold initiate contact with them. Jarrold proposed that the Kirbys contribute a book to a series to be titled the "Observing Eye."[46] However, many women were not as fortunate as the Kirbys, Buckley, and Gatty, especially when they first tried to connect with a publisher. They had to take the initiative in order to find a firm willing to publish their books. Ward had to concoct a clever scheme to draw the attention of publishers to her work. She privately printed 250 copies of her *Sketches with the Microscope* in 1857 and sold them from Ballylin by subscription at three shillings a copy. When the book was oversubscribed, her brother-in-law, George Mahon, took a copy to London and persuaded Groombridge of Paternoster Row to purchase the copyright. A year later it was republished, with a few corrections but otherwise unaltered, as *The World of Wonders Revealed by the Microscope*, and later retitled *Microscope Teachings*.[47]

43. Mumm 1991, 160.

44. Kirby 1888, 145.

45. Bell 1924, 5, 46.

46. Kirby 1888, 144.

47. Kavanagh 1997, 61; Harry 1995, 39.

Publishers often wanted to be involved in every aspect of the production of a book. For some, this proved to be a happy arrangement. The Kirbys were glad to receive advice from Jarrold. He thought of the title for their *Plants of Land and Water.* "We were very pleased with it," Mary recalled, "and complimented him on his skill."[48] When Murray recommended to Buckley that she simplify a chapter on refraction in her *Short History of Natural Science,* she replied that it was her "great wish" to make "everything perfectly clear and I am very glad of any hints on this point."[49] Loudon thanked Murray for his suggestions, which had improved the book she was working on. She wrote, "I am very glad to see the running title 'Loudon's Trees, and Shrubs of Britain,' as many persons would be misled by the Latin title."[50] At times the decisions and actions of publishers led to friction with the author. Loudon objected when she learned that Murray intended to put her husband's name instead of hers on the title page of a gardening book she had written. She told Murray that this "would entirely destroy its whole purpose and intent," which was to make public what she had been taught by her husband. Murray wanted to capitalize on John Loudon's name, while Jane wanted to give the book "an air of truth and reality," which "never fails to make its way with the public," and which would also make the work "perfectly original."[51] Both John Murray and his son antagonized Somerville when they urged her to write at a less demanding level more appropriate for the general reader.[52]

Margaret Gatty's sometimes tense relationship with Bell and Daldy illustrates the problems that could arise when author and publisher disagreed on the issue of target audience. While Gatty located herself within the maternal tradition of women writers who wrote primarily for a female audience, Bell and Daldy pushed her to address a much broader public. In 1862 Gatty and her publishers became involved in a heated dispute over her *British Sea-Weeds* that raised crucial questions about authorial control and audience. On November 5, 1862, Gatty wrote to Harvey that Bell was ecstatic with her introduction to the book, but that he did not want it to be "so prominently addressed to *ladies.*"[53] Then a disagreement erupted on the nature of the synopsis. Gatty wrote to Bell that the attempts of previous botanists at

48. Kirby 1888, 145.

49. Buckley to Murray, July 14, 1875, John Murray [Publishers] Ltd.

50. Loudon to Murray, August 10, 1838, John Murray [Publishers] Ltd..

51. Loudon to Murray, May 10, 1839, John Murray [Publishers] Ltd.

52. Patterson 1969, 331.

53. Gatty to W. H. Harvey, November 5, 1862, Sheffield Archives, HAS 48/323.

composing an intelligible synopsis had been a "signal failure <u>as far</u> as <u>amateurs are concerned</u> (for whom alone I write)." Daldy wanted a briefer synopsis, but that was impossible, and Gatty begged "you and Mr. Daldy to leave the matter to me."[54] Privately to Harvey she wrote, "I am disgusted with the Synopsis and Bell."[55]

When Gatty received a letter from Bell containing a synopsis composed by Daldy, she exploded. Writing directly to Daldy, she protested that had she known that they would make the final decision on the nature of the synopsis she would never have sent them the manuscript for publication. "You have announced the Synopsis as <u>mine</u>, and now you seem inclined to rearrange it after your own plans without reference to my expressed opinion." Gatty felt "a sort of right to control this part of the matter." She also objected that the synopsis was too complex for amateurs. She hoped that her seaweed book would extend her readership. Currently, "the aristocracy and clergy seem my supporters but I should like to <u>stretch out</u> a little further," especially to the manufacturers and merchants in Manchester, Liverpool, and Leeds, where, in her opinion, she was virtually unknown. But Daldy's synopsis would limit her "circle of readers." Threatening to withdraw the manuscript, she suggested a compromise: let Harvey examine both synopses and judge which was the most suitable.[56]

On the same day Gatty wrote to Harvey, complaining about Bell and Daldy's obstinacy. She did not believe that their synopsis, based on internal structure, could be a guide to amateurs. Having the family name in capitals on one side and generic names opposite, as Daldy proposed, was attractive. But "those A.B. beginnings and the a.b.c. subdivisions of the genera and the bouleversement of order were enough to craze and mislead St. Guy Faux himself." Bell and Daldy could "make out a Bentham synopsis in the old language," and announce it as Daldy's or they could use her superior plan. Although she had argued in her letter to Daldy that she wanted to expand her readership, to Harvey she presented her synopsis as a "lady's synopsis," that was "from a lady to a lady and intelligible to the 'softer' sex in more senses than one."[57] Harvey supported Gatty, and she wrote to express her gratitude, hoping that the debate was now at an end. However, she could not resist

54. Gatty to Bell, November 12, 1862, Reading University Library, Archives, George Bell Uncatalogued Series.

55. Gatty to W. H. Harvey, November 12, 1862, Sheffield Archives, HAS 48/325.

56. Gatty to Bell, November 18, 1862, Reading University Library, Archives, George Bell Uncatalogued Series.

57. Gatty to W. H. Harvey, November 18, 1862, Sheffield Archives, HAS 48/331.

a parting shot at Daldy's synopsis. "Families as well as genera were tossed about ad lib," she declared, "it made my head go round to try and follow his arrangements." She described her synopsis as "an entirely amateur's Synopsis of nothing but appearances," that used colors and leaf shape as the determining factors. Gatty's stand on the synopsis also, she thought, spelled out the nature of her audience to her publishers. "I am writing for the Ladies," she insisted, "and if B & D never knew it before they may now."[58]

Gatty clashed with her publishers on other issues as well, from the titles of her books to marketing strategies. After receiving some suggestions from Bell for titling her seaweed book, she wrote that she could "not hear of any of those alliterative" titles sent. Satirizing the fashion for alliterative titles, she offered two of her own, "Sound Sense made Sensible in Seaweeds," or "Seaweed Science Sensibly Simplified Suited to both Sexes and Sages as well as Simpletons."[59] When Bell requested that she write a preface for one of her books, Gatty refused, on the grounds that the notes contained any information that would have gone into a preface. "'No preface' is as essential as 'No popery,'" she half-jokingly wrote to Bell.[60] During their correspondence about a new edition of the *Parables*, Gatty objected to the planned design. "Pretty and tasteful as this is," she wrote, "Mr. Gatty and I are troubled about he size of the book in thickness." Too thin to be priced at seven pounds and sixpence, Gatty recommended the use of larger type, which would also enhance the "Scriptural tone about the Parables."[61] As her books went through the publication process, Gatty became frustrated with the errors she found in the proofs and with Bell's inability to meet publication deadlines. To Harvey she grumbled that Bell sent proof in fits and starts, and mentioned having received the last set now that Bell had "woke up from his last nap." When a book to be sent to Harvey had not arrived, she told him that Bell is "very careless." In November 1862, she worried that Bell's tardiness would lose them the Christmas market for her seaweed book. Later, she began referring disdainfully to Daldy and Bell as "Dildrum and Doldrum" in her letters to Harvey.[62]

---

58. Gatty to W. H. Harvey, November 18, 1862, Sheffield Archives, HAS 48/333.

59. Gatty to Bell, n.d., Reading University Library, Archives, George Bell Uncatalogued Series.

60. Gatty to Bell, n.d., Reading University Library, Archives, George Bell Uncatalogued Series.

61. Gatty to Bell, [August 1860], Reading University Library, Archives, George Bell Uncatalogued Series.

62. Gatty to W. H. Harvey, June 14, 1862, HAS 48/274; June 29, 1862, HAS 48/280; November 1, 1862, HAS 48/321; [February] 1863, HAS 48/363, Sheffield Archives.

Publishers tried to keep their successful authors happy and receptive to accepting more work. Pleased with their first contribution to the "Observing Eye" series, Jarrold quickly encouraged the Kirbys to consider a volume for the same series on insects, which resulted in *Caterpillars, Butterflies, and Moths* (1857).[63] Some publishers were willing to loan or even buy books for their authors that would help them in their work. Loudon expressed her gratitude to Murray in a letter written in 1838 when he loaned her William Buckland's Bridgewater Treatise.[64] Mary Kirby recalled that on several occasions Nelson sent her and her sister a present of books, both in French and English, which "he thought might give us any assistance."[65] Some authors turned to their publishers for financial assistance when they were in dire straits. Some requested that their publisher buy a copyright that they held or that they receive an advance on a manuscript in progress. In June 1856, Sarah Bowdich Lee sold her share of the stock and copyright for *Elements of Natural History* to Longman for £35.[66] Loudon, always pressed for money, requested that Murray take a work on ornithology and entomology of gardens for £150.[67] Bell often came to Gatty's aid in times of financial distress, and she was forced to undertake writing projects for him to work off debt after debt.[68]

As a result of the efforts of publishers to retain the services of their popular authors, some women established long-term relationships with one press rather than pursue the Kirbys' strategy of working with a number of publishers. When to his surprise Somerville's *Mechanism of the Heavens* sold well, Murray refused to take any profit for himself. The generous agreement he had offered Somerville for her first work—she was to receive two-thirds of the profits and the copyright—was adhered to for all of her subsequent publications.[69] Despite her arguments with Bell and Daldy, Gatty built an enduring relationship with them. In 1862, after publishing several works with Bell, she half-jokingly wrote, "I wish you would make me a Partner in the Co. I hate working at so much a line."[70] After a spate of complaints by

63. Kirby 1888, 148.

64. Loudon to Murray, October 30, 1838, John Murray [Publishers] Ltd.

65. Kirby 1888, 221.

66. Beaver 1999, 27.

67. Loudon to Murray, n.d., John Murray [Publishers] Ltd.

68. Katz 1993, 62.

69. Patterson 1969, 321–22.

70. Gatty to Bell, March 23, 1862, Reading University Library, Archives, George Bell Uncatalogued Series.

Gatty about errors, Bell wrote to apologize and wanted reassurance that she would continue to work with them. "Let me now assure you," Gatty replied, "that even when I believed you had left <u>eleven errors in 18 pages</u>, it never crossed my mind to lose confidence in the <u>good Bell and Daldy firm</u>." Gatty's confidence, she affirmed, was grounded on the firm's "utter <u>trustworthiness</u> and the personal friendship I have for long had for its head—including much kindness received."[71]

Like the Kirbys, some female popularizers believed that it was in their best interests to work with more than one publisher. In her dealings with Murray, Loudon threatened more than once to take her books to another publisher. At one point, according to Loudon, she and Murray had a mutual understanding that he would "take a series of my publications, if I would write only for you." As a result of the agreement, she had refused several proposals to undertake works from other presses. If Murray did not give her an advance on works he had agreed to take, she threatened to find employment elsewhere.[72] In 1840 Loudon offered Murray's son a proposed work titled "Instructions in Modern Botany for Ladies, according to the classification of Prof. De Candolle." If he would not take it, she stated that she would go to another publisher.[73] In this case, Murray published her book as *Botany for Ladies* (1842). But in 1842, Murray turned down a book titled "Vegetable Physiology."[74] Loudon ended up working with a series of publishers, including Longman, W. S. Orr, Henry Bohn, William Smith, and even Bell.

The amounts earned by women for their science writing varied from author to author, and depended on whether or not their previous works had been financial successes. Applications to the Royal Literary Fund show that women earned far less than the £100 often assumed to be the lowest price for a book in the midcentury period. Publishers typically paid women £50 for the copyright of their books, though the median value was only £30. The records of the Royal Literary Fund reveal that authorship was precarious and not very lucrative for the overwhelming majority of women.[75] Figures for the earnings of women who wrote about science in particular, and drawn primarily from publishers' archives, are in the same range. Ward sold the

71. Gatty to Bell, December 10, 1863. Reading University Library, Archives, George Bell Uncatalogued Series.

72. Loudon to Murray, n.d., John Murray [Publishers] Ltd.

73. Loudon to Murray, Aug. 25, 1840, John Murray [Publishers] Ltd.

74. Loudon to Robert Cooke, Apr. 23, 1842, John Murray [Publishers] Ltd.

75. Mumm 1990, 33, 35, 44.

copyright to her *Sketches with the Microscope* to Groombridge for £15, but as a wealthy aristocrat she did not need to depend on her earnings for her livelihood.[76] Pratt sold all of her copyright and interest in her *Haunts of the Wild Flowers* (1863) to Routledge for £45 in 1863.[77] Gatty was among the most successful of the women due to the gigantic sales of the *Parables*. She received only £25 from Bell and Daldy for writing the introduction and descriptions for her *British Sea-Weeds*, but she was paid handsomely for illustrated editions of the *Parables*. Excited by the news, she wrote to Harvey, "What do you think! Bell is a-going to give me 150£ for the Illustrated Edition of 3rd and 4th Series of Parables! And for the 2nd Editions of the Illustrated 1st and 2nd Series another 150£ at two payments."[78] But Gatty was more the exception than the rule.

DEFINING AN AUDIENCE

In 1851 Jane Loudon's *Botany for Ladies* was retitled and published by Murray as *Modern Botany*. Ann Shteir interprets this as part of the disappearance by midcentury of deliberately female-specific and gender-conscious science writings by women.[79] Gatty once wrote to Bell that "I write avowedly for the ladies. Let the gentlemen benefit if they choose."[80] Gatty was in the minority among female popularizers in the latter half of the nineteenth century.[81] Opposition to science writing designated specifically for women came from male practitioners, but also from women reformers, educators, and feminists. Many female popularizers of science did not want to limit themselves to a female audience. Loudon's *Modern Botany* opened up a "broader intellectual

76. Kavanagh 1997, 61.

77. University College, London, The Archives of Routledge and Kegan Paul Ltd., Routledge Contracts, I–Q 1853–73, Item 2, October 23, 1863.

78. Gatty to W. H. Harvey, April 1, 1862, HAS 48/240; Gatty to Harvey, January 25, 1864, HAS 48/420, Sheffield Archives.

79. Shteir 1997a, 248–49.

80. Gatty to Bell, n.d., Reading University Library, Archives, George Bell Uncatalogued Series.

81. At first glance Becker's approach to this issue would seem to be identical to Gatty's. When she sent a copy of her *Botany for Novices* to Darwin, Becker told him "it is intended chiefly for young ladies" (University of Cambridge, Becker to Darwin, March 30, 1864, MS.DAR.160: 112). But Becker's strategy was to write in the standardized, impersonal "male" style for women in order to undermine sex segregation in science teaching. As a first-wave feminist she pushed for sameness rather than difference. The harmful result of essentialism, in her mind, was the existence of separate spheres for women (Shteir 1996, 228–29). Gatty was not a feminist and had no interest in challenging essentialism.

space for women while also inviting male readers into this introductory text."[82] She dropped gender-tagged titles in order to reach out to a wider mass reading audience. This represented a fundamental change for female popularizers working in the old maternal tradition, which had defined its audience as being composed of women and children. Most women began to see themselves as writing for the new audience that had developed in the middle of the century that included adult males as an integral component.

Not that female popularizers of this period ignored the younger reader. Wright's *The Observing Eye; or, Letters to Children on the Three Lowest Divisions of Animal Life*, like many of her books, was addressed specifically to the "youthful mind."[83] In the preface to *Stories about Birds of Land and Water*, the Kirbys declared, "It is to you, dear children, we offer this little volume."[84] Like many of their sisters in science, the Kirbys wrote books for adults of both sexes, in addition to works geared to children of various ages. Gatty's *British Sea-Weeds* was written for adults, while her *Parables from Nature* contained short tales apparently for children. However, Gatty did not aim her *Parables* solely at children. By writing for children Gatty believed she was reaching across audiences and generations. "Conscripted by the act of reading," Alan Rauch remarks, "adults would both absorb and transmit Gatty's views on religion and science while the audience of children were themselves amused and taught by the *Parables*." Rauch broadens out his observations to include most works for children, as they included, alongside the conventional adults and children, a number of other categories of readers, such as the child who is being read to, the adult who is reading for themselves, and adults rereading works from their youth.[85] Annie Carey, the author of *The Wonders of Common Things* (1873, Cassell, Petter, and Galpin), instructed her young readers to avoid skimming through her book for mere amusement by reading with an adult. "If carefully read with an intelligent teacher," she wrote, "then it is hoped that it may serve, not only to increase the sum of actual information, but to create an earnest desire for more."[86]

The audience of curious but uninformed readers imagined by women who popularized science cut across class as well as across generations. Ward, for example, began her *Telescope Teachings* by making a series of suppositions about her readers. "We will suppose the reader," she declared "[is] already interested in the appearance of the starry heavens, and acquainted,

82. Shteir 1996, 225–27.
83. [Wright] n.d., preface.
84. Kirby and Kirby n.d., v.

85. Rauch 1997, 137, 140.
86. A. Carey n.d., v.

perhaps, with one or two constellations, inquiring how he may learn more, and what apparatus will be necessary for making observations." Taking the reader by the hand, Ward recommends that they obtain a set of star maps, an almanac with tables of the positions of the planets and other astronomical information, and a small, inexpensive telescope, all affordable by artisans and members of the middle class. She suggests that her readers train their telescopes on Saturn and on the stars. She explains how constellations rise and set each night, assuming an interested reader with little or no knowledge, but not necessarily a child. Jokingly, she remarks that she has "disrespectfully supposed [our reader] to be familiar with no movement of the heavenly bodies, except the rising and setting of the sun."[87] Although Ward was writing about astronomy, rather than botany, her self-conscious attempt to define her own audience is similar to Jane Loudon's. Ann Shteir has observed that Loudon, "like Anne Pratt and other professional writers of popular botany and natural history books . . . situated herself on the threshold of knowledge, helping her readers cross over into scientific study."[88] Shteir's insight about Loudon, Pratt, and other women who wrote about natural history can be extended to most female popularizers from the 1850s to the 1870s.[89] In outlining the simple facts she was about to offer her audience in the preface to her *The Wonders of Common Things*, Carey used the term "threshold" to describe where her readers stood in relation to the body of scientific knowledge. "Elementary knowledge," she declared, "—meaning by that phrase a knowledge of the facts that stand on the threshold of every department of Science and Art—needs most especially to be accurately presented and carefully instilled in early life."[90]

Over and over again, female popularizers defined their audiences, usually in the prefaces or introductions to their works, as those who desired

87. M. Ward 1859, 1, 12.

88. Shteir 1996, 221.

89. Shteir has discussed the notion of "threshold" in science writing by women in several of her works. In an essay on women's magazines she has pointed to how they "gave elementary instruction in systematic botany and guided readers over the threshold into introductory knowledge but provided no access to more complex botanical material." By contrast, she asserts, the typical magazine for men during the same period "indicates a higher intellectual threshold" (Shteir 2004b, 18). In another essay she has referred to Bowdich Lee's *Elements of Natural History* as one of the works of the early Victorian years that introduced readers to the elementary stages of science learning. Bowdich Lee's book offered information on vertebrate zoology for schools and the young as a "stepping-stone" to more complex productions of "deep science" (Shteir 1997a, 244).

90. A. Carey n.d., iii.

to cross over the "threshold of knowledge," whatever their class or gender. Lankester stated that her *Plain and Easy Account of the British Ferns* was "intended as a guide to the lover of Nature, who, though not perhaps scientifically acquainted with botany, may partake of the desire so natural in all minds, to possess, in the best way circumstances will allow, some shadow of the green country lanes and lovely scenes, so refreshing even in the remembrance."[91] Pratt insisted in her *The Flowering Plants, Grasses, Sedges, and Ferns of Great Britain* that "one of the chief objects of this work is to aid those who have not hitherto studied Botany."[92] Ward echoed Lankester and Pratt in the preface to her *Entomology in Sport*, where she announced, "the following pages are intended, not so much for the scientific, as for the young or the comparatively uninstructed reader."[93] Becker's *Botany for Novices* was "adapted for the use of those who have no previous acquaintance with botanical science, but it has been written in the hope that it may introduce its readers to a more extended study of the subject."[94]

Female popularizers of science often saw their books as helping their readers to bridge the gap between ignorance and the daunting works by male practitioners that were too sophisticated for beginners. Gatty began work on an introductory book on seaweeds because she realized that nothing suitable existed for beginners. Writing to Harvey on December 21, 1857, she asserted that David Landsborough's *Popular History of British Sea-Weeds* did not simplify the study, and that even those who owned it and Harvey's *Phycologia Britannica* wrote to her for explanations. Gatty's idea, she wrote Harvey, was to explain the rudimentary matters connected with the study of seaweeds and "so pave the way to the use of your great work by a preliminary A.B.C. book."[95] Interestingly, Pratt had the same object in mind earlier in writing *Chapters on the Common Things of the Sea-Side* (1850, SPCK). "The main object of this little book," she affirmed, "is to enable the reader unacquainted with Natural History, to recognize some of the different objects frequent on our shores." Pratt also hoped that her book would lead the reader to study the works of "our great naturalists," including Harvey, Johnston, Edward Forbes, and Rymer Jones.[96] The Kirbys agreed that demanding books written

91. Lankester n.d., ix.

92. Pratt [1873], 1: 1.

93. Ward and Mahon n.d., 1.

94. Becker 1864, vii.

95. Gatty to W. H. Harvey, December 21, 1857, Sheffield Archives, HAS 48/15.

96. Pratt 1850, v–vi.

by practitioners were often beyond the common reader, but they reminded their audience that their preparatory works were based upon them. "To put into the hands of young persons the scientific works from which the following information has been gleaned," they wrote in the preface to *Caterpillars, Butterflies, and Moths,* "and to expect them to understand them, would be very much like giving a labouring man the Greek Testament and expecting him to read it."[97] Gleaning information from these sophisticated texts in order to write a book on the subject for a wide audience was a complicated process. It was similar to translating ancient books written in a foreign language.

Like many of the Anglican clergymen who popularized science, women writers attempted to make their works accessible to their audience. Two major obstacles prevented a successful translation: the lack of familiarity of the reading audience with systems of scientific classification and their ignorance of the meaning of scientific terms. Many female popularizers solved the classification problem by presenting a simplified system or none at all.[98] For the Kirbys, teaching their readers how to classify was a low priority. In *The Sea and Its Wonders* (1871, Nelson), they declared, "the object aimed at has been to allure him [the reader] to the study of the great book of Nature, rather than to perplex him with a strictly scientific arrangement."[99] In her *Chapters on the Common Things of the Sea-Side,* Pratt told her readers that she would make no attempt at a detailed classification of seaweeds, opting to "simply divide the sea-weeds into the three great groups into which botanists arrange, which are the olive-green, the red, and the green sea-weeds."[100] Even Jane Loudon, whose most important botanical work, *Botany for Ladies,* was intended as an introduction to the natural system, and who saw the Linnaean system as unfit for women, was not adverse to violating the rules of classification in considering the needs of her audience.[101] In her *The Entertaining Naturalist* (1843, Bohn), a revised edition of an older work, she decided not to arrange the orders "quite scientifically; as I could not persuade myself either to displace the Lion from the situation he has held so long at the commencement of the book, or to remove the Whales

---

97. Kirby and Kirby [1861], 5.

98. Decisions on how to handle classification were fraught with significance. Lindley's rejection of the Linnaean system was tied to his attempt to break the link between botany and genteel femininity (Shteir 1997b, 33).

99. Kirby and Kirby 1871, vii.

100. Pratt 1850, 96.

101. Shteir 1996, 221.

and other Cetacea from the place they have near the fishes."[102] To be an entertaining naturalist meant making concessions to the common reader.

Scientific terms were used sparingly, especially in botanical works by female popularizers. Twining believed that botany had until recently been "so peculiarly enveloped in technical scientific language that young persons were almost entirely deterred from learning it." She therefore "avoided as much as possible" all difficult technical and scientific words.[103] Pratt followed a similar strategy in many of her works. In *The Poisonous, Noxious, and Suspected Plants of Our Fields and Woods* ([1857], SPCK) she stated that "as this little book is not intended for the botanist, care has been taken to avoid any terms not readily understood."[104] Bowdich Lee agreed, "the uncertainty and extent of botanical nomenclature, the jargon that it offers of anglicized Latin words, utterly preclude the superficial glance of the mere reader."[105] Loudon recalled that when she began her study of botany, she was in despair, "for I thought it quite impossible that I ever could remember all the hard names that seemed to stand on the very threshold of the science, as if to forbid the entrance of any but the initiated."[106] In her *First Book of Botany* (1841, Bell), after pointing out that all botanical works contained terms that "only serve to perplex the young student," she proposed to select only "the most essential" for inclusion in her book. Armed with these terms, "which may be called the alphabet of the science," the young student could go on to "study either the Linnean or the Natural System of Botany."[107] Becker went further. She promised those readers who felt "deterred by the long words" and who were afraid that they could not learn much of botany "without burdening the memory with a great many long, hard names of plants," that they could make progress without learning any new nomenclature. "Let such be assured," she declared, "that without needing to trouble themselves with any other names than the sweet familiar ones they have known from infancy, they may learn the principles of the science, which have reference, not to the names, but to the structure of the plants, and are to be acquired by a careful examination of living specimens, guided by a few simple rules of classification."[108] By avoiding scientific jargon and complex classification systems, many female scientific popularizers attempted to establish a close relationship with their audience.

102. Loudon 1850, v.
103. Twining 1866, iv.
104. Pratt n.d. [*Poisonous*], xi.
105. Bowdich Lee 1854, iii.

106. Loudon 1842, iv.
107. Loudon 1841, preface.
108. Becker 1864, iii–iv.

LITERARY EXPERIMENTS IN NARRATIVE FORMAT

Female popularizers could also form a bond with their readers through the modification of old, and the creation of new, narrative formats in scientific writing. Scholars have pointed to the importance of studying the creation and evolution of genres used by male practitioners, such as the journal essay, the experimental report, the speculative article, and the literature review.[109] Just as male practitioners consciously constructed a range of distinctive genres, so did popularizers, including many of the women. The familiar format, with its emphasis on a maternal figure who imparted knowledge in a domestic setting, was not, in its early nineteenth-century form, effective for communicating with the emerging mass reading audience composed of men and women of all ages. Women whose careers had begun in the twenties or thirties, like Zornlin, Loudon, and Roberts, and who had adopted the familiar format initially, had already dropped it by 1850. Loudon broke with the previous generation of female botanical writers in her rejection of the familiar format and her use of a first-person discussion that did not include sisterly conversations in the home.[110] Like Johns and Kingsley, many female popularizers experimented with different narrative formats after 1850.

Some women adopted the narrative format of common things. By using familiar objects as a springboard for discussing natural history, female popularizers could begin at the level of their audience while showing that even the commonplace could be fascinating. Pratt's *Chapters on the Common Things of the Sea-Side* was organized around "objects frequent on our shores," such as seaside plants, seaweeds, mollusks, and zoophytes.[111] Carey's *The Wonders of Common Things* dealt with such everyday objects as a lump of coal, a grain of salt, or a sheet of paper, bringing to bear the appropriate science to explain the natural origins of each item and their uses. A number of these books drew on some of the conventions of the familiar format. In Carey's *Wonders of Common Things,* for example, each everyday object is being

109. Bazerman 1988; Dear 1991.

110. Shteir 1996, 222. The familiar format did not die out completely in the midcentury. Among male authors, Ruskin and Kingsley tried to revive the didactic dialogue in the 1860s (Myers 1989). *Arcturus; or, The Bright Star in Bootes: An Easy Guide to Science* (1865), by Miss Sedgwick, presented a dialogue between Harry Wildfire and his mother, in the traditional teacher's role, that covered the seasons and the planets, among other astronomical topics, as well as some geology and chemistry at a very elementary level. Wright's *Observing Eye* (1850) contained letters addressed to "My Dear Young Friends." Mary Ward's *A World of Wonders Revealed by the Microscope* was written in 1858 in the epistolary form, as a series of letters sent to a friend.

111. Pratt 1850, v.

discussed by four children sitting around the fire on a cold winter afternoon. In the Kirbys' *Aunt Martha's Corner Cupboard: A Story for Little Boys and Girls* (1875, Nelson), two idle schoolboys visit their aunt, who tells them entertaining stories about the china, tea, coffee, sugar, and needles in her kitchen cupboard. In the first story, "The Story of the Tea-Cup," she discusses the making of china by the Chinese and how cups and pots were made in England. Aunt Martha, in the best tradition of maternal figures from the first half of the century, aims not only to educate her lazy nephews but also to inspire them to become industrious scholars.

Another narrative format that drew on previous traditions centered on the anecdote, a staple of natural history. Many female popularizers sprinkled their works with anecdotes from other natural history writers, but Bowdich Lee published two books, *Anecdotes of the Habits and Instincts of Animals* and *Anecdotes of the Habits and Instincts of Birds, Reptiles, and Fishes* (1853, Grant and Griffith), composed almost entirely of them. Requiring assistance in finding new material, and unsure of the current conventions surrounding the anecdote, she wrote to Richard Owen while working on the first book. Declaring her wish to "supersede the hackneyed stories now current," she inquired if he would loan her those books that he owned that contained juicy quotes. She also asked him "what sources do you advise me to draw from when the anecdotes are not private?" Finally, she solicited his opinion on the issue of whether or not it would "injure my reputation as an author if I repeat my own anecdotes?"[112] Bowdich Lee decided to include anecdotes about her own interactions with animals, anecdotes told to her by acquaintances, and anecdotes from books and journals. In *Anecdotes of the Habits and Instincts of Birds, Reptiles, and Fishes* she included stories from natural history journals, such as the *Naturalist* and *Naturalists' Magazine,* and from the general periodical press, for example, the *Northampton Mercury, Saturday Magazine, Chambers's Edinburgh Journal,* and the *Edinburgh Literary Gazette.* Among the naturalists quoted were Philip Gosse, Darwin, and Charles Waterton. One of her own anecdotes recounted the story of how she caught a twelve foot shark by baiting a large chain and hook with a piece of salt pork. Divided into three parts, anecdotes on birds, on reptiles, and on fishes, each section was further subdivided by specific types of animals. The pastiche of anecdotes was glued together by Bowdich Lee's short narrative, usually consisting of a brief description of each animal, its geographical location, its food, and its behavior. The anecdotes were often used to provide evidence

112. Lee to Owen, n.d., Natural History Museum, General Library.

for a naturalist's theory about animal habits or instincts. In the section of the book on birds, two anecdotes were related as a demonstration of the "sagacity" and "sociability" in goldfinches. The reptiles section included an anecdote illustrating the "ferocity" of alligators. Like Morris and Johns and other popularizers writing for a broad audience, Bowdich Lee viewed the anecdote as an important source of evidence in natural history.[113]

Female popularizers also wrote books that drew on structures used by male practitioners, though adapted for a broad audience, such as natural history guidebooks, that offered introductions to the classification system of a particular area of study. Variable in length and scope, these guidebooks ranged from the elementary introduction to detailed examinations of each family in the system. The botanists provide the best examples. In her *Botany for Novices,* a small book of about sixty pages, Becker attempted only to explain the principles of the Natural System, rather than provide detailed descriptions and names of individual plants. More sophisticated and detailed, Lankester's *Wild Flowers Worth Notice* focused on flowers that were representative of particular families and that were remarkable for their beauty of appearance or for their usefulness. She acknowledged that her book could not be an exhaustive treatise on British flora such as those large tomes written by "learned botanists."[114] Pratt's massive five-volume *Flowering Plants* and Loudon's *British Wild Flowers* (1844) presented botanical information even more thoroughly. Both are arranged systematically according to the Natural System, and they illustrate each family of plants.

Another narrative format used by female popularizers adopted a chronological structure, depending on the seasons of the year, or on the life cycle, as the organizing principle. The Kirbys' *Caterpillars, Butterflies, and Moths* began with the egg, moved through the chrysalis, and then on to the transformation of the butterfly. Variations on a seasonal structure were more common. Starting with January, Loudon's *The Amateur Gardener's Calendar* (1847, Longman) provided a month by month account of the weather to be expected, of the work to be done, and of the garden enemies at that particular time of year, whether it be birds, insects, or molluscous animals. Loudon claimed that she gave definite and clear directions on when to plant and on other garden operations as compared to other gardener's calendars.[115] Pratt's *Wild Flowers of the Year* (1846, Religious Tract Society) also began in January, discussing a flower in the month when it first appears. She included a

113. Bowdich Lee 1853, 85, 242, 384.
114. Lankester 1905, v.

115. Loudon 1857, 2.

description of each flower, its uses in Britain and abroad, and where it was to be found. These books were indebted to almanacs and to earlier works, such as Mary Roberts's *The Annals of My Village* (1831, J. Hatchard and Son).[116]

Yet another narrative format drew on the conventions of travel literature or on the idea of a ramble through nature, similar to Johns's descriptions of his expeditions. The journey was a replacement for the dialogue format, and was adopted by Loudon in her *Young Naturalist's Journey* (1840, William Smith) and its reissue *The Young Naturalist; or, The Travels of Agnes Merton and Her Mama* (1863, William Smith).[117] According to Loudon, the idea for the book came to her as she was reading the *Magazine of Natural History*. If "stripped of their technicalities," many of the papers could be "rendered both interesting and amusing to children."[118] Loudon decided to combine anecdotes from the journal papers with an imaginary journey. *The Young Naturalist's Journey* is composed of a series of vignettes about Mrs. Merton and her seven-year-old daughter, Agnes, as they visit relations and friends who have an enthusiasm for natural history. Loudon's book has obvious connections with the maternal tradition, though Mrs. Merton is not the sole source of knowledge for her daughter. When they board the train, they meet a woman with a marmoset, which provides the opportunity for Agnes to learn about its native country. In Birmingham, they visit Mrs. Merton's cousin, who has a pair of Virginian partridges from America. After playing

116. Several changes to almanacs in the early nineteenth century may have inspired ideas for new narrative formats in science writing. Charles Knight's *British Almanac*, first published by the Society for the Diffusion of Knowledge in 1828, undercut the older astrological almanacs. In his efforts to counteract superstition, Knight designed the *British Almanac* to emphasize science and rationality. This new association of almanacs with science made possible the notion of creating a genre in science writing that imitated the organization of the almanac. The addition in many almanacs in the early nineteenth century of gardening notes and columns constitutes another new feature that may have suggested to scientific authors that a calendar-like structure could serve well as the basis for books aimed at a popular audience (Perkins 1996, 58, 60, 85–86). Knight's *British Almanac*, and the accompanying *Companion to the Almanac*, contained a variety of types of information, some of it with no connection to science, such as observations on weather, a calendar of important dates, astronomical facts, useful directions for each month, a list of the members of the royal family, directories of the officers of state, commerce, education, and law, useful tables (e.g., for calculating the stamp tax), and general facts about weights and measures. The books by Kirby, Loudon, Pratt, and Roberts resembled the short ten-page section in the *British Almanac* on "Useful Directions for Each Month," which covered the topics of the preservation of health and the management of a garden and a farm in a straightforward, factual manner. The books by these scientific authors were far more literary and poetical in nature.

117. Gates 1998, 44.

118. Loudon 1840, ix.

with the birds, Agnes runs to her mother to find out more about their history and habits. Then it is on to Somersetshire to visit Sir Edward Peregrine, whose falcons become the focus of the discussion. Later, Mrs. Merton and her daughter visit a friend who lives on the banks of the Dart, near Dartmouth, allowing Loudon to explore the topic of fishes. This journey is not a ramble through the country; rather, it is a railway trip that has the potential to expose the reader to various regions within England, as well as exotic locales around the world.[119] Other books by female popularizers that draw on this narrative format include Margaret Plues's *Rambles in Search of Ferns and Mosses* (1861, Houlston), a first-person account of fictional travels to observe ferns indigenous to different parts of England; and Wright's *The Globe Prepared for Man*, which introduces a discussion of geology by describing an imaginary family traveling through England and Wales whose curiosity is piqued by the diverse soil colors and "the changing aspect of the land, from plains to hills and from hills to dales."[120]

Other narrative formats pushed the fictional dimension of science writing even further. In some books, the authors imagined what nature would say were it endowed with the ability to speak. In the Kirbys' *Chapters on Trees* (1873, Cassell), they impress upon the reader the height and age of "the Giant Pine of California." "What a history these old giants might relate," the Kirbys declare, "had they but the gift of speech!"[121] The Kirbys' flight of fancy was the basis of at least two books that take an anthropomorphic turn. In Carey's *The Wonders of Common Things*, everyday objects tell their own stories, like Kingsley's pebble. In the first chapter, a lump of coal promises to tell a group of children a story about its history that will be as "wonderful as any fairy tale can be." The piece of coal discusses its family, its origins, and reminiscences about the grand creatures that played about its feet. For those children who want to hear more, he advises them to ask Sir Charles Lyell about "the annals of our race." Then it moves on to the various uses of its tribe, such as heating houses, running railways, and providing power for steamships and other great inventions.[122] Although Mary Roberts belonged to an older generation than the Kirbys, and though she had made use of the familiar format in several of her earlier works, her *Voices from the Woodlands, Descriptive of Forest Trees, Ferns, Mosses, and Lichens* (1850, Reeve, Benham, and Reeve), written near the end of her career as a scientific author, granted trees the gift of speech. In the preface to the book she anticipated the readers' objections to

119. Gates 1998, 46.
120. [Wright] 1853, 1–2.

121. Kirby and Kirby [1873], 196.
122. A. Carey n.d., 10–18.

this fantasy. "Methinks I hear some one say, Why are things inanimate thus fabled to speak? Why not rather tell concerning their properties and uses, and the places of their growth, without having recourse to fiction?" Roberts pleaded with her readers not to be "chafed," reminding them, "poets in all ages have preferred to instruct mankind after the same manner." A third-person narrator intervenes from time to time to guide the reader from one plant to another, but the main speakers are the lichens, mosses, ferns, and over eighteen different types of trees, that describe their special place in the scheme of things, their historical associations, and their distinctive uses.[123]

Gatty's short stories in her *Parables from Nature* also presented an intriguing mix of fact and fiction while anthropomorphizing both flora and fauna. Humans rarely appear in her stories. The main characters are more often a caterpillar (in "A Lesson of Faith"); a worker bee (in "The Law of Authority and Obedience"); a family of birds (in "The Unknown Land"); a zoophyte, a seaweed, and a bookworm (in "Knowledge Not the Limit of Belief"); flowers (in "Training and Restraining"); crickets and a mole ("Waiting"); spruce firs ("The Law of the Wood"); a dragonfly (in "Not Lost, But Gone Before"); or rooks (in "Inferior Animals"). Gatty attempted to make her children's tales as scientifically accurate as possible and even added in later editions of the *Parables* a lengthy section of notes that included detailed information on the scientific theories informing each of the stories. Even though Gatty's stories contained talking animals and plants, they were based on the observable and the empirical. She consulted her friend Harvey on points relating to his field of expertise. On November 21, 1859, she wrote to ask him how long the *Protococcus nivalis* lasted on the mountains when it made its appearance. "I am trying to get a Parable out of it," she explained to him, "and am anxious not to be incorrect in statistical facts."[124] She wrote to other authorities for advice, including the eminent entomologist Henry Stainton. On September 12, 1860, she asked him for help in describing the shape of butterfly scales for one of the notes and inquired if her description of the death of a butterfly in one of the parables was accurate. "You will be astonished at seeing the proofs again," she wrote, "but I want to know whether 'drooped her wings' (and died) would be correct—as if so it is prettier than "<u>fell down</u>" or <u>dropped</u> her wings, but I <u>should</u> think they did droop them anyhow."[125] Precision

123. Roberts 1850, vii.

124. Gatty to W. H. Harvey, November 21, 1859, Sheffield Archives, HAS 48/51.

125. Gatty to Stainton, September 12, 1860, Natural History Museum, Entomology Library.

on the tiniest detail was important to Gatty, as she conceived of herself as a scientific author conveying important information as well as a storyteller.

"THE WONDERS AND BEAUTIES OF EARTH"

Female popularizers developed an impressive array of narrative formats for communicating with their newly defined audience. Whatever genre they used, they all made aesthetic, moral, and religious themes central to their work. Like the women of the earlier maternal tradition, they presented themselves as guides to understanding the larger significance of scientific theories, though they rarely featured a female character in a motherly role. In her *Flowers and Their Associations* (1840, Knight), Pratt blissfully mused about the beauty of nature and how it could be perceived by one and all. Whether it be the child "treasuring his daisies and cowslips" or the "artisan tending his auriculas," it was a welcome sight. "It is an indication of a perception of beauty—of an awakened love of nature," Pratt asserted, "which will not be satisfied with the object before it, but will comprise, in its regard the wonders and beauties of earth, and bear with it an intellectual joy and improvement." The pleasure derived from wildflowers "lies open to the youngest and the poorest of mankind." Just as birds are "the poor man's music" so are flowers "the poor man's poetry."[126] Female popularizers appealed to their reader's aesthetic sensibilities through their use of language and visual images in order to elicit a sense of wonder. From the microscopic world to the awesome heavens, nature was depicted by these women as a spectacle and as a feast for the senses.

Like natural historians before them, female popularizers introduced poetry and literature into their work. For Pratt, singing birds were only one aspect of how "all nature is full of music." There was music in "gentle winds" as they "rustle through the summer leaves," in autumnal gales' "louder and wilder melodies," and in "the trickling waterfall and the pattering raindrops." It seemed to Pratt as if "all inanimate nature seems pouring forth its anthem from earth to heaven."[127] To appeal to the reader's ear, Pratt began each chapter of her *Our Native Songsters* (1852, SPCK) with a poetic, pastoral rhapsody. She included poetry by Coleridge, Keble, and Wordsworth. Poetry and vivid literary passages are scattered throughout her other works and in the books of other female popularizers. In the *Entertaining Naturalist,*

126. Pratt 1846, 9–10.   127. Pratt 1853, 1.

Loudon quotes from Milton, Shakespeare, Pope, Byron, and Wordsworth. The Kirbys discussed poetic allusions to the many trees they feature in *Chapters on Trees*. In her *Telescope Teachings,* Ward referred to a passage from Humboldt's *Cosmos* on the beauty of solar rays, discussed how the subject had been eloquently treated in English poetry, and then ended with a poem titled "The Sunbeam."[128] In her *Voices from the Woodlands* the talking plants not only used poetic language to describe their own beauty and their important place in nature, but Roberts also included poetry by Coleridge and Southey.[129]

Female popularizers more often appealed to the readers' eyes, rather than their ears, to elicit a sense of the beauty of nature. Lankester emphasized the importance of training and teaching the eye to see nature properly, both as a wondrous world of beauty and as amenable to scientific analysis. "The naturalist—he who has thought and worked amongst the wonderful and curious things of this beautiful world," she declared, "has his eyes sharpened and educated to observe; by constant habit he sees at a glance the arrangement of the parts of a flower, and thereby recognizes its class and order."[130] Pratt painted vivid pictures of nature with her words. In her *Ferns of Great Britain* (1855, SPCK), she constantly drew the attention of the reader to the exquisite beauty of the fern. The fine-leaved Gymnogramma was referred to as a "pretty fragile little fern"; the curled rock brake as an "elegant little fern" that was "so beautiful in outline"; the Willdenow's fern as a "beautiful plant, gracefully waving to every summer wind"; and the lady fern as having a "graceful attitude and elegant outline" (Fig. 3.9).[131] In their *The Sea and Its Wonders,* the Kirbys treated the ocean as a "world in itself" that was "subject to its own laws" where "wonders abound." Enraptured by the beauty of seaweeds and the fantastic creatures that lived among them, the Kirbys offered a lush literary description of a world that "is almost like fairy-land." They compared seaweeds to the beautiful flowers so lovingly described by botanical writers. They were "flowers of [the] ocean, that vie in loveliness with the lily and the rose."[132]

In their efforts to raise the reader's awareness of the beauty of nature, female popularizers often argued that there were exquisite worlds in nature that had completely escaped the reader's notice. When it came to the ocean, the Kirbys believed that landlubbing readers could "form little idea of its

128. M. Ward 1859, 38.
129. Roberts 1850, 47.
130. Lankester 1861, 122.

131. Pratt n.d. [Ferns], 31–32, 59, 66.
132. Kirby and Kirby 1871, vii, 80, 253.

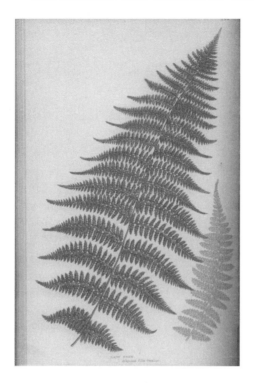

FIGURE 3.9 The lady fern, which Pratt referred to as "the loveliest of all our larger ferns" (Pratt n.d., 66). (Anne Pratt, *The Ferns of Great Britain, and Their Allies the Club-Mosses, Pepperworts, and Horsetails* [London: Society for Promoting Christian Knowledge, n.d.], opposite p. 66.)

hidden beauty."[133] Women writers therefore relied on several techniques to allow their readers to see these beautiful, but hidden, domains of nature. Like Johns, the Kirbys made use of the you-are-there literary technique as a means of encouraging the British reader to imagine themselves in nature. In *Things in the Forest* they transport the reader to far-off exotic locales to observe the gorgeous plumage of various birds. After describing a cypress forest, they address their audience directly, as if they were at the scene spying out a snakebird. "Do you see that pond yonder," they asked, "enclosed by trees, and trees rising out of it? If you watch for a minute, you will see

133. Ibid., 80.

her seated upon one of the branches."[134] Hunting for a bird of paradise in New Guinea in *Beautiful Birds in Far-Off Lands* (1872, Nelson), the Kirbys describe the forest as being "alive with brilliant creatures." They pause in their search to point out some unusual vegetation. "Do you notice yonder tree," they ask, "with great spreading boughs that grow from the top of the stem, and form a kind of leafy plateau?"[135] Another literary technique for uncovering the hidden wonders of nature relied on detailed descriptions provided by sense-enhancing instruments. In her *Chapters on the Common Things of the Sea-Side*, Pratt impressed on her readers the beauty of coral reefs and jellyfish, but to unveil the splendor of smaller organisms she had to resort to the microscope. Observing common zoophytes in the microscope showed that they are "wondrous" and "their beautiful cup-like, or bell-shaped cells" led her to feel "how much there is of beauty lying unperceived by us even in common things." Similarly, many of the British seaweeds needed "the aid of a microscope to discover their beauty of structure."[136]

Conveying the beauty of nature to readers was also accomplished through the use of vivid illustrations, especially those in color, not just via literary techniques. Women in the post-1850 era included far more illustrations and color in their books than their predecessors in the maternal tradition.[137] Beginning in the 1830s, advances in printing technology made diverse imagery widely available and affordable for the first time, and visual images came to play an important role in the work of female popularizers. In order to reach a popular audience, women engaged in science writing recognized that they would have to take advantage of the developing mass visual culture. They became more sophisticated in their use of visual images than their predecessors, consciously manipulating the images in their texts. By contrast, male practitioners did not always use visual images when trying to communicate to a popular audience. In the first half of the nineteenth century the growing middle-class appetite for illustrated novels and weekly periodicals led practitioners of science to fear that pictures would have an adverse effect on the mental faculties of the reading audience. Many male practitioners were reluctant to use illustrations since they could appeal to the senses or the emotions, and not to the rational faculties. The debate in 1838 concerning the use of plates in natural history books revolved around the question of whether or not they would merely foster a superficial appreciation of the beauties

134. Kirby and Kirby 1862, 108.
135. Kirby and Kirby 1872, 21.

136. Pratt 1850, 94, 239.
137. Lightman 2006.

of nature.[138] Similar concerns later in the century may have led would-be professional scientists to use visual images sparingly and to rely mostly on certain types of images such as diagrams, maps, and cross-sections.

However, women made extensive use of visual images in their books, most of which depicted plants, animals, or a particular scene in nature. Although their books did not contain many color illustrations, if any at all, Loudon, Wright, and Zornlin all included a large number of black-and-white woodcuts. Loudon's *Amateur Gardener's Calendar* contained 122 engravings, while her *Botany for Ladies* included 151 figures. Her *Entertaining Naturalist* had even more, close to five hundred engravings.[139] Wright's *Observing Eye* and *Our World* were peppered with small woodcuts, but, in addition to them, her *Globe Prepared for Man* contained a color frontispiece of an erupting volcano and a large pullout of the "Order of Rocks Covering the Globe, 8 or 10 Miles in Depth." Zornlin also relied on woodcuts in her books, some, such as *Outlines of Physical Geography, for Families and Schools* (1851, Parker) contained only nine illustrations (many of them maps and diagrams), while *The World of Waters* and *Recreations in Physical Geography* presented thirty-five and fifty images respectively (Fig. 3.10). Books by Gatty, Lankester, Pratt, Twining, Roberts, Lee, the Kirbys, and Ward featured color depictions of nature. Gatty's *British Seaweeds* used Harvey's colored plates from the *Phycologia Britannica*. Women used color whether they dealt with ferns, trees, flowers, plants in general, birds, or even mollusks.[140] Only a handful of the male popularizers, such as Wood, Johns, Houghton, and Gosse, offered their readers such sumptuously colored books.

Mary Ward's books offer a good example of how female popularizers drew their reader's attention toward the beauty of nature through their use of color images. Like Gatty, Loudon, Pratt, and Twining, Ward was an artist herself. Not only did she draw all of the illustrations in her own books, she also provided male practitioners with illustrations for their works. The

---

138. A. Secord 2002, 28–57.

139. Fewer illustrations appeared in her other books, such as *The First Book of Botany* (thirty) and *The Young Naturalist's Journey* (twenty-three).

140. For ferns see Lankester's *British Ferns* (1880) with its sixteen color plates or Pratt's *Ferns of Great Britain* (forty full-page color plates); for trees see Bowdich Lee's *Trees, Plants, and Flowers* (1854) and its eight color plates and Roberts's *Voices from the Woodlands* (nineteen plates); for flowers see Pratt's *Flowering Plants* (220); for plants in general see Twining's *Illustrations of the Natural Orders of Plants* (160 full page color plates); for birds see the Kirbys' *Beautiful Birds in Far-Off Lands* (numerous full-page color plates of exotic birds); and for mollusks see Roberts's *Popular History of the Mollusca* (1851) with its eighteen illustrations.

FIGURE 3.10 Wilberforce Falls is just one example offered by Zornlin of the beautiful waterfalls and cataracts around the world. It is a "fine example of a double waterfall," which is "one of the most magnificent objects in nature" (Zornlin 1843, 248). (Rosina M. Zornlin *The World of Waters; or, Recreations in Hydrology* [London: John W. Parker, 1843], 249.)

illustrations of Newton's and Rosse's telescopes in David Brewster's *Life of Newton* (1855) are by her hand.[141] Brewster's daughter claimed that she also illustrated Brewster's paper "On the Structure and Optical Phenomena of Ancient Decomposed Glass," and referred to her as having "an exquisite gift of drawing."[142] Ward's colored illustrations were considered so beautiful that two of her works were chosen to go on display in 1862 in the book section of the International Exhibition at the Crystal Palace.[143] Fifteen plates were included in her *Telescope Teachings,* some in color, which were designed from Ward's own drawings. Her "Falling or Shooting Star" depicts a beautiful, starry night enveloped in darkness, complemented by the

141. Harry 1984a, 193.
142. Gordon 1869, 241, 363.

143. Harry 1984b, 471.

A FALLING OR SHOOTING STAR.

FIGURE 3.11 "A Falling or Shooting Star," a plate designed from Ward's own draw-ing intended to illustrate a beautiful night sky, has the quality of a Romantic painting. (Mrs. Ward, *Telescope Teachings* [London: Groombridge and Sons, 1859], opposite p. 166 [in color].)

adjoining discussion of the mystery pervading the subject of shooting stars (Fig. 3.11).[144] *Microscope Teachings* (1864) contained twenty-seven colored il-lustrations of insect wings, scales, and eyes; animal hairs; petals; seeds and pollen of flowers; and the circulation of blood in fish, frog, newt, and bat. The microscope transformed things usually seen as disgusting into objects of resplendent beauty. Insect wings "are among the most lovely spectacles

144. M. Ward 1859, 167.

presented to us by the microscope."[145] As beautiful as her illustrations may be, Ward emphasized that, in the case of insect scales, "the best painted representation could scarcely do justice."[146] She makes the same point about astronomical phenomena. Seeing Saturn for the first time is a vivid experience. "The beauty, too," Ward declared, "of the real planet never fails to strike the beholder in a way no pictured representation can."[147] In her *Microscope Teachings,* where Ward had tried to produce a combination of the guidebook and the panorama, she emphasized the beauty of the scenes produced by the microscope. In the tradition of the panorama, she presented a succession of wonders.[148] In this reference to one of the hallmarks of the London entertainment scene, the panorama, Ward acknowledged her attempt to reach her audience through an appeal to the same aesthetic sensibility at work in the mass visual culture of the period.

MORAL AND SOCIAL TEACHERS

Female popularizers of science in the third quarter of the century may have developed innovative genres and presented more visual images in order to reach their newly defined audience, but they retained the role of moral and social teacher. By doing so, they claimed to possess the authority granted to women in the earlier maternal tradition.[149] Nature provided a model for human behavior and clues to the proper organization of society. Women dealt with such issues as the value of the family, the importance of industriousness, the need for obedience to authority, the dangers of alcohol, the propriety of slavery, and the benefits of imperialism. Of course, the lessons drawn from nature by these women reflected their middle-class background and their British nationality. Male practitioners projected Victorian middle-class society onto nature and then found scientific justification for middle-class values. Female popularizers played the same game.

145. M. Ward 1864, 48.

146. Ibid., 49.

147. M. Ward 1859, 4. Compare this to the Kirbys' handling of color images in *Beautiful Birds in Far-Off Lands.* In the preface they insist that a verbal description of the birds "would have failed to convey any idea of their loveliness; but the Coloured Illustrations which embellish the work bring them before our eyes with life-like reality." Later in the book the descriptions of the stunning Sun-Bird, "or even pictures themselves, would fail to impart an adequate idea of these gems of nature—these beautiful birds!" (Kirby and Kirby 1872, v, 145).

148. M. Ward 1864, vii–viii.

149. Gates 1998, 50–51.

For Pratt, nature spoke of the value of domesticity. Pratt's *British Grasses and Sedges* ([1859]), an SPCK publication, examined the two orders of grasses and sedges, providing information on their stems, fruits, flowers, uses, and location. Pratt opened the book with a discussion of the role of grasses in the vegetable kingdom and offered a description of early spring meadows, emphasizing their beauty. "Wherever now we see a corn-field waving in beauty," she declared, "whether in the climes of east or west, or by the quiet homesteads which lie among the hills and valleys of our native land, it tells of peace, civilization, and domestic happiness; it tells of homes." Pratt then explained that corn was historically sowed by people who had settled down, not by "wild wanderers." Those who tilled the earth gained "softer manner, and gradual improvement in the arts and sciences of civilized life. The house is reared, and children learn beneath its roof the love of kindred, of neighbours, and of country; and agriculture proves alike the source of individual and national prosperity."[150] This ode to domesticity is intended to provide the reader with the proper understanding of the importance of grasses and sedges in the social economy of nature.

Other women drew the reader's attention to the import of industriousness. In the concluding pages of her *Observing Eye*, Wright insisted that there was a lesson to be learned from the accomplishments of insects. Although small in size, their combined labors provided humans with useful items, such as honey and wax, and as scavengers they eliminated harmful waste. Their patient industry teaches us "to be cheerfully active in all that our hands find to do."[151] The Kirbys depicted the entire animal world as one of industriousness. If the reader were "gifted with the power of seeing all that is going on in the animal world, and were to look round you, one fine summer morning, what a busy scene would be spread out before your eyes! Every creature is at work," including the wasp gathering materials for his nest, ants running in and out of their homes, and caterpillars hard at work.[152] According to the Kirbys, enthusiasm for natural history itself taught readers to be industrious. By the end of the Kirbys' *Aunt Martha's Corner Cupboard*, the two lazy boys, "who had before been so idle and ignorant, grew up industrious and learned men" after being inspired by their aunt's stories about the objects in her corner cupboard.[153]

In addition to lessons on domesticity and industriousness, an examination of nature could provide warnings on the dangers of drunkenness.

150. Pratt n.d. [*British Grasses*], 5–6.
151. Wright n.d., 131.
152. Kirby and Kirby [1861], 40.
153. Kirby and Kirby 1875, 175.

Although Bowdich Lee's *Trees, Plants, and Flowers* (1854, Grant and Griffith) was intended to acquaint readers with "the bounties of God," she also pointed to abuses of these gifts. In civilized countries "the culture of vegetables is carried to high perfection, either to supply us with articles of necessity or luxury; for which the whole world is ransacked, and made to yield its treasures to those who seek them." Processing nature into human goods provided employment for thousands, increased wealth through the imposition of duties, produced valuable medicines and fuels, and loaded "our tables with delicacies." Bowdich Lee then proceeded to work her way through the tribes within each order, using Lindley's Natural System as the classification scheme. She presented a physical description of each tribe, some historical information, its location around the globe, and its uses. When she arrived at the "Grass Tribe," the emphasis on the bountiful gifts of God gave way to a stern warning about the abuse of these gifts. Grasses, Bowdich Lee argued, were necessary to all life, and, therefore through divine foresight they were most widely spread over the surface of the earth. "The seeds yield the most nutritious food," Bowdich Lee asserted, "and from many of them are extracted liquids, which, moderately and judiciously used, refresh the failing powers; but which, immoderately and blindly partaken of, destroy life, and plunge the highest work of God into the most abject debasement of intellect and morality." Beer, for example, was a nourishing and wholesome liquor that could restore an individual's strength. But if "taken in excess, the intoxication produced by it is said to brutalize the mind more than any other." God had provided us with a great gift, but "owing to his own sinfulness," a human being could transform a blessing into a curse.[154]

Another important lesson taught by surprisingly many female popularizers concerned the need for submission to authority. Gatty's *Parables from Nature* is, perhaps, the best example of how this theme was handled, though she counseled obedience to secular authority in her other works as well.[155] The *Parables* present ways in which a reader could learn about his or her moral duties and responsibilities through a scientific study of nature.[156] In "The Law of Authority and Obedience," a worker bee becomes rebellious after learning that he and the queen were physically identical when first born and that subsequent differences are merely a result of the food they eat and the shape of the house they live in. After convincing some young bees that this is unfair, they decide to form a hive of equals, but with no leader,

154. Bowdich Lee 1854, v, 1, 2, 75, 80.   156. Le-May Sheffield 2001, 47.
155. Katz 1993, 98, 122.

they cannot decide where to build the new hive. Although they begin to see the wisdom behind a hierarchical society, and although they now perceive that nature provides a way to determine who will rule by giving individuals superior abilities suited to rule, the consequences of their initial rebellion are disastrous. While they were searching for a new hive a young queen escaped from her cell too early and was killed by the old queen. The moral of the story is contained in the conclusion, "and thus the instincts of nature confirm the reasoning conclusions of man."[157] In Gatty's story, the instinctual hierarchical nature of the hive is revealed by science, and confirms the conclusions of those who argue for a political and social system where there are rulers and the ruled. Crickets in "Waiting" are unhappy with their lot in life and yearn for something better. A mole counsels patience, because if they wait "everything will fit in and be perfect at last." After generations the crickets discover their purpose, to sing by the side of hearthstones in human houses.[158] Unlike the impatient worker bees, who disrupt the prevailing natural and social order, the crickets' patience with the nature of things eventually leads to happiness and fulfillment.

In addition to teaching lessons about domesticity, industriousness, teetotalism, and submission, the arrangement of the natural world justified British imperialism. In her *Plant World*, Twining argued that since plants were adapted to their respective countries, "it is intended that man by his power of traveling about to various countries, and of transporting their produce, should distribute it to his fellow-creatures." The exchange of vegetable produce of different countries was mutually advantageous. In the West Indies, for example, the natives did not have the means to make anything useful out of the abundant cotton plant. European ships visited West Indies seaports and received the great bales of cotton the natives had gathered, then it was brought to English manufacturing towns to be made into clothing. "Some are made neat and suitable to English taste," Twining asserted, "but the manufacturer knows the desire of the poor black women for gay-coloured garments, and he has always a portion printed in bright, gaudy patterns." Without the aid of the English manufacturer, the "freed slave" would not have the cotton pods she helped to gather "transformed into the gay dress she is so pleased to wear." Twining presents this as an equitable arrangement, sanctioned by nature and nature's maker. "Thus we see that plants," she declared, "even the humble and lowly, assist in the one great

157. Gatty 1855, 33.   158. Ibid., 121.

and divine law that must even be fulfilled in this world, that man is born to labour for his own benefit and happiness, and for that of his fellow-creatures. Each much work in the lot appointed him by the heavenly Master; and each does in thus working bring profit to himself and to his brethren." Despite Twining's justification of British imperialism, she attacks slavery, designating it as extremely cruel.[159] The tension between supporting imperialism and condemning one of its by-products, slavery, eludes Twining.

Bowdich Lee's lesson on the sanction of nature for British imperialism parallels Twining's. Like Twining, she approved of the "culture of vegetables" undertaken by civilized countries. This kept Britain supplied with articles of necessity or luxury, for which "the whole world is ransacked." Securing "the farthest nations as our colonies" was part and parcel of the "culture of vegetables." Bowdich Lee singled out the discovery of the Gutta-percha tree, essential to the development of cable telegraphy, as an example of how colonization was advantageous. "Time and space are conquered by this one vegetable substance"; she wrote, "and the mighty effect of the submarine electrical telegraph appears like a misty dream of wonder, the entire realization of which is, at present, incalculable." But a discussion of sugar leads Bowdich Lee to an attack on slavery. The cultivation of sugar, she stated, "has been the principal excuse for that wicked and degrading commerce in human flesh, to which the African nations have for many years been exposed." Seemingly oblivious to the relationship between colonization and slavery, Bowdich Lee believed that nature sanctioned one but not the other.[160]

RELIGIOUS TEACHERS: "NO ENTOMOLOGIST CAN BE AN ATHEIST"

Most female popularizers of this period—Lankester and Bodington are among the few exceptions—offered themselves as religious as well as moral teachers.[161] Nature contained clues to proper human behavior precisely because God had created it. Nature was fraught with religious meaning. As

159. Twining 1866, 15-18, 250.

160. Bowdich Lee 1854, 2, 89.

161. Lankester was one of the few female scientific authors who did not pay significant attention to religious themes in her work. Her son, E. Ray Lankester, physician and scientist, was one of the leading scientific naturalists. Bodington, who wrote near the end of the century, identified herself as an agnostic (see chapter 8).

Becker put it, "In nature, nothing is trivial or unimportant, the smallest and most ephemeral of beings, owes its origin to the working of the same laws, and the force of the same Power that produces the greatest and mightiest on the earth."[162] In their works, women treated the bearing of scientific knowledge on religious issues as legitimate subjects of examination. They discussed the relationship between the Bible and science and the significance of natural phenomena for important Christian doctrines. Most of all, they drew upon a discourse of design to present a compelling theology of nature. Many shared a devotion to Christianity with the Anglican parsons. Many of them were Anglican, including Becker, Loudon, and Twining, and some were evangelical Anglicans, including Gatty and Zornlin. Mary Ward was a member of the Church of Ireland, the Irish equivalent of the Church of England. Several came from Nonconformist backgrounds, such as Bowdich Lee, the Kirbys, and Roberts.[163] Although by the midcentury women had moved away from the "narrative of natural theology" and embraced the "narrative of natural history," religious themes still played an important role in their work, even though only a handful of them wrote for Christian publishers such as the Religious Tract Society or the Society for the Promotion of Christian Knowledge.[164]

For some female popularizers, it was important to demonstrate that religious truths communicated in the Bible were in no way at odds with current scientific theories. The title page of Wright's *The Globe Prepared for Man* included two quotes from Psalms on how God laid the foundation of the Earth resting comfortably alongside a quote from De La Beche, one of the leading practical geologists of the period. This set the scene for her attempt to reconcile revelation with the findings of contemporary geology. In her preface, Wright discussed how "discoveries made in Natural History by the study of GEOLOGY, excited a few years ago much alarm in the public mind" as they appeared to "lead to statements at variance with the revealed account of the Creation." She believed that "this alarm has subsided" as a close investigation of the facts revealed the harmony between the Bible and the works of God. It would always be so, according to Wright, as the study

162. [Becker] 1864, 41.

163. I have chosen to focus on what these writers shared rather than how denominational differences translated into divergences in the handling of doctrinal and theological matters. The pervasiveness of religious themes in the works of female popularizers of science, even after the publication of the *Origin of Species,* is the key point here.

164. The Kirbys and Pratt wrote books published by the RTS, as did Brightwen and Giberne later in the century. Pratt and Giberne also worked with the SPCK.

of geology teemed with endless examples of divine order and design and was therefore "calculated to confirm our assurance in the existence of one supreme Creator." However, for the reconciliation to be permanent, Wright insisted that her readers understand that the Bible was intended to deal with humanity's "moral and spiritual relationships to his Maker," and "not to teach man natural history." The Bible affirmed that the earth's surface came to be as it is by God's will, but it "throws no light upon the modes by which the Lord God formed the rocky crusts of the globe." To "gain any idea of the process" we have to study nature.[165]

Other women, in the course of a discussion about a particular animal or plant, pointed to its appearance in the Bible, highlighting the religious meaning of their work. In her *Entertaining Naturalist,* a zoological guidebook, Loudon presented a short section on each animal, a page in length on average. Each section contained information on the animal's geographical location, physical characteristics, habits, and how humans used it. Loudon also discussed biblical references. For example, she declared that "the Hippopotamus is supposed to be the Behemoth of the Scripture. See Job, chap. xl."[166] This was information that Loudon believed was relevant to the discussion and of interest to her readers. It also helped them to see the bearing of natural history on the Bible. Wright explored the symbolical meaning of certain animals in the scriptures. In her discussion of worms in *The Observing Eye,* Wright told her readers that "worms are often mentioned in scripture" where they were "sometimes used as emblems to point out the weakness of man, his earthly mindedness, his exposure to danger, and his liability to corruption."[167] Twining was also interested in using natural history to illuminate the meaning of the Bible. In her *Plant World* she puts her examination of botany into a biblical framework by beginning the book with the first verse of Genesis and considering the significance of God's creation of vegetation on the third day. Later, while analyzing the various parts of a plant, she raises the issue of how the biblical writers used the seed and the root as emblems for the gradual growth of Christianity. The leaf was also an important biblical symbol. "Throughout the whole volume of the Holy Scriptures," she declared, "a leaf is employed as an emblem of some higher signification." But it was impossible to understand biblical allusions to leaves unless "we first learn the nature, office, and properties of leaves." The study of botany, she argued, not only provided "many testimonies to the truth of the Holy Scriptures," it also allowed the reader to "perceive more

165. [Wright] 1853, preface, 10–11.   167. Wright n.d., 20.
166. Loudon 1850, 100.

clearly the full meaning of many circumstances mentioned in the Bible, when we have learnt something of the nature and uses of plants."[168]

Christian doctrine could also be illuminated by a study of the natural world. Resurrection and immortality were favorites with female popularizers. In *Caterpillars, Butterflies, and Moths*, the Kirbys drew an analogy between resurrection and the "miraculous" transformation of the caterpillar into a butterfly. Just as the butterfly went through three stages in its life, from groveling worm, to chrysalis, to perfect insect, so too did humans move from being creatures of the earth, through death, to the "final change."[169] "Does not the study of natural objects teach us deeper lessons than we at first imagine?" the Kirbys asked in *Sketches of Insect Life* (1874, RTS), after a discussion of the metamorphosis of the butterfly as "the shadow of a great and solemn truth."[170] In a similar vein, Wright concluded her *Observing Eye* by pointing out that one lesson to be learned from a study of insects was that God could overcome the sleep of death. Just as the butterfly burst forth into the beauty of a higher life, we can too if God "changes our hearts" and washes us clean of sin.[171] Gatty devoted two stories in her *Parables of Faith* to the analogy between insect transformations and resurrection. In "A Lesson of Faith" a caterpillar does not, at first, believe that she will become a butterfly one day. After experiencing the miracle of this transformation, she learns that with faith anything is possible, including life after death.[172] A later story, "Not Lost, But Gone Before," recounts the tale of a grub who is told by a frog that he will turn into a dragonfly, leave the water, and enter a new world. He promises his grub friends to return after the change has taken place, but after he becomes a dragonfly he can no longer enter the water and he is unable to get a message of hope through to them. In her "scientific" notes, Gatty discusses how separation between the water and air are analogous to the gulf between earth and heaven. "The reader must ponder these things for himself," she wrote. "Similitudes and analogies between physical and spiritual things will not bear pressing too closely; but nevertheless, here and there, Nature seems almost to hold out to us wonderful adumbrations of divine truths, of which the twofold career of the Dragon-fly is a remarkable instance."[173]

168. Twining 1866, 9, 34, 49, 66, 143-44.

169. Kirby and Kirby [1861], 124-27.

170. Kirby and Kirby [1874], 59-60.

171. [Wright] n.d., 131-33.

172. Gatty 1855, 13.

173. Gatty 1861, 142-44, 192. Pratt would have welcomed Gatty's caution and gone further. She insisted that there was no atonement or knowledge of eternal life to be gathered from nature. She

An examination of nature also could be used to teach the reader about the theological notion of providence. The Kirbys believed that instances of divine providence were to be found in the sea. Even though the polyp had nothing but a mouth, sac, and feelers, God had equipped it with enough to survive. They asserted that "the same kind Providence has them in His care, and even the little polyp is not passed over, or neglected, in the vast region over which He reigns."[174] The Kirbys were not the only ones to draw attention to examples of divine providence in the operation of nature. In the case of limpets, Roberts asked her readers, "Saw you not that the Almighty Creator of the universe (without whose permission a single hair does not fall from our heads, nor a lonely sparrow to the ground, neither is a shell or pebble rolled by the billows upon the shore,) provides against their utter extinction through the depredations of sea-birds and rapacious fishes" by giving them simple colors?[175] Loudon praised the "wisdom of Providence" for creating the immense elephant as an herbivore rather than a carnivore.[176]

However, the most pervasive religious theme running throughout the works of many of these women was, undoubtedly, what nature taught us about divine design. Some referred explicitly to Paley as a source of inspiration. In 1887, Mary Kirby recalled in her autobiography that when her father read books to her and the other children, a favorite was a large printed edition of Paley's *Natural Theology*. She asserted, "Dr. Paley's argument about the watch has never been forgotten. If the watch with its mainspring, its chain, and its delicate wheels must have had a maker, how much more the universe, with its mechanism, and its wonderful contrivances must have had a Creator?"[177] In her description of the continual activity of insects, Roberts pointed to their joy and exultation "as Paley beautifully observes."[178] Ward may have been thinking of the same passage when she referred to a chapter "in Paley's 'Natural Theology' devoted to insects, as a proof of design in creation."[179]

Female popularizers ranged across the whole of creation, uncovering traces of God's wisdom, power, and goodness in the tiniest creatures, in

---

feared that natural history enthusiasts might believe that so much of God could be learned through an examination of his works that a study of the Bible was not necessary ([Pratt] n.d. [*Wild*], 12–13).

174. Kirby and Kirby 1871, 59.

175. Roberts 1851, 24.

176. Loudon 1850, 97.

177. Kirby 1888, 8.

178. [Roberts] 1831, 36.

179. Ward and Mahon n.d., 42.

the largest heavenly bodies, and in everything in between. They did not intend to persuade their readers through philosophical argument that an omniscient and omnipotent God existed, though they did rely at times on the discourse of design to illustrate that this was so. Rather than demonstrating God's existence, they aimed to display the design in nature. In her *Telescope Teachings,* Ward emphasized the vastness of the universe revealed by the telescope, and wondered with her readers "what wonders of creation may be contained in those distant worlds? what amazing scenes of grandeur and glory, what triumphs of power unbounded, what monuments of loving-kindness no less infinite!" Yet, she reminded her audience, the greatest telescopes allowed us to see only "one portion of boundless space" or "but a part of the Almighty's works."[180] Whereas Ward stressed the power of God signified in the immensity of space, Wright pointed to the vastness of time. The study of geology, Wright affirmed, inspired "reverence towards our Great Creator." Looking into the formation of the earth's crust "extends our thoughts into immensity; teaching us that the revolutions of nature have occupied immense periods of time; and lastly, it stands forward as an abiding witness of the marvellous wisdom and goodness of that Almighty Being, who from Eternity has formed all things, and still upholds the universe by the power of his own creative will." The main message of both of Wright's geology books, *Our World* and *The Globe Prepared for Man,* is that the creator used natural processes over eons of time to make the earth ready to support life.[181] Zornlin maintained that geology excited "our admiration and delight," whether we regarded it as a key to finding the treasures of the natural world "or in the higher light of one of the various means which lead us to the contemplation of the wisdom, power, and goodness of the beneficent Creator of all things."[182]

Natural history, whether it dealt with plants or animals, provided even greater potential for teaching readers to see contemporary science through the eyes of the theologian of nature. As Twining remarked in her *Plant World,* the study of plants offered a "striking illustration" of the "might and wisdom of the Creator."[183] Loudon agreed. In her *First Book of Botany* she discussed the adaptation of lichens and seaweeds to their environments. This led her to observe that "the meanest objects in the vegetable kingdom have been as wonderfully contrived, and shew as evidently the Divine wisdom by which they have been organized and fitted for the stations they are to fill, as the noblest forest tree; and the more we study to unfold the secrets of

180. M. Ward 1859, 205–6.
181. [Wright] 1853, 22, 94.

182. Zornlin 1852, 97.
183. Twining 1866, 10.

Nature, the more we shall feel impressed with awe and admiration of that beneficent Being, who has made all these wonders."[184] In her *Wild Flowers of the Year*, Pratt declared that the design in the vegetable world was so clear and evident that only the fool could ignore it. "The argument so often applied to the various works of creation," she declared, "that an instance of design necessarily implies a designer, is so obvious, that a child can understand it."[185] Even the specific parts of a plant provided fertile ground for a discourse of design. For Loudon it was the flower that displayed "Divine care and wisdom in its construction," while Roberts pointed to the intricate design of the seed as an illustration of the "wisdom and beneficence of the Creator."[186]

The animal world offered no less of an opportunity to teach religious lessons about the design in nature. There was no end to divine cleverness when it came to finding specific instances of design in unusual body structures. According to Loudon, the pelican's "pouch Providence has allotted to the bird, that he may bring to his eyrie sufficient food for several days."[187] Pratt was fascinated by how the cormorants were "provided by Almighty skill, for the purpose of feeding on larger fishes than can be swallowed by most birds, with a very large gullet or tube, between the mouth and stomach, and it is said even to be able to swallow a flat fish."[188] But even common organs possessed by many creatures displayed design. For Ward, the eyes of vertebrate animals were "constructed on a plan exceedingly similar to that of the camera-obscura." This led her to describe the eye as "that most beautiful piece of mechanism."[189] Of course, Paley singled out the eye as an organ of extreme perfection, and it became a staple example of design that could only be explained by referring to a divine designer. Darwin was therefore compelled in his *Origin of Species* to explain how natural selection could provide a naturalistic explanation for the development of the eye. Ward was still treating the eye in 1864 as a powerful display of divine design in nature.

Sea creatures and insects provided numerous illustrations of design. Some female popularizers invited their readers to comb the seashore. In her *British Sea-Weeds*, Gatty insisted that studying the seashore was just like "other investigations of the wonderful works of God."[190] Roberts's *Popular History of the Mollusca* (1851, Reeve and Benham) revolved around the important role of the lowly mollusk in the economy of nature. They "are as exquisitely

---

184. Loudon 1841, 30.
185. [Pratt] n.d. [*Wild*], 8.
186. Loudon 1842, 7; Roberts 1850, 257.
187. Loudon 1850, 330.
188. Pratt 1850, 340.
189. M. Ward 1864, 82.
190. Gatty 1872, viii.

contrived, and as carefully wrought for the place and station which they are designed to fill, as the higher orders of creation."[191] The theological themes in the study of insects also did not escape the notice of female popularizers, especially Ward, Wright, and the Kirbys. According to Ward the marks of design were so obvious that "no Entomologist can be an Atheist: he is brought nearer, if possible, than any other student of nature to the forming and guiding Hand."[192] Wright argued that due to the "usual wise arrangements of our kind God" insects were, on the whole, a boon rather than a misery. As "the mighty scavengers of the land" they kept the air fresh and the ground clean. They also supplied us with honey, wax, and silk. An examination of insect bodies, especially their wings and eyes, revealed how wondrously the creator had made them. Even the hooks on the end of the hair of a mite were "a great wonder." She concluded the *Observing Eye* with a rhapsody on how insects were "wonderful little beings," whose perfect "light airy bodies" bore the "stamp of wisdom and power."[193] With two books devoted to the subject, the Kirbys were no less enthused about the wonders of insects. The patterns on caterpillar eggs, they point out, are surprisingly beautiful. "We cannot tell why these minute atoms should be finished with such care and beauty," they declare. "We can only admire them as the work of an Almighty hand" (Fig. 3.12).[194] Like Wright, the Kirbys discussed the design of insect bodies. The manner in which the rings are connected in the head, thorax, and abdomen allows the insect ease in its movements. "In every particular," the Kirbys declare, "we trace a design and a skill far beyond our limited capacity."[195]

Even creatures so small that they could not be seen without a microscope could testify to the design in nature. To the Kirbys, the microscopic world of the sea was "a world full of wonders, and where we find traces, on every hand, of the goodness and the skill of our Creator."[196] Pratt turned the microscope on the coralline-like Griffithsia, and found that its blackish purple tufts displayed "strings of small pear-like substances, most beautifully and symmetrically disposed, each marked with a white cross, surrounded by a rich red colour." She concluded, "all nature has its hidden wonders"that

191. Roberts 1851, 2; see also Kirby and Kirby 1871, 91, 290; Pratt 1850, 92, 238, 296, 319.

192. Ward and Mahon n.d., 42.

193. [Wright] n.d., 94–97, 123, 130.

194. Kirby and Kirby [1861], 11.

195. Kirby and Kirby [1874], 11, 13–14, 16, 49.

196. Kirby and Kirby 1871, 46.

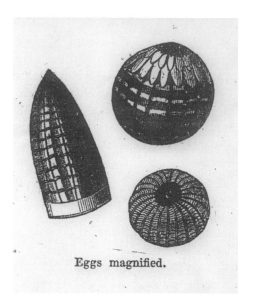

Eggs magnified.

FIGURE 3.12 The intricate patterns on the eggs of caterpillars as illustrations of divine design. (Mary and Elizabeth Kirby, *Caterpillars, Butterflies, and Moths* [London: Jarrold and Sons, 1861], 12.)

"bid us search more deeply into the works of God."[197] Ward devoted an entire book to the microscope. Using this instrument brought the observer "face to face with the minuter parts of God's creation." Ward likened this experience to "visiting a rich, but hitherto undiscovered region" that gave the observer "a new sense of the unfailing power and infinite wisdom of the Great Creator." Inviting the reader to follow her on a journey through the wonders of the microscope, they are amazed by insect wings, scales of insects and fish, hairs and feathers, vegetables, animalcules, and blood. "Is it not truly said," she declared, "that a close scrutiny of God's works conveys with it an awful, overpowering sense of a presence and power more than human? and evinced no less in the smallest than in the greatest of His works."[198]

THE FACADE OF DEFERENCE

Many of the female popularizers of the third quarter of the nineteenth century derived their authority to speak about nature by presenting themselves

197. Pratt 1850, 153.   198. M. Ward 1864, ix, 46.

as members of a long line of women who, in their role as religious and moral teachers, wrote about scientific issues since the eighteenth century. But by treating science within a larger religious framework they were in opposition to the secular agenda of scientific naturalists. Like their clerical counterparts, they represented a challenge to the new locus of authority being constructed by Huxley, Tyndall, and their allies. For Victorian audiences, their work therefore raised the question, who speaks for nature? Female popularizers claimed, as did male practitioners, to speak on behalf of a mute nature, to act as interpreters of the larger meaning of scientific theories.[199] Barbara Gates has compared the attempt to speak in nature's name to a ventriloquist's act. Those who wrote about science created a "dummy" nature through which they could pass on their own pronouncements. Yet essentialist definitions of women as closer to nature made it difficult for female popularizers to be effective ventriloquists. They could not sustain the illusion as well as male practitioners. As a result, women avoided controversial issues in order to maintain their authority.[200] Not only did women avoid public disputes with male practitioners, they also referred positively to their works. Female popularizers erected a facade of deference toward male practitioners as it became increasingly difficult for them, in an era of when practitioners were touting professionalization, to withstand challenges to their authority.

Deference took the form of positive references to the writings of male practitioners, just as it did in the works of male clerical popularizers. As we have seen, women sometimes positioned their works as intermediary preparations for more sophisticated books. Another way to link their work to eminent males was in the form of the dedication. Ward's *Telescope Teachings* was dedicated to the Earl of Rosse, and it informed the reader that her cousin had commended her book as being "good, so far as its limited powers extended."[201] Others pointed out that the basic organization of their book was taken from a male practitioner. Lankester, for example, declared in her *British Ferns*, that she was using William Hooker's book on British ferns as her model, while Loudon noted in the introduction to her *British Wild Flowers* that her systematic arrangement of plants followed Lindley's.[202]

Women writers often mentioned the names of the authors of the chief sources consulted in the opening pages of their books. In the preface to her *Globe Prepared for Man*, Wright acknowledged Lyell, Gideon Mantell, William Buckland, Hugh Miller, and Henry Thomas De La Beche, among

---

199. Lightman 1997d, 207.
200. Gates 1998, 11–12, 36.

201. M. Ward 1859, "To the Earl of Rosse."
202. Lankester 1903, 7; Loudon [1846], 2.

others. Some women also peppered their books with references to specific facts gleaned from the works of male practitioners. In her *Telescope Teaching* Ward declared that she had drawn on books by John Pringle Nichol, Humboldt, John Herschel, Robert Grant, and George Airy. Twining confirmed that she had received assistance from volumes by Lindley and John Forbes Royle while working on the *Illustrations of the Natural Orders of Plants*.[203] Paying deference to male practitioners was a common practice.

Some women deferred to male practitioners while participating in the longstanding tradition of finding moral and religious meanings in nature, seemingly with no concern about Huxley and his allies. But others aired their resistance to the agenda of scientific naturalism in public. Muted signs of disagreement could be glimpsed behind the facade of deference especially when it came to such issues as the status of evolutionary theory, the gender politics of science, and the role of those who were not members of the elite in the production of knowledge. Like Morris, Gatty was a die-hard opponent of evolutionary theory. She often questioned the authority of arrogant male doubters in her parables by elevating the perspectives of animals, plants, and children.[204] A low church Anglican, she told her publisher Bell that she refused to buy a copy of the liberal Anglican manifesto, *Essays and Reviews* (1860), because it was an "unclean thing."[205] In private correspondence she raged against Darwin. To her publisher Bell, she asked for news of the *Origin of Species,* remarking, "what a madness has seized on the Naturalist world!" She marveled that "14/ is no bar to a book selling" that was in opposition to the Bible. On March 19, 1860, she again wrote to Bell, prophesizing that the *Origin* would be "*found out* by somebody, and exposed, as a great man's *blunder*. With Dr. Hooker, Sir Charles Lyell, Professor Huxley & etc. against one, this is much to say—but I *cannot* take any other view."[206] While Kingsley attempted to attach moral and theological meanings to the workings of the evolutionary process in order to accommodate Darwinian theory within Christianity, Gatty would have none of it.

Gatty's outrage over Darwin found its way into her correspondence with William Harvey.[207] She was appalled by the religious and moral implications

203. Twining 1868, 2: 165.

204. Cosslett 2003.

205. Gatty to Bell, April 23, 1861, University of Reading, Library Archives, The George Bell Uncatalogued Series.

206. Gatty to Bell, n.d. and March 19, 1860, University of Reading, Library Archives, The George Bell Uncatalogued Series.

207. Le-May Sheffield 2001, 57–59.

of evolutionary theory but she also told Harvey that the scientific evidence was weak and sarcastically questioned Darwin's mental abilities. Referring to Darwin's depiction of how new species developed over time in the *Origin*, she asked Harvey, "How can that diagram of dotted lines prove any one single thing. Surely there is a soft place somewhere in the learned man's head."[208] Gatty's resistance to evolution put pressure on her friendship with Harvey and, at times, led her to drop the facade of deference she had so carefully built in her correspondence with him. When Harvey refused to take Darwin on publicly, and worse still from her point of view, seemed to side with him on some issues, Gatty expressed her annoyance. "Over Darwin now I can hardly bear the sight of your hand or write," she wrote on August 18, 1860, "so annoyed and discomposed and grieved was I at your last letter." She expressed her hope that someone may yet arise to tackle the Darwin affair "from higher ground," attempting to egg Harvey on.[209] She became alarmed when Harvey showed any sign of weakening in his opposition to Darwin, and in the same month invited him to Ecclesfield for a long parley, declaring "Dr. Joseph Hooker shall *not* convert you."[210] But Gatty stepped over the line at least once. In 1863 Harvey sent her a copy of Darwin's *On the Various Contrivances by Which British and Foreign Orchids are Fertilised by Insects* (1862) as a gift. Gatty refused to read it at first, and criticized Darwin for lacking philosophical insight. Apparently Harvey replied to her criticisms of Darwin and chastised her. Gatty's response was to pull back deferentially. "You have come down heavy enough for once," she wrote, "but I brought it on myself so there's an end."[211]

Since Harvey declined to take on Darwin, Gatty decided to air her opposition publicly, although she recognized that her impact on the debate would be minimal. She wrote to a friend on March 13, 1862, that Darwin might have some "curious muddle-headed confusion of brain which enables him to believe in a great Creator and a great creative 'Natural Selection' (capitals of course) at the same time." She then referred her friend to her story "Inferior Animals" (1861), where "I have combated the Darwin presumption as far as I could in a small way without entering lists which I was not qualified for."[212] In "Inferior Animals" Gatty tells a story about a gathering of rooks that are

208. Gatty to Harvey, March 12, 1860, Sheffield Archives, HAS 48/69.

209. Gatty to Harvey, August 18, 1860, Sheffield Archives, HAS 48/88.

210. Gatty to Harvey, [August 1860], Sheffield Archives., HAS 48/93.

211. Gatty to Harvey, December 29, 1862 and January 11, 1863, HAS 48/340 and HAS 48/347, Sheffield Archives.

212. Gatty to Mrs. Carter, March 13, 1862, Sheffield Archives, HAS 58/18.

discussing the origins of humanity. One rook declares that humans are not superior; rather, they are "neither more nor less than a degenerated brother of our own race." Satirizing the inferential logic of evolutionary argument, the rook argues that if on this proposition you can account for things otherwise unaccountable, then you must accept it. The story ends with the narrator realizing that this was all a dream caused by reading an unnamed book lying on their desk. This leads the narrator to remark on the "first temptation," the sin of desiring to "be as God in knowledge."[213] Although Gatty's "Inferior Animals" is the only story to critique Darwin directly, the entire *Parables from Nature* was created to resist the growing materialism in science. By writing fictional short stories, she managed to find a way to participate in the debate over the validity of scientific naturalism without calling attention to herself or her aims.[214]

For some women like Gatty, resistance to scientific naturalism was based primarily on religious grounds. For others it may have had more to do with the attitude of scientific naturalists toward women. Female popularizers were not, with few exceptions, vocal supporters of the women's movement. Some of them were active as popularizers before women's rights became a prominent issue on the political scene in the late 1860s. Lankester and Becker were the exceptions. Lankester was one of the signatories to the 1866 suffrage petition, and she wrote periodical articles on such issues as women's health and employment.[215] Becker was one of the leaders of the suffragette movement, and she worked with such Victorian feminists as Emily Davies, Elizabeth Wolstenholme, and Josephine Butler.[216] But the rest were, to all appearances, fairly conservative in their views on women's role in society. According to Maxwell, Gatty, for example, did not believe it was proper for women to speak in public and to dare to instruct men on the issues of the day.[217] In 1861, while on a trip to Ireland, Gatty ran across Frances Power Cobbe and the Social Science Congress. She recognized that "there *are* grievances and *abuses* in workhouses and elsewhere, and agreed with Cobbe that they should be "ventilated," but not by "*the ladies.*" "Perhaps," she wrote to Harvey, "I am too old to take up anything so new as the ladies

213. Gatty 1954, 203, 212; Rauch 1997, 144–45.

214. Rauch 1997, 140–41.

215. Shteir 2004a, 1182.

216. Bernstein 2004, 166.

217. Maxwell 1949, 138–39.

teaching the men what to do."[218] However, although female popularizers presented their scientific activities as an extension of the "womanly" duties sanctioned by their society, some nevertheless led lives that contradicted the conventions of their gender.[219]

Although some female popularizers may have questioned the gender status quo, even the most radical among them did not challenge eminent scientific naturalists directly on this issue. Becker's *Botany for Novices* does not deal explicitly with women's issues. However, she adopted the style of the standard, impersonal male botany book in order to reject the notion that female scientific education should take place in a "textual separate sphere," as in the gender-specific botany books using the familiar format.[220] After Becker's *Botany for Novices* appeared in 1864, she wrote a series of periodical articles where she began to address the relationship between the "woman question" and science more directly. Her overall strategy as a first-wave feminist was to undermine essentialism and the notion of separate spheres for women that had been based upon it. She believed that this would give women access to science and other traditional male preserves, like the political world. In "Female Suffrage" (1867), she argued that "sentimental," rather than "scientific grounds," provided the rationale for asserting, "women have nothing to do with politics."[221] In "Is There Any Specific Distinction between Male and Female Intellect?" (1868), Becker systematically undermined the notion that "because women are weaker in body than men they must be weaker in mind." Those who maintained intellectual difference also believed that the intellects of men and women "should be cultivated in a different fashion and directed in different ways, and that there is a 'sphere' or 'province'

218. Gatty to Harvey, August 29, 1861, Sheffield Archives, HAS 48/170.

219. Le-May Sheffield 2001, 217. In the case of Gatty, Sheffield has shown that the female characters in her *Parables* do not conform to the essentialist conception espoused by Darwin and other scientific naturalists of women as defined by their emotions. Gatty presents them as having deeper insight into nature through their use of both thought and feeling (Le-May Sheffield 2001, 56). Rauch reads Loudon's futuristic *The Mummy* as making a powerful statement about the "possible efficacy of women as leaders." Not only does a matriarchal figure bring order to England, the women now wear pants (Loudon 1994, xxiv). If a rejection of the traditional public dress code is to be considered a sign of advanced views on the woman issue, then Gatty's instructions to female readers in her *British Sea-Weeds* on clothing is positively radical. Here she recommends that when undertaking "shore-hunting" women must "lay aside for a time all thought of conventional appearances" (Gatty 1872, viii).

220. Shteir 1996, 228–29.

221. Becker 1867, 316.

assigned to each." Becker therefore perceived the entire gendered system of separate spheres to be based upon the issue of intellectual difference. She put forward three propositions. First, there was no distinction between the intellects of men and women that corresponded to, or was dependent on, the organization of their bodies. Here Becker turned her knowledge of science to account. Males were not universally superior in strength throughout the natural world. Second, current intellectual distinctions between men and women were a result of education and other social circumstances, not some inherent natural difference. Third, despite the impact of external circumstances leading to intellectual difference between men and women, these "do not differ more among persons of opposite sexes than they do among persons of the same."[222]

Whereas Becker's 1868 article drew on her scientific knowledge to undermine essentialism, her next article, "On the Study of Science By Women" (1869), discussed the consequences of using essentialist reasoning to enforce women's "exclusion from the pale of scientific society." Without the intellectual activity provided by scientific pursuits, women fall prey "to morbid religious excitement" or sink into "a weary kind of resigned apathy." But, Becker affirmed, women had little access to scientific institutions and societies. She argued that if educational institutions were reformed in order to allow the same opportunities for training as men, the result would be a larger cadre of researchers and an explosive impetus to scientific progress. "The rate of advancement will be far more than doubled," she claimed, "because the untrained and stationary half of mankind necessarily acts as a drag on the other."[223] Becker's papers anticipated and rejected both Darwin's position on the intellectual inferiority of women in his *Descent of Man* (1871) and Huxley's exclusion of women from the Anthropological Society in the same year.[224] Yet she never spoke out publicly against them.

Perhaps it was Becker's previous relationship with Darwin that led her to hold her tongue and pen in the 1870s. In May 1863 she wrote to Darwin for the first time, sending him a peculiar form of *Lychnis dioica*, later determined to be diseased. This led to a series of letters during the same year on parasites and botanical hermaphrodites. Subsequently, Becker wrote a report on this topic and read it at the British Association meeting in Exeter in 1869, arguing that her observations on how a parasitic fungus had caused flowers to assume a bisexual form could be seen as a confirmation of Darwin's

222. Becker 1868, 483-85, 487, 490.   224. Richards 1983, 57-111; 1989, 253-84.
223. Becker 1869, 388, 391, 404.

theory of pangenesis.[225] In 1864 Becker sent Darwin a copy of her *Botany for Novices,* hoping that Darwin, as one who had "attained the greatest eminence in the pursuit of science," might "feel pleasure in the thought that others, however far removed from them, should be led to share in the same degree, the happiness which the study of nature is capable of affording."[226]

Two years later she was writing Darwin again, uncertain if he remembered her and thanking him for his previous kindness. Now she asked him if he would lend a "helping hand" to the newly formed Manchester Ladies' Literary Society by sending a paper that could be read to the first meeting. Darwin complied with a paper on climbing plants, and Becker's letter, dated February 6, 1867, contained thanks for contributing to the success of the fledgling society.[227] Over the years Becker began to feel indebted to Darwin. She wrote again on January 13, 1869, praising his *Variation of Animals and Plants under Domestication* (1868) and reported her observations on parrots and cockatoos. Later that same year she sent him an abstract of her BAAS paper on parasitic fungus and bisexuality and asked him for advice on which journal to send it to for publication.[228] Becker may therefore have refrained from opposing Darwin's views on the intellectual abilities of women in the *Descent* out of respect for their friendship. In 1877 she was still corresponding with Darwin, writing to him about botanical issues on Manchester National Society for Women's Suffrage letterhead.[229] Although Becker never confronted Darwin, Huxley, or any of the other scientific naturalists in the public domain, she nevertheless used what she had learned from their work and that of other male practitioners in her fight for the vote. In the course of studying science by sitting at Darwin's feet she had developed accuracy and attention to detail that prepared her for assuming a leadership role in the women's movement.[230]

In addition to opposing the secular and sexist views of scientific naturalists, albeit in subtle ways, some female popularizers resisted their claims to a monopoly on the production of knowledge. They could be enthusiastic

---

225. Burkhardt et al. 1999, 424–25, 435, 457, 527–28, 571, 578, 884; Shteir 1996, 230.

226. Becker to Darwin, March 30, 1864, University of Cambridge, MS.DAR.160.

227. Becker to Darwin, December 22, 1866; December 28, [1866]; February 6, 1867, University of Cambridge, MS.DAR.160.

228. Becker to Darwin, January 13, 1869; October 14, 1869; December 29, 1869, University of Cambridge, MS.DAR.160.

229. Becker to Darwin, January 16, 1877, University of Cambridge, MS.DAR.160.

230. Holmes 1912/13, 7.

about the potential of their readers to make important scientific discoveries. Zornlin told her readers that if a student observed carefully they "may even lend his aid to the furtherance of this science. Instances are not wanting, both in our own day and in former times, of individuals from every class of society raising themselves to eminence by the pursuit of science; and the path is open to all." Everyone had his or her part to play, according to Zornlin. Country dwellers could take note of the minerals and plants in their own neighborhood or record observations of the habits of insects, thereby contributing "his quota to the general mass of knowledge." The city dweller could observe cloud formations and temperature.[231]

Becker was no less optimistic about her reader's chances of making new scientific discoveries. "Every accurate observer who records a new fact," she declared, "every patient observer who records a new fact, every patient watcher who traces the working of an obscure law of nature, contributes a brick to the great pyramid."[232] Becker called on Darwin and Newton to help her make the point that anyone could make an important scientific discovery. Quoting Darwin on how some of his most interesting observations on climbing plants occurred when he was confined to a sick room, Becker told her readers that "such an example should encourage others to do likewise." Since many aspects of the most common plants and animals were still imperfectly understood, "any woman who might select one of these creatures, and begin a series of patient observations on its habits, manner of feeding, of taking care of its young" might produce "something of real, if not of great, scientific value."[233] In her presidential address to the initial meeting of the Manchester Ladies' Literary Society on January 30, 1867, Becker argued that nobody should be deterred from "making or reporting original observations, by a feeling that they are trifling or unimportant." She reminded her readers that the most apparently insignificant observations "have led to results which have turned the whole current of scientific thought. What would be a more trifling circumstance than the fall of an apple from a tree?" Yet, in this instance Newton "saw the law of gravitation," while in the connection between flies and flowers Darwin perceived "some of the most important facts which support the theory he has promulgated respecting the modification of specific forms in animated beings." Although she acknowledged, "we are not Darwins nor Newtons, and cannot expect to make surprising discoveries," Becker nevertheless preserved a role for

231. Zornlin 1855, x–xi.
232. [Becker] 1864, 42.

233. Becker 1869, 389.

her readers in the scientific community.[234] In opposition to the hierarchical vision of science championed by would-be professionalizers, Zornlin and Becker offered an egalitarian conception.

RESISTING SCIENTIFIC NATURALISM

Women and Anglican clergymen who wished to be popularizers in the second half of the nineteenth century had much in common. Their traditional authority to speak out on these matters was being questioned by male practitioners, in particular the scientific naturalists. A key part of Huxley and Tyndall's strategy to win acceptance of evolutionary theory was to deny that members of the Anglican clergy had the expertise to determine scientific truth. The authority of women to write on scientific topics was threatened by their exclusion from scientific societies and by the claims by some male practitioners that by their very nature they were intellectually inferior. Both groups wrestled with the problems raised by evolutionary theory and the hierarchical vision of would-be professionalizers of science. Both aimed to address a rapidly developing mass reading audience. They made science more accessible by avoiding jargon and complex classification systems, and conveyed scientific information in an attractive literary style. They both emphasized the moral and religious lessons to be learned through an understanding of nature, and they stressed the sense of wonder to be experienced when faced with its beauty. Together, women and Anglican clergymen who wrote about science formed a formidable group whose common agenda could frustrate the goals of scientific naturalists. Since female popularizers came out of a distinct tradition of scientific writing, and since they were confronted with a different set of obstacles than those facing Anglican clergymen, I have dealt with them separately in this chapter. But as we move into the latter half of the nineteenth century, they can be discussed in a more integrated fashion.

This is not to suggest that the Anglican clergymen and female popularizers formed a cohesive group like the scientific naturalists. Huxley and his allies were centered in London and had numerous opportunities to undertake coordinated action. The monthly meetings of the X Club, to which Huxley, Tyndall, Hirst, Frankland, Spencer, and others belonged, provided an ongoing forum for strategizing about their goals. Women and Anglican

234. As quoted in Blackburn 1971, 36–37.

clergymen had nothing remotely similar. But they knew one another's work and occasionally referred approvingly to it. Gatty, for example, admired Wood's work. On July 10, 1866, she wrote to Bell that "Wood's 'Anecdotes' and Illust. Nat. His. . . . by Routledge are the best *modern* things I have seen."[235] However, women and the Anglican clergymen could be critical of one another's work. Gatty disliked the work of Kingsley. She told Bell not to refer her to *Glaucus* as an example of a good natural history book. "It is a *pretty popular* book—*worthless for use*—utter trumpery," she insisted. "Kingsley is not scientific tho' a clever man can put on anything and dress it nicely up."[236] Kingsley reviewed Gatty's *British Sea-Weeds* favorably in the *Reader* when it first appeared. "This is a beautiful book," he announced, "and, if it be only as good as it is pretty, it must needs be a valuable one." But he declined to evaluate its "scientific merit," saying that "would involve long and laborious collations with other books—a task necessary and suitable only for a professedly scientific writer, in a professedly scientific review."[237] Gatty was cautious about Kingsley's praise. His reluctance to address the scientific value of her book led her to exclaim, "I *smell a rat* that Glaucus is not an algologist." Gatty suspected that Kingsley knew little about natural history in general. Gatty was also hostile toward Kingsley because of his receptiveness to Darwin and his liberal Christianity. She pointed out to Harvey that in *Water Babies* Mother Carey, "who would seem to represent Nature (or "Natural Selection") in one place—and the Irishwoman who more resembles Conscience—and the two Fairies Doasyouwouldbedoneby and Bedonebyasyoushouldo—all seem to be one and the same at the end. Is this *Truth* or Gospel Charity (Love)—or Nature or God's law or what?" Gatty objected to the way that Kingsley conflated natural selection and the divine nature in which she found her religious and moral lessons. But she thought, "separately the little allegories are exquisite."[238] There could be serious points of disagreement between women writers and the Anglican parsons, but there were also disputes between members of the X Club, such as Spencer and Huxley's famous falling out in the late 1880s.

235. Gatty to Bell, July 10, 1866, University of Reading, Library Archives, The George Bell Uncatalogued Series.

236. Gatty to Bell, n.d., University of Reading, Library Archives, The George Bell Uncatalogued Series.

237. [Kingsley] 1863, 162.

238. Gatty to Harvey, August 20, 1863 and August 31, 1863, Sheffield Archives, HAS 48/388 and HAS 48/392.

The convergence of the heirs to the maternal tradition and the theological tradition championed by Anglican parsons may have contributed to the perception in the closing decades of the century that mass culture had become "feminized" and the cause of the decline of civilization. During this period, when socialism and the women's movement challenged traditional male dominated culture, mass culture became associated with women, while genuine culture was considered to remain the prerogative of men. By the end of the century, political, psychological, and aesthetic discourse consistently gendered mass culture and the masses as feminine, while high culture was still seen as the privileged realm of male activities. This was a European-wide phenomenon, as defenders of a masculine high culture included the Goncourt brothers, Nietzsche, and Gustave Le Bon.[239] The participation of Anglican clergymen in the mass market of science therefore could affect the perception of their authority.[240] The very notion of "popular science" acquired some of its negative connotations as a result of being a part of mass culture.

239. Huyssen 1986, 191–96.

240. I am indebted to Jennifer Tucker for this point.

✳

CHAPTER FOUR

The Showmen of Science

Wood, Pepper, and Visual Spectacle

JUST BEFORE Christmas Day, 1862, John Henry Pepper invited a small group of literary and scientific friends, and members of the press, to his Royal Polytechnic Institution to see a performance of Edward Bulwer-Lytton's "A Strange Story." His plan to surprise his visitors with a preview of a new optical illusion worked better than he could possibly have imagined. The audience was so startled by the ghost illusion that Pepper decided not to explain how it worked. The following day he hurriedly took out a provisional patent, sensing its almost unlimited potential.[1] Pepper then prepared a companion lecture for the play, "A Strange Lecture," where he explained the wonders produced by the "Photodrome," an optical apparatus that caused phantoms to appear at will.[2] One periodical recommended that "everybody should go and hear this lecture" and praised the Photodrome for its ability to create "the most beautiful effects we have ever witnessed." But the crowning triumph of the lecture was Pepper's production of a "real veritable spectre, so real that the spectator hardly believes the Professor when he states that it is a mere illusion, a fact, however, which he establishes by walking clean through it."[3] Pepper's ghost caused a sensation and drew thousands of

---

1. Pepper 1890, 3.

2. "Polytechnic Institution" 1863, 19.

3. University of Westminster, Archives, Press Cuttings, Book of Press Cuttings Relating to RPI, 1863 [January 4], "Polytechnic Institution."

visitors to the Royal Polytechnic Institution, including Prince Albert and other members of the royal family.[4]

Nearly twenty-one years later, John George Wood, another showman of science, also startled his audiences with a carefully managed spectacle. To illustrate key points in his Lowell Lectures in Boston, he drew "rapid impromptu sketches" of creatures that gradually took shape before the eyes of those attending. Audiences were particularly impressed by their magnitude. The sheet of black canvas that he used to draw on was stretched on a wooden frame that gave him a surface of eleven feet by five feet six inches. Close up, the drawings appeared coarse and clumsy, but when viewed from thirty or forty feet away they were elegant pictures that were clearly visible in every part of the largest hall. The impact on the audience was electrifying. In one of his Lowell Lectures Wood spoke on the whale to a packed room. "When I opened the lecture," he reported to his family, "by drawing the whale, eleven feet long, in two strokes, there was first dead silence, and then such a thunder of applause that I had to wait." Then Wood drew a little sailor on the whale's back to illustrate its gigantic size, and the crowd "laughed and cheered in the heartiest manner."[5] Wood's larger-than-life sketches catered to the popular audience's taste for spectacle. He realized that if science lectures were to become a popular form of entertainment, and if he were to succeed as a public lecturer, he had to satisfy the craving for visual images that was the hallmark of mass culture in this period.

Pepper and Wood were among the most well known popularizers of science in the second half of the century. Both lectured and published extensively. Both turned their lectures into spectacles and incorporated a multitude of visual images in their books. In this chapter I will examine Wood and Pepper as influential showmen of science. Many popularizers of science capitalized on the pictorial turn in British culture. A significant number of female popularizers, as we have seen, offered their readers sumptuously illustrated books, some in glorious color, but Wood and Pepper took the exploitation of their audiences' hunger for spectacle and vivid visual images to another level. Like Brewer, Wood was among the first of a new breed of science writer who established careers for themselves, taking advantage of the new market conditions. For Wood, making a living as a popularizer of science was a struggle. Appealing to the eye of his audience gave him a competitive edge. It also provided him with a powerful vehicle for presenting his theology of nature. Pepper lectured at, and later managed, the Royal

4. "Polytechnic" 1863b, 9.   5. T. Wood [1890], 203.

Polytechnic Institution, one of the new institutions of science established in the first half of the nineteenth century to reach out to a broad audience. His optical illusions helped the Polytechnic to survive the impact of the Great Exhibition of 1851, which had drastically reduced attendance during the early 1850s, and to compete with the other London entertainment attractions. Although Pepper deferred to the authority of evolutionary naturalists, he did not hesitate to outline the larger religious themes informing science. Together, Wood and Pepper demonstrated the potential of science to attract vast, new audiences by incorporating visual spectacle.

JOHN GEORGE WOOD AND HIS RISKY VENTURE

Shortly after John George Wood's death in 1889, a debate began concerning his proper place in the history of science for the public. To some, Wood was among the preeminent scientific writers of his day. The *Times* obituary credited Wood with doing "more to popularize the study of natural history than any writer of the present age."[6] In his hagiographic biography published a year later, Theodore Wood claimed that his father was "the first to popularize natural history, and to render it interesting, and even intelligible, to non-scientific minds." As a "pioneer in the work of popularizing natural history," Wood "had many subsequent imitators, but he himself imitated no one."[7] Twenty years later, John Upton again thrust greatness upon Wood. In Upton's *Three Great Naturalists*, Wood was included alongside Darwin as one of the three most important naturalists of his age. "Though not perhaps to be ranked with the most famous naturalists," Upton declared, "[Wood] did more than any other man to popularize the science of natural history."[8]

However, praise for Wood's accomplishments was not universal. In the *Saturday Review* an anonymous reviewer of Theodore Wood's biography mercilessly attacked J. G. Wood's scientific credentials and denied him a special role in the history of science. Not only was Wood's "scientific equipment sadly deficient," the reviewer declared, "even as a popularizer, he was neither the first nor in the first rank." According to the reviewer, the earliest writer on natural history for a popular audience was Gilbert White, followed by Charles Waterton. The "first systematic popularizer of general natural history was P. H. Gosse, whose *Canadian Naturalist* was published when J. G. Wood was a child," the reviewer asserted. It was therefore "a grave

6. "Obituary" [Wood] 1889b, 9.   8. Upton [1910], 190.
7. T. Wood [1890], vii, 125-26.

historical mistake" to claim for Wood "any species of initiative." Theodore Wood, the reviewer declared, "has done his father no good turn by this dangerous excess of laudation." Misplaced filial loyalty had led Wood's son to present a false picture of his father's importance as a popularizer of science.[9]

A more nuanced assessment of Wood's significance appeared in F. A. Mumby's *The House of Routledge*, twenty-four years after Upton's study. Mumby recognized what previous commentators had missed—that Wood's importance as popularizer was inextricably connected to the market conditions of the midcentury period. Wood could not "popularize" science on his own. He needed to ally himself with a publisher who also perceived that there was a growing reading audience for science. Mumby stated that "encouraged by the success of one or two early books by the Rev. J. G. Wood, George Routledge now set about popularizing natural history" by planning a shilling series of handbooks and asking him to write several of the volumes. According to Mumby, "these books gave the first impulse in the fifties and sixties of the nineteenth century to many young readers who afterwards became distinguished naturalists, and marked the beginning of that intelligent interest in the living universe and outdoor life which has grown steadily and surely with each succeeding generation." One of Wood's contributions was the stunningly successful *Common Objects of the Country* (1858). Mumby reported that the book "carried the general public by storm" and that the first edition of a hundred thousand sold completely within a week. Although the sales figures for the first edition are almost certainly an exaggeration, Mumby astutely pointed out that Wood was able to reach a large group of readers thanks to Routledge's initiative in popularizing his works.[10] This was an audience previously tapped only by Robert Chambers. Unlike Chambers, Wood published under his own name, avoided contentious topics, and wrote well over thirty natural history books. Wood was a "Victorian sensation," not on the basis of one or two controversial texts, but due to his sustained efforts over the course of a long career as scientific author and lecturer that began in the 1850s.[11] Moreover, and here Mumby was not as astute, Wood's popularity can be attributed, at least in part, to his lavish use of visual images in his books and lectures.

9. "The Rev. J. G. Wood" 1890, 479.

10. Mumby 1934, 75–78.

11. Although Chambers also had a long career as writer, he did not devote himself exclusively to writing scientific works, though he published several books on topics relating to science and wrote many of the articles on science in *Chambers's Journal*.

Wood could easily be placed alongside the other Anglican clergymen who popularized science. He shared many aspects of their agenda, especially their articulation of a theology of nature. But he is distinct from that group in two important ways. First, he gave up a full time clerical post to pursue a career as a popularizer of science. In this respect, he resembled Brewer, though the latter never attempted to undertake ambitious lecture tours. Second, he was a showman of science who was attuned to the visual culture of the period. John George Wood (1827–1889) was the son of John Freeman Wood, a surgeon and chemical lecturer at Middlesex Hospital, and his German wife Juliana Lisetta Arntz (Fig. 4.1). Wood was the eldest of a large family of fourteen. He was a voracious reader with a powerful memory. His interest in natural history was encouraged by his father, who spent Sunday afternoons with his children peering through a microscope to examine objects found in their garden. The family also kept a wide variety of pets. In 1838 Wood was sent to Ashbourne Grammar School in Derbyshire, where his uncle Reverend George Jepp was headmaster. He returned to Oxford in 1844, matriculated at Merton College, and graduated in 1847 at the age of twenty with a BA. At Oxford he became enamored with classical learning and continued his natural history studies. He kept a variety of pets in his rooms, including caterpillars, bats, insects, and snakes. For several years Wood tutored at a school in Wiltshire as he had matriculated so early he had to wait before he could apply for ordination. He returned to Oxford in 1850 to prepare for ordination, but spent much of his time at the Anatomical Museum at Christ Church working with Henry Acland on comparative anatomy, specializing in insects. Bishop Samuel Wilberforce ordained him as deacon in 1852. He was ordained as priest in 1854, but already his interest in natural history was leading him away from a clerical career.[12]

In 1851 Routledge commissioned him to write a popular survey of natural history, published in 1853 as *The Illustrated Natural History*. This was the first in a series of successful books for the reading public undertaken for Routledge during the 1850s. For ten years Wood juggled clerical duties and his writing. He resigned his curacy in Oxford in 1854 in order to devote himself to his writing, but in 1856 he accepted the position of chaplain to St. Bartholomew's Hospital in London. The light duties left him time to write. In 1859 he married Jane Eleanor Ellis, with whom he had three children. By 1862, having established himself as one of the most widely read English authors on natural history, Wood left St. Bartholomew's and moved

12. T. Wood [1890], 1, 2, 7, 8, 13, 21; Gilbert 2004b, 2193–96.

FIGURE 4.1 The Reverend J. G. Wood, a new breed of popularizer of science.
(Theodore Wood, *The Rev. J. G. Wood: His Life and Work* [New York: Cassell Publishing
Company, 1890], frontispiece.)

to Belvedere, near Woolwich. Although he acted as an unpaid curate in the
parish of Erith, Kent, until 1873,[13] and assumed the position of precentor

13. Twice in 1869 the Anglican authorities questioned Wood's voluntary work for the Church.
When he filled in for an ailing vicar, the Archbishop of Canterbury, A. C. Tait, pointed out that he
was officiating as curate without any authority. Wood explained that he had simply carried on the
services during the vicar's absence and that he had had no way of knowing that the illness would
last so long. Later that year, members of Wood's flock accused him of "unchristian and unfeeling"
behavior when he refused to officiate at a funeral. On November 11, 1869, he wrote to Tait, saying
that the funeral party had arrived long after the appointed time and that Wood had to leave to
perform the most important duty of all, the baptism of a dying child. Wood was keenly hurt by
the attacks on him and claimed, "while acting in lieu of the Vicar, I have never spared myself, and

of the Canterbury Diocesan Choral Union, organizing the annual festivals from 1869 to 1875, Wood now began to build a career as a scientific writer.[14] Leaving the security of a clerical position for writing was a risky move, particularly in the early 1860s. It was difficult, if not impossible, to earn enough money to make a living as a science writer for the public unless a lucrative journal editorship was available. Nigel Cross asserts that during this period journalists in general lived in the shadow of poverty.[15]

Wood's long relationship with Routledge, which lasted over thirty-five years, was one key to his success. Just as Wood represented a new breed of popularizer, George Routledge represented a new generation of publishers who recognized the advantages to be gained by exploiting the developing mass market. In the 1840s he developed several strategies for reaching this market, all revolving around attracting readers with volumes at unprecedented low prices, sometimes as little as one shilling or eighteen pence. He sold remainders from other firms at cheap prices, making his money off the large volume of trade; he pirated books by American authors; and he inaugurated several library series, especially the very successful Railway Library. In the 1850s and 1860s, Routledge was among the first to exploit the illustrated publications genre. He hired the Dalziel brothers and other engravers to illustrate a series of volumes that included the works of Shakespeare and natural history books.[16]

Routledge commissioned Wood to write seven of the eleven volumes in the "Common Objects" series published between 1857 and 1875. Designed as a series of inexpensive handbooks on natural history for a popular audience, Common Objects was a huge success for Routledge. The series was set up to capitalize on the enthusiasm of leading proponents of English popular education, such as Kay-Shuttleworth, for teaching science through an examination of familiar, everyday objects. It was this same emphasis on common things that earlier had propelled Brewer's *Guide* to immense sales. Wood's contributions to Routledge's series were especially popular. His first, *Common Objects of the Sea Shore* (1857), sold extremely well (Fig. 4.2). Routledge printed 1,000 copies of the first edition in April (priced at one shilling), but by 1860 four more editions of 19,000 copies each were needed

indeed, have exceeded the duties assigned to me" (Lambeth Palace Library, Tait Papers, vol. 162, ff. 270, 297).

14. T. Wood [1890], 25-26; Gilbert 2004b, 2193-96.

15. Cross 1985, 123.

16. Barnes and Barnes 1991, 261-64.

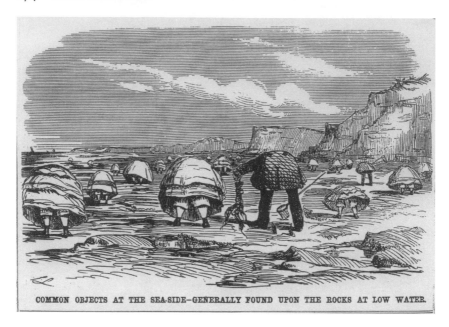

COMMON OBJECTS AT THE SEA-SIDE—GENERALLY FOUND UPON THE ROCKS AT LOW WATER.

FIGURE 4.2 It was a sign of Wood's popularity when *Punch* poked fun at his book and those who read it by depicting them as if they were the common objects of the sea when they combed the shore for marine flora and fauna. ("Common Objects at the Sea-Side—Generally Found upon the Rocks at Low Water," *Punch* 35 [August 21, 1858]: 76.)

to keep up with the demand.[17] Wood's next contribution, *Common Objects of the Country* (1858), was even more successful—it became one of the best-selling science books of the second half of the century. Wood's son reported that the first edition of 100,000 copies was completely sold out within a week, but this is inaccurate.[18] According to the Routledge Publication Books, the first edition of 6,000 copies, which appeared in February 1858, priced at one shilling, was followed in March by a different edition of 3,000 copies priced at 3 shilling, 6 pence. Routledge printed 64,000 copies of both edi-

17. *Archives of George Routledge & Co. 1853-1902*, 1973, 2: 373-74.

18. T. Wood [1890], 61. Other scholars have stated that 100,000 copies of the first edition of *Common Objects of the Country* were sold. Allen and Barber give no reference for the assertion, while Altick footnotes Mumby. But Mumby offers no reference as proof (D. Allen 1976, 139; Barber 1980, 14; Altick 1983, 389; Mumby 1934, 78.) Theodore Wood's number appears to be the source of the confusion.

tions within a decade of the book's first appearance, and 86,000 copies by 1889.[19] The sales of Wood's two books in the Common Objects series were simply phenomenal, and his success in the late 1850s working with Routledge was likely the crucial factor in his risky decision to pursue a career in popularizing science.

During his career as a popularizer, Wood published over two dozen natural history books, the majority of them with Routledge, and he became involved in other types of projects. In addition to writing for a wide variety of journals, he edited some of the natural history classics, such as Charles Waterton's *Wanderings in South America* (1879, Macmillan) and Gilbert White's *Natural History of Selborne* (1853, Routledge). His son recalled that Wood was always busy writing in his study, working twelve hours a day.[20] Although he was such a prolific writer, Wood found it difficult to survive on the income earned through his books and essays. His son acknowledged that he lacked "business qualities" and that as a result he did not demand enough from publishers for his work.[21] Wood sold the copyright of *Common Objects of the Country* to Routledge for only £40 in 1857, while his publisher earned £3,000 for the same title.[22] For authors like Wood who wanted the money right away and who had no way of knowing how successful their book would be, selling the copyright seemed like the best option.[23] Wood became

19. *Archives of George Routledge & Co. 1853–1902*, 1973, Publication Books, vol. 2, 424; vol. 3, 87, 96; vol. 5, 324.

20. T. Wood [1890], 270–72.

21. Ibid., 234.

22. *Archives of George Routledge & Co. 1853–1902*, 1973, Contracts c1850/78, reel 2, vol. 3 R–Z, 316; E. Wells 1990, 60. According to other sources, the amount of the copyright was slightly lower or higher. Crosland recalled that Wood received £30 for the copyright while Wells claims that he was paid £60 (N. Crosland 1898, 118; E. Wells 1990, 60).

23. There were basically two options for payment, shared profits between publisher and author or outright purchase of the manuscript. In the first option, publishers paid for the production and distribution of the book and shared the profits or losses with the writer. The author received a half or a third share. From the author's point of view, there were two disadvantages to this option. The publisher could "cook" the books and make it look as though there was no profit and the payments came in slowly. For authors at the subsistence margin it was not an attractive option due to the latter disadvantage. The second option, selling the copyright outright, gave the publisher complete discretion to print the book in whatever format they wished and as many times as they wished. A drawback of this option was that the amount received for the copyright did not reflect the later success of the work. If, as in Wood's case, the book sold extremely well, the author was not paid one pence more (Fyfe 2004, 211–12). According to Weedon, the most common practice until midcentury was the outright sale of copyrights. Under pressure from authors, publishers moved to royalty agreements (Weedon 2003, 159).

more business savvy in later years. In 1860 he signed an agreement with Routledge to receive £30 for each of forty-eight monthly parts of a revised edition of the *Illustrated Natural History*, and in 1867 thirty-two monthly parts of *Routledge's Illustrated Natural History of Man in all Countries of the World* paid him £30 each.[24] Later, for shorter works, Wood received £25 for *Common Moths of England* and the same amount for *Common British Beetles* from Routledge in 1877.[25]

However, for Wood, a career as a popularizer proved to be financially precarious, especially in the 1870s. When he fractured his right hand in an accident in 1874 and could barely hold a pen, he was forced to apply to the Royal Literary Fund. Writing to the committee on December 30, 1875, he confessed that this was the third time he had had to use up all of his savings and now "I suddenly find myself with a sick wife, six young children, no income, no funds, and no work which is worst of all." Wood revealed that in more prosperous times he averaged an annual net income of £450, but there had been years in which he had earned nothing. He believed this was but a temporary difficulty, and that he was slowly regaining his health and would soon be able to work again. All he asked for was immediate help to enable him to save his library and museum, "which are most costly to purchase, sell for little, and are in fact my stock in trade." Wood received £125 from the Royal Literary Fund in January 1876. In the same year his friend Newton Crosland also persuaded Disraeli to give Wood a Civil List Pension of £100 per annum. In 1877 Wood was compelled to apply again to the Royal Literary Fund. "The book trade, on which I almost entirely depend," he wrote, "has been so depressed for the last three years, that my copyrights yielded a merely nominal sum, and my large and growing family literally ate up my earnings and savings." Due to the state of the publishing trade, Wood could find no work. "Perpetual anxieties" had driven him into a "severe illness" from which he was rapidly recovering, and "much against the wish of the medical men" he was preparing two books in the hopes that the book market would be recovered by the time he had finished them. This time Wood received £50.[26]

24. *Archives of George Routledge & Co., 1853–1902*, 1973, Contracts c1850/78, reel 2, vol. 3 R–Z, 1853–73, 355, 376.

25. University College, London, The Archives of Routledge and Kegan Paul Ltd., Routledge Contracts R–Z, Item 5, 562, 564.

26. Cambridge University Library, Royal Literary Fund, File No. 1982, Letters 2, 11, 12, 21; Crosland 1898, 121.

At this point Wood began to consider lecturing as a possible means of supplementing his income.[27] He had already been lecturing sporadically. He delivered his first lecture many years before, in Oxford Town Hall on March 11, 1856.[28] However, from 1879 until his death Wood conducted ambitious lecturing tours of England, averaging about ninety lectures in a season. During his most extensive tour during the 1881–82 season, Wood gave more than 120 lectures. He spoke on a wide variety of topics. In September 1880, Wood presented a series of lectures at Forest School, Walthamstow, on insect transformation, bees, ants, wasps, spiders, reptiles, and the horse.[29] In 1881 he lectured on "Unappreciated Insects" at the Literary Institute in Altrincham and at the Mechanics' Institute at Mawdsley, on "Ant Life" at the Temperance Hall in Leek, on jellyfish for the Weymouth Lecture Society, and on "The Horse" at Marlborough College.[30]

Wood was in competition with professional scientists like John Tyndall who had tremendous reputations as powerful speakers and great showmen. Tyndall's theatrics during his Royal Institution lectures were legendary. Once, he accidentally knocked a flask off a desk while preparing for a lecture, but quickly jumped over the desk and caught the flask before it hit the ground. He then rehearsed the "accident" and included it in his lecture. Another time, in 1862, Tyndall had the chairman at his spring lecture light a cigar at the invisible focus of a beam of infrared radiation.[31] Realizing that he was entering a competitive field, Wood created an innovative speaking style that won him scores of invitations to lecture. Rather than illustrating his lectures with prepared diagrams or magic lantern slides, he drew "rapid impromptu sketches" of battling ants, gigantic whales, and other creatures. Wood used color pastels, imported from Paris, which positively glowed on the large black canvas drawing surface that he had specially designed.[32] A reporter for the *Altrincham and Bowdon Guardian* was deeply impressed by Wood's sketches, which were "not mere diagrams, but finished pictures in colours

27. Upton [1910], 165.

28. T. Wood [1890], 128.

29. Guy 1880, 4.

30. "Interesting Lecture at Altrincham" 1882, 4; "Lecture Last Night on 'Unappreciated Insects'" 1881, 8; "Lecture on Ants" 1881, 4; "Lecture on Jelly Fish" 1881, 5; "Marlborough" 1884, 8.

31. McMillan and Meehan 1980, 49.

32. T. Wood [1890], 156.

of great beauty." In his estimation, "Mr. Wood's method of lecturing is in fact we believe, unique."[33]

According to Wood's son, the sketches were "always perfectly exact in every particular," and "no line was ever rubbed out or alteration ever made." Wood's seemingly spontaneous sketches were the outcome of long and careful prior preparation. First, he made a tracing of the object he wished to draw from some trustworthy woodcut. He would copy this two or three times upon a slate, always attempting to do so with the fewest possible lines. Next, he would make a very careful sketch in color upon the back of a small paper strip. Then, finally, he would stand before his black canvas and practice drawing the sketch over and over until he could execute it without hesitation and without mistake. Like Tyndall's total control over his experimental apparatus, Wood's command of his pastels in the accurate depiction of the natural world served as proof of his scientific expertise.[34]

Wood's innovative speaking style won him an invitation to give the prestigious Lowell Lectures in Boston in 1883-84. The trip was a success in terms of the enthusiastic response Wood received from his audiences. But Wood's failure to hire an agency to organize engagements in addition to the Lowell Lectures doomed the enterprise financially, and he would have made just as much had he stayed in England. His second transatlantic tour in 1884-85 was a disaster and had to be cut short. This time he had hired an agent, but an incompetent one who lied to Wood about the number of firm engagements. After that experience Wood decided never again to undertake another trip across the Atlantic. Wood cleared about £300 a year through his lecturing activities. His usual fee was five guineas, from which he had to deduct the cost of railway fares, cabs, tips, agents' fees, and other expenses. Lecturing was not as lucrative as Wood had hoped.[35]

Since Wood began to support his family with his earnings as author and lecturer, he had to be more productive than those clergymen who had elected to popularize science while holding a paying position within the Church. The pressure to earn a living took its toll on Wood's health. Crosland recalled that "Wood's multifarious occupations and the consequent strain upon his nervous system became so great . . . that at one time he was tempted to resort to stimulants as a means of keeping his faculties in active working order."

---

33. "Lectures for Altrincham and Bowdon" 1881, 5.

34. T. Wood [1890], 159.

35. Ibid., 234-35, 249-50, 266; Upton [1910], 171.

The result, Crosland judged, was detrimental to his health.[36] Once Wood started lecturing, the physical demands of the circuit undermined his constitution. Wood caught a severe chill while on a lecturing tour and then died shortly thereafter, on March 3, 1889. "The poor fellow literally died in harness," Alfred Whitehead wrote in the *Times*. Whitehead, vicar of St. Peter's, Kent, and rural dean of Westbere, was making a plea on behalf of Wood's family, whom he had left "totally unprovided for." Acting as treasurer for the "J. G. Wood Fund," he invited potential donors to send the money to his bankers.[37] Thirteen days later, Wood's widow wrote to the Royal Literary Fund for help. "So impossible did he find it to maintain his family by his pen alone," she explained, "that in 1879 he adopted public lecturing as a supplementary profession and yet after working almost without a days' holiday for nearly 37 years, he has been able to leave behind him nothing but the proceeds of his life insurance policy." Crosland also wrote to the committee, listing a series of reasons why "this industrious author and good man failed to make provision for his family." The list included his financial commitments to his mother and siblings, illness and the accident to his writing hand, "improvident bargains with publishers" because of the pressure of economic circumstances, the cost of his research on natural history, and the loss of income due to the unsuccessful lecturing tour of the United States.[38] Wood's financial difficulties illustrate just how risky it was for him to have embarked on a career as popularizer of science in the early 1860s, despite the new opportunities that had arisen on account of a rapidly growing reading public and improvements in printing technologies.

MAKING SCIENCE ACCESSIBLE TO THE PUBLIC

Whatever Wood's financial problems, his natural history books sold well, and he believed that they were distinctive in their accuracy and clarity. In his *Boy's Own Book of Natural History* (1861, Routledge) he declared, "there is at present no work of a really popular character in which accuracy of information and systematic arrangement are united with brevity and simplicity of treatment." The "best-known popular works on Natural History" were lacking in several respects, Wood maintained, including incorrect

36. Crosland 1898, 120.
37. Whitehead 1889, 15.
38. Cambridge University Library, Royal Literary Fund, File No. 1982, Letters 24, 28.

classification schemes, inaccurate illustrations, ill-informed accounts of animals, and, most especially, "the absence of explanations of the meanings and derivations of scientific words."[39] Wood repeatedly criticized those who rendered science inaccessible through the use of an unnecessarily complicated nomenclature. In his *Homes without Hands* (1865, Longman) he condemned systematic zoologists who invented fanciful names for newly discovered species instead of relying on Greek words familiar to anyone moderately versed in the classics.[40] In the opening pages of *Animal Characteristics* (1860, Routledge), he made fun of the scientist's penchant for coining strange terms, such as "Proboscidian Pachydermatous Mammifer" for elephant. Although he recognized the use of technical terms as an economical aid, he charged that some scientists intentionally misused them to "mystify their readers, and to take a pride in throwing a veil of impenetrable language over their descriptions." By doing so, they reversed "the real duty of an author, by puzzling people with easy matters, instead of rendering puzzling matters easy."[41]

In addition to avoiding scientific terms where possible, Wood also made his work more accessible to his readers by drawing on one of the important conventions of natural history writing: the use of anecdotes. In the preface to *Boy's Own Book of Natural History*, Wood pointed to the inclusion of many "new anecdotes" as a strength of the book. "In many cases," Wood stated, "the anecdotes related have never been published before, and in many more, they have been extracted from works which, either from their scarcity, their cost, or their nature, would be very unlikely to be placed in the hands of general readers."[42] Wood often drew on the first-hand observations of other naturalists or explorers to enliven his work. To Wood, anecdotes were important narrative tools for conveying knowledge, not just a means to spice up his writing. In *Man and Beast* (1874, Daldy, Isbister), where he argued that the lower animals possessed many human qualities, Wood cited as proof "more than three hundred original anecdotes, all being authenticated by the writers, and the documents themselves remaining in my possession."[43] Anecdotes were valid forms of evidence.[44] Wood also believed they were

39. J. Wood 1861a, iii.

40. J. Wood 1870, 283–84.

41. J. Wood [1860], 2–4.

42. J. Wood 1861a, iv.

43. J. Wood 1874, I, vii.

44. Interestingly enough, one of Wood's anecdotes from *Out of Doors*, recounting his painful encounter with a stinging jellyfish, was later used as evidence by Sherlock Holmes to solve a puzzling case in "The Adventure of the Lion's Mane" (Doyle 1974, 210–11).

the most appropriate form for communicating information about living beings. In his *Illustrated Natural History* (1853, Routledge) he insisted that the "true object of Zoology" was to study the "Life-nature" of organisms rather than "to arrange, to number, and to ticket animals in a formal inventory." Although acquaintance with the material framework of any creature was useful, it was far more important to "know something of the principle which gave animation to that structure." Wood claimed that his emphasis on "animating spirit" rather than "outward form" informed the entire shape of his *Illustrated Natural History*. "In accordance with this principle," he declared, "it has been my endeavor to make the work rather anecdotal and vital than merely anatomical and scientific."[45]

By avoiding elaborate scientific terms while incorporating colorful anecdotes, Wood adopted a writing style designed to appeal to a popular audience. He also made vivid illustrations an important feature of his books. Wood did not draw the illustrations for his books, partly because he was too busy writing and partly because the skills needed to draw huge pictures for lectures were different from those needed for book illustrations. He could make more money letting the publisher hire an illustrator, or a team of illustrators, while he wrote more books. No matter which publisher Wood worked with, he ensured that first-rate scientific illustrators were involved in the production of his books. Tuffen West, one of the most famous microscopical illustrators of the second half of the nineteenth century, was hired to create the twenty plates for Wood's *Common Objects of the Microscope* (1861, Routledge). For the *Boy's Own Book of Natural History*, Wood assured his readers that the illustrations had "all been designed expressly for the present work, and that the combined abilities of Messrs. Harvey and Dalziel, as artist and engravers, are a guarantee for their accuracy and perfect execution." Wood drew some of the anatomical and microscopical illustrations himself from "actual specimens."[46] The Dalziel brothers recalled that Wood kept a close watch on the production of his illustrations. For the *Illustrated Natural History*, issued in monthly parts over the course of nearly four years, they went to see Wood every Monday morning to "receive new lists of subjects, to report progress of those in hand, and to discuss the matter generally." Wood was very picky when it came to illustrators. He considered Harrison Weir's drawings to be "always picturesque, but never correct." It was the opposite problem with T. W. Wood (unrelated to J. G. Wood), who produced drawings that were technically correct but artistically

45. J. Wood n.d., 1: vi.   46. J. Wood 1861a, iv.

deficient. Joseph Wolf's work met with his approval, as did W. S. Coleman, who, when living specimens could not be obtained, endeared himself to Wood by searching for the most reliable representation.[47] Routledge was willing to invest huge sums of money for the lavish illustrations in Wood's books. Mumby estimates that before the end of the 1850s Routledge had paid £50,000 to the Dalziel brothers for the wood engravings alone.[48]

Later, when Wood began working with the Longman publishing house, he produced books brimming with illustrations. In *Insects at Home* (1872, Longman) there were seven hundred figures drawn by E. A. Smith and J. B. Zwecker and engraved by G. Pearson, including twenty plates and seventy-nine woodcuts. For *Insects Abroad* (1874, Longman) Wood used the same artistic team. Since here he was dealing with insects that could only be found outside of England, he took special precautions to obtain accurate drawings. His preface assured the reader that the six hundred illustrations had been made from actual specimens. Wood came up with a startlingly novel idea for the illustrations in *Insects at Home*. In the preface he remarked that the reader would likely notice that "these figures of insects are but slightly shaded, and in many cases are little but outline." Wood said that this was "intentional, and the shading is omitted in order that the reader may supply its place by colour." Since the insects were described fully in the text, Wood believed that the reader would have no problem in identifying the proper color to be used. "I very strongly recommend the possessor of the work to colour these illustrations," Wood affirmed, "as he will thus fix the insects firmly in his mind, and quadruple the value of the volume to other readers."[49] Wood gave his readers the opportunity to feel as though they were actively engaged in the drawing of illustrations.[50]

Wood's approach to writing made his work particularly appealing to children. If the father of a family introduced Wood's *Common Objects of the Country* into the household, the *Literary Gazette* believed that he would see "his children exercising their faculties of observation, and obtaining wholesome entertainment from the commonest objects by which they are surrounded."[51] The *English Woman's Journal* recommended putting this "simple and bright spirited little book" into the hands of "young people, especially of girls, as it could not fail to interest them in minute observation of

47. Dalziel and Dalziel 1978, 266, 270-71.
48. Mumby 1934, 79.
49. J. Wood 1872, vi.

50. Lightman 2000, 657-61.
51. "Publications Received" 1858, 373.

nature."[52] Wood often addressed the young reader as his intended audience. In the preface to his *Common Moths of England* (1870, Routledge), he explained that this book was meant to be an "introduction" to entomology designed to match the experience of the young English reader. Too numerous to mention the names of all the moths of England, Wood made a selection of them "so that the young Entomologist will find in the following pages a figure and description of nearly every Moth that he is likely to find. As a rule, the commonest and most conspicuous species have been selected." He assured his juvenile reader that he would describe the moths "as simply as possible" and that the "dry scientific technicality" would be "considerably simplified."[53]

In attempting to make science accessible to a mass reading audience, Wood saw himself as opposing those who were resolved to professionalize science. In his *Illustrated Natural History*, he was critical of the "inordinate use of pseudo-classical phraseology" by naturalists that led the study of animal life to be "considered as a profession or a science restricted to a favored few, and interdicted to the many until they have undergone a long apprenticeship to its preliminary formulae." Even worse, the need for expert training was accepted outside the scientific community, in that "the popular notion of a scientific man is of one who possess a fund of words, and not of one who has gathered a mass of ideas." Wood believed that "any one of ordinary capabilities and moderate memory" could be "acquainted with the general outlines of zoology."[54] In some of his works, though, Wood asserted that the ordinary person could go beyond mere familiarity with natural history to the discovery of new knowledge. The whole point of the Common Objects series, for Wood, was to invite his readers to participate in the making of science. In his *Common Objects of the Microscope* he denied that only those who traveled to the ends of the earth to gather knowledge or those who owned expensive scientific instruments could practice science. Wood recommended microscopes to his readers that were affordable yet suitable for serious scientific work. Many of his most "valuable original observations" in his notebook were made by "an old lady in her daily perambulation of a little scrap of a back yard in the suburbs of London, barely twelve yards long." Wood had no doubt that "if any one with an observant mind were

52. "Notices of Books. *The Common Objects of the Country*. By the Rev. J. G. Wood" 1858, 347.

53. J. Wood 187?, preface, 1.

54. J. Wood n.d., 1: v.

to set himself to work determinately merely at the study of the commonest weed or the most familiar insect, he would, in the course of some years' patient labour, produce a work that would be most valuable to science and enroll the name of the investigator among the most honoured sons of knowledge."[55] Wood held that through the study of common objects, common readers could discover important knowledge. He insisted that his readers did not need sophisticated scientific instruments, a laboratory, or even expert training in order to play a significant role in science.

Wood's resistance to the goals of professionalizing scientists was related to his attraction to spiritualism and freemasonry, and to his ambivalent attitude toward evolutionary theory. There is no mention of Wood's Masonic activities or his devotion to spiritualism in his son's biography, perhaps because it would have damaged his reputation as a popularizer of science. Routledge, a Freemason himself, may have encouraged Wood to join the secret society.[56] Wood belonged at least as early as 1854. In an article on "Masonic Symbols" for the *Freemasons' Quarterly Magazine* in 1854, signed as "Bro. The Rev. J. G. Wood," he proposed that those Masons who had made some progress in any of the sciences "should illustrate the various symbols of our Order, throwing light on them by means of their scientific knowledge." Then, based on what he had learned writing his book *Bees: Their Habits, Management, and Treatment* (1853, Routledge), he discussed the parallels between a beehive (one of the emblems of the Craft) and a lodge of Freemasons.[57] A year later, Wood wrote a report of his visit to a French lodge in order to show that members of the society were received with as fraternal a welcome in another land as in their own. Describing how the French customs officer, a fellow Mason, facilitated the processing of his documents, Wood remarked "the mystic powers of the Craft began their work immediately on our landing on French ground." Impressed by the ceremony at the French lodge, he wrote that it was "magnificently performed, and I never witnessed anything more striking." Wood's love of spectacle, evident in the choral festivals he organized as precentor of the Canterbury Diocesan Choral Union and in his sketch-lectures, take on new meaning in this light. However, Wood was upset by two aspects of the ceremony, which detracted from the grandeur. The hand organ was out of tune and participants in the ceremony had the unpleasant habit of spitting.[58] Wood's affiliation with the Masons lasted until his death, yet it is difficult

55. J. Wood 1861b, 4–5, 7.
56. E. Wells 1990, 58.

57. J. Wood 1854, 45–50.
58. J. Wood 1856, 136, 222.

to detect Wood's connection with the Masons in any of his natural history works.[59]

Similarly, traces of Wood's spiritualism are not easy to find in any of his books, but according to an obituary in *Light*, he was one of the earliest members of the spiritualist movement.[60] Newton Crosland claimed to have introduced Wood to spiritualism in 1856, and he maintained that Wood's son had repressed this important dimension of his life because he could not see how it could be reconciled with Anglicanism. When Wood first became interested in spiritualism, Crosland quotes him as exclaiming that "little glimpses of such vast and deep meaning have been given, that I feel much like a little chicken that has just chipped the shell, and for the first time seen the sky."[61] At least for a time, Wood became excited by spiritualism. Edmund Evans, an engraver and printer, recalled that soon after he began working for Routledge in 1856 or 1857, Wood often wanted him to come with him to witness a séance. "He promised I should see and hear wonderful things," Evans remembered, "massive pieces of furniture moved by unseen powers, instruments of music play untouched, and mysterious doings generally."[62] Crosland's wife asserted that Wood became a "seeing medium as well as a rapping one, occasionally discerning the atmosphere about people." In 1857 he wrote to her that he heard continuous raps as he was translating from St. Paul's Epistles. Later, she recalled, "lying spirits" deceived Wood and he began to resist their influence. However, she affirmed that Wood believed in the existence of animal, as well as human, apparitions. She saw manifestations of Wood's spiritualism in his book *Man and Beast*, where he "boldly argues, and on Biblical authority, in favour of the immortality of all creatures."[63] The fact that Wood argued for the existence of a heaven for animals in this work on biblical authority—and not by virtue of an encounter with bestial ghosts—indicates how much he kept his spiritualist beliefs in the background in his published works.

Likewise, Wood rarely discussed his attitude toward evolution in public. His son claimed that he rejected it at first, but subsequently modified his

---

59. "Obituary" [Wood] 1889a, 80. I am indebted to Robert Gilbert for drawing my attention to Wood's connection to the Masons.

60. "Sudden Decease of the Rev. J. G. Wood, F.L.S." 1889, 115.

61. Crosland 1898, 119, 123, 153.

62. McLean 1967, 36.

63. Crosland 1893, 248-49. I am grateful to James Secord for pointing out the Croslands' recollections on Wood as a spiritualist.

views to accept the notion that evolution was responsible for some, though not all, animal development. Wood came to believe that evolutionary theory was in no way opposed to religion. However, his son acknowledged, "it was far from easy to gather his views regarding a subject upon which he was so reticent, and upon which he evidently considered that it was as yet premature to pronounce a decided opinion."[64] Crosland confirmed that Wood began as an opponent of Darwinism but later softened his position. He complained that Wood's about-face was due to the influence of "that materialist Professor Flower, and for a time he seemed inclined to coquet with the Darwinian hypotheses. I argued and remonstrated with him on the subject, and finally he allowed it quietly to drop."[65]

Given Wood's reticence on the subject of evolution, it is difficult to find evidence for his changing views in his published works. His *Man and Beast*, published just three years after the appearance of Darwin's *Descent of Man* (1871), is almost certainly intended to reject human evolution, at least in conjunction with materialism. Here he argued that the "lower animals share immortality with man in the next world." By insisting that animals had a soul, Wood, of course, granted humans one as well. He began by arguing that the scriptures did not deny a future life to the lower animals and then discussed how they shared with humans the attributes of "Reason, Language, Memory, a sense of Moral Responsibility, Unselfishness, and Love, all of which belong to the spirit and not to the body." Although Darwin's name and the word "evolution" are never mentioned, Wood's strategy is the mirror opposite of the one adopted in the *Descent of Man*. Darwin stressed the human qualities of animals in order to place humans within the animal kingdom, which made them subject to the natural law of evolution. Wood conferred human attributes to animals to raise them to heaven's gates. Of course, in Wood's scheme, humans, as well as animals, were endowed with a soul and were not mere products of natural law.[66]

In one of his last published works, *Romance of Animal Life* ([1887], Isbister), Wood referred positively to Darwin in the introduction, drawing on the evolutionist's work on worms to illustrate the dependence of humanity on one of the smallest of God's creatures. "As the late Charles Darwin has shown us," Wood declared, "man could not have found a place upon the earth had not the worm prepared the way for him." Without the lowly earthworm there would be no trees or grasses—"is not this physical fact"

64. T. Wood [1890], 113–14.
65. Crosland 1898, 125.

66. J. Wood 1874, 1: v; 2: 345.

Wood asked, "as romantic as any production of the novelist's brain?" In his discussion of the ape, Wood asserts that it is a "universal rule of nature" that the structure of an animal is that "best suited to its life and habits." Yet, again, Wood cannot bring himself to use the term "evolution."[67] Wood's reticence is significant, especially in light of Francis Morris's explicit attacks on evolutionary theory. Whereas Morris publicly attacked Darwinism, Wood made the strategic decision to mute his opposition.

VISUALIZING A THEOLOGY OF NATURE

Wood's handling of his commitment to the religious dimension of science is in line with those popularizers of science who presented a subtle "theology of nature" accommodated to the more secular sensibilities of a mid-nineteenth-century audience. In his works all of the heavy-handed references to God's wisdom, goodness, and power in classical natural theology as formulated by Paley were excluded. Wood, as his son states, "never seasoned his writings with texts, and seldom even quoted Scripture" and "abstained on principle" from saying that "Creation was God's work." He preferred to "point his moral indirectly, and to leave it rather to be drawn by inference than gathered from direct statement." Wood "had the most utter detestation of that form of writing generally described as 'goody,' which causes many a reader to throw aside many a book in disgust." More than once Wood was infuriated when the editor of a religious magazine inserted scriptural quotations into one of his articles after the proofs had passed through his hands.[68] Wood was not alone in despising the unnecessary inclusion of theological themes in natural history works. In 1866, an anonymous critic in the *Popular Science Review* censured a practice that was "too common in Natural History treatises," the habit "which certain naturalists display of dabbling in Divine matters. It appears to them that, unless they drag the Creator into every second paragraph, their essay will not possess the necessary religious veneering for the public taste."[69]

The relationship of Wood's theology of nature to natural theology is most clearly presented in his *Nature's Teachings* (1877, Daldy, Isbister). Even though Wood never used the word "God," the book is based on the analogical reasoning at the heart of natural theology. Wood stated that the aim of the book is "to show the close connection between Nature and human

---

67. J. Wood 1889, 14, 167.
68. T. Wood [1890], 98, 68.

69. "Reviews. Popular Zoology" 1866, 215.

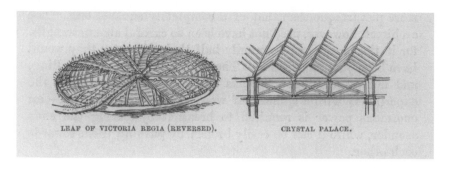

LEAF OF VICTORIA REGIA (REVERSED).   CRYSTAL PALACE.

FIGURE 4.3 A depiction of how architecture resulted from gardening. Wood demonstrates the parallels between the design of the Crystal Palace by Paxton and the unique vegetable cellular structure of the Victoria Regia plant. (John George Wood, *Nature's Teachings* [London: Daldy, Isbister, 1877], 196.)

inventions, and that there is scarcely an invention of man that has not its prototype in Nature."[70] Wood ran through the gamut of human invention, including nautical, military, architectural, instrumental, optical, acoustic, and artful, to demonstrate the parallels between nature and human arts. The fixed nets used in fishing and hunting have their natural analogy in the web of the common garden spider. The prototype of plate armor lies in the bodies of the lobster, crayfish, or shrimp. Even the most modern technology, such as the network of telegraphic wires, has its parallel in the nervous system of the human body, while the suspension bridge is modeled on the creepers of tropical climates. Sometimes the parallel is the result of unconscious imitation of nature, as in the case of the pulley and the human hand. But other times an inventor deliberately copies a structure in nature, for example the novel architectural plan used in the building of the Crystal Palace based on vegetable cellular patterns (Fig. 4.3).[71]

As John Brooke and Geoffrey Cantor have shown, the use of analogical reasoning is one of the main strategies in natural theology.[72] Four terms are arranged in two pairs, one pair included a man-made artifact (such as a telescope or a watch) and its human maker while the other a part of the natural world (such as the eye) and its maker, God. In both cases design implied a designer and craftsman. Whereas Paley demonstrated that artifact is to artisan as natural object is to God, Wood leaves out any explicit reference

70. J. Wood 1877, v.   72. Brooke and Cantor 1998, 190.
71. Ibid., 196, 452.

in *Nature's Teachings* to the fourth term, God. He always begins with the human invention and then looks to nature for a similar, previously existing structure, in the same way that Paley moved from the contrived quality of a watch to a designed natural world. Paley explicitly pushed the reader to recognize that just as the existence of a watch demanded the existence of a human contriver, the designed quality of nature proved the existence of a divine creator. Wood leaves it to the reading audience to fill in the missing term. In many of his other works Wood did refer to God, but usually limited his discussion of the religious meaning of his books to the introduction or preface. In this way he provided a religious framework without constantly harping on marks of design throughout the body of the text.

Wood's theology of nature plays a crucial role in all of his works, from the earliest to those produced later in his career. His discussion of religious themes is often reinforced by the inclusion of vivid visual images. His best-selling *Common Objects of the Country* opens by drawing an analogy between learning to read the Bible and learning to read nature. "To one who has not learned to read," Wood declared, "the Bible itself is but a series of senseless black marks; and similarly, the unwritten Word that lies around, below, and above us, is unmeaning to those who cannot read it." Wood declared that the "following pages are written" for those who would like to read nature ("the unwritten Word") but cannot.[73] As the reader moves through the book and begins to read "the unwritten Word" with Wood's help, understanding depends on seeing nature as a world of purpose, beauty, and wonder. In this book Wood discussed bats, mice, rats, and other mammals, then moved on to reptiles and insects, stressing that each creature has its place in the "scale of creation." Whenever Wood comes across a creature that is commonly reviled, he shows his readers that it served some purpose and that it is aesthetically pleasing. Bats are "very pretty" creatures and they "thin the host of flies and other insects" in the air. The toad, a "much-calumniated creature," devours great numbers of garden pests while possessing a beautifully "brilliant eye." When he reaches the insects, Wood dwells on the unearthly beauty of moths and butterflies. He describes one little moth as "the epitome of every fragile, fairy-like beauty, and seems fitter for fairy tale, 'once upon a time,' than for this nineteenth century." At times Wood has to rely on the microscope to reveal the magnificence of insect bodies. The British diamond beetle looks ordinary enough to the unassisted eye, but when put under the microscope it "glows with jewels and gold." Wood's detailed descriptions,

73. J. Wood 1858, 1–2.

and his sumptuous color plates, were intended to elicit a sense of wonder in the reader. "Few people have any idea of the wonders that they will find inside even so lowly a creature as a caterpillar," Wood insisted, "—wonders, too, that only increase in number and beauty the more closely they are examined." After explaining the intricate anatomy of a silkworm, an imaginary friend utters "exclamations of astonishment." Whereas the friend had previously thought "caterpillars were nothing but skin and squash," now they have become a "new world" to him. This is the reaction that Wood is hoping to draw from his reader (Fig. 4.4).[74]

Wood presented a similar theology of nature, enhanced by vivid visual images, in many of his subsequent natural history studies. In the preface to his *Illustrated Natural History for Young People* (1882, Routledge), Wood stated that his "end" was to lead the reader to "notice how wonderfully each creature is adapted for its peculiar station by Him who has appointed to each its proper position." But Wood recognized that some readers had prejudices that led them to a "causeless abhorrence of certain creatures" such as snakes, spiders, or toads. Wood urged them to examine such creatures with "a more reverent eye," because they were beings that the "Maker saw at the beginning of the world and declared very good." The true naturalist, Wood declared, "will see as much beauty in a snake, spider, or toad, as in any of those animals which we are accustomed to consider models of beauty." When the organizational details of these supposedly ugly and frightful animals are studied under the microscope, we become "lost in wonder and amaze at the vastness of creation, which, even in one little, apparently insignificant animal, presents to our eyes innumerable marvels—marvels which increase in number and beauty as our power for perceiving them increases."[75]

Wood's attempt to train his readers to see nature through a "reverent eye" is most striking when he is dealing with tiny objects put under the microscope. Unlike would-be professional scientists such as Huxley, who claimed that proper use of the instrument was confined to those few trained to the task in modern laboratories, Wood believed that anyone could make important scientific discoveries with a microscope. His *Common Objects of the Microscope* discussed different types of microscopes, explained how to prepare cells, offered suggestions on interesting objects to view, and invited readers to compare what they saw with the twelve plates of illustrations at the back of the book. Plate III (frontispiece) contains fifty-two images of various vegetable and flower structures disembodied from the original

74. Ibid., 3–5, 44, 59, 65, 108, 153, 171.   75. J. Wood [1882], vi–vii.

FIGURE 4.4 *Common Objects of the Country* featured a dozen colored plates. Wood referred to them frequently throughout the book to elicit a sense of wonder in the reader. The daddy long legs or crane fly depicted in figure 1 "shows a very pretty species, covered with yellow rings." Commenting on figure 2, a bee, Wood said that it was "remarkable for the beautifully tufted extremities of its middle pair of legs. Figure 7 features the common black cocktail beetle, a carrion eater that has an unpleasant smell. "Repulsive as it is," Wood wrote, "its wings are very beautiful, and the mode in which these organs are packed away under their small cases is most wonderful" (J. Wood 1858, 167, 172, 175). (John George Wood, *Common Objects of the Country* [London: Routledge, 1858], plate H.)

plants. Although Wood admitted that the drawings could not fully capture "the lovely structures," the symmetrical and geometrical shapes illustrate his point about the hidden beauty of the microscopic world. "Their form can be given faithfully enough," he affirmed, "and their colour can be indicated; but no pen, pencil, or brush, however skillfully wielded, can reproduce the soft, glowing radiance, the delicate pearly translucency, or the flashing effulgence of living and ever-changing light with which God wills to imbue even the smallest of his creatures... whose wondrous beauty astonishes and delights the eye, and fills the heart with awe and adoration."[76] While T. H. Huxley's future science teachers were trained to see a fully secularized material world at his lab at South Kensington, Wood's reading audience was expected to see divine beauty through the lens of the microscope.[77]

In *Insects at Home* and *Insects Abroad*, published within two years of each other, Wood offered his readers lavishly illustrated books that transformed the entomologist into a theologian of nature. In the preface to *Insects Abroad* Wood stated his two objects. First, in keeping with the goodness of divine design, he hoped "to show the great and important part played by Insects in the economy of world, and the extreme value to mankind of those insects which we are accustomed to call Destructives." The creatures most detested by humans—mosquitoes, ants, and wood-boring beetles and termites—were actually among our greatest benefactors. According to Wood, insects were "working towards one purpose, namely, the gradual development of the earth and its resources."[78]

Insects were not just good in the larger scheme of things; they were also beautiful. The second object of *Insects Abroad* was to encourage readers to "note the wonderful modifications of structure which enable the insects to fulfil their mission, and the surpassing beauty with which many of them are endowed." *Insects at Home* also emphasized this theme. "We find among insects," Wood asserted, "a variety and brilliancy of colour that not even the most gorgeous tropical flowers can approach, and that some of our dullest and most insignificant little insects are, when placed under the revealing lens of the microscope, absolutely blazing with natural jewellery."[79] Throughout both books Wood drew the reader's attention to the beauty of insect bodies and wings, making good use of his many illustrations. The frontispiece to *Insects Abroad* depicted different species of beetles, including the large goliath beetle that dominated the lower half of the image. In contrast to the

76. J. Wood 1861b, 5, iv.
77. Gooday 1991, 307–41.

78. J. Wood 1883, v, 5.
79. Ibid., v; J. Wood 1872, 2.

disembodied images of *Common Objects of the Microscope*, the illustrations in the insect books presented their subjects in a natural setting filled with lush vegetation and teeming with life. Here was what Darwin termed the "face of nature bright with gladness" so central to the natural theology tradition. The striking frontispiece to *Insects at Home*, rendered in vivid color, presented a variety of different insects, including one of Wood's favorites, the "handsome Great Green Grasshopper" (Fig. 4.5).[80]

It was no challenge for Wood to highlight the beauty of some butterflies, moths, and beetles. But he was not satisfied unless he could persuade his readers that nearly every insect was aesthetically pleasing in some way. Of the dragonfly depicted in Figure 1 of Plate VII of *Insects Abroad*, Wood said, "It is really a most lovely insect, its wings glittering with iridescent hues of metallic purple, green, blue and gold." Even as Wood turned to insects commonly seen as ugly and disgusting, his "reverent eye" found something attractive or wonderful to praise. A giant earwig, on the left near the bottom of Plate VII of *Insects at Home*, was described as a "fine insect," while the common earwig, on the left near the top, was shown with "its beautiful wings extended" (Fig. 4.6). In this illustration Wood also included a field cockroach and its egg case (bottom left), a field cricket (just below the center), and a mole cricket (center bottom), which, he remarked, is "shaped wonderfully like the mole." In some cases Wood was forced to turn to the microscope for help in finding the element of beauty he wanted his readers to see in all insects. "When placed under the microscope," Wood declared, "the Flea really becomes an interesting insect, with some share of beauty about it." Although "singularly unpleasant" in a room, the gnat was "marvelously beautiful under the microscope." Wood recommended that the gnat be examined with a succession of powers, beginning at the lowest and ending at the highest, so that the observer could appreciate its beauties of detail "by degrees." "As dull and colourless as the Gnat may appear to the unaided eye," Wood announced, "it has only to be placed under the revealing glass of the microscope to blaze out in a magnificence which would pale all the fabled glories of Aladdin's fairy palace."[81]

---

80. Near the beginning of chapter 3 of the *Origins of Species*, titled "Struggle for Existence," Darwin declares, "We behold the face of nature bright with gladness, we often see superabundance of good; we do not see, or we forget, that the birds which are idly singing round us mostly live on insects or seeds, and are thus constantly destroying life" (C. Darwin 1985, 116). For Wood's praise of the grasshopper see J. Wood 1872, 247.

81. J. Wood 1883, 356; 1872, 230, 226, 245, 591, 601-2.

FIGURE 4.5 Wood's happy world of beautiful insects, as seen through his "reverent eye." (John George Wood, *Insects at Home* [London: Longmans, Green, 1872], frontispiece.)

FIGURE 4.6 An illustration designed to train the eye of the reader to see the wonder in earwigs and cockroaches. (John George Wood, *Insects at Home* [London: Longmans, Green, 1872], 228, plate VII.)

Wood's success as a popularizer of science, such as it was, owed a great
deal to his use of visual images in both his popular lectures and best-selling
books. It gave him an edge in a competitive field. Working with Routledge
in the 1850s, and after the stunning success of *Common Objects of the Country*,
he witnessed firsthand the growing market for science books in the mid-
Victorian period. By the early 1860s he had the confidence in his writing and
the market to take the risky move of giving up his clerical profession in order
to build a career as a popularizer of science. Sensitive to the reading habits of
his audience, he designed books that presented scientific information in an
accessible way and that appealed to the aesthetic sensibilities of a public with
a voracious appetite for visual images. Recognizing that his reputation as an
author of natural history works could be tarnished, he avoided any public
expression of his connections to the Masons and to spiritualism. Unwilling
to antagonize Darwin or his scientific naturalist allies, he chose to mute his
criticism of evolutionary theory. Finally, Wood rejected an appeal to traditional natural theology, designed for an older generation of readers. He
elected to offer a more subtle theology of nature, depicted in images of harmonious animal life, and addressed to a sophisticated public. As a result, he
became an influential popularizer whose *Common Objects of the Country*
outsold the works of would-be professional scientists, including the often
celebrated and foremost canonical scientific work of the nineteenth century,
Darwin's *Origin of Species*.

JOHN HENRY PEPPER AND THE
THEATRICS OF SCIENCE

In 1887 Wood wrote an article for the *Nineteenth Century* in which he complained about the dullness of museums. "I speak on behalf of the General
Public," Wood declared. "Full of interest to the expert, there is no concealing the fact that to the general public a museum, of whatever nature,
is most intolerably dull, as I know by personal experience." Wood was
critical of existing zoological galleries, botanical collections, and geological
museums. He advocated three different types of museums, one for purely
scientific purposes, one for those trying to learn the rudiments of science
in order to become students, and one for the general public. While the first
two types of museums already existed in England, Wood claimed that there
were none of the third type, and he presented his vision for the ideal public
museum. Wood believed that such an institution must take into account
the "sublimity of ignorance which characterizes the general public." Taking
zoology museums as an example, he insisted that they must be "pre-eminently

attractive" and that they should deal with zoology as "the science of life" by presenting a scene in the life history of the animals exhibited. "In all the large groups there should be a background representing faithfully a local landscape," he wrote, "actual objects being merged gradually into the pictorial representations in the way which has of late years proved so effectual in the various panoramas representing the siege of Paris, the battle of Tel el Kebri, and similar scenes." Lecturers, carefully selected for their ability to place themselves "in the mental condition of their hearers," would explain the animal groupings. "The most learned men are not necessarily the best teachers," Wood warned, as they assumed "that their hearers are already somewhat versed in the subject, and in consequence are unintelligible." If such an institution were in existence, Wood predicted that it would "attract thousands who otherwise would not set foot inside a museum."[82]

A museum of the type that Wood was envisioning had existed six years earlier, though it emphasized the physical sciences. The Royal Polytechnic Institution was one of the new galleries of practical science established in the second quarter of the nineteenth century. Reporting on the opening of the Polytechnic Institution to the public on August 6, 1838, the *Mirror* declared that it had been formed "for the advancement of Practical Science, in connexion with agriculture, arts, and manufactures."[83] The brainchild of Charles Payne, former manager of the Adelaide Gallery, and Sir George Cayley, a wealthy Yorkshireman with an enthusiasm for the mechanical arts, the Polytechnic soon outstripped its main competitor, the Adelaide Gallery.[84] It was equipped with industrial tools and machines, a laboratory, a lecture theater, and a large display room, known as the Great Hall, where the main exhibits were housed. Among the main exhibits were the diving bell and diver, an oxyhydrogen microscope, large electrical machines, and model boats floating in a long canal (Fig. 4.7). The diving bell was a unique feature. Hanging beneath the west end gallery and constructed of cast iron, it weighed three tons. Four to five persons could fit inside while the diving bell was submerged under water. The diving bell was a huge attraction. Costing one shilling for a descent, in 1839 it was asserted that it earned £1,000 per year for the Polytechnic.[85] In his *Frankenstein's Children*, Iwan Morus has already drawn attention to the importance of the galleries of practical science like the Polytechnic and the Adelaide Gallery during the second

82. J. Wood 1887, 384–87, 392, 395.

83. "The Polytechnic Institution, Regent Street" 1838.

84. Weeden 2001, 1; Morus 1998, 80.

85. "Diving Bell" 1839, 98.

FIGURE 4.7 The Great Hall of the Polytechnic with the diving bell in the very front of the picture. (Hermione Hobhouse, *History of Regent Street* [London: Macdonald and Jane's, 1975], plate XVIII. Courtesy Guildhall Library and Guildhall Art Gallery, City of London.)

quarter of the nineteenth century. They provided a site for mechanics and London's instrument makers to present spectacular shows and to appear before the public as men of science. For the middle classes who toured the exhibits, they offered a way to make sense of the factory system as an essential component of the progress of industry.[86] The Polytechnic survived well into the second half of the nineteenth century, and did not close its doors until 1881. When John Henry Pepper became manager of the Polytechnic in 1854, it had already begun to evolve into a new phase of its existence as the kind of public science museum that Wood imagined.

86. Morus 1998, 82–83.

From his home base at the Polytechnic, Pepper established himself as an influential writer and lecturer from the 1850s to the 1870s. Like Wood, he refrained from attacking eminent scientific naturalists, and incorporated elements of the developing mass visual culture into his work. Both made a career for themselves as popularizers of science, though Pepper accomplished this goal differently from Wood. Pepper was employed by an institution devoted to providing science to the public, whereas Wood established and managed his career entirely on his own. While both Wood and Pepper took advantage of the growing market for accessible science books, Pepper was also the beneficiary of the expansion of existing museums and the founding of new ones. Many of Britain's great museums first opened their doors during the period from the 1820s to the early 1880s, including the National Art Gallery (1824), and the South Kensington Museum (its collections went on to be the nuclei of both the Victoria and Albert Museum and the Science Museum, 1857). All told, two hundred metropolitan, provincial, and university museums were founded in Britain, which leads Nicolaas Rupke to refer to this period as "the age of museums." Among them were important science museums, and in his select list Rupke includes the Museum of Practical Geology (1851) and the British Museum (Natural History) in South Kensington (1881).[87] Rupke could have added the two key galleries of practical science, the Adelaide Gallery (1832) and the Polytechnic (1838). Pepper is representative of the growing number of lecturers and demonstrators hired at these two institutions, and others that catered to a popular audience, whose popularizing activities were an integral part of their jobs.

Born to Charles Bailey Pepper, a civil engineer, John Henry Pepper (1821–1900) was educated at King's College School, and then later studied analytical chemistry at the Russell Institution with J. T. Cooper. In 1840 he was appointed assistant chemical lecturer at a private school of medicine run by R. D. Grainger. He married Mary Ann Benwell of Clapham Common, London, in 1845. Pepper gave his first lecture at the Polytechnic in 1847 and impressed the managers so much with his speaking skills that he was hired as lecturer and analytic chemist in 1848 and then as manager of the institution in 1854. He continued in this role, with some short interruptions, until 1872, when he resigned for good after a quarrel with the board of directors over the extent of his autonomy. During his time at the Polytechnic he published five science books for a popular audience, including *The Boy's Playbook of Science*

87. Rupke 1994, 13–15.

(1860, Routledge), and established himself as one of the premier showmen of science. He tried to recreate his successful form of science entertainment at the Egyptian Hall, Piccadilly, but lost money on the venture, and went on tour in the United States, Canada, and Australia from 1874 to 1881. He accepted the post of public analyst in Brisbane, Australia, in 1881, stayed there until 1889, and then returned to England, where he remained until his death in 1900.[88]

When Pepper took over the reins of the Polytechnic in 1854 he became the director of an institution that had already eliminated its major competitor, the Adelaide Gallery, which had closed its doors in 1845 after a thirteen-year run.[89] However, another London exhibition created problems for the Polytechnic. Pepper needed to come up with a strategy for attracting customers whose expectations had been raised by their experiences exploring the Crystal Palace in 1851 on shilling days. The Polytechnic's offerings must have seemed meager in comparison.[90] On October 28, 1854, an advertisement for the Polytechnic in the *Athenaeum* noted that it was "under the sole Direction of J. H. Pepper" and promised, "every NOVELTY in GENERAL SCIENCE will be secured to the public." This signaled Pepper's intention to exploit the relationship between the Polytechnic and the London entertainment scene by bringing in more music and spectacle. But the notice also announced that "GOOD DRAMATIC READINGS are now added to the other attractions of the Institution," an indication that theater was to play a major role in future Polytechnic programs.[91]

In May 1855, the Polytechnic was offering "Dissolving Views of Sinbad the Sailor," a "Diorama Illustrating the Voyage across the Atlantic" and dramatic readings.[92] In November of the following year, the bill of fare included a musical lecture, and juvenile lectures by Pepper on the chemistry of fireworks illustrated "by a complete miniature series, constructed by Mr. Darby, the celebrated Pyrotechnist."[93] In 1858, the *Illustrated London News* judged the Polytechnic program "is now of the most attractive order." A new "musical and pictorial entertainment" was being offered, in which a young visitor to

88. Boase 1965c, 386–87; Cane 1974–75, 116–28; J. Secord 2002, 1648–49; W. Brock 2004, 1572–73.

89. Morus 1998, 81.

90. Altick 1978, 472–73.

91. "Royal Polytechnic Institution" 1854, 1306.

92. "Royal Polytechnic Institution" 1855, 470.

93. "Royal Polytechnic" 1856, 483.

the Crystal Palace falls asleep in the Egyptian Court and voyages to famous sites on the Nile. Illustrated by dissolving views, the oral narrative was "relieved" at times by Mr. Cooper's "capital buffo songs." Also on the program were lectures on natural magic accompanied by illusive phantasmagoria, dissolving views on the mutiny in India, and more strictly scientific lectures on railway signals, fire detectors, and Pepper's talk on "A Scuttle of Coals."[94]

In 1859, the periodical press announced the engagement of Christofor Buono Core, a.k.a. "the Salamander," who walked through a cage of flame impervious to harm, thanks to a clever invention. Also on the program were dissolving views of Don Quixote, a harp performance, and the phantasmagoria. When Pepper on February 26 added a "Wheel of Fortunatus" to "pour its gifts upon the juvenile visitors morning and evening," the press was "at a loss to imagine" what else could be "devised to keep up the popular enthusiasm."[95] Pepper never stopped trying to outdo himself and continued to bring in new entertainments. The Christmas program for 1861 included a lecture on modern magic, the Garibaldi Bell Ringers, a lecture and dissolving views on "Navies, Dockyards, and Iron-Clad War Steamers," a comic pantomime, the phantasmagoria, an infant vocalist, a Russian cat and performing bird, and lectures by Pepper on "The Iron Age" and on the science of the Whitworth and other guns.[96] Pepper was so successful in fashioning the Polytechnic to be a home for his popularizing activities that already, by 1861, he was judged by the *Chemical News* in a review of his *Boy's Playbook of Science* and his *Playbook of Metals* to be the most famous science lecturer of his age. "Perhaps there is no public scientific lecturer whose name is so familiar to the present and rising generations," the *Chemical News* contended, "as that of the author of the above-named works. His presence at the Polytechnic Institution, where most of us have had the opportunity of appreciating the animated style of his delivery, and the cleverness with which he selected the most striking experiments to illustrate his subject, was a tower of strength to that establishment."[97]

---

94. "Royal Polytechnic" 1858, 11.

95. University of Westminster, Archives, Press Cuttings, Book of Press Cuttings Relating to RPI, 1858.

96. University of Westminster, Archives, 1861 Programme for Christmas time entertainments and lectures.

97. "Notices of Books" 1861, 29.

THE GHOST SPECTACLE

Pepper's fame as the public scientific lecturer par excellence received an even greater boost when in 1862 he began to exploit the potential of an optical illusion suggested by the inventor Henry Dircks.[98] "Pepper's Ghost," a surprisingly realistic specter, became the star in countless plays produced at the Polytechnic while Pepper delivered accompanying lectures on the scientific principles of optics and on the shams of spiritualism. Pepper had already been lecturing on "Optical Illusions" in 1856, and on "Remarkable Optical Illusions" in 1857 and saw this as an area that could attract a substantial audience.[99] Just before Christmas Day in 1862, Dircks's invention, vastly improved by Pepper, was used to produce a ghost illusion that stunned a small audience of scientific friends and members of the press previewing a performance of Edward Bulwer-Lytton's "A Strange Story" at the Polytechnic.[100] Pepper prepared a lecture to accompany the play, titled "A Strange Lecture," illustrated by an optical apparatus that produced eerie phantoms.[101] At some point, Pepper began to tell a story in his "Strange Lecture," about a student who sees the apparition of a skeleton late at night and whose sword swings right through it.[102] By February, Pepper had introduced a new lecture, "Burning to Death, and Saving from Death," followed by the still popular ghost scenes from the "Strange Lecture." The "Spectre-Drama" was playing in the morning and the evening, except on Tuesdays and Wednesdays.[103] By Easter, the play had become so popular that it was moved into the larger theater of the Polytechnic where the dissolving views were usually exhibited. "Special written permission" was obtained from Charles Dickens to mount a production of his "The Haunted Man" as a vehicle for exhibiting the ghost illusion and it ran for fifteen months.[104] In the years that followed more plays were mounted featuring the ghost illusion, and they became a regular part of the Polytechnic's program. In December 1864 two new spectral tableaux were presented, "The Indian Widow's Suttee" and "Snow White and Rosy Red."[105] In 1865 the *Times* announced that "Mr. Pepper's

98. Lightman 2007, 119–122.

99. "Royal Polytechnic" 1856, 1612; "Royal Polytechnic Institution" 1857, 35.

100. Pepper 1890, 3.

101. "Polytechnic Institution" 1863, 19.

102. Pepper 1890, 29.

103. "Polytechnic" 1863a, 218.

104. Pepper 1890, 12.

105. "Royal Polytechnic" 1864, 666.

ghost is put to new uses in a dramatic entertainment, devised by Mr. Pepper himself and entitled the 'Poor Author Tested.'"[106]

From 1862 until his resignation from the Polytechnic in 1872, Pepper also devised new illusions that featured spooky apparitions, weird transformations, and mysterious disappearances. Visitors were treated to a rhymed speech from the disembodied head of Socrates, to a Shakespearean soliloquy from the floating heads of Hamlet or Lear, and to a choral song from Joshua Reynolds's cherubs. In 1867, Pepper's daily lecture was titled "The Head of the Decapitated Speaking, and the Eidotrope" (Fig. 4.8).[107] In 1870 the *Times* remarked on Pepper's extraordinary ability to devise new optical illusions. "Were it not notorious that his 'ghosts' are susceptible of infinite improvement," the *Times* reporter declared, "we should say that they had reached their culminating point of perfection in his present illustration of Sir Walter Scott's 'Lay of the Last Minstrel,' in which, by a combination of set scenery and optical illusion, an undoubtedly picturesque approach to the weird and supernatural is attained." In his most recent innovation, "the seeming specters are multiplied, so that while one group remains stationary, two or three reflections of it float around" (Fig. 4.9).[108]

Through his roles as lecturer and manager at the Polytechnic, Pepper's influence was pervasive. His lecturing inspired budding young scientists. Sir Henry Enfield Roscoe recalled hearing Pepper while a student at the high school of the Liverpool Institute during the 1840s. "I shall never forget the impression made upon me by one of Mr. Pepper's lectures," he wrote in his autobiography. Seeing Pepper in action led him to wonder, "shall I ever . . . attain to the position of a scientific lecturer and burn phosphorus in oxygen on a large scale before an admiring audience?"[109] Sir Ambrose Fleming fondly recalled trips to the Polytechnic when he was twelve years old, where Pepper was the "presiding spirit of the place." He remembered thrilling pictures of the Indian Mutiny, an exciting experiment on rendering fabrics noninflammable, the diving bell and diver, and Pepper's lectures on chemistry.[110] Pepper also attracted the attention of Queen Victoria and the royal family. In 1855 the *Illustrated London News* reported that the queen had given the Polytechnic the sum of one hundred pounds "as an acknowledgment of her Majesty's approbation of the various entertainments presented on the recent occasion of the Royal visit to this institution."[111] In May, 1863, when Pepper's "Ghost" was creating a sensation, Pepper presented his ghost

106. "Polytechnic Institution" 1865, 12.
107. "Royal Polytechnic" 1867, 30.
108. "Polytechnic Institution" 1870, 8.

109. Roscoe 1906, 18–19.
110. Fleming [1934], 8–10.
111. "Polytechnic Institution" 1855, 491.

FIGURE 4.8 A poster advertising the attractions at the Royal Polytechnic Institution
sometime in 1866 or 1867. Pepper's optical novelties, the "Christmas Carol" with
the ghost and the speaking decapitated head, are prominently featured. ("The Royal
Polytechnic Institution [1866–67]," London Play Places 4, John Johnson Collection,
Bodleian Library, University of Oxford. Courtesy the Bodleian Library, University of
Oxford.)

lecture to the Prince of Wales and Prince and Princess Louis of Hesse by
royal command. After the performance the special visitors went behind the
scenes "and examined with much interest the machinery and appliances
required to produce the Polytechnic 'ghost.'"[112]

112. "Polytechnic" 1863b, 9.

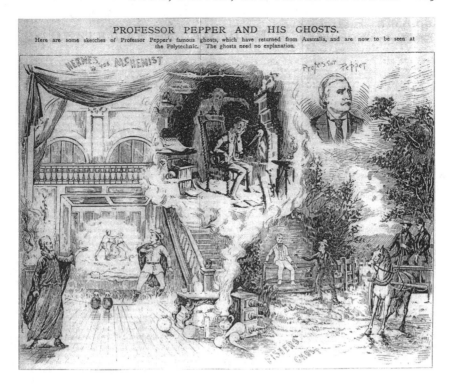

FIGURE 4.9 "Professor Pepper and His Ghosts." Appearing in the *Pall Mall Budget* just after Pepper's return from Australia in 1890, the ghost illusions were so familiar to Victorian audiences that the caption for the illustration asserted that "the ghosts need no explanation." ("Professor Pepper and His Ghosts," *Pall Mall Budget* [January 16, 1890]: 85.)

The Polytechnic's target audience, however, was neither the aristocracy nor the scientific elite. In 1861, when the Polytechnic was almost bought by new owners intending to turn it into a music hall, the *Chemical News* had some advice for the institution's management: improve the "character of the entertainments" by offering "illustrations or expositions of the latest discoveries in science." If this were done, "the Polytechnic would then be to the middle what the Royal Institution is to the upper classes, and would be patronized in as generous a manner."[113] The contrast between the Royal

113. "Miscellaneous. The Polytechnic Institution" 1861, 384.

Institution and the Polytechnic is telling. The missions of the two institutions were quite distinct. Whereas the Royal Institution put its emphasis on research, as well as public lectures, the Polytechnic highlighted teaching and the diffusion of knowledge above all else.[114] Even when the Royal Institution opened its doors for a "public" lecture, during one of the Friday evening discourses first instituted in 1826, the distinguished audience was primarily composed of members and their friends. Since the annual subscription cost five guineas, most members of the Royal Institution belonged to the aristocracy or the upper middle class.[115] In comparison, the Polytechnic was open on a regular basis and to all who could afford the one shilling entrance fee in effect from the time it first opened until it closed in 1881. The one shilling price of admission was cheap enough for skilled laborers in regular employment during the 1850s and 60s, but not for the large proportion of working men who were without a regular and sufficient income.[116]

Shortly after becoming director, Pepper attempted to make the Polytechnic more accessible to workers. He supplied books of tickets to factories allowing admittance to the Polytechnic to workers and their families on Monday evenings for only sixpence each.[117] A comparison of two illustrations from contemporary periodicals points to a change in the social composition of visitors to the Polytechnic. In "the Polytechnic Institution, Easter Week" (1844), men in top hats and women in bonnets from well-to-do families casually examine the exhibits in the Great Hall (Fig. 4.10). In an illustration from 1858 titled "Christmas Holidays at the Polytechnic," however, the audience is more diverse and boisterous (Fig. 4.11). The varied social composition of the Polytechnic's customers during the period when Pepper was working there is apparent in one account of a visit in 1870 by a reporter for *Fun* magazine. He observes among the crowd "the intelligent working men striving hard to master the obstacles which long hours of labour and short wages throw in the way of self-improvement, and the cads and counter-jumpers who try to look like swells and don't succeed; the wondering provincial and his equally astonished chaperone, the family parties

114. Layton 1977, 538. Initially, scientific research was not envisioned by its founders as one of the primary goals of the Royal Institution. But when Humphry Davy was appointed as professor of chemistry in 1802 he established scientific research as an important feature, in addition to making the Royal Institution a popular venue for lecturing (James 2002a, 7–8).

115. James 2002b, 120, 140; Forgan 2002, 31–32.

116. Small 1996, 273.

117. "Patron—H. R. H. Prince Albert" 1854, 945.

FIGURE 4.10 The Polytechnic depicted as the embodiment of rational entertainment for the well-off family. ("The Polytechnic Institution, Easter Week," *Pictorial Times* 13 [1844]: 232.)

and the young lovers."[118] Dependent on entrance fees to ensure financial stability, Pepper's entire operation hinged on attracting large crowds to the Polytechnic. Here he was highly successful. The *Times* reported that "crowds of people, especially of the young" visited the Polytechnic during the Christmas holidays in 1858. On Boxing Day alone, it was estimated that five thousand people were admitted to the Polytechnic.[119] Later, after Pepper enlisted the aid of spirits, the Polytechnic drew sizeable crowds on a regular basis, not just on holidays. On May 26, 1866, an advertisement in the *Illustrated London News* claimed that 109,000 visitors had already seen Pepper's illusion.[120] One of the magic lantern operators, T. C. Hepworth, estimated that during the time of the ghost illusion visitors came "at the rate of two

118. Our Special Sightseer 1870, 223.

119. "Fatal Accident at the Polytechnic Institution" 1859, 12.

120. "Royal Polytechnic" 1866, 511.

FIGURE 4.11 A crowd gathers around the electric machine, amused by the frustration on one patron's face as he interacts with the exhibit. ("Christmas Holidays at the Polytechnic," *Illustrated London News* [December 1858]: 607.)

thousand per diem."[121] In 1860 in his *Boy's Playbook of Science*, Pepper acknowledged that the "'South Kensington Museum' now takes the lead, and surpasses all former scientific institutions by its vastly superior collection of models and works of art."[122] But in its heyday under Pepper, especially after the introduction of the famous ghost illusion, the Polytechnic was one of the favorite destinations in London for those Victorians who wanted to visit an institution of science.[123] It was one of the few genuinely "public"

121. Hepworth 1978, viii.

122. Pepper 2003, 1.

123. Estimating the number of visitors at the Polytechnic is difficult. If Hepworth's two thousand per day claim is to be trusted, then at its peak the Polytechnic attracted approximately one hundred thousand visitors a year. A comparison to attendance at other museums may help to gauge Pepper's success. Alberti has noted that large public museums, which were not specifically focused on science, attracted hundreds of thousands per annum. The Sheffield Public Museum was visited by 350,000 in its first year, while half a million people came to the British Museum (Alberti 2007). Given its size, the Polytechnic seems to have been able to compete with the large public museums.

scientific institutions. As one of the premier science institutions of the mid-nineteenth century in Britain, it introduced countless Victorians to science through Pepper's theatrical spectacles.

PUTTING THE POLYTECHNIC ON THE PAGE

What, exactly, did Pepper say to the crowds who flocked to the Polytechnic? To answer this question we must turn to Pepper's published works, some of which were based on his Polytechnic presentations, as periodical accounts rarely described the contents of his lectures in detail. Although publishing projects were secondary to lecturing in Pepper's career—in contrast to Wood, whose lecturing supplemented his writing—his books were remembered fondly by eminent scientists in their memoirs, in particular his *Playbook of Science*. The *Playbook* was filled with hundreds of experiments and demonstrations, designed for older boys, from the physical sciences, including mechanics, pneumatics, optics, heat, electricity, magnetism, chemistry, and astronomy. It stood within a long tradition of presenting experimental physics as based on lecture demonstration and public display, exemplified in Tom Telescope's *Newtonian System of Philosophy* (1761), Jeremiah Joyce's *Scientific Dialogues* (1800–1803), and John Ayrton Paris's *Philosophy in Sport Made Science in Earnest* (1827). James Secord asserts that the *Playbook* "became the most widely read introduction to physics and chemistry for young people in the mid and late Victorian era."[124] Sir Ambrose Fleming wrote that *The Playbook of Science* was a "much-read volume in my private boyish library."[125] H. E. Armstrong recalled that he had "feasted" upon Pepper's *Playbook of Science* "in its original form, a far better book than any now available for boys."[126]

Like Wood, Pepper worked closely with Routledge, especially when he first began to write science books. Routledge had both Wood and Pepper in his stable of writers in the late 1850s and early 1860s. Wood was still contributing to the Common Objects series when Pepper's *The Playbook of Science* first appeared in September 1859, priced by Routledge at six shillings. It sold well, though not up to the mark established by Wood's *Common Objects of the Sea Shore*. Routledge's first edition of 3,000 was quickly followed in January of 1860 with a second edition of 3,000. By 1869 the book was in its seventh edition and Routledge had printed 16,000 copies. A total of 34,000

124. J. Secord 2003a, v–vi.   126. Armstrong 1973, 61.
125. Fleming [1934], 11.

copies were printed by 1893.[127] Pepper also chose to work with Routledge on two other books, *The Playbook of Metals* and *Scientific Amusements for Young People*, both published in 1861, and both intended to capitalize on the success of *The Playbook of Science*. But for his *Cyclopaedic Science Simplified* (1869), he chose to "jump ship" to Frederick Warne and Company. Pepper would have become acquainted with Warne through Routledge, as both Warne and his brother William worked for Routledge when they first entered the bookselling trade. When Routledge moved into publishing, William (in 1848) and then Frederick (in 1851) became partners in the firm. Although Frederick continued as a partner with Routledge after his brother's death in 1859, he eventually, in 1865, started his own business and immediately began an ambitious program of expansion that lasted almost to the end of the century. Warne's policy of offering wholesome, family entertainment at popular prices, and in particular his attempt earn a reputation as a leading publisher of children's books, made his publishing house a good fit with a popularizer like Pepper who had already written two books for children. Frederick Warne and Company later received a boost by publishing Beatrix Potter's *The Tale of Peter Rabbit* in 1902, as well as her later works that drew on her knowledge of plant and animal life (Golden 1991, 327-33).

Pepper clearly viewed his book projects as extensions of his work at the Polytechnic. In fact, at times, they seem to be nothing more than thinly veiled advertisements for his home institution as well as blatant self-promotion. In his *Playbook of Science*, for example, Pepper took credit for introducing the colorful "Illuminated Cascade" into the Polytechnic's exhibits. "The total internal reflection of light by a column of water," Pepper declared, "is an experiment that admits of great variety as far as colour is concerned, and is one of the most novel and beautiful experiments with light presented to the public within the last few years." Pepper later bragged that at the Polytechnic he had six magic lanterns working at the same time to produce effects in the views illustrating the voyages of Sinbad.[128] In his *Cyclopaedic Science Simplified*, Pepper referred time and again to the scientific instruments being used at the Polytechnic to illustrate points about light, electricity, heat, magnetism, pneumatics, acoustics, and chemistry. Among the Polytechnic apparatus discussed (frequently accompanied by an illustration) were the equipment for the ghost illusion, the phenakistoscope, the Photodrome,

---

127. *Archives of George Routledge & Co., 1853-1902*, 1973, vol. 3, 559; vol. 5, 430; vol. 6, 651.

128. Pepper 2003, 293, 305.

FIGURE 4.12 The Hydro-Electric Machine at the Polytechnic in *Cyclopaedic Science Simplified,* one of the many scientific instruments in the book that represented the important role of Pepper's institution in scientific research. (J. H. Pepper, *Cyclopaedic Science Simplified* [London: Frederich Warne, 1869], 273.)

metallic reflectors, the Plate Electrical Machine, air pumps, and the diving bell. At one point in his discussion of electricity he coyly told the reader that he "abstains from saying anything about a new gigantic coil" being built for the Polytechnic, since it was unproven, but even the brief mention of the "Mammoth Induction Coil" was intended to draw attention to the Polytechnic's future attraction. After a lengthy discussion of the experiments carried out with the hydroelectric machine at the Polytechnic, Pepper declared, almost apologetically, that he was "fearful that our readers may think the writer too prone to talk of Polytechnic doings" (Fig. 4.12).[129]

In addition to publicizing the spectacular Polytechnic, Pepper had good reason to feature its experiments and instruments in his book: he wanted

129. Pepper 1869, 24, 73, 85, 197, 223, 273, 388, 441, 443.

the Polytechnic to be perceived as being more than a science museum where customers came to be entertained. In a discussion of an apparatus used in experiments on polarizing light, he reminded his readers that one of the Polytechnic's employees, J. F. Goddard, had received a silver medal from the Society of Arts for this instrument. Moreover, Pepper insisted that the public demonstrations that were repeated over and over again for the amusement of the public were in reality serious scientific experiments producing knowledge of natural processes. An exhibit in which miniature torpedoes were used to blow model ships in the air was an "experiment," although it was "performed so frequently at the Royal Polytechnic." After reading *Cyclopaedic Science Simplified*, readers came away with the impression that the Polytechnic was at the forefront of scientific research. The periodical press adopted this view. When the Giant Induction Coil was unveiled at the Polytechnic in 1869, both the *Illustrated London News* and *All Year Round* were convinced that the new apparatus would lead to an increase of knowledge.[130]

Given that Pepper's books were designed to promote the Polytechnic as an important scientific institution, it is not surprising to find in them some of the themes and strategies in his lectures. The ghost illusion is used to explain the phenomenon of reflection in *Cyclopaedic Science Simplified*, while a discussion of optical illusions provides Pepper with the opportunity to expose charlatans and wizards.[131] The dramatic spectacles performed at the Polytechnic are matched by Pepper's inclusion of a multitude of visual images in his books. *Cyclopaedic Science Simplified* featured six hundred, *The Playbook of Metals* contained three hundred, and *Scientific Amusements for Young People* included one hundred illustrations. Pepper's desire to instruct and entertain a diverse audience within the walls of the Polytechnic is paralleled by his attempts to write for both adults and children. *Cyclopaedic Science* is pitched at the level of an adult with little scientific knowledge, while the *Boy's Playbook of Science*, *Popular Lectures for Young People*, and *Scientific Amusements for Young People* are geared toward children. In the last book, Pepper asserted that "the primary object of the very elementary scientific articles in this work is to excite *curiosity* in the youthful mind so as to induce a desire to read and study more extended and complete works on the various subjects to which they relate."[132]

130. Ibid., 114, 344, 388, 441, 443; "The Great Induction Coil at the Polytechnic Institution" 1969, 402; "Playing with Lightning" 1869, 620.

131. Pepper 1869, 31; 2003, 277.

132. Pepper 1861b, 77.

Pepper's lectures and his books also shared in common a deferential attitude toward elite scientists. When he was manager, the reputation of the Polytechnic depended on Pepper's status within the scientific community. Since the title "professor" was conferred upon him by the Polytechnic board, rather than a university, his claim to that designation was always somewhat suspect. Like Wood, he was careful not to antagonize eminent scientists, especially those who worked at important scientific institutions. Pepper cultivated a relationship with Faraday while the latter was still alive. When Faraday told Pepper that he did not understand how the ghost illusion worked, Pepper took him behind the scenes and placed his hand on one of the concealed glass plates essential to the deception. "Ah," Faraday remarked, "now I comprehend it; but your glasses are kept so well protected I could not see them even behind your scenes."[133] Shortly after Faraday's death in 1867, Pepper began to speak about his legacy. According to the *Times*, Pepper was lecturing on Faraday's achievements in 1867 at a *conversazione* held at the Polytechnic.[134] In his *Cyclopaedic Science Simplified*, he included a lengthy chapter on Faraday's researches in the electricity section, and then began the chemistry section with a heroic account of his life.[135]

Tyndall, Faraday's successor at the Royal Institution, was also treated favorably in Pepper's lectures and his books. If imitation is the sincerest form of flattery, then Pepper was "buttering up" Tyndall with a vengeance. In his 1862 Spring Lecture Tyndall lit a cigar at the invisible focus of a beam of infrared radiation.[136] In 1866 the *Times* reported that Pepper's new lecture on "Combustion by Invisible Rays" featured the ignition of combustibles, including a cigar, by invisible rays. Pepper also cooked a piece of meat by placing it at the focus of a large, concave, metallic reflector, one hundred feet across the main hall of the Polytechnic from a second reflector with a charcoal fire at its focus. An approving account of Tyndall's apparatus to illustrate the heating power of invisible rays appeared in *Cyclopaedic Science Simplified* with an illustration of the cooking meat.[137] In 1871 Pepper was lecturing on glaciers and on the theory of heat considered as a mode of motion, two topics that Tyndall had written about earlier in *The Glaciers of the*

---

133. Pepper 1890, 35.

134. "Polytechnic Museum" 1867, 6.

135. Pepper 1869, 315, 527.

136. McMillan and Meehan 1980, 49.

137. "Royal Polytechnic" 1866, 1; Pepper 1869, 197, 205.

Alps (1860) and *Heat Considered as a Mode of Motion* (1863).[138] Although Pepper dealt mainly with the physical sciences, he managed to include some quotes from Charles Darwin, the "eminent naturalist," concerning the theory of evolution.[139]

Despite his admiration for Tyndall and Darwin, and his heated opposition to spiritualism, which he shared with Huxley, Pepper's attitude toward evolutionary naturalism was somewhat complicated. At some point in his life, Pepper converted to Roman Catholicism, a decision normally fraught with serious consequences in Protestant England, and usually an indication of strong religious convictions.[140] According to one contemporary, Edmund Wilkie, Pepper was a sincere Christian who "never let slip an opportunity of impressing upon his hearers that the man of science by endeavoring to penetrate deeply into the hidden secrets of nature was guilty of no irreverence, and that the idea that science and unbelief go hand in hand was totally devoid of foundation." Wilkie claimed that Pepper ended his astronomical lectures with arms and eyes raised, and with a quotation from the Psalms, "The Heavens declare the Glory of God, and the firmament showeth His handiwork."[141] Pepper referred to God in his published writings, though even more sparingly than Wood. In the *Playbook of Metals* he pointed out to his readers that coal was the remains of vast forests "which have lived, died, and have been entombed by the all-merciful hand of Providence thousands of years ago, and are now being exhumed for our benefit, to give us health, in the shape of cheap warmth, and wealth more abundant than the imaginary contents of Aladdin's cave."[142] In the *Boy's Playbook of Science*, Pepper again presented the idea that "the earth has been wonderfully prepared for God's highest work—Man," but this time in reference to the expansion of water by cold. The "supreme wisdom" of God's preparation of the earth for humanity was apparent "in the fact that water offers the only known exception to the law 'that bodies expand by heat and contract by cold.'" But Pepper used discretion when it came to expressions of faith in his books, as he deemed it inappropriate for scientific writers to speak about the relationship between God and His creation. He believed that the perfection of nature should "check the most eloquent speaker or brilliant writer who attempted to offer in appropriate language, the praises due to that first great

---

138. "Polytechnic" 1871, 9.
139. Pepper 1861a, 9.
140. Boase 1965c, 386.

141. Wilkie n.d., 74.
142. Pepper 1861a, 5.

creation of the Almighty, when the Spirit of God moved upon the face of the waters and said, 'let there be light.'"[143]

However, like Wood, Pepper worked to undermine the hardening boundaries between the practitioner and the layman. He believed that scientific knowledge could be made accessible to all and that an expert elite should not control it. In 1865 Pepper published a play titled *The Diamond Maker; or, the Alchymists's Daughter* (M'Gowan and Danks). He is credited on the title page for being responsible for "the Plot, the Scenic, the Optical, and Other Scientific Arrangements," while "Phransonbel" is listed as the author of the dialogue. Described as a "Romantic Drama, in Three Acts," the play provides a showcase for Pepper's illusions, including a skeletal ghost. It is unclear whether or not the play was ever performed on a London stage. Pepper had allowed the ghost to haunt other London theaters. The Haymarket, the Britannia, and Drury Lane were among those theaters that took out licenses to use the illusion.[144] Perhaps Pepper had hoped to mount his own London production. But "The Diamond Maker" was more than just a vehicle for Pepper's optical illusions.

In this play Pepper explored the complex relationship between scientific discovery and public knowledge. The play is set in the indeterminate past in one of the dukedoms in Germany. Marie, the daughter of an alchemist, and the duke have fallen in love. She resolves to end the romance, since her father believes that the duke is an evil tyrant. The alchemist, however, has a secret. He has used his power to create a diamond. But concealing his discovery has alienated him from others, who, he is afraid, will want his knowledge. Having "tracked science up her winding paths" the alchemist looks down "from their dizzy heights on all the petty aims, the low ambitions, the frauds, the ignoble strife that men call living." The alchemist fears the duke most of all, as the combination of political power with alchemical knowledge would give him the means to extend his tyranny. So his secret must die with him:

> Blind insolence sits throned in palaces,
> While knowledge crouches waiting for the spring.

When a new age comes ("the spring") the world will be ready for the knowledge he possesses.[145] But the duke fears the alchemist's powers and tries, unsuccessfully, to have him killed, not knowing that he is Marie's father. The duke is later injured during a hunt and Marie goes to him, thinking that

143. Pepper 2003, 255, 364.   145. Pepper and Phransonbel 1865, 28.
144. Pepper 1890, 30.

he is dying. They marry and the alchemist's heart is broken, though he is reconciled to his daughter just before he dies. Pepper's alchemist is a tragic figure. By hiding his knowledge he cuts himself off from the human race. He cannot understand the love that moves Marie to run to the duke after he is hurt. He misinterprets the intentions of the duke, who is not the cruel tyrant he imagines. Although Pepper is condemning ancient alchemy, the play can be read as a criticism of any group or individual attempting to monopolize science. Knowledge should not be hidden; rather it should be made accessible to the public, as it was in the Polytechnic. "The Diamond Maker" hints at tensions between Pepper's commitment to the accessibility of science and the scientific naturalist's professionalizing agenda. Whereas Pepper insisted that knowledge was produced in the public space of the Polytechnic, scientific naturalists asserted that discovery took place largely in the private space of the laboratory. Elitist scientific naturalists sometimes behaved as if they were alchemists, jealously guarding their secrets so that they could maintain the sharp distinction between expert and layman.

REMEMBERING THE SHOWMEN

H. G. Wells once asked his readers the question, what was the "greatest influence upon your mental growth, what book, what teacher, what experiences?" Wells then considered how to respond to that question in his own case, admitting that it could be answered in thousands of ways. Scholars have generally pointed to T. H. Huxley, Wells's teacher at the Imperial College, as the greatest influence in his life.[146] However, in this introduction to a book on natural history, Wells identified Wood as his most important teacher, and pointed to the reading of one of Wood's books as having had the greatest influence on his intellectual development. As a child age seven, Wells was laid up in bed with a broken leg when he became absorbed in Wood's *Natural History*. He recalled, "it does seem to me that the day I opened that once popular favourite, Wood's *Natural History*, was in its way exceptional, and that I did come upon something then that made a distinctive, new and fruitful beginning in my thoughts."[147]

The visual images sparked his youthful imagination. His mind was "born anew" as he "turned over the pages, looking at the plentiful woodcuts, reading here and there, and a multitude of new creatures, in groups and

---

146. Mackenzie and Mackenzie 1973, 57; Foot 1996, 16; Hammond 2001, 31; D. Smith 1986, 11.
147. H. Wells 1937, xv.

families and regiments, came marching into my imagination." In comparison to other natural histories, which presented the natural world as one of disordered wonders and marvels, Wood's illustrations revealed an "orderly arrangement of living things." Although Wood offered a theology of nature, Wells perceived the *Natural History* to be in step with the most recent scientific discoveries. To the youthful Wells, it marked "the progress of the better half of a century." The book included Australian animals as well as the gorilla, which scared him so much that "for a time I went upstairs to bed in grievous fear of meeting it." One scholar has suggested that the picture of the gorilla in Wood's *Natural History* left such an impression on Wells that it was the inspiration for the Morlocks, the underground troglodytes of his *Time Machine*. Images of the gorilla and the notion of livings things bound up "in these interesting clusters, these classes and orders" led Wells to think about evolution, although Wood never used the term in his book.[148]

While the vivid illustrations stood out when Wells remembered the powerful impact of Wood's *Natural History* on his young mind, recollections of the Royal Polytechnic Institution often focused on the spectacular exhibits and demonstrations. One visitor in 1870, a journalist who wrote for *Fun*, was "swept away by the magic of remembrance" just before entering the Polytechnic. He gazed on the building "which contains those wondrous triumphs of science and ingenuity, the diving-bell and the Zoëtrope, the dissolving views and the doubling up perambulator." As he looked, "visions long since hidden in the past rise before me" and he was again "a young boy—being led by one now long departed, to be electrified, horrified, and charmed." Overwhelmed by nostalgia, the visitor admits that he almost shed a tear. After composing himself, he walks into the building and recovers his humorous demeanor. When he is directed to the minor theater, "where I was told I could see Pepper's ghost," he becomes annoyed as he had "ventured in for the purpose of seeing and hearing the Professor in the flesh." The journalist is impressed by Pepper's lecture on the Franco-Prussian war, though he complained that Pepper "goes into ecstasies a little too often." He enthusiastically praised the panoramic views of the Rhine and the battlefields that accompanied the lecture. The Polytechnic is described as a visual, and aural, banquet. After having "feasted my eyes and ears till I could no more, I went away . . . very much pleased with myself and with the Polytechnic."[149]

148. Ibid., xv–xvi; Geduld 1987, 2.   149. Our Special Sightseer 1870, 223.

The emphasis on visual spectacle served Wood and Pepper well. Gigantic freehand sketches became the hallmark of Wood's reputation as a popular lecturer. Vivid illustrations helped Wood become one of the most successful popularizers of science in terms of book sales. They also facilitated Wood's attempts to make natural history more accessible to Victorian audiences, and they enhanced his presentation of a theology of nature. Pepper's use of theatrical spectacle, in particular his optical illusions, ensured that the Polytechnic survived well into the post–Crystal Palace era. Visitors flocked to the Polytechnic to see his ghost illusion. Using the Polytechnic as his base of operations, Pepper established himself as one of the foremost science lecturers of the period. When he resigned after twenty-four years at the Polytechnic, he took his optical illusions to Canada, the United States, and Australia. He "peppered" his books with illustrations of Polytechnic instruments and their spectacular effects, ensuring that they would be attractive to potential readers. Capitalizing on the Victorians' hunger for spectacular visual images, Wood and Pepper were among the first to build careers for themselves as popularizers in the competitive marketplace of science.

———— ✳ ————

CHAPTER FIVE

The Evolution of the Evolutionary Epic

IN THE preface to his *The Evolutionist at Large* (1881, Chatto and Windus), Grant Allen raised the issue of how to communicate "the general principles and methods of evolutionists" to "unscientific readers." Since "ordinary people care little" for "minute anatomical and physiological details," the focus on "underlying points of structure" in biological works had to be avoided. The reading public "cannot be expected to interest themselves in the *flexor pollicis longus*, or the *hippocampus major* about whose very existence they are ignorant, and whose names suggest to them nothing but unpleasant ideas." Instead, Allen believed that his readers would be most interested in "how the outward and visible forms of plants and animals were produced" through the process of evolution. Allen's plan in this book was therefore to take a simple and familiar natural object "and give such an explanation as evolutionary principles afford of its most striking external features." Just as Wood had focused on common objects of the seashore or the country, Allen presented his readers with short evolutionary studies of everyday things, such as a strawberry, a snail shell, a tadpole, a wayside flower, or even a walnut.[1]

The story of the origin of walnuts is typical of how Allen dealt with evolutionary themes. He began the essay by pointing to the importance of examining the evolutionary process at work in the commonest objects. Allen acknowledged that Darwin had had to write two bulky volumes on human evolution in his *Descent of Man*, since "we would have refused to listen to him had he given us two volumes instead on the Descent of Walnuts." But

---

1. G. Allen 1881a, vii–viii.

viewed "as a question merely of biological sciences," Allen insisted, "the one subject is just as important as the other." An understanding of evolution was possible through the study of virtually anything, no matter how mundane. The origin and development of walnuts for example, was "a subject upon which we may profitably reflect, not wholly without gratification and inter-est." Allen's account of the walnut stressed the different survival strategies of fruits and nuts. Whereas fruits wanted to be eaten so that their seeds would be dispersed, nuts had developed a hard shell so that their seeds would not be eaten. This led Allen to some reflections on the evolutionary process writ large. Nature was a "continuous game of cross-purposes," where animals perpetually outwitted plants and plants in return continually attempted to outwit animals. "Or," Allen declared, "to chop the metaphor, those animals alone survive which manage to get a living in spite of the protections adopted by plants; and those plants alone survive whose peculiarities happen suc-cessfully to defy the attack of animals. There you have the Darwinian Iliad in a nutshell."[2] In this essay Allen skillfully drew the entire evolutionary epic—the "Darwinian Iliad"—out of something as small and common as a nut.

Allen was not the only popularizer who adapted the evolutionary epic as a vehicle for conveying contemporary scientific ideas to a general reading audience. The evolutionary epic became one of the most important narrative formats in the second half of the nineteenth century. It derived its scientific legitimacy from the concept of evolution as a gradual, lawful, and progres-sive development in the natural world. It assumed epic status by moving through vast expanses of time, by ranging across a series of scientific dis-ciplines, or even by presenting heroes who performed deeds of great valor. The evolutionary epic proved to be a versatile genre for popularizers. Not only did it offer a gripping cosmic story to readers, it could also provide a grand synthesis of scientific knowledge that seemed to surpass Somerville's synoptic overviews of the physical sciences, though the emphasis was really on biology and geology. Moving forward in time from the creation of the earth, as set forth in the nebular hypothesis, to the evolution of higher life forms, provided a superb organizing principle for conveying masses of scientific information to diverse audiences. The evolutionary epic allowed popularizers to range across astronomy, geology, biology, and many other sciences. The idea of evolution became the key to finding the connections between the various branches of scientific knowledge. Evolutionary stories

2. Ibid., 161–62, 171.

that were cosmic in scale, then, drew on the association of "epic" with vastness of scope across time or subject matter, or across both. But some popularizers drew on the notion of "epic" connected with literary traditions. As a narrative format, the evolutionary epic provided the opportunity to create evolutionary heroes who could capture the imagination of readers, as had Odysseus and Achilles in Homer's classical epics. Both forward-looking evolutionary theorists, locked in a battle with religious bigotry, and clever and courageous animals, fighting with their natural enemies and a hostile environment, could be portrayed as heroes engaged in an epic war.

Robert Chambers, the anonymous author of the *Vestiges of the Natural History of Creation* (1844, John Churchill), created the evolutionary epic in its modern form. Chambers took the monad-to-human style cosmic evolutionary narrative, which had its origins in Lucretius, and combined it with the new sciences of the nineteenth century for the first time. In his *Victorian Sensation,* James Secord has referred to the *Vestiges* as a hybrid or generic monster that shared with other new genres of "reflective science" the theme of progress. To create this monstrous new epic, Chambers drew on the narratives of progress to be found in astronomy, based on sanitized versions of Pierre-Simon Laplace's nebular hypothesis; in geology, as expressed in William Buckland's Bridgewater Treatise, *Geology and Mineralogy* (1836), and Gideon Mantell's *Wonders of Geology* (1838); and in biology, as found in French naturalists like Jean-Baptiste Lamarck. Chambers, Secord asserts, reshaped the evolutionary cosmology of the Enlightenment by adopting the epic conventions of historical fiction, Sir Walter Scott in particular, in order to domesticate what had been seen as a dangerous scientific theory. Secord argues that *Vestiges'* most important influence over the long term was to "provide a template for the evolutionary epic-book-length works that covered all the sciences in a progressive synthesis."[3]

The fascination of Victorian reading audiences with the evolutionary epic was linked to the popularity of spectacles in this period and therefore to the activity of showmen of science like Wood and Pepper. A rhetoric of spectacular display was built into the *Vestiges,* although there were no illustrations. The *Vestiges* offered readers a guided tour through a museum of creation. Other authors presented a tour of creation, such as the Rev. Thomas Milner in his *Gallery of Nature* (1846), likely the best-selling scientific part work of the decade. Milner did include striking illustrations, though

3. J. Secord 2000, 41–42, 56–57, 59, 90, 461.

he did not tell a story of development like Chambers.[4] The link between the *Vestiges* and Victorian visual culture can be taken one step further, by comparing the evolutionary epic to the panorama. Both attempted to provide the viewer with a privileged bird's-eye perspective of a scene of grandeur. In many ways, the evolutionary epic can be seen as the literary twin of the panorama, and this perhaps explains why the narrative format initiated by Chambers was so influential in the second half of the nineteenth century. It resonated with important developments within nineteenth-century visual culture, especially when evolution was put on display or when it supplied the guiding theme in a heavily illustrated book.

Besides Allen, David Page, Arabella Buckley, and Edward Clodd were among the most important exponents of the evolutionary epic in the second half of the nineteenth century.[5] They were prolific and widely read. Many members of the Victorian reading audience encountered Lamarck, Chambers, Darwin, and Spencer secondhand through reading the works of these authors. They constituted a group of influential popularizers of science distinct from the Anglican clergymen, the women of the modified maternal tradition, and the showmen of science. As a fourth group of popularizers, they shared some of the aims of evolutionary naturalists such as Huxley and Tyndall. But they were by no means slavish imitators or uncritical disciples. For Page and Buckley, their adaptation of the evolutionary epic provided a means for subverting the secularizing goals of Huxley and his allies. Although Clodd and Allen admired Huxley, Spencer, and Darwin, and adopted their more secular approach to evolution, they maintained their independence. These four popularizers represented the two major interpretations of the evolutionary epic, the religious and the secular. Although Darwin is important for the construction of the evolutionary epic after 1859, as might be expected, surprisingly, he was not the decisive influence. None of these four popularizers looked to Darwin as the chief inspiration behind their evolutionary epic. For Page, Darwin veered too much in the direction of materialism. Buckley looked to Wallace to understand the spiritual dimensions of evolution. Clodd affirmed that he was more indebted to Huxley and

---

4. Ibid., 439–40, 463.

5. None used the phrase "evolutionary epic" to describe the narrative format of their works. One of the earliest uses of the term was in *Naturalism and Agnosticism* (1899) by the psychologist James Ward. He used it to describe Spencer's account of evolution in his *First Principles* (Ward 1906, 2: 269). The term appeared more frequently in the twentieth century when a variety of intellectuals used it to convey the grand scope of the evolutionary process.

Tylor for his views on evolution, while Allen was enamored with Spencer. Allen asserted that while Darwin's work had been restricted to the biological realm, Spencer had been the first to bring all of the sciences together in an evolutionary synthesis. For these popularizers, Darwin's name was not synonymous with evolution, and they portrayed him as only one of a number of individuals who had contributed to the making of the evolutionary epic in its modern incarnation.

DAVID PAGE: THE EVOLUTIONARY EPIC OF COSMIC DESIGN

In the final chapter of his *The Earth's Crust* (1864, 6th ed., 1872), titled "General Inferences as to World-History," David Page (1814–1879) tried to sum up the broader implications of his outline of geology. Although geology was a young science, and "imperfect as are many of her methods, she has established the fact that the crust of this earth is undergoing incessant change." Geology did not merely reveal "a panorama of incessant change," it also disclosed "a progress from lower to higher—each ascending stage in time being characterized by higher forms of life, built up after a uniform plan." Geology threw no light on the origin of life, yet it could trace "the course of creation, and form a knowledge of that course to indicate the secondary causes employed by the Creator for its regulation and development." Page's aim as popularizer was to present an evolutionary epic that illustrated, in spectacular fashion, the design in geological processes.[6] Although Page used visual illustrations sparingly, he followed in the footsteps of geological authors such as Hugh Miller and Thomas Hawkins, who used language to spark the visual imaginations of their readers and to evoke a sense of the sublime in panorama-like scenes of the earth's past.[7] In the case of Page, the evolutionary epic was used to subvert the secular meaning that scientific naturalists had found in Darwinian evolution.

Page's interpretation of evolution was read by thousands of readers. When he died in 1879, Henry Clifton Sorby, then president of the Geological Society of London, described him as "being one of the earliest, and throughout his life one of the most successful, popularizers of geological ideas."[8] The author of the obituary in *Nature* similarly stated, "few names

---

6. Page 1868, 118–19.
7. O'Connor 2002; 2003.

8. Sorby 1879–80, 40.

FIGURE 5.1 David Page, popularizer of geology. (Picture Library, National Museums of Scotland. Courtesy of The Trustees of the National Museums of Scotland. Courtesy the Picture Library of the National Museum of Scotland.)

have been more familiar to general readers in geology than that of this practiced writer."[9] Page spent much of his life carving out a career as a freelance science writer and lecturer (Fig. 5.1). Page's father, a mason and builder, sent him at age fourteen to study for the ministry in the Church of Scotland at the University of St. Andrews. Although he did not pursue a clerical position, he remained an adherent of the Moderate party for his entire life. The Edinburgh publisher W. and R. Chambers employed him between 1843 and 1851, and while there he produced a series of geological primers for the firm and functioned as editorial assistant on *Chambers's Journal*. Page must have been considered a valuable and loyal employee, as he was one of the few people

9. "David Page" 1879, 444.

entrusted with the secret that Robert Chambers was the author of the *Vestiges*. After the Chambers brothers refused to make him a partner, Page quit the firm and attempted to out Robert as the anonymous author after giving a lecture in 1854. James Secord refers to this as the "most serious attempt to break the anonymity." After leaving W. and R. Chambers, Page wrote over fifteen books on geology and physical geography published under his own name that were geared toward a broad readership. William Blackwood and Sons published many of them. The trajectory of Page's career was unusual. In 1863 he was elected president of the Geological Society of Edinburgh, a post he occupied until 1868. Then, in 1871, he was appointed professor of geology in the Durham University College of Physical Science in Newcastle, an indication of how fluid the lines were, even as late as the 1870s, between popularizer and practitioner.[10]

Page's evolutionary epic was conveyed to audiences in his lectures, in textbooks, and in books. According to contemporaries, Page was a skilled and dynamic lecturer. Ralph Richardson, honorary secretary of the Geological Society of Edinburgh at the time of Page's death, owed his introduction to geology to Page. He attended Page's geological class at Edinburgh University in the summer of 1866. Richardson asserted that "his lectures were at once graphic and concise; his systematic mind permitted no chaos; everything was lucid and well arranged." Richardson also vividly remembered two of Page's lectures at *conversazioni* held by the Edinburgh Geological Society in the Museum of Science and Art. On December 6, 1866, Page lectured to 1,700 ladies and gentlemen on "Geological Life Periods," illustrated by diagrams illuminated by the oxyhydrogen light. Page used a similar method of illustration at the second *conversazione* on January 23, 1868, when a large crowd listened to him speak on "Ice Action."[11]

Page's textbooks were also popular. His *Introductory Text-Book of Geology* (1854, Blackwood), which reached a sixth edition within a decade, and a twelfth edition in 1888, aimed, as Page put it, to "furnish an elementary outline of the science of Geology." Page claimed that in preparing this book "the utmost care has been taken to present a simple but accurate view of the subject, to lead the learner from things familiar to facts less obvious, and from a knowledge of facts to the consideration of the laws by which they are governed." Page believed that by adopting such a method he transformed

---

10. Ford 2004, 1521–23; J. Secord 2000, 368, 395–97; 2004f, 322–23.

11. Richardson 1871, 220.

geology from a "dry accumulation of facts" into "one of the most attractive departments of natural science." He began by defining the object of geology, and then moved to discussions of geological agencies; the general arrangement, structure, and composition of the materials that composed the earth's crust; a classification of those materials; and then a series of chapters on the different stratified systems from the oldest to the newest. In his final chapter, Page offered a review of these systems and the "general deductions" to be drawn, the chief of which was evidence of progress from "humbler to more highly organized forms" in each system.[12] Page's *Advanced Text-Book of Geology* (1856, Blackwood), intended as a "sequel" to the *Introductory Text-Book of Geology,* reached a sixth edition by 1876 and was customized for "senior pupils" and those who wanted to study geological subjects in more detail. The main object of the book was to help the student become a "practical observer" in the field and to enable them "to read with appreciation the higher treatises, special monographs, papers, and new discoveries of others." The advanced textbook dealt in more detail with the economic aspects of geology and put greater emphasis on such themes as the importance of observation, the danger of using conjecture, and the problems with catastrophism.[13] Producing textbooks must have been a profitable business for Page. In addition to the two textbooks in geology, Page also wrote the *Introductory Text-Book of Physical Geography* (1863, Blackwood), which reached twelve editions by 1888, and the *Advanced Text-Book of Physical Geography* (1864, Blackwood), which passed through three editions by 1883.

Page's science writing activities were by no means limited to textbooks. In books aimed at a broad reading audience, such as his *Geology for General Readers* (1866, Blackwood), he attempted to present "a simple and familiar exposition of its [geological] leading truths and principles." Reaching a twelfth edition in 1888, *Geology for General Readers* most likely exceeded print runs of 10,000 copies (as did two of his textbooks). Page gave notice to the reader that he was "discarding technicalities as much as possible and avoiding the formality of a text-book." He took the time to explain basic terms by using analogies to familiar, everyday objects. For example, he referred to the earth's exterior as crust "much in the same way as the housewife speaks of the crust of her loaf, or the schoolboy of the crust of ice that forms on the stagnant pool during the frosts of winter."[14] In his *Chips and Chapters* (1869, Blackwood), Page declared that he saw no reason why the principles

12. Page 1854, 3, 136.
13. Page 1856, iii, v.

14. Page 1870, v–vi, 17.

of geology, "with slight modification," could not be "made attractive to the many" who had "no need, and indeed no time, for systematic training in science."[15]

His *Past and Present Life of the Globe* (1861, Blackwood), which presented a "sketch in outline of the World's Life-System," aimed to link the "remote to the recent—the living to the extinct—that the general reader may be enabled to form some intelligible conception of the whole as a great and continuously evolving scheme of vegetable and animal existences." By stressing the big picture, rather than attempting to "teach anatomical details or point out specific distinctions," Page made geology and paleontology appealing to his readers.[16] Similarly, in the *Earth's Crust*, Page explained, "twenty years' experience in lecturing to miscellaneous audiences has convinced me that what is primarily required for the diffusion of knowledge is less a full and systematic explanation than a pleasant and perspicuous outline." In the preface to this book Page identified the readers for whom he wrote. The audience included "young men striving after self-instruction," "men in business," the "leisurely, who seek information simply as an accomplishment," and "the gentler sex, unprepared for technicalities." All of these readers were "anxious to read and learn, were the facts only presented to them in a handy and intelligible form."[17] Although Page saw geology as "deserving the study of every cultivated mind," he strongly advocated it as an integral part of the education of all women. He remarked that women were already well represented among geological observers and collectors. Page was willing to broaden the education of women so as to include science, but he assured his readers that such activities represented no challenge to the gender status quo. Women, he believed, could pursue their enthusiasm for geology "without interfering, more than any other relaxation, with the discharge of those domestic and social duties which, after all, form woman's noblest and most natural function."[18]

Page's approach to science writing appealed to Victorian readers, and over the years he earned a sizeable sum as a result. Throughout the 1850s Page was receiving money from Blackwood and Sons for writing the first editions of books and for revising subsequent editions. His total earnings in that decade came out to about £275 and thirty guineas. In 1854 he received thirty guineas for the copyright to the *Introductory Textbook of Geology* and another thirty pounds for the copyright to the *Advanced Text-Book of Geology*.

15. Page 1869, 5.
16. Page 1861, 7-8.

17. Page 1868, iii.
18. Page 1869, 91-92, 101.

In 1857 he was sent twenty-two pounds, five shillings, and seven pence, two-thirds of the estimated profits on the third edition of the *Introductory Textbook,* as per his agreement with Blackwood and Sons. Two years later he received a little more than £11 for the same book, £100 as final payment of the copyright on *Handbook of Geological Tours and Geology,* and over £78 for editing and preparing the second edition of the *Advanced Text-Book of Geology.* The following year he was receiving over £34 for the fourth edition of the *Introductory Text-Book of Geology.*[19] Throughout the 1860s he continued to earn money from Blackwood and Sons, at least £263 and 177 guineas.[20] Although Page was no longer as dependent on his earnings as a popularizer once he was appointed to his professorship at Durham University College, the money kept rolling in during the early 1870s, when he cleared at least £235.[21]

Throughout his works, Page directed his audience's attention to both the economic and intellectual advantages of geological study. Dealing with the former in his *Geology and Its Influence on Modern Beliefs* (1876, Blackwood), Page maintained that the earth was a "great storehouse of wealth," but in order to determine the "place, position, abundance, and accessibility" of valuable geological formations required scientific skill. Page insisted that geological knowledge was useful to farmers, land valuators, civil engineers,

19. National Library of Scotland, Blackwood Archive, September 18, 1854 and November 10, 1854 [MS 4106 f206, f208]; November 5, 1857 [MS 4126, f47]; November 14, 1859, September 10, 1859, September 3, 1859, and December 6, 1859 [MS 4142 f7, f6, f3, f8].

20. Earnings for 1861: £39 for revising the fifth edition of the *Introductory Text-Book* and over £80 for revising the third edition of the *Advanced Text-Book.* Earnings for 1863: 150 guineas for the copyright of the *Introductory Text-Book of Physical Geography* and the *Advanced Text-Book of Physical Geography.* Earnings for 1865: five guineas for the copyright to *Examinations in Physical Geography.* Earnings for 1867: over £31 for his *Scientific Reading Book.* Earnings for 1868: over £87 for the profits on the fourth edition of *Advanced Text-Book of Geology,* £11 for profits on the third edition of *Introductory Text-Book of Physical Geography,* two guineas for editing the third edition of *Geological Examination,* twenty guineas for the copyright of *Chips and Chapters,* and £15 for profits on the seventh edition of *Introductory Text-Book of Geology* (National Library of Scotland, Blackwood Archive, June 8, 1860 [MS 4153, f5]; December 20, 1861 [MS4163]; December 20, 1861 [MS 4163, f198, f199]; January 13, 1863 [MS 4184, f131]; February 6, 1865 [MS 4203, f7]; July 15, 1867 [MS 4225, f163]; January 7, 1868, January 15, 1868, August 1868, and January 7, 1868 [MS 4238, f131, f132, f133, f134]).

21. Earnings for 1871: just over £53 for revising and editing the ninth edition of the *Introductory Text-Book of Geology* and the third edition of *Geology for General Readers.* Earnings for 1872: over £132 for his share of the profits on the fifth edition of the *Advanced Text-Book of Geology* and on the fifth edition of the *Introductory Text-Book of Physical Geography.* Earnings for 1873: £25 for the second edition of the *Advanced Text-Book of Physical Geography* and the same amount for the tenth edition of the *Introductory Text-Book of Geology* (National Library of Scotland, Blackwood Archive, September 25, 1871 [MS 4281, f1, f2]; March 1872 and May 18, 1872 [MS 4296, f1, f2]; February 21, 1873, May 19, 1873 [MS 4310, f1, f2]).

and mining engineers. He argued that geology applied to the work of the potter and glassmaker, to the manufacturer of mineral pigments and dyes, to the metallurgist and chemist, to the jeweler, and to the mechanical engineer and machinist. The entire "mechanical and commercial supremacy" of Britain depended, Page pointed out, "on our coalfields: our steam-engines, steamships, railways, telegraphs, and endless machinery are their direct outcome, and would have been simply impossible without them." In sum, geology directed our attention "to those minerals and metals whose applications are so intimately interwoven with the character and progress of our modern civilization."[22] Although the utility of geology was an important theme in Page's works, he dealt far more with the intellectual, or theoretical, advantages of studying geology. The aim of theoretical geology was "to arrive at a rational history of the successive phases of the globe."[23] Not only was there an inherent "intellectual interest" in such a study, but also the end result was to "increase our admiration of the means employed by the Creator to alter, to diversify, and to sustain." Looking into the structure of the earth's crust provided "newer and deeper insight into the laws and ordainings of nature, and from all deeper insight of nature the human intellect arises wiser, happier, and more exalted."[24] In his lengthy discussions of the results of theoretical geology in many of his works Page presents his version of the evolutionary epic.

Page begins his examination of the history of the successive phases of the globe in *The Present and Past Life of the Globe* by drawing the reader's attention to a pile of rocks on a table (Fig. 5.2). These "fragments of rock," which the "feet of the ignorant might spurn from their path" are actually "invested with as high an interest as the obelisks of Egypt, or the sculptures of Nineveh" in the eye of the geologist. Like Wood, Kingsley, Allen, and other popularizers, Page used commonplace objects to involve the reader in his exploration of science. He invested them with a mystery designed to entice the reader to accompany him further. "Rough and mutilated as these fragments may appear—obscure as are the forms impressed on their surfaces," Page declared, "they embody a tale of the world's PAST as legible to the eye of Science—and often far more connected—than these sculptures on this slate, or those hieroglyphics graven on that sarcophagus." The tale is, of course, the evolutionary epic, and it was heard only with the help of the paleontologist, who "resuscitates as it were the life of former epochs." Page emphasized the spectacular vision that comes into view if the paleontologist uses his knowledge in conjunction with the other sciences to bring the

22. Page 1876, 39, 41, 43–46, 54, 60.   24. Page 1870, 32.
23. Page 1856, 7.

FIGURE 5.2 The commonplace objects whose mysteries conceal the evolutionary epic. (David Page, *The Past and Present Life of the Globe* [Edinburgh and London: William Blackwood and Sons, 1861], 17.)

past to life. "Lifting the veil from the Past," he asserted, "he displays the terraqueous aspects of the globe at the successive stages of its history; even as now, through the combined labours of the geographers, the botanist, and the zoologist, we are enabled to present a panorama of existing lands and seas with all their exuberant and varied vitality." Page then proceeds to paint a panorama for his readers as he moves from one geological period to the next, illustrating the "one truth that geology has established more clearly than another . . . that of the progressive evolution of life on this globe."[25]

After examining the order existing in all present life, Page transports the reader back into the far, middle, and recent past in a search for a unity of plan. This leads to his chapter on "The Law," where he draws a series of conclusions. A uniformity of type and pattern has persisted throughout time as radiate, articulate, molluscan, and vertebrate "range side by side as distinctly now, each within its own typical idea, as when they first clothed the land and peopled the waters." Quoting American naturalist Louis Agassiz approvingly, Page stated that the evolution of life is "but the realization of a pre-determined plan." Page stressed the continuity within the "one great system or COSMOS," and rejected the notion of great catastrophes leading to the extinction of the flora and fauna of one period and the re-creation of new forms. A true uniformitarian, Page asserted that all life forms "are inseparably interwoven into one gradual and continuous sequence."[26] Even in his textbooks, Page emphasized the uniformitarian principle that "the operations of nature appear to be fixed and uniform within ascertainable limits" and argued that beyond these limits "there seems to lie some great law of *cosmical progression*, clearly indicated in the geological history of the past" (Fig. 5.3).[27]

Page explicitly linked the notion of cosmical progression to theological themes, and made the search for divine design central to the objectives of geology. "The highest aim of our science," Page declared, "is to discover the Creative Plan which binds the whole into one unbroken and harmonious life-system."[28] Seen in this light, Page presented geology as a revelation of divine activity through the ages and as a guarantee of future progress. "The great COSMICAL DESIGN which geology now labours to reveal," he insisted, "will be steadily upheld by the Omniscient omnipotence of Him 'with whom is no variableness, neither shadow of turning.'"[29] More than any of the other natural sciences, Page believed, geology "is calculated to impress with convictions of Divine intelligence and design." Here Page

25. Page 1861, 17, 19, 21, 26, 112.

26. Ibid., 183, 196.

27. Page 1856, 270.

28. Page 1861, 21.

29. Page 1856, 280. The phrase "with whom is no variableness, neither shadow of turning," is a quotation from James 1:17. I am indebted to Paul Fayter for identifying this and other biblical references in this chapter.

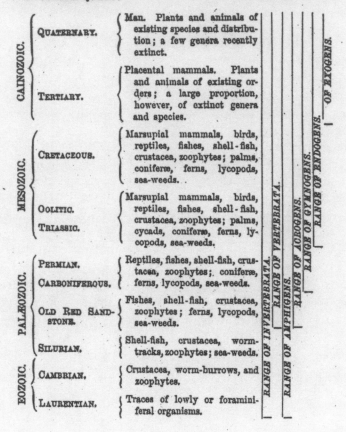

FIGURE 5.3 Page's diagrammatic representation of the "Order and Succession of Life," which is designed to illustrate the cosmic progression from the earliest to the most recent geological age. (David Page, *Geology for General Readers* [Edinburgh and London: William Blackwood and Sons, 1870], xv.)

was responding to those who "attempted to fix on geology . . . the ban and odium of materialism."[30] In their study of "existing nature," zoologists and botanists ennobled our conceptions of God by providing the "theologian with irrefragable proofs of unity of plan and design throughout Creation." But these conceptions would be even more exalted when the theologian discovered "the same harmonies of design and the same unity of plan running through untold ages, and spreading and ramifying through forms so numerous and varied that, varied and rife as existing Life may be, it constitutes but the merest fraction of the Life that has been, and of the forms that have passed away."[31] In comparing astronomy to geology, Page acknowledged that the former "may be a loftier theme." However, Page pointed out that the "loftiness of its topics only renders them the colder and more remote," while Geology "has ever an immediate and human interest" since the earth's past is inseparable from its present. The main point of the evolutionary epic, that we are part of a divine plan still in process, exalts "our conceptions of creation" and "can never tend to weaken our reverence for the power, wisdom, and goodness by which it is directed and sustained."[32] Contrary to uninformed views of geology as encouraging materialism, Page argued that it offered "the theologian new evidences at every turning, of creative wisdom, goodness, and design."[33]

For Page, the design was embedded in natural law, rather than Paleyian notions of contrived organisms. Since geology's role as "the main promoter of this idea of natural law" had led it to be branded as materialistic, Page discussed the link between law and design at length in his works.[34] In the preface to his *Past and Present Life of the Globe*, he avowed that his main object was to impress the general reader "with the universality and uniformity of natural law." He denied that "by recognizing in every instance the fixity and unerring operation of Law" he was placing "a wider distance between the Creator and his works" or that knowledge of natural law encouraged irreverence. "On the contrary," he declared, "he who knows most of creational law, and that the most intimately, stands generally the least in need of the injunction—'Put off thy shoes from off thy feet, for the place whereon thou treadest is holy.'"[35] Law in nature was "but the mode in which the Creator has chosen to manifest himself in his works, and the highest attainment of reason is to give intelligible expression to these modes,

---

30. Page 1869, 112.
31. Page 1861, 25.
32. Page 1870, 142.

33. Page 1868, 121.
34. Page 1869, 112.
35. The biblical quote is from Exodus 3:5.

so that we may be enabled to determine their courses and anticipate their results." Page argued that geology's existence as a science depended on the notion of uniformity. "Destroy this belief in the continuous operation of natural law," Page affirmed, "and appeal to 'revolutions' and 'cataclysms,' and you present a world of disorder, a Creator without a plan, and the human reason striving in vain to elaborate a system from phenomena over which no system prevails." But if the earth's crust were subject to uniform natural law, then the geologist could be confident that the present plan resembled the past plan, and that in tracing the development of that plan from past to present it was possible to uncover "some conception, however faint, of the divine idea of its Creator."[36] No wonder that the reviewer for the *British Quarterly Review* said of Page that he was "a philosopher who thinks it is no disgrace to be a Christian—a man of science who believes in an Author of science."[37]

Page's attempt to enshrine design within natural law, and to present knowledge of it as the goal of geology, must have been irksome to those who sympathized with the goals of scientific naturalism. One of Page's most persistent critics, William Sweetland Dallas, reviewed his works in the *Westminster Review* from the late 1850s to the late 60s. A natural historian, with a particular interest in entomology, Dallas prepared lists of insects for the British Museum from about 1847 to 1858. He was curator of the Yorkshire Philosophical Society's museum from 1858 to 1868, and thereafter, until his death, the assistant secretary to the Geological Society of London.[38] Author of the *Natural History of the Animal Kingdom* (1856) and *Elements of Entomology* (1857), Dallas developed close ties with Darwin in the late 1860s. He prepared the index for Darwin's *Variation of Animals and Plants under Domestication* (1868) and in 1868 undertook translation work for Darwin. With Darwin's support Dallas won the election to be assistant secretary of the Geological Society. Later, Dallas compiled the index for the *Descent of Man* and delighted Darwin in 1872 with his work on the glossary of scientific terms in the sixth edition of the *Origin of Species*.[39] In his review of Page's *Advanced Text-Book*, Dallas acknowledged that the book would be useful to the student "who has mastered the elementary doctrines of the science," but he complained that

36. Page 1861, 8, 72, 242.

37. "Past and Present Life of the Globe" 1861, 218.

38. Boase 1965a, 8.

39. Burkhardt and Smith 1985, 254, 264, 285, 288, 326, 348, 354, 366.

it was "not, however, a very interesting book," since it was "crowded with dry details which have little connexion one with another." Moreover, Page did not discuss many "topics of the highest importance." As a result, "the student will be but little trained by it in that use of his higher reason which geological inquiry, if rightly directed, tends to promote." Dallas then raised questions about Page's expertise. Some topics, such as the origin of coal and of the carboniferous limestone formation, "are treated in the manner of a man who has got up his information from books, and has not thought out the study for himself."[40]

Three years later, Dallas was reviewing Page's *Handbook of Geological Terms and Geology* (1859, Blackwood) and pointing to "serious defects in it, as well as some mistakes." If it reached a new edition, he recommended that Page "obtain the assistance of experts in the principal departments of the subject, instead of trusting to his own general knowledge of the whole range of it."[41] Dallas was also critical a year later of the final chapter on "The Law" of Page's *Past and Present Life of the Globe,* referring to it as the "most unsatisfactory part of his treatise" as it contained "vague rhetorical platitudes."[42] This was the key chapter in which Page deduced the cosmical progression in the geological record. Perhaps Page had in mind reviews such as those by Dallas when he remarked in the preface to his *Geology for General Readers* that it had become "fashionable in some quarters to sneer at popular sketches."[43] But Dallas continued to voice his criticisms. In his review of *Geology for General Readers* he pointed to the repetitiveness in Page's books. "One of the conditions of existence of popular scientific writers seems to be," Dallas declared, "that they should from time to time hash up their materials as it were, and furnish them to the public under a new form." Page's geological writings, Dallas complained, were "very similar in the general nature of their contents."[44]

Dallas was also puzzled by Page's reluctance to commit himself to the "development" or the "natural-selection" hypothesis.[45] Page's bitterness over his break with Chambers can be read into his rejection of premature speculation about the nature of evolutionary law. Although Page asserted that geological science had revealed the existence of a cosmical design in the fossil record, he maintained in his *Advanced Text-Book of Geology* that there was insufficient evidence to determine whether or not there was a law of

40. [Dallas] 1857, 282.
41. [Dallas] 1860, 302.
42. [Dallas] 1861, 257.

43. Page 1870, v.
44. [Dallas] 1866, 243.
45. [Dallas] 1861, 257.

development whereby lower species evolved into higher ones. He insisted, "there is little to be gained by indulging in surmise and hypotheses, however curious and ingenious, unless they are based on fact and observation." Then he referred the reader to *Vestiges* and the critical reviews of it by Adam Sedgwick and David Brewster.[46] In later works Page was less coy. In his *Past and Present Life of the Globe* he proclaimed that the *Vestiges* "stands bastardized by the moral cowardice that shrinks from avowing its paternity." He included *Vestiges* among a list of speculative theories that threatened to undermine the credibility of science. "'Theories of the Earth,' 'Vestiges of Creation,' 'Untieings of the Geological Knot,' 'Pre-Adamite Sketches,' and 'Scriptural Reconciliations,'" Page warned, "are ever crowding thick upon us—enough to destroy the reputation of any science not founded on the sure and ample bases of Truth and Philosophy."[47]

Page was also hostile toward Darwin's theory of natural selection. In his *Past and Present Life of the Globe* he discussed his objections to the theory at length. Natural selection was too materialistic to account for the designed nature of the "great cosmical plan." The plan was "surely more than mere 'physical development'—something higher than the 'transmutation of specific forms under the force of external conditions'—something more precise and definite than 'natural selection in the struggle for existence,' or any other of the materialistic hypotheses that have been recently advanced to account for the great chronological elimination of vitality." Page grouped Darwin, Chambers, and Lamarck together as theorists who had espoused a materialistic hypothesis. Any theory that emphasized "'physical' transmutation," whether it be Lamarck's appeal to "the modifying influence of new external conditions," or the *Vestiges'* conception of the "force of internal volition on the embryotic organisms," or Darwin's "gradual accumulation of minute beneficial changes," they all adopted "the same blind-chance process." Page denied that there was any "direct evidence" of "such a process" in "existing nature or in that which has become extinct." Moreover, any hypothesis that relied on "the force of mere external conditions" was unable to explain order. Page was willing to accept the notion that natural selection had an influence on the diversity of life, but it was only a small part "of the Creator's plan," and it functioned as a secondary activity limited in its power and range of applicability. Page quoted the final paragraph of the *Origin of Species,* and pointed out that here Darwin allowed for factors other

46. Page 1856, 278-80.   47. Page 1861, 209, 245.

than natural selection, such as inheritance and use and disuse. According to Page's reading, Darwin also admitted that life was originally breathed into a few forms or one form. Yet, Darwin inconsistently referred throughout the *Origin* to chance for all subsequent development. "If science is constrained to admit a Divine origination of life," Page declared, "why should she be ashamed to confess to an equally Divine sustaining of its subsequent manifestations?"[48] Page's rejection of the theory of natural selection as speculative and materialistic represented a serious challenge to the agenda of scientific naturalism.

Huxley and his allies would also have been concerned about Page's views on the authority of scientific experts. Although Page's bitter dispute with Chambers had led him to criticize the theory of evolution found in the *Vestiges,* Page nonetheless shared Chambers's more egalitarian conception of science. The issue of authority was central to the controversy surrounding the *Vestiges.* Chambers refused to accept the notion that the negative judgment of practitioners on specific sections of the book was enough to discredit his main theory. He presented himself as someone who had gained insight into the workings of nature by going beyond the specialized horizons of the experts, and he appealed to ordinary readers as the final arbiters of the truth of his theory. In response, practicing scientists asserted their authority and touted the superiority of specialization.[49] Page shared Chambers's rejection of the authority of specialists. In his *Advanced Text-Book of Geology* he emphasized that the objects of geological research were scattered everywhere around us. "Not a quarry by the wayside," he declared, "not a railway-cutting through which we are carried, not a mountain-glen up which we climb, nor a sea cliff under which we wonder, but furnishes, when duly observed, important lessons in geology." All that was required to "proceed in the study as a practical observer" was a hammer, a bag for specimens, a sketchbook, and an observing eye, "a pair of willing limbs," and the "diligent use of his text book." Later in the book he again insisted that it was not difficult to "acquire a knowledge" of geology.[50] Special expertise and training was not necessary to become a practicing geologist. Page was critical of narrow specialists who objected to his attempts to make geology intelligible to the majority of general readers. "To those who may sneer at 'smatterings of science,'" Page replied, "or grow facetious on the 'dangers of a little learning' (and these are generally the mere technical tradesmen

48. Ibid., 117, 197, 200, 210–11.
49. Yeo 1984, 6, 24–27.

50. Page 1856, 7, 289.

of some narrow department), I have only to answer, that a beginning must be made somewhere . . . and that the mind is more likely to be stimulated to further inquiry by the generalizations of a vivid outline than by an array of details, the very nomenclature of which is often a puzzle and perplexity."[51]

For Page, a "vivid outline" of the evolutionary epic was enough for the reader to grasp the cosmic design throughout the geological ages. Emphasizing "an array of details" in the geological record could lead the reader to be blinded by the "busy panorama of life, growth, and decay." Instead, Page aimed to provide readers with the meaning of the big picture. In *Geology for General Readers*, Page discussed how our limited perspective did not allow us to see how over time the earth's crust was incessantly modified by atmospheric, aqueous, organic, chemical, and igneous agencies. However, the current state of the earth's crust had been formed by the action of these agencies over the eons. "The plains of the Old World," he wrote, "the historic fulfilments of the past—China, Hindoostan, Mesopotamia and Egypt—were borne down from the mountains of Asia and Africa; just as the prairies, the Ilanos, and pampas of the New World, the hopes of the advancing future, are the gifts of the Andes and the Rocky Cordilleras." The comfort and development of humanity was bound up with this process of degradation and reconstruction, and Page marveled at "this system of interdependence between the organic and inorganic—between the mechanical processes of nature and the social development of man!" Few took the trouble to try to understand this system of incessant change, and they deprived themselves "of much rational enjoyment" while paying "little regard to the system of nature of which they form so prominent a part, and how little reverence for Him who has given them eyes to see, and understanding to understand." Page pointed to the admirable "system of compensation by which decay in one part is balanced by renovation in another." He drew the reader's attention to the "beauty and consistency" of nature. His works provided his readers with a glimpse of the panorama of life, growth, and decay in order to reveal the "unity of design" in the evolutionary epic.[52]

ARABELLA BUCKLEY: SPIRITUAL EVOLUTIONISM

While the evolutionary epic could be conveyed as a story about cosmic design, it could also be given a spiritualist gloss. Alfred Russel Wallace, the cofounder of the theory of natural selection with Darwin, was among the

---

51. Page 1861, 241.   52. Page 1870, 48–49, 198.

foremost British spiritualists after publicly announcing his conversion in 1866.[53] But the writer who played a leading role in bringing together spiritualism with the evolutionary epic was actually Arabella Buckley, who, Wallace recalled, was "my most intimate and confidential friend" in the early 1880s.[54] Wallace and Buckley had become close friends because of their shared interest in spiritualism. She concealed her fascination with spiritualism from friends like Lyell, Darwin, and Huxley. Like Wood, she kept it buried in her published works in order to avoid raising questions about her credibility as a scientific author. Scholars have virtually ignored this aspect of Buckley's life and thought. If the subtext of Buckley's works is connected to her spiritualist leanings, then her entire conception of the evolutionary epic will be seen in a different light.

The youngest daughter of the Rev. J. W. Buckley, vicar of St Mary's, Paddington, Arabella Buckley (1840–1929) became Charles Lyell's secretary in 1864. Through him she came into contact with important publishers and eminent men of science such as Darwin, Huxley, and Wallace. After Lyell's death in 1875, she embarked on a career as a popularizer of science. Her first major work, *A Short History of Natural Science* (1876, 5th ed., 1894, Murray), was designed to be the first book to treat "the difficult subject of the History of Science in a short and simple way."[55] Periodical reviewers and practicing scientists received it well. In its notice of the second edition, the reviewer in the *Westminster Review* asserted that the book was "now too well known to need any detailed analysis."[56] Darwin wrote to Buckley on February 11, 1876, that the idea behind the book was "a capital one, and as far as I can judge very well carried out. There is much fascination in taking a bird's eye view of all the grand leading steps in the progress of science."[57] Buckley authored over ten books on science, many for children, between 1876 and 1901, working primarily with publishers such as Cassell and Company, Edward Stanford, and John Murray. From 1876 to 1888 she also lectured on scientific topics. While in the midst of her career as popularizer of science, Buckley married Dr. Thomas Fisher, of Christchurch, New Zealand, in 1884.[58]

53. Fichman 2004, 172.

54. Wallace 1905, 2: 378. Wallace explored the link between evolution and spiritualism in works that appeared well before Buckley started writing, but *Man's Place in the Universe* (1903) was his first major published work containing a discussion of cosmic evolution (Fichman 2004, 293–99).

55. Buckley 1876, vii.

56. "Contemporary Literature" 1880, 582.

57. F. Darwin 1887, 3: 229.

58. Gates 2004a, 337–39; "Mrs. A. B. Fisher" 1929, 9.

Buckley had connections to Darwin and his inner circle, and her books contain positive references to their scientific research. She pointed to John Lubbock's investigations into ants in *Life and Her Children*.[59] In her *Fairy-Land of Science* (1879, Stanford), she noted Tyndall's work on crystals in an approving tone. She also quoted from Huxley's penny lecture on "Coral and Coral Reefs" as an example of going beyond dry facts and bringing a scientific subject to life.[60] Buckley was friendly with Huxley and on occasion she was a teatime guest of his wife, Henrietta.[61] But she was alarmed by Huxley's heterodox beliefs. In 1871, after hearing Huxley's Royal Institution lecture on the "Metaphysics of Sensation," she wrote to him asking for clarification on his views. "Will you excuse me for troubling you about the end of your Friday evening's lecture," she asked. "My remark that 'I could not believe it' was not quite so impertinent as it must have appeared and it would be a great satisfaction to me to know whether I can have misunderstood you." Buckley explained that in the past she had often defended Huxley from charges of heterodoxy on the grounds that "you do not deny us a power of conception of God if only we will allow that it is imperfect and not talk of Him as if he were 'a man in the next street about whose actions we were perfect judges.'" The conclusion of the lecture, which implied that it was improper to even form a conception of God, "pained" her, as "this defence seemed to me destroyed the other evening."[62] In 1873, Buckley visited with Henrietta and they chatted about Huxley's reputation. After the visit, Henrietta wrote to her husband on July 7, that Buckley "wished that you could write some book or books that should leave its mark behind, for that yours would always be a contested name." Then Henrietta stated her hope that Huxley would become a "leader of a new school of thought where the materialists and spiritualists shall be united in brotherhood." It is unclear if Buckley expressed the same expectations as Henrietta, but Huxley's response of August 8 indicates that he was annoyed with Buckley for encouraging his wife to entertain such hopes. "I don't quite understand what Miss Buckley meant about the 'disputed' reputation," Huxley declared, "unless it is a reputation for getting into disputes." But Huxley was not concerned about any reputation except that of being honest and straightforward. He did not see his role as reconciling the "antagonisms of the old schools" and did not

59. Buckley 1881, 274.   60. Buckley 1879a, 21–23, 87.
61. L. Huxley 1902, 1: 427.

62. Buckley to Huxley, May 22, 1871, Huxley Papers, Imperial College Archive, 11.182.

think that rapprochement between free thought and traditional authority was a possibility in any case.[63]

Buckley had a more intimate bond with Darwin, and she visited him at his Down home in the late 1870s and early 80s.[64] Francis Darwin asserted, "Miss Buckley was one of the few women who could be regarded as *his* friend—though there were many women whose society he enjoyed very much." With Buckley, he could "talk about her books in which he interested himself" and her "success was pleasant to him."[65] Over the years Darwin responded with encouragement whenever Buckley sent him her work. After reading the first two chapters of *Life and Her Children,* Darwin wrote to Buckley on November 14, 1880, saying that she had "treated evolution with much dexterity and truthfulness." He praised the "plan" for her work and remarked, "who can tell how many naturalists may spring up from the seed sown by you. I heartily wish your book all success."[66] Darwin's friendship with Buckley was also based on their shared reverence for Lyell. Francis Darwin remarked that one reason for the bond between Buckley and Darwin was that "she belonged to the scientific world too through her connection with Lyell."[67]

Buckley's enormous respect for Lyell was evident in her works and elicited reactions from Darwin. In her *Short History of Natural Science* she remarked that Lyell, "like all other great men, . . . was humble and reverent in his study of nature." She extolled his love of truth and believed that "by his conscientious and dispassionate writings he did much to persuade people to study geology calmly and wisely, instead of mixing it up with angry disputes, like those which, in the time of Galileo, disfigured astronomy."[68] Darwin wrote on February 11, 1876, "you have done full justice, and not more than justice, to our dear old master, Lyell."[69] In her entry on Lyell in the ninth edition (1878) of the *Encyclopaedia Britannica,* she praised his "gentle nature, his intense love of truth, his anxiety to help and encourage those who cultivated his favourite science." His "extreme freshness of mind" protected him

---

63. Darwin and Desmond, forthcoming. Transcription supplied by Adrian Desmond.

64. Colp 1992, 7.

65. F. Darwin, "Reminiscences of My Father's Everyday Life," 64.

66. Darwin to Buckley, November 14, 1880, Cambridge University Library, Charles Darwin Papers, MS.DAR.143: 184.

67. F. Darwin, "Reminiscences of My Father's Everyday Life," 64.

68. Buckley 1876, 409.

69. F. Darwin 1887, 3: 229.

from dogmatism in old age and allowed him to accept the work of younger men. This flexibility was a key factor in his acceptance of natural selection even though he had earlier rejected Lamarckian evolution.[70] Darwin wrote to Buckley in 1881 to express his enthusiasm for the entry. "I do not think that it could have been better done," he told her, "and you have brought out clearly and forcibly his high merits." Darwin believed that "all of Lyell's admirers ought to feel grateful to you."[71]

Barbara Gates has already demonstrated that Buckley was indebted to Lyell for many of her narrative strategies. In the course of working with him for eleven years she learned how to focus on the big picture through the use of vivid literary images that resembled panoramas or cross sections. Like Lyell, she helped the reader to see back through time, or downward through the earth's crust, or through the eyes of other creatures.[72] It could also be argued that Buckley's intellectual debt to Lyell involved his more religious interpretation of evolution. Whereas Darwin's theory of natural selection was nested in a framework that leaned toward agnosticism, Lyell remained a Christian. A committed Unitarian, he accepted evolution, but only as a mode of God's immanent, creative activity in providential law. However, for Lyell, there was a sharp distinction between humans and lower animals, and he never accepted human evolution.[73] Buckley's debt to Lyell on this score is contained in the dedication to the memory of Lyell and his wife in *A Short History of Natural Science.* Here she proclaimed that she owed "more than I can ever express" to the Lyells. She hoped that her book would "help develop in those who read it that earnest and truth-seeking spirit in the study of God's works and laws which was the guiding principle of their lives."[74]

Despite her immense respect for Lyell, and the important impact that he had upon her writing, Buckley formed the deepest intellectual bond with Wallace. Wallace recollected that he met Buckley in the summer of 1863. At the evening receptions at the Lyells' from 1863 to 1872, attended by eminent scientists and intellectuals, Buckley befriended the socially awkward Wallace. According to Wallace, she pointed out to him "the various celebrities who happened to be present, and thus began a cordial friendship which

70. [Buckley] 1890, 102.

71. Darwin to Buckley, July 11, 1881, Cambridge University Library, Charles Darwin Papers, MS.DAR.143: 187.

72. Gates 1998, 53.

73. Bartholomew 1973.

74. Buckley 1876, v.

has continued unbroken, and has been a mutual pleasure and advantage." Buckley was among the friends with whom Wallace investigated spiritualism.[75] In 1874 they corresponded at length on Buckley's experiences with mediums. Initially, Buckley visited a medium to deal with a serious case of writer's block. During the third visit the medium mesmerized her, and she was sent into a trance "almost the whole time I was there." After each visit "writing has been easier," she told Wallace, and "yesterday I wrote five large pages of perfectly coherent writing in less than twenty minutes." Even more astonishing, Buckley was told that she was "a strong medium" herself and she believed that she was getting messages from dead relatives. The following day she acknowledged that critics of spiritualism like William Carpenter would "say I was a victim to hysteria or mania." Although she could not explain her experiences to a skeptic, she was glad that her reason had shown her "that I am not excited mentally in the least and can reason upon it as if it were someone else while at the same time being the agent I am able to convince myself that there is no deception."[76]

Following the death of Wallace's eldest child, Herbert, Buckley wrote a letter of condolence to him on April 25, 1874. "How wonderful it is how *completely* Spiritualism alters one's idea of death," she exclaimed, "but I think it increases one's wish to know what they are doing." Buckley reflected on Wallace's good fortune of having many spiritualist friends who could get information for him on his son. In a "P. S.," Buckley hinted that she might have received a message from Herbert. Since writing the first part of the letter she had "had a communication which I should hesitate to send so soon after your loss if I did not know that you are able to balance probabilities and take it for what it is worth. I wish I could get rid of this feeling that it may be partly my own imagination."[77]

Buckley was deeply affected by the apparent communication from Wallace's late son and from other spirits. A month later, she was still trying to determine whether or not they were genuine. She wrote to Wallace on May 26, 1874, that both of her "tests" had failed to confirm their authenticity and now she was becoming suspicious about her potential for becoming a medium. Her investigations into spiritualism seemed to have hit a brick wall. "I have now been at work for *five months*," she declared, "giving myself

75. Wallace 1905, 1: 433-35; 2: 296.

76. Buckley to Wallace, February 19 and 20, 1874, British Library, Alfred Russel Wallace Papers, MS Add 46439, f.82.

77. Buckley to Wallace, April 25, 1874, Natural History Museum, Wallace Papers.

to it as trustfully as I could doing all which I could gather or even guess was required of me, and asking to be told further ways of advance and yet I am not one whit further advanced in any tangible way than I was in the beginning." Since Buckley believed that some spirit communications had come from her dead sister, her mother had "such a happy faith in it as uniting her with my sister that I cannot bear to shake it." She also reported to Wallace that Lyell had told her that Huxley had recently proved a medium to be an impostor and that she planned to see the Huxleys soon to hear the entire story. Buckley seems to have kept her fascination with spiritualism from Lyell and the other members of the Darwin circle. In an almost conspiratorial tone, she told Wallace that Lyell had read an article of Wallace's on spiritualism, "but of course is not a bit convinced by it." Buckley expressed her disappointment with Lyell's closed mindedness.[78]

Buckley was still thinking about spiritualism in the late 1870s, and in 1879 she wrote an anonymous essay for the *University Magazine* on "The Soul, and the Theory of Evolution" that summed up her position.[79] Buckley began by ruling out "pure materialism" as "inadequate to account for the facts of human life" and appealed to John Tyndall's point that it was impossible to explain how molecular action can produce consciousness. Then she examined different types of spiritualism by the light of evolutionary theory. Metempsychosis, or the Eastern doctrine of transmigration of souls, and Christian notions that souls came direct from the hand of God, were both undermined by evolutionary theory. Although she rejected the materialist's notion that humanity was mortal, and the idea that immortality could evolve out of mortality, she was prepared to accept the proposition that the whole human series was immortal. This proposition merely implied that the power referred to as life is "something distinct from ordinary inorganic forces," and Buckley believed it was "quite scientific to assume such a power" since "when we have weighed and measured all the mechanical forces of the body . . . we obtain no clue to the power which combined these forces into a living being." We therefore "are compelled to assume something behind them which we call life." If this life-principle, or "spirit," exists, Buckley argued, "we must suppose, on the theory of evolution, that it is passed on from flower to seed, from animals to their offspring, from parent to child" and that during each lifetime it draws in fresh supplies from the general

78. Buckley to Wallace, May 26, 1874, British Library, Alfred Russel Wallace Papers, MS Add 46439, f.95.

79. The article is attributed to Buckley in the *Wellesley Index* because it was signed "A. B." like an article by her in *Macmillan's Magazine* (W. Houghton 1987, 366).

fund of spirit. This notion of spirit was scientific in Buckley's eyes because it could be related to theories about the ether and about the indestructibility of energy. Spirit was never "localized" in material substance, but permeated the organic form "in the same way as ether is supposed to pass between the grosser atoms of matter." She also pointed out that "science has amply proved that there is no such thing as destruction of force in our portion of the universe" and if spirit was not "convertible into material forces, then there is nothing in the dissolution of the body to affect it."[80]

Buckley believed that her conception of spirit, combined with evolutionary theory, explained how habits developed—due to the "life-force permeating the organism," not because of the evolution of material bodies. Moreover, it gave purpose and meaning to the life of the race and the individual throughout the entire evolutionary process. "If, then, we can conceive permanent impressions accumulating through countless generations of animals," Buckley insisted, "leading to developed instincts, emotions, and passions, and thus on to the complex nature of man, who through savage life gains new experiences; then the upward struggle, with all its difficulties and pain, finds an explanation and a moral justification." Buckley gave this evolutionary theodicy a progressive and individualistic spin. Every newly born individual gained some experience and moved in some degree in the right direction and therefore must at the end of life "have gained something in its passage through the world." The result was a constant "individualization of spirit." Referring to this as the position of the "spiritual evolutionist," Buckley argued that "the life-principle is gradually individualising itself, and going back with certain qualities impressed upon it, to carry on in a future existence the development of its powers, whatever these may be." Metempsychosis involved a fixed quantity of souls appearing again and again on the earth and led to a doctrine of Nirvana as well as a longing for annihilation since "the old weary round should be trodden" continuously. Spiritual evolutionism opened the way "to a new existence fraught with hopes and freed from the trammels of grosser matter." All humans could look upon themselves and all creatures as the "germs of future beings." Man "is immortal because the life-principle within him is eternal and indestructible," Buckley declared, ". . . he is individually immortal because that principle has here received the stamp of individuality."[81] In this article Buckley laid out a position on spiritualism informed by science with which she was now comfortable. Buckley was working on her *Fairy-Land of Science*

---

80. [Buckley] 1879b, 1, 4, 7–9.   81. Ibid., 8–10.

at the same time that she wrote this article. Although the analogy she draws between invisible natural forces and fairies in this book can be interpreted as a device to entice children to become interested in the wondrous world of nature while retaining a naturalistic perspective, it could also reflect her fascination with spiritualism. I will focus on how her spiritual evolutionism found its way into the two books on the evolutionary epic written shortly after the article appeared.

In her *Life and Her Children* (1880, Stanford), which was in print up until 1904, Buckley aimed to "acquaint young people with the structure and habits of the lower forms of life; and to do this in a more systematic way than is usual in ordinary works on Natural History, and more simply than in text-books on Zoology."[82] Starting with the simplest forms of life, tiny slime animals, Buckley worked her way through six divisions of animal life. The second division included animals with simple weapons of attack and defense (e.g., sponges), while the third dealt with prickly skinned animals (e.g., the star-fish). Buckley then moved on to the shell-inhabiting animals of the fourth division, the worms of the fifth division, and the jointed-foot animals of the sixth and largest division, which encompassed crabs, centipedes, spiders, and six-legged air-breathing insects. Buckley told her readers that the seventh division, the backboned animals, would be dealt with in a separate volume.[83] *Winners in Life's Race* was published by Edward Stanford in 1882 and was in print up until 1901. Buckley referred to it in the preface as the natural sequel to her earlier book on invertebrates.[84]

Buckley's evolutionary epic in these two books is populated with unlikely heroes who courageously use their natural advantages to survive the struggle for existence. Each animal group possesses some special advantage that enables them to "spread their children over the world; the sponges had their co-operative life and their protecting skeletons, the lasso-throwers their poisonous weapons, the prickly-skinned animals their tube feet and stony casing, the mollusca their wonder-working mantle." Buckley calls upon the reader to admire these plucky beings. Animals with ringed bodies and jointed feet, the arthropods, succeed in the battle of life when they "hold their own and fright bravely" using the "tools and weapons" supplied to them by nature. The jellyfish, part of that group of beings who use a weapon "as simple, as deadly, and far more wonderful in its action than the lasso of the American hunter," are to be respected for their willingness to "open out

82. Buckley 1881, v.   84. Buckley 1882, v.
83. Ibid., 12.

so freely and eagerly in the depths of the quiet ocean." All forms of life in the six divisions preceding the more highly evolved vertebrates become an inspiration to the human reader for their heroism under fire. Glimpsing the labors of these beings "may lead us to wish to fight our own battle bravely and to work, and to strive and bear patiently, if only that we may be worthy to stand at the head of the vast family of Life's children."[85]

Two additional themes dominate *Life and Her Children* and *Winners in Life's Race*: the family bond uniting all life and the religious dimensions of the evolutionary process. Barbara Gates has discussed the first theme at length in her treatment of Buckley as one of several women who retold the story of science. Buckley's emphasis on mutuality was, in Gates's words, "her favorite improvement upon Darwin."[86] The title "Life and Her Children," Buckley declared in the preface, was selected in order to "express the family bond uniting all *living* things."[87] Buckley begins the first book by asking if most people ever considered "how brimful our world is of life, and what a different place it would be if no living thing had ever been upon it." This leads her into a disquisition on the abundance of life. "There is no spot on the surface of the earth," she declared, "in the depths of the ocean, or in the lower currents of the air, which is not filled with life whenever and wherever there is room" (Fig. 5.4). This vision of life as an "invisible mother ever taking shape in her children"—a superproductive mother nature—is then balanced against the notion of struggle. The tendency of nature to produce life leads to overcrowding, which forces plants and animals to improve by developing weapons and defense mechanisms. But Buckley stressed that by studying how animals managed the struggle for existence the reader would come to see how humans were related to "Life's other children," though the development of authentic mutual sympathy was not to be found among the invertebrates. By recognizing that the invertebrates "have as much a real history as you or I have, with real struggles and difficulties which they can only overcome by using all their powers," the reader can sympathize with their labors and feel a sense of kinship.[88] In *Winners*, Buckley traces the evolution of true sympathy, the love of parent for child, starting with the bony race of fish, through reptiles, birds, and mammals. Several of Buckley's illustrations depict various vertebrates protecting, nurturing, or feeding their young. The sticklebacks tend to their nest; the eagle brings food to its young; a buffalo cow defends her calf; and the humpback whale

85. Buckley 1881, 142, 158, 51, 76, 13.   87. Buckley 1881, v.
86. Gates 1998, 60.   88. Ibid., 1, 4, 6, 300, 101.

LIFE IN THE DEEP SEA.
(for description see list of illustrations)

FIGURE 5.4 A crowded ocean view of life in the deep sea forms the frontispiece of Buckley's *Life and Her Children*. (Arabella Buckley, *Life and Her Children* [London: Edward Stanford, 1881], frontispiece.)

FIGURE 5.5 A woolly monkey comforts her child in a very humanlike gesture. (Arabella Buckley, *Winners in Life's Race* [London: Edward Stanford, 1882], 247.)

suckles her young (Fig. 5.5).[89] Life teaches her children that to win the race they must learn that "unity is strength." More important than the study of how different forms have become physically fitted for their lives "is the great moral lesson taught at every step in the history of the development of the animal world, that amidst toil and suffering, struggle and death, the supreme law of life of the law of SELF-DEVOTION AND LOVE."[90]

Buckley's emphasis on moral evolution was grounded on religious considerations, related to her commitment to spiritualism, the second major theme dominating the two books that presented her evolutionary epic. Even before Buckley had become deeply involved with spiritualism, she had explored the larger religious implications of Darwin's theory of evolution. In her "Darwinism and Religion" (1871), published in *Macmillan's Magazine*, Buckley denied that Darwin's views on the origins of the social instincts, as stated in the *Descent of Man*, would lead to gross materialism.[91] On May 14, 1871, Wallace wrote to Darwin, asking, "do you not admire our friend Miss Buckley's admirable article in *Macmillan?* It seems to me the best and most original that has been written on your book."[92] In her later, postspiritualist,

89. Buckley 1882, figs. 14, 47, 71, 85.
90. Ibid., 352-53.
91. [Buckley] 1871, 46.
92. Marchant 1916, 216-17.

works, Buckley emphasized the design in nature and the notion that an unseen power guided the evolutionary process toward a moral end. Buckley often used a discourse of design, with its stress on wonder and contrivance, to describe the intricacies of animal bodies. In *Life and Her Children,* Buckley referred to even the simplest creatures, such as the tiny slime animals, as being "beautiful and wonderful." Upon examining the "curious and beautiful" sponge she asked her readers, "what architect has laid the fibres so skillfully, and formed such a wonderful and intricate structure?" The sea urchin is a "quaint, clever, wonderful, and skilful piece of mechanism."[93] Buckley's evolutionary epic is infused with a theology of nature.

The religious and spiritualistic dimension of Buckley's evolutionary theory goes even deeper. The capitalization of "Life" in *Life and Her Children,* when interpreted in light of Buckley's essay on "The Soul and the Theory of Evolution," takes on added significance. More than just the superproductive Mother Nature, "Life" implied to Buckley the life-principle or divine spirit that informed the entire evolutionary process. This theme is developed more fully in *Winners in Life's Race,* particularly in the concluding chapter, though in the preface Buckley announced that her book would have accomplished its purpose "if it only awakens in young minds a sense of the wonderful interweaving of life upon the earth, and a desire to trace out the ever-continuous action of the great Creator in the development of living beings." Titled "A Bird's Eye View of the Rise and Progress of Backboned Life," the final chapter provides the key to the evolutionary epic unfolded in both books. As a whole, the history of the various branches of the great backboned family "has been one of a gradual rise from lower to higher forms of life." A study of a chapter in the gradual development of any invertebrate will show that "this is the mode in which the Great Power works," not by "sudden and violent new creations." But a glimpse into the workings of spirit over the course of the whole process of evolution, which is what Buckley aimed to produce in her two books, provided the best evidence of a guiding divine force. "Could we but see the whole," she wrote, "we should surely bend in reverence and awe before a scheme so grand, so immutable, so irresistible in its action, and yet so still, so silent, and so imperceptible, because everywhere and always at work." Instead of seeking "marvels of spasmodic power," attention should be focused on "the greatest proof of a mighty wisdom in an all-embracing and never-wavering scheme, the scope

93. Buckley 1881, 32, 35, 98.

of which is indeed beyond out intelligence, but the partial working of which is daily shown before our daily eyes."[94]

Buckley's notion of an individualization of spirit within a progressive evolutionary process lay behind the grand, organic vision she offered her readers in the next section of the conclusion. By means of the "facts" collected by Darwin, "our great countryman," and the "careful conclusions which he drew from them, we have learned to see that there has been a gradual unfolding of life upon the globe, just as a plant unfolds first the seed-leaves, then the stem, then the leaves, then the bud, the flower and the fruit; so that though each plant has its own beauties and its own appointed work, we cannot say that any stands alone, or could exist without the whole." Looking at the evolutionary process in nature in this way gave natural history "quite a new charm," for it set as its task the study of living forms and the remains of those now gone in order to understand how the different branches of life developed "so as to lead to the greatest amount of widespreading life upon the globe, each having its own duty to perform." With this "great thought before us that every bone, every hair, every small peculiarity . . . has its meaning, and has, or has had, its use in the life of each animal," we can never weary of the study of "tracing out the working of Nature's laws, which are the expression to us of the mind of the great Creator." Buckley then returned to her point about the centrality of mutuality in nature and tied it in to her notion of a divine spirit at work. When we look upon "the whole animal creation as the result of the long working out of nature's laws as laid down from the first by the Great Power of the Universe, what new pleasure we find in every sign of intelligence, affection, and devotion in the lower creatures!" These signs showed that the struggle for existence led not only to "wonderfully formed bodies, but also to higher and more sensitive natures," and taught us that "intelligence and love are often as useful weapons in fighting the battle of life as brute force and ferocity." Family love, which we thought was a gift only to humanity, "an exception only found in the human race," has been "gradually developing throughout the whole animal world."[95]

Buckley remained in touch with Wallace and continued to think through her notion of spiritual evolution after the publication of the two books of the early 1880s. When Buckley decided to accept a proposal of marriage from Dr. Fisher, she notified Wallace through his wife on October 9, 1883. She informed the Wallaces that her home would be in Devonshire after she was

94. Buckley 1882, viii, 334, 336.   95. Ibid., 346, 348, 351.

married. She asked, "Do you remember my asking you if you thought I could keep up work in the country? Tell Mr. Wallace I mean to go on working."[96] Buckley continued her career as a popularizer of science after her marriage, and she maintained her friendly correspondence with Wallace. On February 16, 1888, Wallace responded to Buckley's question about the physiology of ferns and mosses and why they have persisted so long in competition with flowering plants. Wallace also discussed his work on a popular sketch of Darwinism and told Buckley that he believed that he could arrange the chapter on hybridity "more intelligently than Darwin did, and simplify it immensely by leaving out the endless discussion of collateral details and difficulties which in the 'Origin of Species' confuse the main issue."[97] In their correspondence, Buckley and Wallace were exchanging views on how to explain evolution to a general reading audience.

Buckley returned to her exploration of the spiritual dimensions of evolution in her *Moral Teachings of Science* (1891, Stanford). Again, she presented her evolutionary epic of the development of sympathy and love in the vertebrates, culminating in humans. The sense of duty in humanity was the "outcome of evolution, or the unfolding by natural law of the will of the Creator." In the concluding chapter of the book Buckley included a discussion of immortality. She could not conceive of the possibility that humans passed away into nothingness, as it was an unjust scheme unworthy of a great power. Ultimate goodness existed for each individual as well as for the universe. But she envisioned a time when, "through the action of ever-widening sympathy, the narrow boundary of self will break down more and more till our own individuality will survive only to be merged by sympathy in that of others." When that time arrived, "we shall realise at last that we are indeed but individual fractions of Our Universal Life."[98]

Wallace was still trying to guide Buckley in her study of spiritualism in 1896 and a year later wrote to her about his admiration of Oliver Lodge's address to the Spiritualists' Association.[99] Buckley's dialogue with Wallace on spiritualism lasted until just before his death in 1913. In 1910, Wallace sent Buckley a copy of his new book, *The World of Life: A Manifestation of Creative Power, Directive Mind and Ultimate Purpose*. Buckley thanked him for the gift, and told him how much she liked the book. "Whether people agree with you or not," she wrote, "(and I do agree with the main part of

96. Buckley to Mrs. Wallace, October 9, 1883, Natural History Museum, Wallace Papers.

97. Marchant 1916, 295–96.

98. Buckley 1891, 98, 108, 118.

99. Fichman 2004, 302–3.

it) I think it will be very beneficial in the present age of materialism and pessimism."[100]

Buckley's correspondence with Darwin was not nearly as frank. On November 14, 1880, Darwin wrote to Buckley that he had read the first two chapters of *Life and Her Children*. Since she had "treated evolution with much dexterity" he expected that she would "escape" persecution from heretic hunters. "You will not be called a dangerous woman," he predicted.[101] Ironically, Darwin missed the entire spiritualist subtext of Buckley's work. He thought that Buckley had included religious themes to blunt potential criticism of her book as materialistic. In his own work, Darwin had been careful to conceal his own ambivalent position from his reading audience when it came to the issue of religious belief. He assumed that Buckley had adopted the same strategy. In the opening pages of *Life and Her Children* he would have seen a passage reminiscent of a key sentence in his *Origin of Species*. Here Buckley declared that all living beings obeyed the law to increase, multiply, and replenish the earth from the "day when first into our planet from the bosom of the great Creator was breathed the breath of life."[102] As he was revising the first edition of the *Origin*, Darwin altered the final sentence of the book concerning the relation of creator and creation: "There is grandeur in this view of life, with its several powers, having been originally breathed by the Creator into a few forms or into one." Darwin inserted the phrase "by the Creator" into the second edition of December 1859 and it was thereafter retained.[103] Despite the resemblance between the two passages, Darwin was wrong about Buckley's true religious beliefs, tempered, as they were, by her encounter with spiritualism. But Buckley had concealed her interest in spiritualism from Darwin and her other scientific naturalist friends. Buckley, no less than Darwin, was a master of ambiguity.

EDWARD CLODD: EVOLUTION FROM GAS TO GENIUS

On July 18, 1881, Buckley wrote to Edward Clodd, thanking him for inviting her to his house one Sunday evening. "Unfortunately," she replied, "my father being a clergyman, I can very rarely get away from home on the Sunday and shall not be able to come." But Buckley was eager to meet Clodd

---

100. Buckley to Wallace, December 16, 1910, Natural History Museum, Wallace Papers.

101. Darwin to Buckley, November 14, 1880, Cambridge University Library, Darwin Papers, MS.DAR.143: 184.

102. Buckley 1881, 4.

103. C. Darwin 1959, 759.

and invited him to drop by her house any Tuesday afternoon from three to six o'clock. She also thanked him for "the kind way in which you speak of my work."[104] By 1881, Clodd had established himself as an important author of books on religion in light of modern science that had brought him into close contact with Huxley and other scientific naturalists. Buckley's desire to meet him was based on her belief that they had much in common. Both were writing for a general audience and both had connections to important evolutionists. Only seven years later, Clodd would present his version of the evolutionary epic in *The Story of Creation*. Yet, Clodd was an outspoken critic of spiritualism and over thirty years later wrote several books in which he attacked it. Although Huxley and the other scientific naturalists would have applauded Clodd's stance on spiritualism, they would have questioned his friendship with important rationalists. Unlike Huxley, who strategically kept his distance from secularists, Clodd became deeply involved in their activities. If, from Huxley's perspective, Buckley erred by indulging the religious sentiments of spiritualists, Clodd was equally mistaken in consorting with a movement that lacked respectability due to its associations with atheistic working-class unbelief.

Edward Clodd (1840–1930) became a popularizer of science only after a long struggle to escape the influence of his devoutly religious parents. He came from a humble family of farmers and sailors (Fig. 5.6). His maternal grandfather had been a Greenland whaler and his father was a brig master. When Clodd was a baby his family moved to Aldeburgh, an old fishing and smuggling port at the time. His father, a pious Baptist, tried to instill a fear of God in his son and pushed him toward becoming a Gospel preacher. He attended the Aldeburgh Grammar School, a private school that emphasized learning for its own sake. Encouraged by his schoolmaster and his mother, Clodd began to read widely. Although no science was taught at the school, he won Maria Hack's *Lectures at Home* as an academic prize, and it inspired him to build a primitive telescope. When his mother took him to London in 1851 to see the Great Exhibition, the impact was profound. To a boy of eleven, Clodd recalled, "this was to enter a wonderland which surpassed all that his mind could conceive."[105] Clodd began to disdain small-town life, rejected his parents' wish that he become a minister, and secretly decided to go to London to seek his fame and fortune as soon as he completed his schooling.

104. Buckley to Clodd, July 18, 1881, Leeds University Library, Clodd Correspondence.
105. Clodd 1926, 8.

FIGURE 5.6 Based on the painting of Clodd by John Collier placed in the library of the Rationalist Press Association in 1914. (Edward Clodd, *Memories* [London: Watts, 1926], frontispiece.)

Clodd obtained a job as an accountant in 1855 while he was visiting his uncle in London. He worked at a series of similar positions at various accountancy firms and in 1862 became a clerk at the Joint Stock Bank, of which he later became secretary in 1872, a post he held until he retired. Clodd devoted his leisure time to reading, primarily books on science and history, and to attending science lectures. During his teens he left his strict Baptist faith behind, and before he was twenty he had become a liberal Christian. He accepted Darwin's evolutionary theory when the *Origin of Species* appeared in 1859 and the intellectual ferment of the 1860s drove him even further away from the orthodoxy of his youth. According to Clodd, two books "set him free," Huxley's *Man's Place in Nature* (1863), as it brought humans into evolution, and Edward Tylor's *Primitive Culture* (1871), since it applied the theory of evolution to every branch of knowledge.[106] For Clodd, anthropology

106. Ibid., 16.

proved that Christian doctrines of the fall and redemption were based on myth and that biblical stories of paradise, creation, and the deluge could be traced to their birthplaces in the Euphrates valley or the uplands of Persia.[107] Clodd had been studying astronomy during the late 1860s and early 70s, and his election as fellow of the Royal Astronomical Society in 1869 had led him to become friendly with William Huggins and Richard Proctor. But his encounter with Tylor drew his attention away from astronomy and toward anthropology, and it inspired him to become a popularizer of science. Intellectual, rather than economic, motives lay behind his desire to write for a broad audience. He was by no means dependent on the money he made as an author since his career as a banker provided a healthy annual income.

In his first book, the *Childhood of the World* (1873, Macmillan), written for children, he explained how the Bible should be understood in light of new scientific developments, particularly in anthropology. He discussed the life of prehistoric humans and the evolution of the human mind. The success of the *Childhood of the World*, which reached a fourth edition within two years and sold 20,000 copies within six years, encouraged Clodd to expand the second part of the book into the *Childhood of Religions* (1875, H. S. King), which also sold well.[108] During the latter half of the 1870s Clodd became part of an intellectual circle that included prominent scientific naturalists such as Grant Allen, W. K. Clifford, T. H. Huxley, and Leslie Stephen. By 1880 Clodd could no longer accept a liberal theistic position and adopted a form of agnosticism. He believed that the idea of God, a futile attempt to explain the mysteries of existence, had been inherited from earlier, more primitive ages. The heterodoxy contained in his *Jesus of Nazareth* (1880, Kegan Paul) elicited strong criticism. Tylor warned Clodd about a tract, titled *A Caution against the Educational Writings of Edward Clodd* (1880) by Catholicus, which attacked him as a freethinker who was poisoning the minds of children.[109] John Ruskin, with whom he had been corresponding, wrote to Clodd that *Jesus of Nazareth* "gave me more pain, and caused me more deadly discouragement, than any book I ever yet opened." Although well intended and temperate, Clodd's book was like a "dose of arsenic or of strychnine" to Ruskin, who berated Clodd for being insensitive toward believers like himself and for his lack of understanding of Christ.[110] Clodd

---

107. Ibid., 17.   108. McCabe 1932, 31–32.
109. Tylor to Clodd, Nov. 24, 1880, Leeds University Library, Clodd Correspondence.
110. Ruskin to Clodd, Nov. 11, 1881, Leeds University Library, Clodd Correspondence.

decided to avoid all direct criticism of religious beliefs in the future and to devote himself to expositions of evolution for a general audience.[111]

Over the next twenty years Clodd produced a series of books on evolutionary theory and evolutionists, including *Story of Creation* (1888, Longman), *The Story of Primitive Man* (1895, Newnes), *A Primer of Evolution* (1895, Longman), *Pioneers of Evolution* (1897, Grant Richards), *Grant Allen: A Memoir* (1900, Grant Richards), and *Thomas Henry Huxley* (1902, Blackwood). During the 1880s he was also a regular contributor to Richard Proctor's journal *Knowledge,* and acted as assistant editor during Proctor's lecture tours in America and Australia.[112] At first Clodd worked with a number of established publishers, such as Macmillan, Kegan Paul, and Longman, but in the late 1890s he took a chance on a new firm run by Grant Richards, the nephew of his good friend Grant Allen. Starting in 1890 Richards worked for W. T. Stead in an editorial and clerical capacity, thanks to an introduction from Allen. Here Richards became acquainted with many of the prominent London publishers and authors. When he opened his publishing house in 1897, backed financially by Allen, he drew on his earlier contacts and approached established writers like Clodd to publish their books with him. Clodd's *Pioneers of Evolution* was actually the first book published by Richards. Gross profits for *Pioneers,* including both the sales of the book and the sale of the American publishing rights to Appleton, came to £100. Richards later recruited Allen, George Bernard Shaw, and A. E. Housman, among others, to contribute their work to his publishing house.[113]

Clodd's treatment of the evolutionary epic can be divided into two main types: accounts of the cosmic evolutionary process in its various manifestations, and studies of those intellectuals and scientists who contributed to the growth of evolutionary thinking. An example of the first type, the *Story of Creation,* priced at six shillings, was presented by Clodd as a "brief and handy" exposition of the theory of evolution to those who did not have the "time or courage" to grapple with Spencer's "bulky volumes."[114] McCabe, Clodd's biographer, claimed that 2,000 copies sold within two weeks, and that 5,000 copies were sold within three months.[115] The Longman records confirm that Clodd's book was a very successful best seller. The first edition of 4,000 copies printed early in 1888 was nearly all gone by June, when Longman produced another 2,000 copies. In 1890 yet another 4,004

111. McCabe 1932, 69.
112. Lightman 2004b, 450–52.
113. Brockman 1991, 272–73.

114. Clodd 1890, [xi].
115. McCabe 1932, 72–73.

copies were printed to keep up with the demand.[116] New editions of the *Story of Creation* were produced right up until 1925. The book also attained a wider circulation through its appearance in 1895 in an abridged form under the title *A Primer of Evolution* and in 1904 as part of the Rationalist Press Association's cheap reprints series.

Although Clodd began the *Story of Creation* with a discussion of the importance of Darwin, he very quickly pointed out that the *Origin of Species* dealt only with organic evolution, a small part of an "all-embracing cosmic philosophy" that encompassed "whatever lies within the phenomenal—the seen or felt—and therefore within the sphere of observation, experiment and comparison." Clodd's aim of outlining a cosmic evolutionism that would draw on all of the sciences was encapsulated in the frontispiece, a photograph of the Orion nebula taken by the astronomer Thomas Common (Fig. 5.7). The Orion nebula had been at the center of debates between astronomers from the 1840s to the 1860s over the legitimacy of the nebular hypothesis, with its links to theories of cosmic progress. Supporters of the nebular hypothesis, such as radical astronomer John Pringle Nichol, argued that telescopic observations of the Orion nebula provided proof of the existence of true nebulosity, the kind of material that Laplace's theory presupposed.[117] As a symbol of the law of progress throughout the universe, the frontispiece of the Orion nebula embodied Clodd's wish to combine the nebular hypothesis with organic evolution to form a single, coherent story of the progressive development of the universe. In the religiously inflected evolutionary epics of Page and Buckley the nebular hypothesis, with its materialistic connotations, played little or no role at all. But for Clodd the nebular hypothesis was a key part of his presentation of cosmic evolution. The first part of the book dealt with force and energy as the raw materials of the universe, as manifested in stars, nebulae, the sun and the planets, the past life history of the earth, and current life forms. Clodd referred to this part of the book as a "description of the things evolved." In the second part of the book, titled "Explanatory," he moved on to an elucidation of the theory of evolution, which included chapters on the evolution of stellar systems, the solar system, and the earth, the origins of life and life forms, the origin of species, and social evolution. Clodd acknowledged that his account of inorganic evolution was indebted to "the 'nebular theory' of Kant and Laplace," but modified to take into account the doctrine of the conservation of energy. For his discussion of the origin of species, he concentrated on Darwin's theory of

116. *Archives of the House of Longman, 1794–1914* 1978, E2, 459–60.

117. Lightman 2000, 679.

NEBULA OF ORION

Enlarged from a photograph taken direct by Mr. Common

FIGURE 5.7 Common's photograph of 1883 was among the first to capture details of celestial phenomena that could not be seen by the human eye even with the aid of the best telescopes of the time. (Edward Clodd, *The Story of Creation* [London: Longmans, Green, 1890], frontispiece.)

natural selection. At the beginning of the second part of the book he offered a diagram intended to illustrate the ascent of the higher life forms from the lower, a reworking of Darwin's conception of a treelike arrangement to describe the evolution of species.[118]

To Clodd, the last chapter of the "Explanatory" section on "Social Evolution" was crucial for the significance of the entire book. He wrote to his cousin that the *Story of Creation* "is the most important thing that I have

118. Clodd 1890, 2, 133, 139.

ever done or can ever expect to do, for one has sought to apply the great theory of our time to human conduct and motives for right conduct."[119] In this chapter Clodd brought to bear all of his work on human evolution and anthropology. He argued that a truly universal theory of evolution would include the development of the human mind. If mind lay outside the range of causation, he believed, then the entire doctrine of the conservation of energy fell to pieces, for that would imply that humans had the power to add to something, which, according to the physicist, could neither be increased nor lessened. Clodd maintained that humanity was "one in ultimate beginnings, and in the stuff of which he is made, with the meanest flower." He covered the evolution of society, language, morals, and theology. In summing up the "Explanatory" section of the book, Clodd pointed out he had told an epic story with a single plot. "We began with the primitive nebula," he declared, "we end with the highest forms of consciousness; the story of creation is shown to be the unbroken record of the evolution of gas into genius."[120]

The second type of evolutionary epic produced by Clodd focused on the struggles and triumphs of key evolutionary figures—the geniuses produced by the evolutionary process—rather than the history of the universe from beginning to end. In his *Pioneers of Evolution: From Thales to Huxley*, Clodd established the existence of a long line of evolutionary pioneers. Clodd defined a pioneer as one who is a "foot-soldier; one who goes before an army to clear the road of obstructions" and to "cut a pathway through jungles of myth and legend to the realities of things." Just as Homer's epic the *Odyssey* contained its heroes who performed exceptional deeds of valor, Clodd's evolutionary epic had its champions. The first pioneers of evolution, "the first on record to doubt the truth of the theory of special creation, whether as the work of departmental gods or of one Supreme Deity, matters not—lived in Greece about the time already mentioned; six centuries before Christ." In this history of scientific heroism, the promising inquiry into evolutionary truth in the ancient period was cut short by the rise of Christianity, revived during the Renaissance, and then put on firm footing by modern evolutionists.[121]

In his section on "Modern Evolution," which takes up half of the book, Clodd discussed Darwin and Wallace first. To Clodd, Darwin was one of the great pioneers, so much so that the book's frontispiece featured a picture of him. Wallace, however, was criticized for exempting human spiritual and

119. McCabe 1932, 72.
120. Clodd 1890, 206, 228.

121. Clodd 1907, 3.

intellectual nature from natural selection and for his spiritualism. Clodd proposed to drop him "out of the ranks of Pioneers of Evolution" for his transgressions. Spencer was credited with formulating a forward-looking cosmic evolution while Huxley was praised for going beyond Darwin and applying the latter's theory to humans. "Huxley was the Apostle Paul of the Darwinian movement," Clodd proclaimed, "and one main result of his active propagandism was to so effectively prepare the way for the reception of the profounder issues involved in the theory of the origin of species, that the publication of Darwin's *Descent of Man* in 1871 created mild excitement." Although Clodd may have intended his analogy between Huxley and Paul to be ironical, his book is a hagiographical study of the important evolutionists. The book climaxed with Clodd's assertion of the inevitability of extending evolution to all facets of human existence, including our intellectual and spiritual nature. Whereas the "old theologies of civilised races, useful in their day, because answering, however imperfectly, to permanent needs of human nature," were now in reality "lineal descendants of barbaric conceptions," the pioneers of evolution had led humanity to see that the abiding value of evolution was "in the extension of its processes as explanation of all that appertains to mankind."[122] The progressive heroes of evolution had almost won their epic struggle with the forces of barbaric theology by helping the human race to embrace cosmic evolutionism.

Not only a commercial success, a number of Clodd's eminent friends also praised the *Pioneers of Evolution*. It was in print up until 1921. Originally published by Grant Richards, Watts and Company published several editions, some in the Rationalist Press Association cheap reprints series. Two of his literary friends had nothing but praise for *Pioneers*. George Gissing wrote to Clodd, "the universe may say with truth that there is 'not a dull page in the book.'"[123] In a letter dated January 17, 1897, Thomas Hardy wrote that he was impressed by the way that Clodd had been able to capture the panorama of the evolutionary epic in such a compressed form. He thought the book showed "a breadth of grasp, and a power of condensing the stupendous ideas scattered over Time in fragments, which you have never before equaled. It is just as when one sees a landscape of miles length reproduced in a charming miniature picture inside a camera."[124] Herbert Spencer thanked Clodd for

122. Ibid., 133, 184, 197, 233, 245.

123. Gissing to Clodd, January 19, 1897, Leeds University Library, Clodd Correspondence.

124. Purdy and Millgate 1980, 143.

sending him a copy of the book and congratulated him "on having done a good piece of work." He believed that readers would obtain "a much better grasp" of the doctrine of evolution by understanding the stages through which it "has itself been evolved." Spencer was also pleased with the treatment he received in *Pioneers*, as it corrected the notion that Darwin created the theory of evolution ex nihilo. "Your description of my own share in the matter will go some way towards rectifying the general misapprehension," Spencer declared.[125] The Russian geographer and anarchist Peter Kropotkin wrote Clodd on February 9, 1897, that the book "reads like a novel" and yet still "conveys such a mass of information." Kropotkin suggested that Clodd follow it up with "more and more books of the sort" as "they are badly wanted."[126] Whether or not Clodd was following Kropotkin's advice, he later wrote biographical studies of Huxley and Allen that were also informed by the hagiographical approach to the history of evolution contained in *Pioneers*.

Clodd's magnetic personality led him to forge friendships with many of the leading liberal scientists and literary figures of his time, and some of his closest friends were important scientific naturalists. Among his correspondents were Henry Walter Bates, Annie Besant, Matthew Arnold, Edmund Gosse, Frederic Harrison, William Huggins, Max Müller, Richard Proctor, George Bernard Shaw, and W. B. Yeats. He exchanged letters with Thomas Hardy from 1891 until 1923. Impressed with Clodd's intellectual energy, H. G. Wells wrote in 1905, "You might like [to be] a Fabian you know."[127] But Clodd shared many ideals in common with the scientific naturalists, many of whom were his friends. Indeed, in his *Between Science and Religion*, historian Frank Turner included Clodd in his list of important scientific naturalists, placing him alongside Leslie Stephen and Grant Allen as one of the essayists of the group.[128] Clodd corresponded with James G. Frazer, Francis Galton, John Lubbock, Herbert Spencer, and Edward Tylor. He counted several of the scientific naturalists as being among his closest friends. Upon his return from Jamaica in 1876, Grant Allen called upon Clodd, which began a lifelong friendship.[129]

Huxley was another important comrade. In his autobiography Clodd wrote, "it was worth being born to have known Huxley." They first met at

125. Spencer to Clodd, January 17, 1897, Leeds University Library, Clodd Correspondence.
126. Kropotkin to Clodd, February 9, 1897, Leeds University Library, Clodd Correspondence.
127. Wells to Clodd, December 10, 1905, Leeds University Library, Clodd Correspondence.
128. Turner 1974, 9.
129. Clodd 1926, 21.

Clifford's. Although just slender acquaintances in 1879, Clodd sent Huxley his *Jesus of Nazareth*. Huxley responded with enthusiasm that the book was just what "I have been longing to see; in spirit, matter and form it appears to me to be exactly what people like myself have long been wanting." Huxley had been fighting "to oppose and destroy the idolatrous accretions of Judaism and Christianity," but he had no sympathy with those who wanted to remove the Bible entirely from the schools, as it contained praiseworthy ethical ideals.[130] During this period of his life, Clodd shared much with Huxley, including the attempt to preserve what was still valuable within organized religion. Gissing, after reading the *Pioneers of Evolution,* remarked on the intellectual similarities between the two. "You are very happy in your section on Huxley," he told Clodd, "which, by the bye, reminded me how strongly you resemble him in the blending of literary culture with scientific attainment."[131]

Clodd dedicated his *Primer of Evolution* to Huxley, which drew an appreciative letter from Huxley's wife Henrietta shortly before his death. "Your kind letters and the 'Primer of Evolution' with its dedication to my husband have indeed been valued by us," she wrote. "He was able to read the dedication and gave a pleased smile."[132] Clodd also sent Henrietta a copy of his *Pioneers of Evolution* in 1897, and told her that "one chief pleasure in writing it has been the chance thus afforded of paying loving tribute to the illustrious man whom it is your pride to have called husband, and whose friendship was one of the rare gifts which life has brought me."[133] Later, after the publication of Clodd's *Thomas Henry Huxley,* Henrietta again conveyed her gratitude. "How shall I adequately thank you for it," she declared, "for your fine tribute to his genius, his nobility of character, his active rectitude and tender humanity toward his fellows." After reading a few pages, she was so moved that "I could hardly see the words for blinding tears." She had written to tell him "how deeply what I did read touched me, and to give you my thanks."[134] About a month later, when she had been able to read more of the book, she contacted Clodd once more, affirming how much she valued his biography of her husband. The public, and not just the family, was indebted to Clodd for this book since its reasonable price put it within

130. Ibid., 37, 40–41.

131. Gissing to Clodd, January 19, 1897, Leeds University Library, Clodd Correspondence.

132. Henrietta Huxley to Clodd, April 2, 1895, Leeds University Library, Clodd Correspondence.

133. Clodd to Henrietta Huxley, January 13, 1897, Imperial College, Huxley Collection, 12.255.

134. Henrietta Huxley to Clodd, February 21, 1902, Leeds University Library, Clodd Correspondence.

reach of "the masses who cannot afford to buy 'The Life.'" She observed, "your affection for my husband shines through its pages."[135]

Despite his high regard for Huxley, in the early twentieth century Clodd entered into a close working relationship with a group of freethinkers that Huxley despised, the secularists. The coarse atheistic philosophy of Charles Bradlaugh and his secularists had always repelled Huxley and many of his scientific naturalist colleagues. When a group of dissident secularists arose in the mid-1880s who disassociated themselves from Bradlaugh's militant secularism, an alliance between Huxley and this branch of secularism became a possibility. Led by the publisher Charles Albert Watts, they began to promulgate a more respectable form of agnosticism. In 1890 he established the Propaganda Press Committee, the seeds of an ambitious publishing program designed to turn the Watts and Company headquarters at Johnson's Court into a propaganda machine for free thought and agnosticism that would outdo both Bradlaugh's publication efforts and those of the Society for Promoting Christian Knowledge or the Religious Tract Society. Renamed the Rationalist Press Committee in 1893, in 1899 it was incorporated under the Companies Acts and became known as the Rationalist Press Association (RPI).[136]

When Watts first broke away from Bradlaugh's group, he attempted to enlist the support of Huxley and some of the other elite scientific naturalists. But he made the mistake of offending Huxley, and the eminent biologist never forgave him. Huxley wrote to Tyndall on November 25, 1883, that he had never imagined "that any-one could play such a dishonourable trick as this Watts has done." Watts had written asking for his views on agnosticism and Huxley innocently thought that his reply was a private communication. Watts "not only printed this without asking leave or sending a proof, but paraded me as a 'contributor'" in the first issue of the *Agnostic Annual* in 1884. Then, Huxley told Tyndall, Watts "had the impudence to say that he was going to issue a second edition and would be obliged if I would enlarge what I said!" Huxley then wrote him a letter "that he will recollect,—and not print."[137] Huxley advised some of his friends not to get involved with the unscrupulous Watts. Huxley's warnings must have made an impression and must have been widely known. Eleven years later, Karl Pearson still remembered the incident. Writing to Huxley in 1894, Pearson told him

---

135. Henrietta Huxley to Clodd, March 24, 1902, Leeds University Library, Clodd Correspondence.

136. Lightman 1989, 289, 303.

137. Huxley to Tyndall, November 25, 1883, Tyndall Papers [Correspondence], RI MS JT/1/TYP/9, p. 3106.

that Watts had asked him to submit a piece to the *Agnostic Annual,* and claimed that Huxley would probably contribute to the same issue. Pearson recalled that "some years ago the Editor drew forth a remonstrance from you by using your name in an unauthorised manner," and Watts was now at it again. Pearson suggested that somebody ought to give Watts "a little correctional talking to."[138]

Even while Huxley was alive, Clodd was more receptive to interacting with radical secularists. G. W. Foote asked him in 1878 to speak at South Place Institute one Sunday evening at a lecture series organized by the Council of the British Secular Union. Clodd accepted the invitation and spoke on the antiquity of man.[139] After 1900 Clodd reconsidered his earlier resolution to produce only constructive works, and he began to write attacks on occultism and spiritualism.[140] In 1906 he became chairman of the RPA, a position he held until 1913. McCabe asserted that Clodd tried to restrain the RPA. Clodd believed that a program of rational culture could be implemented without direct criticism of religious beliefs.[141] When he resigned, Watts wrote to him that his resignation was "the heaviest blow the R.P.A. has sustained since it has been established." He expressed his appreciation for Clodd's "many and great services to the Association," and in Clodd's honor a painting of him was placed in the Association library in 1914.[142] Clodd became a rationalist hero, celebrated by Frederick James Gould in his *Pioneers of Johnson's Court* (1929), just as Clodd had turned Huxley into a scientific hero in his *Pioneers of Evolution.*[143] But this was the same group of freethinkers that Huxley had warned his friends to steer clear of. Although Clodd admired Huxley and his scientific naturalist friends, he did not hesitate to assert his intellectual independence.

Robust in old age, Clodd was active well into the twentieth century. In 1915, when he was seventy-five years of age, he retired from the Joint Stock Bank and moved permanently to Aldeburgh, the town of his youth. He was still lecturing at the Royal Institution in 1917 and 1921, and writing

138. Pearson to Huxley, July 20, [1894], Huxley Papers, 28.199.

139. G. W. Foote to Clodd, October 14 and October 25, 1878, Leeds University Library, Clodd Correspondence.

140. McCabe 1932, 83.

141. Ibid., 128–29.

142. Leeds University Library, Clodd Correspondence, Watts to Clodd, January 17, 1913; F. Gould 1929, 43.

143. F. Gould 1929, 43–44.

on folklore and occultism. He was eighty years old when he published his last book, *Magic in Names and Other Things* (1920, Chapman and Hall), an attack on revivals of ancient animism in spiritualism and Christianity. In the post–World War I era, Clodd became more militant as he saw dangerous signs of decay.[144] But Clodd's importance lies in his appeal to the late Victorian reading audience. He was among those popularizers of science who fascinated the public with his stories of a cosmic evolutionary process that encompassed all of the products of human intelligence. He was one of the first to incorporate the study of anthropology into the evolutionary epic and to present it to a popular audience.

GRANT ALLEN: THE EVOLUTIONIST AT LARGE

Clodd's *Story of Creation* was dedicated to his fellow popularizer of science, Grant Allen. The dedication thanked Allen for his constant interest in the book, for his novel conceptions of force and energy, and for his friendship.[145] Allen returned the compliment in 1894 when he dedicated his collection of poems, *The Lower Slopes,* to Clodd.[146] After Allen's death in 1899, the grieving Clodd wrote his *Grant Allen: A Memoir.* It was Allen who had initiated the friendship. He stumbled across a copy of Clodd's *Childhood of the World* sometime while in Jamaica from 1872 to 1876. Impressed, he noted Clodd's address and later they corresponded.[147] Allen finally met Clodd in 1882. Based on a shared interest in exploring how evolution impacted on a broad range of intellectual issues, they became close comrades. Clodd recalled that their friendship "had its beginning in the knowledge of community of taste in scientific pursuits, and of large, although not complete, agreement on social questions, while its growth into affectionate relationship was fostered by intercourse as frequent as circumstances permitted."[148] But whereas Clodd's evolutionary hero was Huxley, Allen looked to Spencer.

Charles Grant Blairfindie Allen (1848–1899) was born in Kingston, Canada, the son of J. Antisell Allen, then incumbent of the Holy Trinity Church on Wolfe Island. In 1867 he went to Merton College, Oxford, from which he graduated in 1871 (Fig. 5.8). Allen then accepted a position at the new Queen's College at Spanish Town, Jamaica, but it closed after only three years. Upon returning to England, he began a career as a prolific popularizer

144. McCabe 1932, 153.
145. Clodd 1890, [vii].
146. G. Allen 1894b, v.

147. Clodd 1926, 21.
148. Clodd 1900, 113.

FIGURE 5.8 Grant Allen, scientific author, novelist, and Spencerian disciple.
(Edward Clodd, *Memories* [London: Watts, 1926], opposite p. 20.)

of science, producing countless periodical articles and over eighteen science
books, many of which contained collections of his essays. Allen's first two
books, *Physiological Aesthetics* (1877, Henry S. King) and *The Colour-Sense*
(1879, Trübner), were intended as original scientific works that applied
evolutionary theory to new domains. Although they were financial failures,
they were well received by critics and brought Allen to the attention of
Darwin and other evolutionists, as well as to journal editors. Allen became
the master of the short scientific essay, and he could really churn them out.
He contributed to *Belgravia, Cornhill Magazine, Fortnightly Review, Longman's Magazine, Pall Mall Gazette,* and to *Knowledge.*[149] In addition to King
and Trübner, he also established a working relationship with a number of
publishers, including Chatto and Windus, Longman, George Newnes, and

149. Morton 2004, 36–39.

John Lane. Nevertheless, he was unable to earn enough to support his family through scientific journalism, and beginning in 1880 he began to shift his energies into writing novels and short stories. Sensational novels such as *The Woman Who Did* (1895, John Lane), the tragic story of an advanced Girton graduate who would rather "live in sin" than be enslaved by marriage, sold well. Allen's literary works were the key to his subsequent financial success, and they brought him the respect of such writers as Arthur Conan Doyle and George Meredith.[150] But he continued to pen scientific works, and David Cowie has persuasively argued that even Allen's fiction, as well as his writings on travel, aesthetics, and religion, served as a vehicle for pushing forward his evolutionary naturalist agenda.[151]

For Allen, the most important evolutionary naturalist was Herbert Spencer. In 1897 Allen praised Clodd's *Pioneers of Evolution,* particularly since it showed the central importance of Spencer's position in the "evolutionary advance."[152] Allen had been exposed to Spencer at an early age by his father, who was also an admirer. As an undergraduate at Oxford Allen had devoured Spencer's *First Principles* and the *Principles of Biology,* and while in Jamaica he read the *Psychology* several times.[153] Fired up by his reading of the *Psychology,* Allen wrote to Spencer on November 19, 1874, to tell him that he agreed "in the main with your chief speculations" and to express his gratitude for the guidance they had given him in making sense of human existence. He enclosed an ode to Spencer's greatness in order to "render you thanks for the personal assistance you have rendered <u>me</u> in interpreting the phenomena of the universe."[154] Later published in his *Lower Slopes,* "To Herbert Spencer" described the synthetic philosopher as the "deepest and mightiest of our later seers" who had "read the universal plan." The poem is a tribute to Spencer as a system builder, who, using the blocks supplied by lesser builders, "Rears high a stately fane, a grand harmonious

---

150. Rozendal 1988, 13.

151. Cowie traces the general trajectory of Allen's work, asserting that there was a pattern to his career as an author. The first period was dominated by serious work on physiology, followed by periods that focused on science, fiction, and then travel writing and religion. Cowie maintains that "within the diversity of Allen's work there was a uniformity supplied by his evolutionism and science, and in each of his works Allen attempted to consolidate or expand the authority of science and to advance the public understanding of evolution" (Cowie 2000, 19).

152. G. Allen 1897, 252.

153. G. Allen 1904, 612.

154. Allen to Spencer, November 10, 1874, University of London Library, Spencer Papers, MS 791/102[i].

whole."[155] Not expecting a reply to "a self-introduced letter," Allen was surprised to receive a letter from Spencer on December 10, 1874, expressing his gratification to find one who recognized the "meaning and scope of the work to which I have devoted my life."[156]

Allen responded to Spencer on February 9, 1875, by sending a paper on "Idealism and Evolution" and asking him to "use his influence with the Editor of the Contemporary or the Fortnightly to get it published." Allen acknowledged that in making this request he was "taking a great liberty," but he hoped that a "common interest in the highest subjects may form some link between us; even though I am an unknown beginner, and you have already made your name known to all genuine thinkers through the English speaking world at least."[157] Allen's self-deprecation and his fulsome flattery of Spencer had the desired effect, as Spencer brought the article to the attention of the editor of the *Contemporary Review*.[158] When Allen returned to England in 1876, he resolved to visit Spencer. Forgetting the exact street number of Spencer's residence, Allen was shocked to find that Spencer's neighbors did not know who was living among them. "The greatest philosopher that every drew breath," he recalled, "...the maximum brain on earth, is living in this square—and not a soul in the place has ever heard of him." Allen was even more surprised when he met Spencer and saw that he looked like "the confidential clerk of an old house in the City."[159] Despite his disappointment, Allen continued to cultivate their relationship. He asked on February 26, 1877, if he could dedicate his book *Physiological Aesthetics* to Spencer, insisting that "everything which I say in it is strictly in accordance with your views of psychological evolution."[160] The dedication read: "The Greatest of Living Philosophers, Herbert Spencer, I Dedicate (By Permission) This Slight Attempt to Extend in a Single Direction the General Principles Which He Has Laid Down."[161]

155. G. Allen 1894b, 45–46.

156. G. Allen 1904, 612-13. Spencer liked the poem and sent it to Edward Youmens who had it published in *Popular Science Monthly* (G. Allen 1875, 628; 1904, 613).

157. Allen to Spencer, February 9, 1875, University of London Library, Spencer Papers, MS 791/104.

158. Allen to Spencer, May 23, 1875, University of London Library, Spencer Papers, MS 791/108.

159. Allen 1904, 613-14.

160. Allen to Spencer, February 16, 1877, University of London Library, Spencer Papers, MS 791/117.

161. G. Allen 1877.

Allen paid public tribute to Spencer repeatedly in his scientific writings, presenting him as the key figure in the development of the concept of cosmic evolution. In his *Charles Darwin* (1885, Longman), Allen credited Darwin with showing that evolution was scientifically valid and that it was applicable to the human sciences. However, "on the other hand," Allen declared, "the total esoteric philosophic conception of evolution as a cosmical process, one and continuous from nebula to man, from star to soul, from atom to society, we owe rather to the other great prophet of the evolutionary creed, Herbert Spencer."[162] In the pages of a book purporting to celebrate Darwin's achievements, Spencer emerges as the more significant thinker when Allen assigns them their relative places in the pantheon of evolutionary heroes. Spencer wrote an appreciative letter to Allen in October 1885. He thanked him for "setting forth in various places the relations in which I stand towards the evolutionary doctrine, because it is a thing which I have not been able to do myself, and which none of my friends have hitherto taken occasion to do for me."[163] Two years later, in an article summing up the recent progress of science, Allen discussed the penetration of evolutionism into all dimensions of human life. The evolutionary movement "as a whole sums itself up most fully of all in the person and writings of Herbert Spencer," as to him humanity owed "the general concept of evolution as a single all-pervading natural process." Spencer had been evolutionism's "prophet, its priest, its architect, and its builder."[164]

In 1897, Allen offered an extensive comparison of the two evolutionists in his article "Spencer and Darwin." Darwin's primary contribution to the advance of evolution had been his theory of natural selection. He had been falsely credited with the idea of descent with modification in plants and animals, and he did not originate the idea of evolution as a cosmical process. Erasmus Darwin, Lamarck, and others first put the former forward, according to Allen, while the latter originated with Spencer, "and I venture to say from Herbert Spencer alone." After stripping Darwin of much of the glory often granted to him, Allen attempted to establish Spencer's status as an independent thinker. In regard to cosmic evolution, Spencer was "not in the remotest degree beholden for the origin of his ideas to Darwin." He was indebted to Immanuel Kant, Laplace, and the English geologists. "Thus," Allen concluded, "so far is it from being true that Mr. Spencer is a disciple of

162. G. Allen 1885, 191.
163. Clodd 1900, 126.
164. G. Allen 1887, 876-77; Atchison 2005, 55-64.

Darwin, that he had actually arrived at the idea of Organic Evolution, and of Evolution in General, including Cosmic Evolution, Planetary Evolution, Geological Evolution, Organic Evolution, Human Evolution, Psychological Evolution, Sociological Evolution, and Linguistic Evolution, before Darwin had published one word upon the subject."[165] Although Spencer did not reciprocate Allen's lavish praise in his published works, he did appreciate Allen's abilities. He wrote to Clodd on June 11, 1900, thanking him for a copy of his memoir of Allen. "I was often surprised by his versatility," Spencer told Clodd, "but now that the facts are brought together it is clear to me that I was not sufficiently surprised." Spencer added that he had always been impressed with Allen's "immense quickness of perception."[166]

Allen presented the Spencerian evolutionary epic to his reading audience in two different forms: in the expository essay and in the style of the natural historian. In his essay "Evolution"(1888) in the *Cornhill Magazine,* an example of the first form, Allen set out to correct common misconceptions about evolutionary theory. Contrary to popular belief, Darwin's *Origin of Species* had nothing to do "with evolution at large." Long before Darwin's book had appeared, evolutionary principles had been recognized in astronomy by Kant and Laplace, in geology by Lyell, and in biology by Lamarck. But, Allen explained, evolution, "according to the evolutionists, does not even stop there. Psychology as well as biology has also its evolutionary explanation." This is Allen's cue to launch into a discussion of Spencer's evolutionary psychology. From here, Allen moved to an examination of how evolutionists dealt with the application of evolutionary theory to human culture and society, as illustrated in Tylor's and Lubbock's work. "Having shown us entirely to their own satisfaction the growth of suns, and systems, and worlds, and continents, and oceans, and plants, and animals, and minds," Allen declared, "they proceed to show us the exactly analogous and parallel growth of communities and nations, and languages, and religions, and customs, and arts, and institutions, and literatures." The grand conception of the uniform origin and development of the universe, "summed up for us in the one word evolution," was the joint product of "innumerable workers," and not just Darwin, who worked toward "a grand final unified philosophy of the cosmos." Allen singled out Spencer for special mention as the one who established the use of the term "evolution" and who founded a system of philosophy applicable to the entire universe based on evolutionary theory.

165. G. Allen 1897, 254, 259, 261.

166. Spencer to Clodd, June 11, 1900, Leeds University Library, Clodd Correspondence.

"It is a strange proof of how little people know about their own ideas," Allen remarked, "that among the thousands who talk so glibly every day of evolution, not ten per cent are probably aware that both word and conception are alike due to the commanding intelligence and vast generalizing power of Herbert Spencer."[167] In this essay, as elsewhere, Allen tried to push his readers to reinterpret the significance of Darwin's contributions to science while at the same time outlining the cosmic evolutionism tied so closely to Spencer.

Allen's second method of presenting the evolutionary epic was more novel and important. The master of the short natural history essay, he had the ability to cram the story of cosmic evolution into a small literary space, in comparison to Page, Buckley, and Clodd, who relied more on the book as a vehicle for telling their grand story. Of course his master, Spencer, had required ten volumes to convey his evolutionary epic in his monumental *Synthetic Philosophy*. Allen's debt to natural history is evident in his edition of Gilbert White's *Natural History of Selbourne* ([1899], John Lane). Although White's book was a product of the past, filled with erroneous ideas, Allen claimed that it was far in advance of those of his time. Most late eighteenth-century books were filled with medieval fables and marvelous survivals of folk tales. White's method was more important than his results as he taught how to observe. "He shows us," Allen asserted, "by an object-lesson of patience and watchfulness how we ought to proceed in the investigation of nature." White represented to Allen the "dawn of the philosophic spirit in science" and he saw him as a forerunner of the "generation of colossal thinkers" that included Lyell, Darwin, Spencer, and Huxley. Allen praised the "universality of his broad interests" and found in the natural history tradition the scope for exploring the universality of the evolutionary process.[168]

Allen's periodical articles can be described as evolutionary natural history essays. In many of these essays he presented his important innovation: combining a focus on a common object in nature with the evolutionary epic. In his "Pleased with a Feather" (1879), Allen is sulking in his London drawing room during a murky winter afternoon. He admits that it is not a good opportunity for the "pursuit of natural history" since he cannot go out for an excursion. But he begins to study a feather sticking out of an Indian cushion on the sofa. The feather would serve as "the text for a humble lay-sermon concerning the nature and development of feathers in general, and the birds or human beings who wear them." What interested Allen about the feather is that it grew and that this was true of all plumage in the history of animal

167. G. Allen 1888a, 35, 45–47.   168. G. Allen [1899], xxxvi, xxxix.

evolution. This point quickly leads Allen to conclude that an examination of the feather, like any other piece of organic nature, could lead to grand conclusions about the cosmos. The fact that "everything has grown, throws a fresh and wonderful interest into every little object which we can pick up about our fields or our houses." Whereas the old view of creation as single and instantaneous made each creature and its organs seem like a "mere piece of moulded mechanism, with no history, no puzzle, and no recognizable relation to its like elsewhere," the new view of creation as continuous, progressive, and regular "teaches us to see in every species or every structure a result of previous causes, an adaptation to pre-existing needs." This allowed us to find "in a flower, a fruit, or a feather, innumerable clues which lead us back to its ultimate origin, and give delightful exercise to our intelligence in tracing out the probable steps by which this complex whole has been produced." The hidden hints in flowers, fruits, or feathers pointed to "some strange fact in the past history of the species." Whether it be an "unobtrusive spur or knob," or a "tuft, a spot, or a streak," they teemed with "information for the seeker who has found out the method of seeking aright." Like Kingsley, who could use the tale of a pebble to unravel the geological record, Allen could focus on one object and find an entire world in it, though one formed by cosmic evolution. In the case of feathers, Allen believed that they had been developed and selected through the habit of flight. Once in existence, further development—in particular the infinite variety of coloring and tone—had taken place through the process of sexual selection. The use of feathers by human beings for aesthetic decoration showed the similarity of artistic feeling running through the whole animal series.[169] Allen dealt with evolutionary aesthetics in a number of his essays and books. He argued that aesthetic feelings had a physical basis and that they are the product of natural and sexual selection. They were therefore not unique to humans.[170]

Allen often gathered his evolutionary natural history pieces together to form a book collection. His aptly titled *Vignettes from Nature* (1881, Chatto and Windus) contained essays that first appeared in the *Pall Mall Gazette*. In the preface he explained that they were written from an "easy-going, half-scientific, half-aesthetic standpoint" and he hoped that they would "do a little good in spreading more widely a knowledge of those great biological and cosmical doctrines which are now revolutionizing the European mind." In essays on fallow deer, butterfly hunting, a big fossil bone, carp, seaside

169. G. Allen 1879c, 712-13, 718, 721-22.   170. J. Smith 2004.

weeds, and a mountain tarn, Allen found unexpected illustrations of natural and sexual selection. The essays relied on natural history conventions, as he conducted nature rambles and interacted with the wildlife. In "Fallow Deer," Allen used the first-person voice to begin his examination of the evolutionary pedigree of the deer. "To-day I have brought out a few scraps of bread in my pocket," he wrote, "and the fawns are tame enough to come and eat it from my hand on the open." Thinking about how these fallow deer are part of "an old indigenous fauna" led Allen to a discussion of how the evolutionist was now similar to an archaeologist. Every animal contained within it the telltale clues as to its evolutionary ancestry, and these could be observed now that Darwin and Spencer had opened our eyes. "We have all been living all our lives in the midst of a veritable prehistoric Ilium," Allen asserted, drawing on the epic connotations of the Latin name for Troy, "with all its successive deposits and precious relics lying loose about us." For the deer, the clue to its development was contained in the bony projections on the forehead as they later helped in the struggle for existence when they evolved into antlers.[171]

Allen was at his most cosmic in his *Evolutionist at Large,* in which he reprinted essays originally from the *St. James Gazette.* It was moderately successful in terms of sales. Chatto and Windus produced a first print run of 1,000 copies and then another run of the same number for the second edition in 1884.[172] In 1889 it was published in the United States as part of the Humboldt Library series. Allen's essays focused on a wide variety of well-known natural objects, such as ants, butterflies, slugs, snails, tadpoles, berries, and nuts, to provide an evolutionary explanation for their most striking external features. In the essay "Distant Relations," for example, Allen opens with a rustic description of an old mill and an adjacent pond. By beginning with this familiar scene, Allen ensures that he and his readers start at the same point in the commonplace, and it allows him to "speak to people in words whose meanings they know." But as Allen and the reader examine the tadpoles in the pond together, they are abruptly thrust into the strange world of evolution. "At the bottom of this shallow pond you may now see a miracle daily taking place," Allen remarked, "which but for its commonness we should regard as an almost incredible marvel. You may there behold evolution actually illustrating the transformation of life under

---

171. G. Allen 1881b, v, 1, 3, 4.

172. University of Reading, Chatto and Windus Ledgers, 3: 379.

your very eyes."[173] Throughout the *Evolutionist at Large*, Allen drew on the same strategy to convey the evolutionary epic. In an essay on berries, Allen begins with some deceptively common observations on the relationship between fruits and nuts. Most fruits, he asserts, are really nuts that disguise themselves in sweet and beautifully colored pulp so that a bird will swallow its seed. But sweet juices and bright colors are of no use to a plant until there are eyes to see and tongues to taste. This prompts Allen to move back in time to tell his cosmic story of fruit and fruit eater. There could not have been "fruit-bearers on the earth until the time when fruit-eaters, actual or potential, arrived upon the scene: or, to put it more correctly, both must inevitably have developed simultaneously and in mutual dependence upon one another," Allen deduced. "So we find no traces of succulent fruits even in so late a formation as that of these lias or cretaceous cliffs." Allen then invited his readers to reflect on the wonderfully complex interdependency of beings in the evolutionary process. "No fruit, no fruit-bird; and no fruit-bird, no fruit . . . fruits and fruit-eaters [are] linked together in origin by the inevitable bond of mutual dependence."[174]

Although living beings were mutually dependent in Allen's evolutionary epic, there was no moral connection between them. In contrast to Buckley, Allen, who tended toward atheism, viewed the evolutionary process as amoral and without any religious, or spiritual, purpose.[175] Allen developed an ironical, whimsical style in order to make the horrors of the struggle for existence palatable to his readers. In his *Flashlights on Nature* (1899, Newnes), Allen handled gruesome accounts of animal behavior with a light touch. In one essay, an attack by warrior ants on another colony of ants "affords one opportunities for endless amusing glimpses into the politics of a community full of comic episodes and tragic dénouements." In another essay, Allen observed a garden spider, which he named Rosalind, through an entire season. Referred to as the most ferocious and bloodthirsty animal on earth, Allen described how the spider lay "concealed like a secret assassin in her nest," waiting for her next victim. But Allen's humorous style defused this "curious drama of blood and treachery," as he moved to a discussion of "the delicate question of the domestic relations of spiders, which are certainly not of a sort to be commended for imitation." In a mocking tone of moral indignation, he declared, "I regret to say it is her reprehensible habit to devour alive her unsuccessful suitors, and sometimes also the father of

173. G. Allen 1881a, vii, x, 96–97.   175. Morton 2005, 183.
174. Ibid., 94–95.

her own children." This helped Allen to drive home the point that nature is "intensely utilitarian" as each kind "regards as little the feelings of other kinds." This is no different, Allen insisted, from the way the fisherman paid no attention to the feelings of herrings. "A race that skins living eels at Billingsgate, and decks its hats with egrets in Hyde Park," Allen wrote, "has no just ground of complaint, after all, against my poor, misguided, husband-eating Rosalind." Allen ironically turns the tables on those readers tempted to condemn Rosalind and the entire evolutionary process. His humor helps readers to see that they implicitly accept the amorality of evolution.[176]

In some cases, Allen demonstrated in a humorous fashion how the factor giving an animal a leg up in the struggle for existence resulted in horrendous suffering for other animals. An essay titled "A Woodland Tragedy" dealt with the butcher-bird and its cruel habit of impaling prey to keep them alive until it was ready to dine (Fig. 5.9). This "ingenious but hateful invention . . . has secured him a place in the struggle for existence." But Allen points out that the butcher-bird treats its prey with the same indifference that a fishmonger treats a lobster. "It is her business to provide for her own young," Allen declared, "and she does it as ruthlessly as if she were a civilized human being." Ironically yoking ruthlessness with civilized behavior, Allen then comically depicts the butcher-bird as an exemplar of domestic and social virtue. The butcher-bird "believes himself to the end to be a model father, a tender husband, an ornament to society, and a useful citizen."[177] The contrast to Buckley could not be starker. Whereas Buckley solemnly regarded parental feeling as the epitome of the moral dimensions at the basis of the evolutionary process, Allen brought out the irony in the fact that devotion to one's offspring meant agony and death for other animals in an amoral natural world.

Just as Allen developed an ironical style to make the amoral evolutionary process more acceptable to his readers, he also retained a sense of wonder to enable readers to face the loss of divine beauty in nature. Beauty was sometimes accidental, due to the same natural causes, and sometimes designed, produced by the aesthetic choice of living beings engaged in a struggle for existence.[178] In neither case was beauty the result of divine

176. G. Allen 1899, 47, 60, 66, 70, 203.

177. Ibid., 76, 80, 89, 93.

178. Levin has argued that Allen began as a Darwinist who depended on mechanistic explanation and language, but that later, upon turning to aesthetics, he saw the need for preserving a place for mind and purpose in human evolution without succumbing to vitalism like Shaw and Wallace (Levin 1984, 77–89).

NO. 3.—PART OF THE BUTCHER-BIRD'S LARDER.

FIGURE 5.9 The butcher-bird's larder. (Grant Allen, *Flashlights on Nature* [London: George Newnes, 1899], 76.)

design.[179] Indeed, Allen insisted that the logical conclusion to draw from Darwin's theory of natural selection was the end of all natural theology. Darwin had shown that nature "does not teem with design and contrivance." Whatever design existed, it was a result of the operation of natural selection, which meant that it served the selfish interests of the species. The object of flowers was not to delight the human eye but to guide the fertilizing insect to the pollen.[180] Allen acknowledged Wood's point in his *Nature's Teachings* that "there is hardly a device invented by man which she [Nature] has not anticipated." But they were produced in nature because of the evolutionary process, not by the hand of God, and then imitated by humans in their struggle to survive. Whereas Wood dealt with the whole range of human invention, Allen emphasized the connection between invention and evolution by focusing on the use of "deception, wiles, and stratagems" in hunting

179. G. Allen 1881a, 215.   180. G. Allen 1888c, 2.

and warfare. The camouflaged traps for wild beasts were but a version of the deceptions used by animals and plants to catch their prey. Coats of mail and ironclads were based on the defensive armor "common in nature" in hard-shelled insects and marine creatures.[181]

While rejecting the notion of design, Allen used the language of wonder drawn from Romantic conceptions of the sublime to describe the sense of awe felt when humans beheld the evolutionary process. Here Allen was following Darwin's lead.[182] After presenting a discussion of the earwig's beauty reminiscent of Wood, Allen playfully admonished the reader that if, "after this, you ever despise those horrid earwigs, I shall think you have no taste for the wonderful in nature."[183] Allen was not to be outdone by Buckley or Wood in his depiction of nature as a world of wonder. He declared that "even Jules Verne's wildest story is comparatively tame and commonplace in the light of that marvellous miniature forest," a patch of English grass and mosses that was the scene of an awful life-and-death struggle between an ant and a mayfly. He compared the intricate and complex evolutionary cosmos to the world of unicorns and dragons invented by "mediaeval fancy," the dreams of the "Greek mythologist,' and the fables of the "Arabian story-teller." As for the story of the pedigree of the daisy, no fairy tale was "more marvellous, and yet certainly no fairy tale was ever half so true."[184] Allen stressed that evolution was not a thing of the past; rather, it continued to operate in the present. He stated that "it is a fatal habit to picture evolution to oneself as a closed chapter; we should think of it rather as a chapter that goes on writing itself continuously for ever."[185] Those who observed and understood the cosmic evolutionary process taking place in nature experienced a sense of awe and wonder. For this reason Allen referred to several of his essays

181. G. Allen 1901, 29, 142.

182. A number of scholars have noticed Darwin's debt to the Romantic notion of the sublime that influenced Burke, Wordsworth, Carlyle, and many others. Paradis has discussed Darwin's fascination with the aesthetic category of the "sublime" or "the emotional sensation of vastness one felt upon viewing an immense landscape from a great distance" (Paradis 1981, 87). Sloan has argued that the Romantic sublime of Goethe, as developed further by Humboldt, forms the main source of Darwin's pantheistic reflections on nature as vital, unifying, and creative, rather than the concept of nature in British natural theology (Sloan 2001, 251-69). Kohn has explored the important role of the Romantic sublime in the aesthetic construction of Darwin's theory (Kohn 1997, 13-48).

183. G. Allen 1899, 134.

184. G. Allen 1884, 8, 10.

185. G. Allen 1881b, 47-48.

in *Flowers and Their Pedigrees* as being a "sermon," because the evolutionary gospel that he was preaching still emphasized this sense of wonder.[186]

ALLEN, SCIENTIFIC NATURALISM, AND THE WORLD OF PUBLISHING

Due to his critical position on orthodox religion and his passion for evolutionary theory, Allen naturally gravitated toward the scientific naturalists when he returned to England from Jamaica in 1876. They also provided him with an entrée into the worlds of science and publishing. In 1879 Tyndall invited Allen to lecture at the Royal Institution on color sense.[187] Allen admired Tyndall. After Tyndall's death, Allen penned a character sketch in which he referred to him as "one of the prime leaders in the great revolution of the nineteenth century," next in importance only to Huxley and Spencer among the first generation of evolutionists.[188] Allen also corresponded with Huxley, and in 1882 he sent him his *Vignettes from Nature* to get an opinion on how well he had balanced accuracy with readability. Huxley wrote that the book was "delightful." He compared Allen to Shakespeare's character Falstaff, whom Huxley considered a great philosopher. "If Falstaff had been soaked in Evolution instead of sack," Huxley wrote, "I think he might have 'babbled o' green fields' in some such general fashion." Huxley had "no fault to find on the score of accuracy"; in fact, he found "much to admire in the way you conjoin precision with popularity." Huxley added that he thought highly of Allen's *Colin Clout's Calendar*. "With a few illustrations to help ignorant people to find what they ought to see," he asserted, "I would not wish for a better lure to the study of nature."[189]

Allen also cultivated Darwin. From the late 1870s to the mid-1880s, Allen drew on Darwin's botanical works to promote Darwinian botany and physiological aesthetics in books and in the periodical press.[190] In 1878 he sent Darwin a paper on the coloration of flowers and fruits, explaining that it filled a "small gap in that portion of your great theory which relates to

---

186. G. Allen 1884, 79, 131.

187. Allen to Tyndall, February 28, 1879, Royal Institution of Great Britain, Tyndall Papers. In that same year Tyndall also wrote to Warren de la Rue to recommend Allen for a job (Tyndall to de la Rue, February 25 or 26, [1879], Royal Institution of Great Britain, Tyndall Papers).

188. G. Allen 1894a, 21, 25.

189. Clodd 1900, 112.

190. J. Smith 2004; 2006, 160–63.

those structures."[191] Allen not only flattered Darwin with his reference to "your great theory," he also presented himself as a humble drone who was working to complete the magnificent edifice fashioned by the master. In 1879 Allen sent Darwin a copy of his new book, *Colour-Sense*. Anticipating criticism, Allen justified his reliance on recorded observations rather than on experiments. He explained that the heavy demands of his journalistic career forced him to "give to science the little leisure which remains to me after the business of bread-winning for my family is finished." But Allen believed that he could "be of some little use to scientific men by throwing out such hints as occur to me, and by working . . . in my own way, with the few materials which come within my reach." If he had the time and money, he would have preferred to take up physiological psychology in a scientific way and to do laboratory work. Allen signed off expressing "all the respect which every evolutionist owes to the founder of his faith."[192] Darwin must have responded quickly and sympathetically, as Allen wrote him nine days later, thanking him for "both your kind letters," and for the helpful criticism that they contained. Continuing the self-deprecating style of the previous letter, he acknowledged the imperfections of his work and apologized "for addressing you at all." But he knew that Darwin's interest in scientific truth was so great that he would forgive "even the bungling guesses of a learner."[193] Darwin did what he could to help Allen when his career as popularizer was floundering owing to ill health and insufficient income. When Allen was seriously ill in 1879, Darwin helped raise enough money to enable Allen to spend a winter in the south of France.[194]

Darwin reacted enthusiastically to Allen's *Evolutionist at Large*, writing, "the whole has pleased me. Who can tell how many young persons your chapters may bring up to be good working evolutionists!" Darwin complimented Allen on his clear and fluent writing style, and praised the novelty of some of his views. However, he did have one reservation: "Some of your statements seemed to me rather too bold," though Darwin acknowledged that in a work of this kind boldness may be an advantage."[195] Allen replied

---

191. Allen to Darwin, March 13, [1878], Cambridge University Library, Manuscripts, Charles Darwin Papers, MS.DAR.159: 41.

192. Allen to Darwin, February 12, 1879, Cambridge University Library, Manuscripts, Charles Darwin Papers, MS.DAR.159: 43.

193. Allen to Darwin, February 21, [1879], Cambridge University Library, Manuscripts, Charles Darwin Papers, MS.DAR.159: 44.

194. G. Richards 1932, 72; Morton 2005, 18.

195. Darwin to Allen, February 17, 1881, Dittrick Museum.

that he was "greatly gratified and not a little flattered by your very kind letter." Allen thanked Darwin for thinking of him, and for taking the "trouble of writing to me, in the midst of so many subjects which occupy your attention." But after expressing his gratitude, Allen defended his boldness by conveying a gentle lesson to Darwin on how to attract readers who were not scientific practitioners. "I was quite aware that the papers were a little bold," he declared, "but of course in writing for a daily paper one is obliged to adapt oneself to a very different audience from that which one addresses in a scientific book."[196] Allen had to be "bold" in his use of the evolutionary epic as a framework for his discussions of flowers, berries, and nuts if he wanted to attract readers. The following year, Darwin sent Allen a microscope as a gift. Allen thanked him and hoped that he would have more time in the future for "original observation, in which I have no doubt the microscope will be of great service to me."[197]

Other practitioners of science were not so enthusiastic about Allen's work over the years, and some voiced their objections in the pages of the journal *Nature*.[198] The anonymous reviewer of *The Evolutionist at Large* criticized Allen for being too superficial. He charged that Allen condensed and exhibited the teachings of Darwin and Spencer "in the most simple gossiping style," leaving the "most puzzling questions" unanswered.[199] Two would-be professional botanists attacked Allen after the appearance in 1882 of a short research paper by him in *Nature* on "The Shapes of Leaves."[200] Frederick Orpen Bower, one of Huxley's demonstrators at South Kensington in the early 1870s, wrote that Allen's article called for an "emphatic protest on behalf of botanists." Articles like Allen's, which contained "blunders of such magnitude, but written with that assurance of style which naturally carries conviction to the mind of the unwary, and disseminated through the country in a widely read journal like NATURE, cannot but produce a rich crop of erroneous impressions." Not only was Allen incapable of writing a proper research paper, Bower asserted that he was not even informed enough to convey scientific knowledge to a nonexpert reading audience, as the "popular writer must, before all things, be master at least of the first rudiments of

---

196. Allen to Darwin, February 19 [1881], University of Cambridge, Charles Darwin Correspondence, MS.DAR.159: 47

197. Allen to Darwin, March 24, 1882, Cambridge University Library, Manuscripts, Charles Darwin Papers, MS.DAR.159: 49.

198. Morton 2004, 37–38.

199. W. O. 1881, 27.

200. G. Allen 1883.

the subject on which he writes."[201] In the same issue Allen was again criticized, this time by William Turner Thiselton-Dyer, another one of Huxley's demonstrators from South Kensington during the early 1870s. Thiselton-Dyer complained of the "growing tendency, especially in writings intended for popular consumption, to explain everything by it [evolution] deductively." Explaining complicated morphological phenomena in a deductive fashion amounted to little more than a "literary exercise," failed to advance scientific understanding, and, in avoiding the necessary inductive work, "the theory of evolution runs a very good chance of being burlesqued."[202]

In 1888, Allen produced another piece of work that he considered a serious contribution to science, *Force and Energy: A Theory of Dynamics*, which argued for a redefinition of some of the most fundamental concepts in physics, such as energy. Unlike most popularizers of science, Allen yearned to be considered an original scientific thinker. The book had remained in manuscript form since 1877, until Clodd had showed him the first rough sketch of his *Story of Creation*. Discovering that they were working along the same lines, Allen had loaned Clodd the manuscript; Clodd decided to use Allen's theory in outline in the dynamical position of his forthcoming volume. After seeing Clodd's position attacked by critics, Allen decided that the whole theory should be published. Although he recognized that his theories were heretical, he asked scientific readers to approach the book with an unbiased mind.[203] However, nothing could have prepared him for the attack in *Nature* launched by the physicist Oliver Lodge. To Lodge, Allen's book "lends itself indeed to the most scathing criticism; blunders and misstatements abound on nearly every page, and the whole structure is simply an emanation of mental fog." Lodge was struck by the "contempt" that Allen, and other "paper philosophers," must "feel for men of science." Allen must see Newton and Thomson as "intolerable fools" for muddling away "their brains in concocting a scheme of dynamics" based on the wrong definition of energy that has to be set right "by an amateur who has devoted a few weeks or months to the subject." Lodge referred to Allen as one of Spencer's disciples, and avowed that both offered "criticisms from the outsider's point of view."[204]

Karl Pearson's attack on *Force and Energy* must have been even more difficult for Allen to bear, since he, unlike Lodge who was attracted to spiritualism, adhered to the principles of scientific naturalism. Pearson cast

201. Bower 1883, 552.

202. Thiselton-Dyer 1883, 554-55.

203. G. Allen 1888b, vii–xiv.

204. Lodge 1889, 289–92.

serious doubt on Allen's grasp of dynamical science, charging, "he has apparently never learnt its catechism." Since his definition of the terms "force" and "energy" were erroneous, and "absolutely unintelligible," every time he used them in his book "he simply introduces complete nonsense into well-known principles." Pearson compared Allen to a stubborn circle-squarer who refused to give up on his heterodox ideas. As a result, Allen had earned a place in the yet to be written "history of pseudo-science." But Allen was no harmless crank. The story of creation that he and Clodd had given to the world professed "to have a scientific basis, and herein is the source of the ill it may possibly give rise to." Allen, Pearson insisted, was spreading "a great deal of error in the lay mind, for Mr. Allen is a popular writer, whom all sorts of folk read."[205]

Occasionally, Allen's resentment of his critics surfaced in his writings, and it took the form of sarcastic comments aimed at narrow specialists. In *Flowers and Their Pedigrees* he discussed a minor point concerning the daisy's development to please that "terrible person, the microscopic critic," a "very learned and tedious being" who proclaimed to everybody "that you don't know something because you don't happen to mention it." In fear of him the science writer was "often obliged to trouble one's readers with petty matters of detail." Allen mocked this critic by naming him "Smelfungus," and was willing to go only so far to satisfy his "prejudices."[206] In 1895 he extended his rejection of narrow specialists to encompass the flaws of scientific orthodoxy as a whole. He complained that an "infallible and unassailable scientific priesthood" had been established, partly as a result of the Germanization of scientific education in England and America. Darwin, one of the great leaders of science in the past, had not been produced by having science drilled into him by methods similar to those used to train Prussian soldiers. Darwin "was not a trained physiologist. He was not a drilled and dragooned South Kensington student." Darwin was "merely an amateur, a lover of truth." Similarly, Spencer had had no specialized training and was the most original thinker "the world has ever known." The orthodox scientist merely followed "closely along accepted lines the accepted notions," and he could add details but he "cannot possibly upset, reconstruct, revolutionize." By contrast, the "task of the inspired outsider" was to come "to the work fresh, and with a fresh impulse" in order to revolutionize. Allen accepted the label conferred upon him by Lodge, and aimed to show that even the outsider

---

205. Pearson 1888, 421–22.   206. G. Allen 1884, 30–32.

had an important role to play in the progress of science.[207] To Allen, the vision of a trained cadre of professional scientists, so important to Huxley and his allies, had produced a rigid orthodoxy. Allen's disdainful reference to the "drilled and dragooned South Kensington student" could only have been a poke at Huxley, who, upon retiring in 1885, left behind the rigorous regime he had established to train biology students at South Kensington.[208]

Allen's edition of Gilbert White's *Natural History of Selbourne,* which he produced near the end of his life, takes on new meaning in light of his growing antagonism for would-be professional scientists. In his introduction, Allen looked back wistfully to White's age, when a "comparatively unoccupied gentleman of cultivated manners and scientific tastes" could make a contribution to science. "Those times have passed away," Allen wrote with regret. "Science has become a matter of special education." In aiming to advance knowledge, scientific education had produced narrowly trained "inventors, discoverers, producers of new chemical compounds, investigators of new and petty peculiarities in the economy of the greenfly that affects roses," rather than producing "many-sided men and women." While the power of the narrow specialist had grown, "the field of the amateur has been sadly curtailed." It was not possible to "attain to new facts or generalizations without the copious aid of libraries, instruments, collections, co-operation, long specialist training." Allen offered a new edition of the *Natural History of Selbourne* to show "how far White differed in the width and universality of his broader interests from the narrow and specialized man of science of to-day."[209]

Allen's increasing dislike for narrow scientific specialists was matched by his growing contempt for the entire publishing system in England. Even with the backing of eminent scientific naturalists Allen found that he could not survive financially as a popularizer of science.[210] His serious scientific works of the late 1870s and 80s, *Physiological Aesthetics, Colour-Sense,* and *Force and Energy,* were all commercial failures. The money he earned from his evolutionary natural history essays, and the books collecting some of

207. G. Allen 1895, 302–7.

208. Desmond discusses how Huxley established a "modern command structure in South Kensington," which swarmed with royal engineers. Desmond asserts that at the "Kensington Barracks" science "acquired real military authority and an air of national purpose" (Desmond 1997, 542, 632–33).

209. G. Allen [1899], xxxi, xxxix.

210. Peter Morton's recent book on Allen and the "socioeconomics of professional authorship" is the best-sustained analysis of how Allen tried to extract a living from the literary marketplace (Morton 2005, 8).

these essays together, was not enough to support his family.[211] It was no easier in the 1880s for popularizers to carve out a career for themselves than it had been in the 1850s. Allen tried to establish a career as a popularizer of science during a time when the world of publishing and journalism was undergoing a radical transformation from 1880 to 1895. Nigel Cross points to the "introduction of syndication, the expansion of the popular press, the founding of the Society of Authors, the rise of the literary agent, the relaxation of mid-Victorian pruderies in fiction, the triumph of the adventure story and of the gossip column" as factors that led to the climate of change that pervaded George Gissing's *New Grub Street* (1891).[212]

Other popularizers of science before him, such as Jane Loudon, Mary Kirby, and Margaret Gatty, had supplemented their income by writing fiction. Allen adopted this strategy in the early 1880s. He continued to produce some scientific works, but at the same time began to write light fiction, establishing a reputation for himself in that area by the middle of the 1880s. Allen churned out a series of successful sensationalist novels, but even so, it was not until the success of *The Woman Who Did* in 1895 that he finally achieved financial security.[213] By adapting to the market and transferring his energies from science writing to fiction, as well as to journal and magazine work, he managed to avoid impoverishment. Unlike J. G. Wood, who only produced natural history works, Allen never had to resort to applying to the Royal Literary Fund. Allen may have saved himself from financial disaster by moving away from popularizing science, and he may have had the opportunity to continue his exploration of evolutionary themes in his fiction, but ultimately, he found novel writing to be drudgery.[214] By 1892 he began to resent that he had been forced into doing such work. He wrote to one correspondent, "I share your inability to read novels, and don't much care for writing them." He complained, "the public will <u>read</u> my scientific articles; but it won't <u>pay</u> me for writing them; so in order to live, I have to take to fiction."[215]

Allen's difficulties in establishing himself as a successful writer led him to voice his criticisms of the entire publishing system and the readers who

211. Morton estimates that he received a total of £500 for his first four volumes of science essays and that this was too little a return for three years of hard work (Morton 2005, 91).

212. Cross 1985, 204.

213. Cowie 2000, 61.

214. For discussions of the evolutionary themes in Allen's fiction see Melchiori 2000 and Cowie 2000.

215. Allen to unknown correspondent, September 13, 1892, Pennsylvania State University Libraries, Mortlake Collection, Album 1.

supported it. In his "Ethics of Copyright" (1880) he insisted that in the current copyright system the rights of the author were subordinated to the interest of the public in getting cheap and good books.[216] Allen's criticisms became even more pointed nine years later in an anonymous piece on "The Trade of Author" (1889). Not only were authors ill-paid in comparison to other learned professions, in order to succeed they also were forced to give the public what it wanted: books that that were "uncritical and unthinking" that reflected the public's "own commonplace banal level." If you did not, "somebody else will survive in the struggle for life, while you go to the wall or into the workhouse." It was "the gospel according to Darwin and Malthus applied to art."[217] Allen's contempt for his readers could, at times, surface in his science essays. In the preface to his *Falling in Love* (1889, Smith, Elder), he contrasted his predilection for science, and champagne, "as dry as I can get them," with the public's yearning for sweeter bubbly. To accommodate the taste of his readers, "I have ventured to sweeten accompanying samples as far as possible to suit the demand."[218] In his *Flashlights on Nature,* he refrained to go into more details on the Hessian fly "lest I weary that fastidious and somewhat lazy person, the 'general reader.'"[219] All of Allen's frustration with the direction his writing career had taken can be read in this insulting reference to his readers.

Allen could also scorn periodical editors who demanded that he convey complex scientific ideas to an uninformed public in too short a space. When W. T. Stead asked him to write a short summary of the state of science in a couple of pages each month for his *Review of Reviews,* Allen set out to prove that it could not be done when "the people will not hear." Allen may have been the master of the short, evolutionary natural history essay, but not as compact as Stead envisioned. After discussing the complexities of Darwin, Spencer, and Wisemann's views on the nature of heredity, Allen exclaimed, "in spite of all my trying, the reader is as much at sea as at the outset. You can't explain those things off-hand in so short a space to the general public." Allen acknowledged that this would be "heresy" to the editor.[220] He could not accept Stead's application of the principles of the new journalism to the communication of scientific knowledge.[221]

Allen's unhappiness with his lot in life in the 1890s may have led him to become more critical of his chief evolutionary hero. In 1891 Allen's new

216. G. Allen 1880, 153–60.
217. [G. Allen], 1889b, 261, 267–69, 273.
218. G. Allen 1889a, preface.
219. G. Allen 1899, 293.
220. G. Allen 1890, 537–38.
221. Dawson 2004, 183–84.

triple-decker novel, *Dumaresq's Daughter,* was published, and it contained a pathetic character strikingly similar to Spencer. Haviland Dumaresq was the originator of the "Encyclopaedic Philosophy," "the profoundest thinker of our age and nation—the greatest metaphysician in all Europe." He had devoted twenty-five years of his life to mastering the infinite mass of detail that formed the groundwork of the "Encyclopaedic Philosophy" before he even began to write his book. But he lives in poverty and, outside of a few disciples, has had little recognition. When Charles Linnell, an admirer, meets him, he realizes that Dumaresq's mind works in an eccentric fashion, as he sees everything, even "the merest small-talk," as "a peg on which to hang some abstract generalization." But Dumaresq has several more serious flaws. His years of excessive toil and privation have taken their toll—he takes pure opium to calm his nerves.[222] Moreover, he has betrayed some of the principles of his own philosophy. Although critical in his writings of aristocrats for their worship of wealth, Dumaresq wants to surround his daughter Psyche with comfort and luxury. When Linnell falls in love with his daughter, Dumaresq will not allow them to marry, as Linnell is a poor artist.[223] Spencer, of course, never married. The novel raised the question, what would have happened if Spencer had had a daughter? Allen provided him with an alternate reality to demonstrate that however much the Synthetic Philosophy might appeal to a recluse like Spencer, its asceticism was too much for those who had families or any substantial human relationships.

Allen's increasingly critical attitude toward Spencer in the 1890s is evident in an extraordinary article that appeared in 1904, five years after Allen's death, and shortly after Spencer's death. Allen had actually written it in 1894, but would not allow publication while Spencer was still alive.[224]

222. G. Allen 1891, 1: 127. Spencer reputedly took drugs to counter his sleeplessness. After visiting Spencer as he lay dying in December 1903, Beatrice Webb wrote in her diary of his pessimism about the world. "Indeed," she remarked, "the last twenty years have been sad—poisoned by morphia and self absorption" (B. Webb [1950], 32). Allen's account of one of Dumaresq's drug induced hazes implies that Spencer's grasp of the connections in nature, the basis of his Synthetic Philosophy, was in part a result of his habit. As Dumaresq walks in a field "the opium was transforming the earth into heaven for him. Space swelled, as it always swelled into infinite abysses for Haviland Dumaresq when the intoxicating drug had once taken full possession of his veins and fibres." Dumaresq's "dilated vision" causes the horizon to "spread boundless in vast perspective" and the hills to rise up "in huge expansive throes" so that they become "high mountains." His own stature doubles and he no longer walks in "our prosaic world: each step appeared to carry him over illimitable space: he trod with Dante the broad floor of Paradise" (G. Allen 1891, 1: 223-24).

223. G. Allen 1891, 1: 17, 53, 109, 208.

224. G. Allen 1904, 610.

Clearly, Allen knew that Spencer would be hurt if he saw what Allen really thought of him. Allen depicted Spencer as a man almost devoid of emotion, a man whose "soul was less richly endowed" than most. "Nature, in making him," Allen asserted, "had concentrated all her energies, so to speak, on intellect. . . .He was pure intellect, and little more: the apotheosis of reason in a human organism." As a result, Spencer possessed "the finest brain and the most marvellous intellect ever yet vouchsafed to human being." Although Allen regarded him on the intellectual side with "the profoundest reverence," he nevertheless did not agree with him on everything. There were "serious errors and lapses in his intellect." Allen acknowledged that a large part of *First Principles,* the first volume of the Synthetic Philosophy, was "vitiated by a false conception of Energy," and that "there are serious misconceptions in parts of the 'Sociology.'"[225]

However, Allen maintained that Spencer's most significant error lay in his attitude toward socialism. Allen's radical interests place him in the middle of late nineteenth-century British progressivism. He spoke out against the iniquities of the capitalist system and the selfishness of restricting access to private land.[226] This eventually drove a wedge between him and his evolutionary hero. "Especially toward the end of his life," Allen declared, "I think he went often grievously wrong, more particularly in his social and political thinking." Influenced by Spencer's *Social Statics* (1850), and his idea of land nationalization, many of Spencer's earlier admirers had become socialists. Allen asserted that Spencer had considered Beatrice Webb and himself as "his chosen disciples," as "his two favorite followers," and that it was a great blow to him when they proclaimed their adherence to socialism. Allen quoted from an angry letter sent by Spencer to him on October 23, 1890, that touched on the political differences dividing them. "I hear that you have turned socialist," Spencer indignantly wrote. Allen believed that Spencer's notion of individualism "did not hang together with the rest of his philosophy" and that it was intellectually inconsistent. Spencer "did not see that an individualism which begins by accepting all the existing inequalities and injustices is no individualism at all; that his own early principle of land nationalization struck the keynote of revolt; and that socialism offers the only real hope to the thorough-going and consistent individualist of the future." While all those whom Spencer had deeply influenced had followed out the logical consequences of his earlier ideas by becoming socialists, Spencer's key supporters in his old age were the Tories and militarists he had earlier

225. Ibid., 610–11, 619.   226. Nottingham 2005, 101, 107.

denounced. Allen received the angry letter from Spencer just a year before *Dumaresq's Daughter* was published. It may have led him to create the character Dumaresq, who embodied all of the striking inconsistencies that he perceived in Spencer's political and social thought.[227] Allen's socialism drove a wedge between him and the other scientific naturalists as well. In his character sketch of Tyndall, Allen pointed out that "all of the men of that first generation who spread the evolutionary doctrine among us are now reactionary in politics" and "bitterly hostile to the Socialism of the future."[228] Indeed, in 1890 Huxley published a series of essays attacking socialism in the *Nineteenth Century,* including his "Capital—The Mother of Labour" and "Natural Inequality of Men."

DEFERENCE AND THE BUTLER AFFAIR

If popularizers of science departed too much from the agenda of scientific naturalism, they risked challenges to their authority. Buckley, Clodd, and Allen had one glaring example of the consequences of attacking a leading scientific naturalist: the sad fate of Samuel Butler (1835-1902). Their reactions to Butler's excommunication from the inner Darwinian circle reveal the complex strategies involved in deferring to scientific practitioners in exchange for furthering one's career in scientific writing. Like Allen, Butler had become enthralled with evolutionary theory while abroad and returned to England determined to cultivate Darwin and other members of his circle in order to further his literary career. After a stint as a sheep farmer in New Zealand, Butler was back in England in 1864 and in correspondence with Darwin shortly after his return. Between 1865 and 1877 he sent most of his essays and all of his books to Darwin to gain his approval. But Butler became increasingly critical of Darwin, starting with the publication of his *Life and Habit* in 1877, where he argued that his theory of evolution, based on the concept of inherited habit, offered a better explanation of biological facts than Darwin's emphasis on natural selection. In this book Butler presented his own version of the evolutionary epic that combined Lamarckian evolution with the notion of inherited habit. The existence of habits in humans indicated that we are born with the memories of our progenitors. Every step forward was passed on to succeeding generations through inherited habit. All living things, according to Butler, possessed the same unconscious memory, and were therefore part of a single organism. Butler's aim in the

---

227. G. Allen 1904, 610, 626-27.   228. G. Allen 1894a, 25.

book was to build his notion of instinct as inherited memory into a grand picture of life as a single, vast compound animal. His evolutionary process was fraught with purpose in that it was the result of will, though not a divine will.

Butler had at first hoped that the Darwinian circle would welcome his entrance into the arena of debate on evolutionary theory. But by the time he wrote his *Evolution, Old and New* (1879), it was clear to him that his concept of inherited memory was not being taken seriously by Darwin or his allies. Butler began to use his considerable satirical powers to criticize Darwin. Butler held Darwin responsible for clouding all discussion of evolutionary issues by introducing his theory of natural selection, which, he argued, was inferior to the teleological evolutionary theories of Georges-Louis Buffon, Erasmus Darwin, and Jean-Baptiste Lamarck. Whereas in the earlier book he presented himself as a friend of Darwin's who offered a different though complementary theory, in this book Butler was intervening in the scientific debates over the nature of evolution by criticizing Darwinian theory and countering it with a theory of his own. But Butler alienated the Darwinians even further in a controversy that arose over the publication of a two-part biography of Erasmus Darwin by Charles Darwin and Ernst Krause. This book appeared shortly after Butler had sent Darwin a copy of his *Evolution, Old and New*. Purporting to be an English translation of a previously published paper by Krause, *Erasmus Darwin* contained several new passages that, according to Butler, borrowed from *Evolution, Old and New* without acknowledgment and, worse still, questioned Butler's sanity in championing the grandfather's version of evolution over the grandson's. When Butler demanded a public apology, Darwin ignored him at the advice of Huxley and other friends, and later, Darwinians such as George John Romanes assailed him.[229] Butler's attacks on Darwin became even nastier in his *Unconscious Memory* (1880) and his *Luck or Cunning?* (1886). They were expanded into a critique of scientific naturalism as overly materialistic and dogmatic.

Butler's decision to cast deference aside and confront evolutionary naturalism head on cost him dearly. In his *Luck, or Cunning?* Butler acknowledged publicly that his anti-Darwinian stance had "got me into the hottest of hot water, made a literary Ishmael of me, lost me friends whom I have been sorry to lose, [and] cost me a good deal of money."[230] Later, in 1901, Butler claimed that taking Darwin on had had a negative impact on the trajectory of his entire literary career. "*Evolution Old and New* and *Unconscious Memory*

---

229. Paradis 2004.   230. Butler 1924, 14.

made a shipwreck of my literary prospects; I am only now beginning to emerge from both the literary and social injury which those two righteous books inflicted on me."[231] For some popularizers, such as Gatty, who could not abide the agenda of scientific naturalists, silence was too high a price to pay in exchange for establishing their authority. But Butler was even bolder than Gatty. He took direct aim at Darwin and eminent scientific naturalists. Butler's fate was an object lesson to other popularizers of science: Challenge the authority of scientific naturalists and you risk losing scientific credibility and face the threat of commercial failure. Buckley, Clodd, and Allen all knew Butler. The lesson was not lost on them.

Buckley stood by Darwin. She saw Butler in the British Museum after having visited Darwin in October 1880, and discussing with him the Krause controversy. Darwin had told her that he knew nothing of what Krause had written and was not responsible for it. She defended the nobleness of Darwin's character and accused Butler of personal malice.[232] Although Butler thought that Buckley was "very nice" when introduced to her for the first time, he later referred to her as a "silly tattling log-rolling mischief-making woman and I dislike her very much."[233] Clodd invited Butler to his home in the late 1870s. But the Krause controversy disrupted their friendship. Butler asserted that "Clodd dropped me when I became so unpopular through my row with Darwin: he met me once or twice and would hardly speak."[234] From Clodd's point of view, Butler "became a man with a grievance" after his "deplorable attack" on Darwin in *Unconscious Memory*. Clodd recalled, "unfortunately he nursed the delusion that every man of science if he defended Darwin was in conspiracy against himself."[235]

Allen came to Darwin's defense publicly. He wrote two reviews of Butler's *Evolution, Old and New*. In a signed review in the *Academy* he complained that Butler's "dazzling flood of epigram, invective, and what appears to be argument" left the reader "without a single idea of what it has all been driving at." Then he objected that Butler had crossed the line in his attack on Darwin. He had allowed himself to "use unseemly and contemptuous language toward an old and honoured scientific chief, whom even his adversaries should respect for his noble devotion to truth and his lifelong pursuit

231. Keynes and Hill 1935, 40.

232. Breuer 1984, 123.

233. Keynes and Hill 1935, 196.

234. St. John's College, Cambridge University, Samuel Butler's Notes, 210.

235. Clodd 1926, 256.

of knowledge."[236] In an unsigned review in the *Examiner,* Allen labeled Butler's notion of the conscious will as "just the sort of mystical nonsense from which we had hoped that Mr. Darwin had for ever saved us," reaffirmed the value of the theory of natural selection, and recommended that Butler "avoid science" and stick to literature.[237] Butler was deeply offended by Allen's reviews, which, he claimed, "led the way of adverse criticism" to *Evolution, Old and New.* By signing one review and not the other, the public, Butler believed, would not suspect that the same person had written both, and it would be concluded that "two independent writers each took the same unfavourable view of the book."[238] Allen, Clodd, and Buckley all followed the lead of the foremost scientific naturalists by punishing Butler for his audacious attacks on Darwin.

But in the mid-1880s, Clodd and Allen became more sympathetic to Butler's situation. When they bumped into each other, Clodd talked up Allen's work, at which point Butler told him that he did not like Allen. Butler was surprised to receive an invitation from Clodd to come to tea one Sunday in 1885. Butler accepted, and then Clodd informed him that Allen would also be coming. Butler thought that this was "rather cheek, for he knew I did not like Grant Allen," but he had already accepted the invitation and decided to go. In one of Butler's accounts of the tea, he claimed that he did his best to be civil and so did Allen. "It all went off very nicely" and Butler shook hands with Allen. He was somewhat disconcerted when Allen told him that *Evolution, Old and New* had been very "useful," as he recalled how one of Allen's reviews asserted that the book left the reader without a single clear idea of his main object. But Butler held his tongue.[239] In another account, Butler appreciated Allen's help in extricating himself from an uncomfortable conversation. Henry Bates, the naturalist, was at the tea and he would not speak to Butler due to his attacks on Darwin. Bates repeatedly spoke of Darwin's brilliant discovery of natural selection. Allen then wondered whether or not Darwin was influenced by the works of his grandfather, a move that Butler interpreted as a ploy to draw him into the conversation, since in recent books he had argued that Darwin had concealed his debts to Erasmus Darwin. With a "wry face," Butler "blurted" out that it was inconceivable that anyone should pay the smallest attention

236. G. Allen 1879a, 426–27.
237. [G. Allen] 1879b, 647.
238. St. John's College, Cambridge University, Samuel Butler's Notes, 209–10.
239. Ibid., 210–11.

to anything written by his grandfather. Everyone laughed and the subject was dropped.[240] Writing to his sister on June 30, 1885, Butler saw the whole occasion as a plan by Clodd to "bring me and one of my particular foes Grant Allen together" and all in all it "went off smoothly." In the end, "Grant Allen wanted to make peace and I let him."[241]

Although Butler's *Luck, or Cunning?* contained criticisms of Allen's extravagant praise of Darwin in *Charles Darwin*, the peace held.[242] Butler appreciated Allen's "handsome acknowledgement of Evolution Old and New" in the preface to *Charles Darwin*.[243] He believed that he and Allen were actually in agreement about the nature of the evolutionary process, since Allen subscribed to Spencerian evolution, which Butler viewed as a form of Lamarckianism.[244] For his part, Allen wrote a sympathetic review of Butler's *Luck, or Cunning?* He recognized that Butler's controversial theory had lost him friends and cost him money. He confessed, "we have all hitherto done an unwitting injustice to Mr. Butler." The world had neglected his clever hint thrown out in *Life and Habit* and had "treated it too hardly." Although Allen did not agree with Butler's harsh criticisms of Darwin, he nevertheless judged *Luck, or Cunning?* to be "a most valuable, original, and suggestive contribution to current evolutionary thought." Allen hoped that biologists would read it and ignore Butler's emotional outbursts directed at Darwin. Butler had something to say that had "often been widely overlooked." Allen warned that the men of science should not "rule him out of court inexorably without even granting him so much as a hearing."[245] Perhaps Allen's rough treatment at the hands of Bower and Thiselton-Dyer had made him more sensitive toward the obstacles that Butler faced in trying to play a role in serious scientific discussion.

Allen's nephew, the publisher Grant Richards, once asserted that Allen "certainly was one of the first and the foremost interpreters of Darwin to the multitude."[246] Although Allen was certainly one of the most important interpreters of evolution to the vast Victorian reading audience, he saw himself as presenting the Spencerian, not the Darwinian, version. Strikingly, none of those who envisioned an exciting evolutionary epic for the public looked to Darwin first and foremost. Page did not accept Darwin's theory of natural selection. Despite his falling out with Robert Chambers, his views

240. H. Jones 1919, 2: 20–21.
241. D. Howard 1962, 141–42.
242. Butler 1924, 196.
243. D. Howard 1962, 152.

244. Butler 1924, 189–92.
245. G. Allen 1886, 413–14.
246. G. Richards 1932, 72.

on evolution were profoundly influenced by the work they did together. Buckley was indebted more to Wallace or Lyell than to Darwin. Although enamored of Darwin, Clodd's cosmic evolution was inspired by Huxley, Tylor, and Spencer. Allen was a self-professed disciple of Spencer's. Moreover, these four evolutionists had complicated relationships with scientific naturalism in general. Page and Buckley can be grouped together as the most resistant to elements of the scientific naturalist's agenda. While Page enshrined design into the heart of his epic, Buckley's endowed the evolutionary process with a moral teleology. Clodd and Allen toed the naturalist line more closely, but they did not slavishly imitate Spencer, Huxley, or Darwin. Both demonstrated their independence in several ways. Clodd, by allying himself with the Rationalists, Allen, by affiliating with socialism, and both by befriending Butler after he had been shunned by the Darwinians for his sins.

CHAPTER SIX

———— ✳ ————

The Science Periodical

Proctor and the Conduct of "Knowledge"

IN 1882, in the pages of the second volume of his newly founded journal *Knowledge,* Richard Proctor wrote an impassioned piece objecting to the recent construction at Kew Gardens. Portions of the wall surrounding the Gardens had been built higher, and the Temperate House Gate was being bricked up. Proctor complained that the national purse paid for the Gardens—they were "the people's own property"—and yet the Gardens were hidden from the public's view. And why, Proctor asked, did the Gardens not open until one o'clock while the director, Joseph Dalton Hooker, entertained guests on a regular basis? Proctor implied that Hooker used the Gardens as a private park for his friends, among whom were fellow X Club members T. H. Huxley, John Tyndall, and Herbert Spencer. Hooker may be an excellent botanist, Proctor acknowledged, but he had insulted and wronged the public. "That long wall," Proctor declared, "is a disgrace to England, a discredit to every Englishman who, having seen it, does not do all that lies in his power to have it replaced by such an enclosure as shall protect without hiding these public gardens."[1] Readers of Proctor's *Knowledge* would have understood that the wall enclosing Kew Gardens was no different in Proctor's eyes from the boundary that practitioners of science maintained to exclude amateurs, women, and the rest of the English public from the world of science. Preventing access to nature allowed the practitioner to claim a monopoly over natural knowledge. In titling his new

---

1. Proctor 1882j, 351–52.

journal *Knowledge,* Proctor staked out his own claim to knowledge on behalf of popularizers such as himself, as well as for his readers.

Proctor's *Knowledge* was founded just after a period of significant growth and development in British periodicals during the middle of the nineteenth century. This was due to many of the same factors that had led to the explosion in the publication of books, such as the demand from increasingly literate and leisured reading audiences for novel and varied sources of knowledge and amusement; new technological developments, including ever more efficient printing machines, that provided ways of catering to this growing and diversifying taste; and reduced costs as a result of changes in taxation. Shilling monthlies, such as *Macmillan's Magazine* (f. 1859) and the *Cornhill Magazine* (f. 1860) arose, attracting educated but not overly affluent readers. More expensive monthly reviews were also established in the 1860s and 70s, including the *Fortnightly Review* (f. 1865), the *Contemporary Review* (f. 1866), and the *Nineteenth Century* (f. 1877), featuring reasoned debate on timely intellectual issues. Scientific issues were included to satisfy the middle-class taste for topical, learned, and entertaining discourse. The new monthlies provided an important forum for professionalizing scientists like Huxley who wished to address a wider audience.[2]

Proctor's *Knowledge* belonged to the growing group of commercial science journals, some catering to popular audiences, that began to appear in the 1840s. These journals were modeled on cheap weekly mechanics' magazines (e.g., the *English Mechanic,* f. 1865), on trade weeklies (e.g., the *Chemical News,* f. 1859), on expensive monthly natural history magazines (e.g., the *Zoologist,* f. 1843), or on periodical genres usually associated with general topics (e.g., the *Popular Science Review,* f. 1862, and the *Reader,* f. 1863).[3] Before 1860, many of the science journals for popular audiences encouraged their readers to participate in the scientific enterprise. These periodicals stressed the universal accessibility of the scientific endeavor based on the proposition that all men possessed the same capacity for understanding nature. An egalitarian vision of a "Republic of Science," the periodical itself was presented as a space for communicating scientific observations to others. Science periodical publishing for the general reader peaked in the 1860s, just as Huxley, Norman Lockyer, and other members of the young guard adopted the professional ideal of the scientific expert. Many science periodicals began to mirror the

2. Cantor et al. 2004, 1, 16, 19–23. For studies of science in the general periodical press in the later period, from 1890 to just before World War I, see Broks 1988, 1990, 1993, 1996.

3. Cantor et al. 2004, 16; W. Brock 1980; Barton 1998b; Sheets-Pyenson 1976; 1985.

goals of the new guard and shifted their emphasis from seeking the participation of their readers to eliciting support for the scientific elite.[4] Like the new
monthlies, the science periodicals founded during this period offered a platform for the would-be scientific professional bent on promoting the cultural
importance of science to the public. The journal *Nature* (f. 1869) was a prime
example.

Both general and commercial science periodicals also provided some
popularizers with opportunities for publishing their scientific work and for
supplementing the income they earned from writing books. Among the clergymen, Kingsley, Webb, Henslow, and Houghton were active journalists.
Kingsley and Houghton published primarily in the general periodical press.
Kingsley wrote over twenty articles for such journals as *North British Review*
(one of the newer quarterlies) and for the monthly *Fraser's Magazine for Town
and Country*, while over fifteen pieces by Houghton appeared in the older established quarterlies, such as *Edinburgh Review* and *Quarterly Review*, as well
as in the *Westminster Review*. Twenty articles by Henslow were published
in both the general periodical press, in journals such as *Nineteenth Century*
and *Modern Review*, as well as in the science periodical *Journal of the Royal
Horticultural Society*. Webb was the most prolific of the four, writing close to
two hundred articles on astronomy, and he placed his work almost exclusively in science periodicals. Many of his pieces appeared in *Monthly Notices
of the Royal Astronomical Society* and *Reports of the British Association for the Advance of Science*, organs of scientific societies, but he also wrote many articles
for periodicals that appealed to a popular audience, including *English Mechanic, Intellectual Observer, Knowledge, Nature*, and *Popular Science Review*.
Few female popularizers contributed scientific articles to periodicals, and
those that did, which included Phebe Lankester and Lydia Becker, were
not particularly prolific. Whereas Lankester wrote a few articles for *Popular
Science Review*, Becker contributed to the general periodical press. A handful of articles by Becker appeared in *English Woman's Review, Contemporary
Review*, and *Westminster Review*.

Turning to the showmen of science, Pepper undertook little work for the
periodical press. His energy went into running the Polytechnic, lecturing,
and writing books. Wood authored over twenty articles for a wide variety
of periodicals, including religious and children's journals, monthlies, magazines, and dailies. His work appeared in *Boy's Own Paper, Cornhill Magazine*,

---

4. Sheets-Pyenson 1976, 57–59; Barton 1998b, 2–3.

Daily Telegraph, Dark Blue, Dublin University Magazine, Good Words, London Society, Longman's Magazine, Nineteenth Century, and *Sunday Magazine,* but not in any science periodicals.[5] The majority of these pieces were on some aspect of natural history. For the *Boy's Own Paper,* for example, he contributed a number of serialized articles on botany, entomology, and other aspects of natural history, sometimes in collaboration with his son Theodore.[6] Although Page engaged in little periodical work, preferring to churn out books, the other three popularizers of the evolutionary epic were more willing to write short pieces. Buckley's *Moral Teachings of Science* was published in installments in *Chautauquan,* in addition to which seven articles came out in journals such as *The Youth's Companion, Macmillan's Magazine,* and *Dublin University Magazine.* Edward Clodd produced over ten articles for such journals as *Knowledge, Modern Review, Longman's Magazine, Folklore,* and *Fortnightly Review,* many on aspects of anthropology, folklore, and evolution. But Grant Allen was more active than any of the clergymen, women, showmen, or evolutionists in this particular area of popularization of science. He produced at least 250 articles in over fourteen different periodicals, from dailies, weeklies, and monthlies to a number of science periodicals. His work appeared in *Belgravia, Cornhill Magazine, Contemporary Review, Eclectic Magazine, Fortnightly Review, Fraser's Magazine, Knowledge, Longman's Magazine, Macmillan's Magazine, Nature, North American Review, Pall Mall Gazette, Popular Science Monthly, Scientific American,* and *Strand Magazine.* For Allen, generating articles at a rapid speed was an integral part of his efforts to make a living as a writer.

As prolific as Allen was, Proctor topped even him. Michael Crowe estimates that Proctor wrote at least five hundred essays, not counting his many technical papers in the *Royal Astronomical Society Monthly Notices.*[7] He wrote for science periodicals aimed at a popular audience, for example, *Intellectual Observer, Popular Science Review, Scientific American,* and *Knowledge,* and for the general periodical press, in journals such as *Belgravia, Contemporary Review, Cornhill Magazine, Eclectic Magazine of Foreign Literature, Fraser's Magazine, The Friend, Littell's Living Age, Longman's Magazine, National Review, Nineteenth Century, North American Review, Open Court, Saint Pauls Magazine,* and *Temple Bar.* Proctor was not merely the author of hundreds of articles, many reprinted in his numerous books on astronomy, he was also the editor of an important science journal. It was due, in part, to his astounding

---

5. Lazell 1972, 3–4.   7. Crowe 1989, 1.
6. Noakes 2004, 159.

productivity, and his editorship of *Knowledge*, that by the time of his death Proctor had become the most widely read astronomical popularizer in the English-speaking world.[8] The London *Times* obituary declared that Proctor had "probably done more than any other man during the present century to promote an interest among the ordinary public in scientific subjects."[9] Proctor's lucid writing style and his willingness to indulge in daring speculations on the existence of extraterrestrial life in his many astronomical writings were also important reasons for his popularity. Proctor was intimately familiar with the worlds of both the practitioner and the popularizer of science. He conducted original research on Venus and Mars, charted the directions and motions of about 1,600 stars, and played a major role in the formulation of the conception of the Milky Way and of the universe as a whole.[10] His scientific papers on various astronomical subjects are to be found chiefly in his eighty-three technical essays in the *Monthly Notices of the Royal Astronomical Society*. Some Royal Astronomical Society members thought highly enough of Proctor's research to nominate him for the Royal Astronomical Society's Gold Medal in 1872, though in the end this honor eluded him. Since he was a liminal figure—one who straddled the worlds of the practitioner and the science writer—an examination of his journal *Knowledge* provides an unusual glimpse into the tensions that existed between would-be professionals, popularizers, and the reading public.

Although I have touched on periodicals from time to time, up until to this point I have focused mainly on books. In this chapter the emphasis will be on periodicals. Barton and Sheets-Pyenson have demonstrated how science journals can be examined to reveal changing conceptions of the boundaries of scientific knowledge and of membership within the scientific community.[11] Here I will offer a case study of Proctor's *Knowledge* in order to illuminate the important role of periodicals in the career of one of the most prolific popularizers of the second half of the century. Periodical work not only provided a steady source of income for Procter, it also allowed him to push an agenda at odds with the goals of practitioners who supported the creation of a professional scientific elite. Fueled by heated controversies with Lockyer and other professionalizing astronomers during the 1870s, Proctor created a weekly journal designed to challenge *Nature* for control of the science periodicals market for the public and to question the role and dominance of practitioners. Proctor aimed to reinvigorate the vision

8. Ibid., 1.
9. "Late Mr. Richard A. Proctor" 1888, 5.
10. Crowe 1989, 9; North 1975, 162–63.
11. Barton 1998b; Sheets-Pyenson 1976; 1985.

of a "Republic of Science," which had animated many science periodicals for popular audiences prior to 1860. Proctor's intention to compete with *Nature* is reflected in his journal's title, objectives, format, content, and the background of the contributors. Proctor's own role as "conductor," rather than editor, of *Knowledge*, his handling of the correspondence columns, and his criticisms of scientific societies and would-be professional astronomers in his own contributions to *Knowledge* reveal his attempt to open science up to full participation. However, Proctor's experiment in science publishing ran into problems only four years after its initial issue. In the process of revamping *Knowledge* into a monthly in 1885, in part as a result of financial pressures, the journal underwent an interesting transformation. Proctor's original vision of a scientific community, in which well-informed members of the public played an integral role, was lost with the move to a monthly format. Now on sounder financial footing, *Knowledge* survived Proctor's sudden death in 1888. A. Cowper Ranyard took over the editorial duties and moved the journal from Longman to W. H. Allen. *Knowledge* continued into the twentieth century, ceasing publication in December 1917—a good run for any periodical in this period.

THE CAREER OF A PROLIFIC WRITER AND LECTURER

Richard Anthony Proctor (1837–1888) became involved in popularizing science when he suffered the tragic loss of a child and soon after faced financial disaster. The youngest of four children of William Proctor, a wealthy solicitor, Proctor was educated by his mother at home until 1848, when he attended a school in Milton-on-Thames for three years (Fig. 6.1). He did not enter King's College, London, until 1855, because of temporary financial problems when his father died in 1850. In 1856 he proceeded to St. John's College, Cambridge. In accordance with his mother's wishes, Proctor intended to enter the Church and read theology and mathematics. But he was distracted by his mother's death in his second year at Cambridge and by his marriage to a young Irish woman. As a result, his academic accomplishments as an undergraduate were disappointing and he graduated in 1860 as twenty-third wrangler. Proctor decided against the ministry after graduating and instead began to study law. He became interested in astronomy a few months after he left Cambridge when he came upon John Pringle Nichol's *Views of the Architecture of the Heavens* (1838) and Ormsby MacKnight Mitchell's *Popular Astronomy* (1860). Fascinated by astronomy, he mounted a small telescope and decided to train himself in the subject so

THE LATE MR. R. A. PROCTOR, ASTRONOMER.

FIGURE 6.1 Richard Anthony Proctor, the "eminent scientific astronomer."
("The Late Mr. R. A. Proctor," *Illustrated London News* 93 [September 29, 1888]: 363.)

he could teach his eldest son. Although he found writing difficult at first, Proctor composed "The Colors of Double Stars," an article of nine pages that took him more than six weeks to complete. It was published in the *Cornhill Magazine* in 1863. When his son died in 1863 Proctor's initial reason for learning astronomy was gone. Distraught over his son's death, Proctor's health began to fail. His doctor advised him to immerse himself in some form of work that would occupy him for at least a year. He chose the study of astronomy, and by 1865 he had written a monograph titled *Saturn and Its System*. The book was a commercial failure, yet it was a critical success that led to his election in 1866 as a Fellow of the Royal Astronomical Society.[12]

The catalyst that actually led Proctor to pursue a career as a popularizer of science was a failed investment. In May 1866, the New Zealand Banking Corporation, in which the bulk of his property had been invested, collapsed. As the second largest shareholder Proctor had a large liability of £13,000. Aged thirty, with not so much as half a crown in his pocket and a family of five to maintain, Proctor decided to move his family from their pleasant

12. Gilbert 2004a, 1641-43; [Ranyard] 1889, 164; Proctor 1895, 393-95.

home near Plymouth to London so he could begin a career as a scientific author and practitioner. In some ways, Proctor was faced with the same challenge as Huxley when the latter returned in 1850 to London in debt after his *Rattlesnake* voyage: how to make science pay when career paths were not well established. Proctor realized that it would not be easy to make a living from his interest in science. Recalling those difficult times in an autobiographical piece, Proctor wrote he was aware that "scientific research is not remunerative" and that "scientific exposition was not likely to be remunerative either." He was "little known to the editors of magazines and journals." His slow progress with his essay on double stars had shown that he "had no particular taste for it." Proctor recalled that he found science writing to be "slow and unprofitable work." In the autumn of 1867, having "entered definitely on literary work," he "passed through experiences enough to age a man ten years in as many weeks. Sickness and death in the family, sorrows and disappointments . . . fell upon me." Moreover, it was a constant grind. "The morning after I heard of the bank's failure," Proctor declared, "and from that day onward, I did not take one day's holiday from the work which I found essential for my family's maintenance."[13]

Proctor slowly established his credentials as an astronomer who could write for the public. Initially, publishers were not eager to accept his book proposals, as none of his early books, such as *Handbook of the Stars* (1866), *Constellation Seasons* (1867), and *Sun Views of the Earth* (1867), sold particularly well. Longman published all three of these works. Proctor finally produced a commercially successful work when Hardwicke paid him £25 to write a small book, *Half-Hours with the Telescope* (1868). The book eventually passed through twenty editions.[14] But Proctor's breakthrough book was his *Other Worlds Than Ours* (1870, Longman), in which he focused, for the first time in his writing, on the theme of life on other planets. Victorian reading audiences were fascinated with Proctor's approach to pluralism. After 1870, Proctor had little difficulty finding a publisher or a journal editor who would accept his work. Astoundingly prolific, Proctor wrote over sixty books during his career—perhaps the most of any nineteenth-century popularizer. Proctor produced a large number of his books by drawing on many of his previously published periodical articles.

Proctor's ability to churn out books and journal articles at a breathtaking speed attracted criticism from some who believed that it compromised the

13. Proctor 1895, 395–97.　　　　14. [Ranyard] 1889, 166.

quality of his work. In a review of one of Proctor's books in 1875, the *New York Times* declared, "this is getting serious. Mr. Proctor's books come so fast that no moderate-sized library will be able to hold them in the course of another year of two." The anonymous reviewer asked, "what is to be done with Mr. Proctor, who produces a new book (as it seems to us) at least four times a year?" Proctor, the reviewer complained, frequently announced changes in his views as a result of the rapid development of astronomical knowledge. "Perhaps if he waited a little longer before writing his books," the critic declared, "his readers would not be obliged at certain intervals to carefully unlearn what he has already taught them." Although the writer acknowledged that Proctor deserved recognition for conveying complex astronomical issues in a form easily understood "by the common mind," he protested that Proctor was now rushing into print a "hasty literary composition."[15]

Success as a popularizer depended on producing a steadily flow of books and articles, as Grant Allen and others learned through painful experience. Allen and Proctor were good friends, and they regretted that they had both been forced into activities dictated by the market and the fickleness of the public. Proctor remarked that Allen was "simply unequalled by all our English writers on popular science for grace and elegance of style and versatility." So it "grieves one to the heart to think that powers such as his should for want of the due appreciation of science in our day be increasingly wasted on those who appreciate only sensational fiction."[16] Allen judged Proctor to have "few equals among modern thinkers" in terms of "width of grasp and breadth of vision," and he could not recall any other contemporary who so impressed him "with a consciousness of intellectual greatness." But Proctor's accomplishments "fell short of his natural powers," and that was "due to the fact that the necessity for earning a living by the work of his brains compelled him to waste upon popularizing results and upon magazine articles a genius capable of the highest efforts."[17]

Early in his career, Proctor attempted to bolster his scientific credentials. In the later 1860s he began submitting articles to the *Monthly Notices of the Royal Astronomical Society* and other scientific journals. He also obtained an appointment as one of the Honorary Secretaries of the Royal Astronomical Society (RAS) and as editor of the *Monthly Notices* in 1872. He resigned the following year as a result of becoming embroiled in controversies with several eminent members of the RAS. As more and more lucrative opportunities

15. "Science By-Ways" 1875, 2.   17. G. Allen 1888d, 193.
16. [Proctor] 1888a, 173.

for lecturing and writing opened up, he did less and less scientific research. Proctor's contributions to the *Monthly Notices* and other research journals fell off sharply after 1873 while his popularizing activities increased. Proctor's first wife died in 1879, and in 1881 he married an American widow, Mrs. Sallie Crawley, and later settled in St. Joseph, Missouri, while continuing his extensive lecturing tours and editing *Knowledge*.[18] He worked for years on an original treatise on astronomy that he hoped would restore his reputation as a researcher, but he died in 1888 in New York of malarial fever before he could finish it. *Old and New Astronomy* finally appeared in 1892 completed by the astronomer A. C. Ranyard and published by Longman.

Proctor's long list of publications suggests that he succeeded, where Allen failed, in making a career as a popularizer of science profitable. Indeed, he was not forced to pen novels or to undertake any other form of nonscientific writing to subsidize his scientific work. Proctor adopted the strategy of diversifying his publishers, and he established a good working relationship with Longman, Kegan Paul, Chatto and Windus, and Smith, Elder, & co., among others. In 1881 he noted, "I have written nearly forty works, which have been published for me by five or six firms, with all of which I have had pleasant relations." Proctor tended to trust the advice of publishers, especially after his painful experience with *Saturn and Its System*. He had included many "long and complex calculations" and elaborate illustrations, thinking that this would sell more copies. When Longman suggested that very few cared for formulas and diagrams, Proctor thought "they underestimated the intelligence of the general reader." He also had not heeded the advice of Longman to print no more than a thousand copies, and as a result the first edition of the book was only now, sixteen years from its first publication, close to selling out. "They knew, and I did not," Proctor admitted, "what was best and wisest." Proctor also recommended to budding authors that they accept the guidance of publishers on advertising, the selection of journals to which a book should be sent for review, and the choice of title.[19]

According to Proctor, a good relationship with a variety of presses translated into a lucrative career as a popularizer. In 1876 he asserted, "of all forms of scientific literature, so-called popular science-writing is the most steadily remunerative." If an author possessed the necessary literary abilities, they could, after a few years of steady work, "very readily achieve an income ranging from £2,000 to £5,000 per annum, whereof a portion equal to the salary of a University professor (say from £500 to £1,000) would come

18. Sarum 1999, 34–54.   19. [Proctor] 1881f, 72.

from the sale of new editions of his works."[20] In Proctor's case, the profits came rolling in once he was established. Among his most successful works, *Other Worlds Than Ours* reached a fourth edition by 1878, by which time 4,500 copies had been printed since it first appeared. It went through at least twenty-nine printings and remained in print up to 1909.[21] In 1870 he received £50 from Longman for his share of the profits of the first edition of 1,250 copies. In addition he was paid £38, half of the profits for the 500 copies sent to America. In 1872 the second edition of 1,000 copies garnered him £75, while the third edition (1872) of 1,250 copies yielded him £110 and the fourth edition (1878) of 1,000 copies brought in £80.[22] This was only one of Proctor's many works that Longman published in the early 1870s. He was also earning money from *The Sun* (1871), *A New Star Atlas* (1872), *The Orbs around Us* (1872), *Essays on Astronomy* (1872), *Expanse of Heaven* (1873), and *The Moon* (1873), among others. Including the payment he received for his work done for other publishers, and for his journal and magazine articles, and it quickly added up to a substantial sum.

Proctor also lectured extensively. During his career he undertook lecture tours in Britain, the United States, Canada, Australia, and New Zealand.[23] He claimed in 1883 to have delivered between 1,500 to 2,000 lectures in the last thirteen years, 386 of them in America and 124 in Australia and Asia.[24] Proctor could cope with the grueling pace of lecture tours since he kept in good shape, continuing athletic activities from his university days. According to Clodd, he was a good oarsman and a first rate fencer.[25] In 1880 the *New York Times* marveled at his stamina and superb physical condition at the age of forty-four. He "never feels unwell, and seems insensible to fatigue," sleeping but five or six hours a day. "But," the journalist asked, "how much longer can he keep it up?" During this particular trip to the United States, he delivered 136 lectures from October 1879 to May 1880, stopping to speak in Washington, Baltimore, Philadelphia, Boston, Brooklyn, and New York, as well as at many educational establishments. Like many of his other lecture tours, it was a financial success, generating gross receipts of $50,000 and a

20. Proctor 1876, 8.

21. Crowe 1989, 4.

22. *Archives of the House of Longman* 1978, A10, 160; C5, 24.

23. Crowe 1986, 369.

24. Proctor 1883d, 25.

25. Clodd 1888, 265.

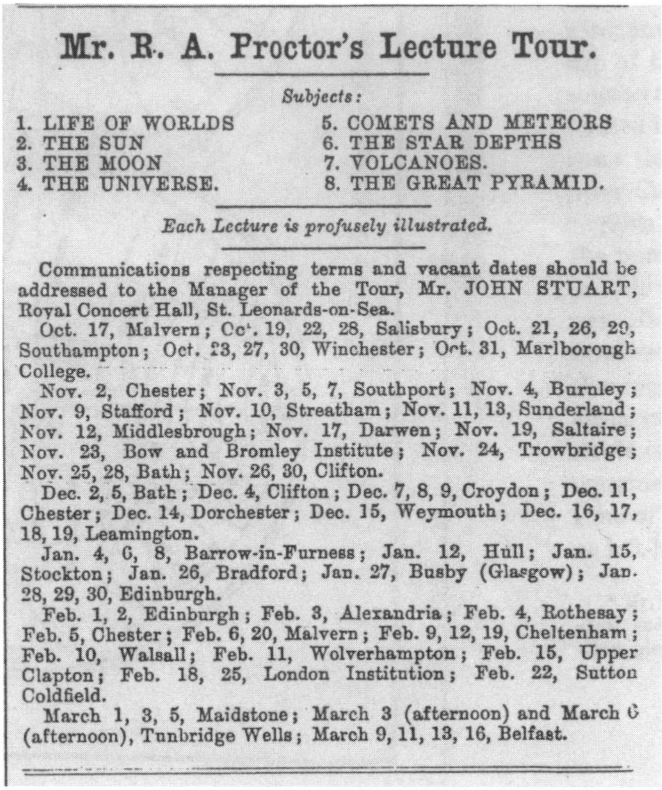

Mr. R. A. Proctor's Lecture Tour.

Subjects:

1. LIFE OF WORLDS
2. THE SUN
3. THE MOON
4. THE UNIVERSE.

5. COMETS AND METEORS
6. THE STAR DEPTHS
7. VOLCANOES.
8. THE GREAT PYRAMID.

Each Lecture is profusely illustrated.

Communications respecting terms and vacant dates should be addressed to the Manager of the Tour, Mr. JOHN STUART, Royal Concert Hall, St. Leonards-on-Sea.

Oct. 17, Malvern; Oct. 19, 22, 28, Salisbury; Oct. 21, 26, 29, Southampton; Oct. 23, 27, 30, Winchester; Oct. 31, Marlborough College.

Nov. 2, Chester; Nov. 3, 5, 7, Southport; Nov. 4, Burnley; Nov. 9, Stafford; Nov. 10, Streatham; Nov. 11, 13, Sunderland; Nov. 12, Middlesbrough; Nov. 17, Darwen; Nov. 19, Saltaire; Nov. 23, Bow and Bromley Institute; Nov. 24, Trowbridge; Nov. 25, 28, Bath; Nov. 26, 30, Clifton.

Dec. 2, 5, Bath; Dec. 4, Clifton; Dec. 7, 8, 9, Croydon; Dec. 11, Chester; Dec. 14, Dorchester; Dec. 15, Weymouth; Dec. 16, 17, 18, 19, Leamington.

Jan. 4, 6, 8, Barrow-in-Furness; Jan. 12, Hull; Jan. 15, Stockton; Jan. 26, Bradford; Jan. 27, Busby (Glasgow); Jan. 28, 29, 30, Edinburgh.

Feb. 1, 2, Edinburgh; Feb. 3, Alexandria; Feb. 4, Rothesay; Feb. 5, Chester; Feb. 6, 20, Malvern; Feb. 9, 12, 19, Cheltenham; Feb. 10, Walsall; Feb. 11, Wolverhampton; Feb. 15, Upper Clapton; Feb. 18, 25, London Institution; Feb. 22, Sutton Coldfield.

March 1, 3, 5, Maidstone; March 3 (afternoon) and March 6 (afternoon), Tunbridge Wells; March 9, 11, 13, 16, Belfast.

FIGURE 6.2 Proctor's advertisement in *Knowledge* for his busy lecture schedule in 1885. ("Mr. R. A. Proctor's Lecture Tour," *Knowledge* 8 [October 16, 1885]: 340.)

clear profit of $15,000 for Proctor.[26] Proctor was able to use his position as editor of *Knowledge* on more than one occasion to advertise his upcoming English tours, listing the dates, places, and titles of his lectures (Fig. 6.2).[27] He complained that London newspapers did not fully report on lectures on scientific topics, and he threatened in 1883 to avoid London in the future as long as the press "limits its reports and notices to theatrical and musical entertainments."[28] Ordinarily, Proctor was able to draw large crowds to his lectures. In 1883 he claimed to have filled the large Corn Exchange at Lincoln with 2,000 people.[29]

26. "Prof. Proctor" 1880, 4.
27. Proctor 1885p, 104.

28. Proctor 1883e, 217.
29. Proctor 1883f, 40.

Yet, even with the payments that Proctor received for writing his many publications, and even with the profits from his lecturing tours, at his death his widow applied to the Royal Literary Fund (RLF) for help. Mrs. Sallie Duffield Proctor estimated that his annual income was over £500 and more when lecturing, but after his debts were paid off there "will be absolutely nothing left for us at present." She hoped to be able to retain royalties for enough of the books to bring in a small income of about £150 per annum in the future. Proctor had died en route to England to undertake more lecturing, and he had allowed his resources to run down, believing that when he arrived his earning power as a speaker would see them through. The RLF awarded £200, half to Mrs. Proctor, and half to the children.[30] The fact that Proctor, who was so prolific as a writer and so active as a lecturer, was in such a precarious state at the time of his death illustrates just how difficult it was for popularizers to maintain a steady source of income to support their families and to provide for unforeseen emergencies, retirement, or even death.

RELIGIOUS PLURALISM AND THE CO-OPTION OF EVOLUTION

If struggling to maintain a decent income was a constant in Proctor's career as popularizer of science, so was his engagement with the theme of extraterrestrial life in his essays, books, and lectures. From 1870 until 1890 he was the most widely read participant in the pluralist debate in Britain and America.[31] Proctor revitalized the debate surrounding the possibility of life on other planets by bringing to bear the results of the "new astronomy," based on the introduction of the spectroscope and the camera into astronomical research. Proctor emphasized the role of the spectroscope in providing evidence that the planets orbiting distant stars were composed of the same materials as the earth. In the late 1860s and early 1870s Proctor believed that Mars offered the best opportunity to demonstrate the existence of extraterrestrial life. He established a new tradition in Martian cartography aimed at convincing his readers that Earth's neighbor was capable of supporting life. His maps of Mars supported his claim that continents and oceans similar to those of Earth could be detected through telescopic observation and confirmed by

---

30. British Library, Royal Literary Fund, File no. 2294.

31. Crowe 1986, 369.

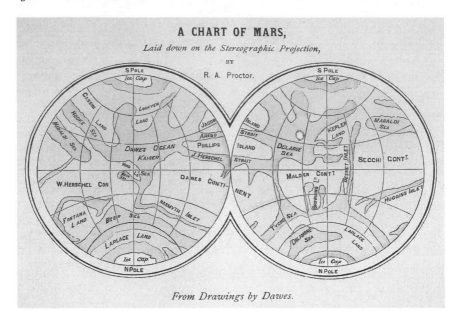

FIGURE 6.3 One of Proctor's maps of Mars, the stereographic projection, with its Eurocentric nomenclature and depiction of earthlike continents and bodies of water. (Richard Proctor, *Other Worlds Than Ours* [London: Longmans, Green, 1870], 92.)

spectroscopic evidence of the existence of earthlike gases and vapors (Fig. 6.3).[32]

To Proctor, the subject of pluralism was inextricable from religious considerations. In his scheme of things, God had created nature to fulfill a certain purpose. The profusion of life on the earth was a sure sign that Nature's "great end" is "to afford scope and room for new forms of life, or to supply the wants of those which already exist."[33] If the purpose of nature was to produce life, then surely it existed on other planets. Therefore, to Proctor, proof of extraterrestrial life confirmed that nature had a purpose and that a designing God was the creative force behind the cosmic arrangement revealed by astronomy. In his *Other Worlds Than Ours*, Proctor was constantly on the lookout for new evidence of purpose in nature. When considering the glowing mass of Jupiter that can sustain no life, readers are invited to find a raison d'être, for Proctor cannot accept the notion that God would create something for no purpose at all. The "wealth of design" in Saturn is so

32. Lightman 2000, 661–71.   33. Proctor 1870a, 18.

striking in Proctor's eyes that we cannot question but "that the great planet *is* designed for purposes of the noblest sort" though we may be unable to fathom them. Proctor enthuses over the recent discoveries of science, which are "well calculated to excite our admiration for the wonderful works of God in His universe."[34] Proctor even structured *Other Worlds Than Ours* along the lines of a cosmic, post-Darwinian theology of nature. The beginning chapters on "What Our Earth Teaches Us" and "What We Learn from the Sun" set the didactic tone for the entire book. Here the telescope, spectroscope, and the other tools of the astronomer's trade reveal nature's lessons concerning God's intentions and will. These first two chapters are part of a nine-chapter section on the solar system, which leads into a series of three chapters on the stars and nebulae, extending the discussion of how God instructs us through nature to the rest of the universe. The concluding chapter, titled "Supervision and Control," deals with the lessons to be learned from an examination of astronomy and the providence of God.[35]

Proctor's emphasis on the religious dimensions of pluralism, and his insistence on purpose in nature, was likely a result of the spiritual turmoil he experienced after the death of his eldest son. Clodd, with whom he was a close friend, recalled, "the death of a darling boy caused him to seek consolation in the Roman Catholic communion." Initially, Clodd was unaware that Proctor had converted. "His perversion came upon me as a surprise," Clodd recollected in his memoirs, "and there resulted a correspondence between us in which I got the worst of it." At the time Clodd was a Unitarian, sitting under James Martineau, and part of a congregation that included Charles Lyell and Arabella Buckley. According to Clodd, he "got the worst of it" in his exchange with Proctor because, as he later found out, there was no halfway house between Catholicism and agnosticism. Whereas Proctor had his feet firmly planted on "the rock of Saint Peter," Clodd's feet "were on the shifting sands of Theism."[36] The correspondence between them on Proctor's Catholicism began on July 11, 1870, when Clodd wrote to Proctor, praising his *Other Worlds Than Ours*. "Your recent work on the Plurality of Worlds," he told Proctor, "has afforded me so much gratification that I must preface this letter by thanking you as the source of that pleasure." Although Clodd was glad that Proctor had added the final chapter on "Supervision and Control," he objected to Proctor's willingness to "abnegate private judgment when I should feel bound to exercise it." Proctor replied on July 12

34. Ibid., 5, 142, 147–48.
35. Lightman 1996, 36–38.

36. Clodd 1926, 58–60.

that fluctuations in his opinions due to the use of private judgment had led him to be suspicious of it. Proctor now looked "on the so-called freedom on private judgment as a delusion and a snare." Clodd replied with a forceful defense of the right of verifying the doctrines accepted by the individual.[37]

Although Proctor denied the validity of the notion of private judgment, the key principle of Protestantism, Clodd still had no idea that his friend had turned Romeward in 1867. He was therefore shocked when Proctor revealed in a letter of July 27 that it had been "nearly 3 years since I became a Cath." Proctor admitted that he was also surprised by the turn of events. "I think 3 years ago I should have laughed outright had any one suggested the possibility of my becoming a Catholic," he wrote to Clodd. Proctor acknowledged that, in one sense, they agreed on the importance of exercising private judgment, for he decided to convert after careful personal deliberations. "I need hardly say that it was only after a long (and in my case somewhat painful) exercise of private judgment," he declared, "that I arrived at a conclusion so importantly affecting all my future." To prove that his renunciation of private judgment could only have come after exercising it to the utmost, Proctor pointed out that no "external circumstances" provided an incentive for him to convert. In fact, there was every reason for him to remain an Anglican. In England it was impossible for a man to become a Catholic "without offending a large circle of friends—nor ordinarily without injuring his prospects."[38]

Clodd wrote back to Proctor on August 2, trying to find common ground between them by using an astronomical metaphor to bridge their differences. Despite their disagreement, Clodd insisted, "the same light is guiding us." Whereas Proctor found clearer guidance "in the bright, visible rays of the spectrum; I am content to be influenced by its ultra-violet rays which though they be invisible, are not unfelt." Clodd maintained that God was found in both private judgment and the Catholic Church's judgment. On August 7, Proctor denied that there was any agreement between them, and he praised the efforts of John Henry Newman, an earnest and sincere individual, to illuminate the Catholic position on the issues they had been discussing. Three days later Clodd brought their correspondence on private judgment and Catholicism to a close. He wrote, "there are many questions raised in your present letter which invite comment but I think that our respective positions have been clearly defined and the grounds on which they

37. Proctor to Clodd, July 12, 1870, Leeds University Library, Clodd Correspondence.

38. Proctor to Clodd, July 27, 1870, Leeds University Library, Clodd Correspondence.

rest sufficiently indicated, so that little wheat remains to be threshed from the shears of controversy."[39] According to one obituary account, Proctor severed his connection to Catholicism in 1875 when he was told that some of his astronomical theories were not in conformity with the teachings of the Catholic Church.[40] Clodd asserted that "the illusions begotten of his un-balanced emotions were dispelled in later years, and he died an Agnostic."[41] However, Proctor's link between religious themes and astronomical topics continued until the end of his life. He referred to himself as an agnostic in the 1880s, yet as Clodd admitted in his obituary on Proctor, he was "a man of deep religious feeling."[42]

Proctor's religious sensibilities led him to co-opt evolutionary theory in support of his powerful mix of pluralism and a discourse of design. Evo-lutionary theory gave his theories more scientific plausibility in at least two ways. First, Proctor used Darwin's theory of natural selection to explain how alien life could exist in extreme environments. In the opening pages of his *Other Worlds Than Ours* he discussed the tendency to believe that other planets would be inhospitable to life in the same way that we might believe that certain parts of our planet—such as the Arctic or the bottom of the ocean—are uninhabitable. "Who would believe," he declared, "for example, that men can live, and not only live but thrive and multiply, in the frost-bound regions within the Arctic circle, if travellers had not visited the Esquimaux races, and witnessed the conditions under which they subsist?" Similarly, if we did not know that living creatures inhabit the depths of the ocean, where land creatures would quickly die, would we not conclude that no life existed there as well? But life survived even in these harsh conditions. Proctor concluded, "even though we could prove that every living creature on this earth would at once perish if removed to another orb, yet we cannot thence conclude that the orb is uninhabited. On the contrary, the lesson conveyed by our earth's analogy leads to the conclusion that many worlds may exist, abundantly supplied with living creatures of many different species, where yet every form of life upon our earth—bird, beast, or fish, reptile, insect, or animalcule—would perish in a few moments." Life could exist in the most extreme environments on earth because of the

39. Clodd to Proctor, August 2, 1870, Leeds University Library, Clodd Correspondence.

40. "Richard A. Proctor Dead" 1888, 1.

41. Clodd 1926, 60.

42. Clodd 1888, 265.

power of natural selection. Over the course of eons of evolution life arose and adapted to the worst conditions on earth. An examination of the Earth taught us that "not only is Nature careful to fill all available space with living forms, but that no time over which our researches extend has found her less prodigal of life." Nature has "a singular power of adapting living creatures to the circumstances which surround them."[43]

Drawing on images of nature as superabundant and adaptive—a concept of nature taken directly from Darwin's *Origin of Species,* though used there to demonstrate that natural selection could produce new species—Proctor pushed his readers toward the notion that nature was powerful enough to produce life on other planets. Why could not the same process lead to the existence of life adapted to the conditions on gas giants or frigid tiny planets that orbited stars from a colossal distance? Proctor developed this same argument in his other works. For example, in an essay titled "Other Inhabited Worlds," first published in 1869 in *St. Paul's Magazine,* and then republished in a collection of essays titled *The Orbs around Us,* also tackled the theme of adaptation and extraterrestrial life. Whatever the nature of the beings living on planets in other solar systems and whatever the peculiarities of the conditions on those planets, "the most perfect adaptation doubtless exists between those unknown living creatures and the structure of the worlds on which they live. This lesson is taught by all that we see around us." Proctor insisted, "adaptation is a fundamental law of nature." To push home the full implications of this point, Proctor then discussed unusual systems where the conditions were almost unimaginable to humans. So far he had been dealing only with systems where planets circled around a central sun similar to our own. But "around the double, triple, and multiple stars there doubtless travel systems of worlds crowded with living creatures." The conditions on these planets were, from an earthly point of view, quite strange, including the existence of complex gravitational forces, remarkable climatic changes, and different colored suns. "The great law of adaptation exerts its influence," Proctor maintained, "however, in these parti-coloured systems as elsewhere; and whatever doubts we may have respecting the actual habitudes prevailing there, we may be sure that they are fully as well suited to the wants of the inhabitants of those systems as are terrestrial habitudes to the wants of the inhabitants of earth."[44] Proctor argued that the universe teemed with life by virtue of the law of natural selection, deftly

43. Proctor 1870a, 10, 15–18.   44. Proctor 1902, 58–59, 61.

turning evolutionary theory into a powerful ally in the establishment of his religiously inflected pluralism.

From the mid-1870s on, when Proctor modified his views on the extent of inhabited planets in the universe, he found a second role for evolutionary theory in his religious pluralism. In the earlier plurality of worlds debate during the 1840s and 50s, the evangelical Scottish physicist David Brewster had upheld the pluralist position. Brewster could not accept the idea of a God who would create a wasteful universe empty of life, except on Earth. Brewster's chief opponent was William Whewell, Master of Trinity, a Bridgewater Treatise author, and considered by his contemporaries to be an authority in numerous scientific fields. Although a liberal Anglican, Whewell refused to be open-minded about the possibility of extraterrestrial life. He defended the notion that earthly life was a special creation of a designing creator. At first, Proctor entered the debate on the side of Brewster and those who believed that there was life on other planets. Yet he increasingly appreciated Whewell's position in the early 1870s as he gave up, one by one, on the inhabitability of planets in the solar system. Proctor began to believe that life might only exist at particular stages in the history of each planet, and he employed evolutionary theory in a larger sense to explain how this was so. Each planet had its own evolutionary history and only at certain times in this history were the conditions right for life. In his *Our Place among Infinities* (1875, King) he declared that, "each planet, according to its dimensions, has a certain length of planetary life, the youth and age of which include the following eras:—a sunlike state; a state like that of Jupiter or Saturn, when much heat but little light is evolved; a condition like that of our earth; and lastly, the stage through which our moon is passing, which may be regarded as planetary decrepitude."[45] There was life on other planets, but not at all times. More than anyone before him who was involved in the pluralist debate, Proctor emphasized the need to take an evolutionary approach by seeing the planets as subject to change and development.[46] Proctor's theory of planetary evolution added yet another evolutionary dimension to his pluralism.

Before Proctor accepted planetary evolution, he had already expressed his support for stellar evolution. In his treatment of Mars as an abode of life, star mapping played an important part in his proof for his position. In 1871 he discussed the tendency of stars to arrange in streams and argued that this

45. Proctor 1889b, 67.   46. Crowe 1986, 377.

was a sign of a more general law determining their aggregation and segregation. "We recognize so clearly within our own solar system such motions and such laws of distribution as suggest a process of evolution," Proctor wrote, "that the mind is led to inquire whether the motion of the stars and their arrangement throughout space may not indicate the action of a yet higher order of evolution." Proctor believed that one key to understanding the evolution of the structure of the universe lay in a study of "star-drift," and he recommended that astronomers adopt a research program that systematically determined the motion of the stars. He contributed two maps based on his own observations, which, in his opinion "fully justify the term 'star-drift,' which I have applied to the stellar proper motions." The maps featured a novel form of charting motion through the use of arrows attached to each star (Fig. 6.4).[47]

Proctor's adherence to biological evolution, to planetary evolution, and to stellar evolution made it easy for him to accept full-blown Spencerian cosmic evolution, as had his friends Allen and Clodd. In the pages of *Knowledge* he pointed to the contradiction of accepting the divine hand in the laws by which pebbles, flowers, insects, and animals are formed while rejecting as irreligious the idea of God's power and wisdom operating on a larger scale. "Evolution on the small scale we may admit without harm," he wrote, "but to see evolution in the development of a world or a world-system, and still more to see evolution throughout the entire universe as revealed to man, this is 'to set God on one side in the name of Universal Evolution.'"[48] In an article on "The Unknowable; or, The Religion of Science," Proctor credited Spencer as being "the first to present fully to the world of thought the doctrine of universal evolution—of which cosmical and biological evolutions are but chapters."[49]

47. Proctor 1878, 19, 119, 156.

48. [Proctor] 1881g, 4.

49. Proctor 1885s, 37. Although Proctor acknowledged in 1885 that he owed much to Spencer, he asserted that he did not blindly accept all of Spencer's teachings. As an example of where they diverged, Proctor pointed to their disagreement on the nebular hypothesis. "I have rejected as unsound the nebular hypothesis of Laplace, which Mr. Spencer values," Proctor declared. Nevertheless, Proctor praised Spencer's philosophy as "at once clearer and profounder, kinder and more considerate, braver in upholding right and resisting injustice, and better calculated if steadily followed—to make men happier and better, than any which hitherto has been propounded to the world" (Proctor 1885f, 273). Proctor rejected Laplace's nebular hypothesis well before the mid-1880s. In his *The Expanse of Heavens* (1873) he listed all of the aspects of the solar system that remained unexplained by Laplace's theory (Proctor 1889a, 183, 186, 189). Proctor accepted

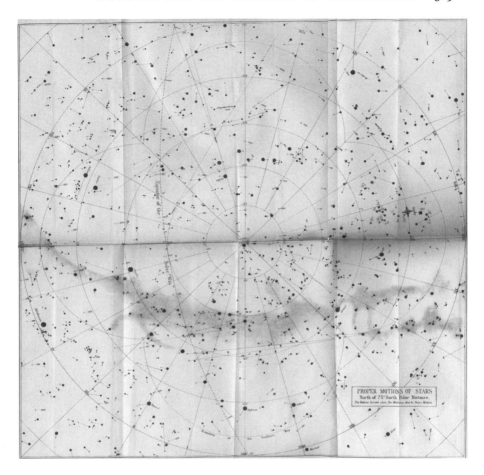

FIGURE 6.4 Proctor's map of the "Proper Motion of the Stars (Northern)." Attached to each star is a little arrow whose direction and length indicated the character and magnitude of the star's proper motion. (Richard Proctor, *The Universe of Stars*, 2nd ed. [London: Longmans, Green, 1878], plate IX, opposite p. 230.)

Proctor was also indebted to Spencer's theistic version of agnosticism, unlike both Allen and Clodd. Proctor embraced Spencer's notion of an un-knowable God as early as the mid-1870s. In his *Our Place among Infinities,*

---

cosmic evolution, but, unlike Allen and Clodd, did not view the nebular hypothesis as scientifically sound enough to stand as the opening stage of the entire process.

he asserted, "so far as Science is concerned, the idea of a personal God is inconceivable, as are all the attributes which religion recognizes in such a being." But, like Spencer he insisted that science could not disprove the existence of an infinite power. Echoing Spencer's notion of the Unknowable, Proctor emphasized the inconceivability and incomprehensibility of the power behind the evolutionary process. "But it is no new thought," he wrote, "no modern discovery, that we are thus utterly powerless to conceive or comprehend the idea of an Infinite Being, Almighty, All-Knowing, Omnipresent, and Eternal, of whose inscrutable purpose the material universe is the unexplained manifestation."[50] In 1883 he was voicing similar ideas in *Knowledge*, though he now identified himself explicitly as an agnostic. In response to a letter from an atheist, Proctor outlined three positions. "On one side are those who think they know what God is and wills," he affirmed, "on the other side those who think and say they know that there is no God that is and wills." Proctor saw both of these positions as mirror images, and told his atheistic correspondent, "your dogmatic denials are as outside our plan as dogmatic assertions of belief." Proctor held to a third position, the one including "they who feel there is a Power at the back of all we know, but do not pretend to know its nature, plans, or purposes."[51]

Although Proctor emphasized the religious dimensions of agnosticism, he nevertheless presented himself as an ally of Huxley and Tyndall. Huxley did not publicly reject Spencer's Unknowable until after Proctor's death, so there was nothing to prevent Proctor from seeing himself as being in line with Darwin's bulldog on religious issues, and he often defended eminent scientific naturalists from the attacks of religious zealots.[52] When Tyndall, Huxley, and Darwin were assailed in the pages of the *Christian Commonwealth* in 1882, he criticized the journal for undermining its own cause. "Tyndall's views about matter and life," Proctor affirmed, "Huxley's about automatism, Darwin's about evolution, are twisted into attacks on religion, with which, in fact, they have nothing whatever to do."[53] As for Darwin's denial

---

50. Proctor 1889b, 2, 34.

51. Proctor 1883i, 338.

52. Lightman 1987, 137.

53. Proctor also defended Huxley in 1886 from daily papers attacking him for expressing his opinions on the Irish question and politics in general. Proctor argued that scientific training provided a scientist a superior grasp of cause and effect, and the action and operation of law. For this reason, Proctor was prepared to give greater weight to the opinions of a Tyndall, a Huxley, or a Spencer, or of any scientific man, "about any political or religious matter, than I would to the opinion of a politician or a priest." Proctor thought it more likely that a man of science, than the

of design, Proctor maintained that the great evolutionist merely indicated "a process which seems to many far more consistent with just ideas of a wise Creator's plan than the ordinary view." On the contrary, Proctor claimed that the "most dangerous enemies religion has in these times are those who teach that the doctrine of evolution is inconsistent with religion."[54] Proctor referred positively to Tyndall's scientific work. In an article on "Professor Tyndall's Theory of Comets" (1869), he judged Tyndall's researches to be "highly ingenious speculation" with great promise.[55] Elsewhere, Tyndall's comments on planetary atmospheres supplied grist for Proctor's pluralistic mill. Tyndall's demonstration that it was impossible to determine a planet's climate merely by considering its distance from the sun gave Proctor the ammunition he needed to assert that "the inhabitants of Venus and Mercury might enjoy a climate as genial as that of our own earth."[56]

Proctor would not have endeared himself to Huxley or Tyndall by co-opting evolution in the service of an updated theology of nature. Moreover, Proctor's criticism of one of the books in Macmillan's Science Primer series in 1876 would have irked Huxley, since the eminent biologist was one of the editors of the series and since the author was his friend Norman Lockyer. Although not mentioning Lockyer by name as the author, Proctor criticized an elementary treatise on astronomy used as a textbook at South Kensington that included so many errors that it was employed by Cambridge examiners to provide "horrid examples" to students. Then Proctor targeted Lockyer's contribution to the Science Primer series. "The author of that book has recently also produced a primer of astronomy," Proctor declared, "in which the worst of those errors have been repeated and mistakes introduced relating to still more elementary matters."[57] In 1888 Proctor was less coy. Naming Lockyer as the author of a recent "absurd production," he also pointed to odd mistakes in his astronomy book in the Science Primer series, "as, for instance, in describing the stars which pass to the zenith of London, which never rise or set at all, as rising and setting 'on a slant.' "[58]

---

politician or the priest, would speak only if he had studied the subject using scientific methods (Proctor 1887c, 43).

54. [Proctor] 1882n, 349.

55. Proctor 1902, 215.

56. Ibid., 32–33.

57. Proctor 1876, 6–7.

58. [Proctor] 1888b, 234.

PROCTOR AND THE CELESTIAL SURVEYORS

Proctor's vendetta with Lockyer played a key role in his decision to establish a new science journal in the 1880s. To understand their animosity, we must examine Proctor's tumultuous relationships with professional astronomers, especially Lockyer, during the 1870s. As a result of his battles with Lockyer and other astronomers, Proctor's attitude toward scientific practitioners in general began to change. According to A. J. Meadows, conflict between Proctor and Lockyer was inevitable. Both, he argues, "were seeking to compensate for an uncertain financial background by achieving public recognition of their scientific eminence."[59] Born in 1835 to a surgeon-apothecary who conducted experiments in electromagnetism, Lockyer became a clerk at the War Office. While still in the civil service he began a serious study of science in 1861 and joined the Royal Astronomical Society in 1862. When the *Reader* was established in 1863 he was asked to be a regular contributor, which brought him into close contact with T. H. Huxley and other metropolitan scientists. Macmillan retained him as the firm's chief scientific expert in 1868. Lockyer's scientific career reached a new peak in 1869, when he was elected Fellow of the Royal Society on the basis of his work in solar spectroscopy, asked to deliver a Friday evening lecture at the prestigious Royal Institution, and then invited to become the editor of Macmillan's new periodical *Nature*. These heady accomplishments were achieved while Lockyer was still languishing in the War Office, having suffered a demoralizing demotion in 1868. However, he was appointed in 1870 as secretary to the Devonshire Commission. While the Cambridge-educated Proctor was churning out books and articles, the self-educated Lockyer was the editor of a new journal and in the thick of the action as a member of an important government commission.[60]

Proctor and Lockyer first clashed at the beginning of the 1870s on the question of the nature of the solar corona. Proctor argued that it was a genuine solar appendage, Lockyer that it was partly caused by the scattering of light in the Earth's atmosphere.[61] Then they disagreed on the bearing of discoveries in solar physics for meteorology, Lockyer maintaining and Proctor rejecting that there was a strong connection between the solar cycle and global weather patterns. It did not help that the June 1870 issue of *Nature* contained an unflattering review of Proctor's *Other Worlds Than Ours*.

59. Meadows 1972, 96.

60. MacLeod 1969, 438; Meadows 1972, 5–6, 12, 17, 22.

61. Meadows 1972, 96.

The author accused Proctor of wild speculations on the issue of extraterrestrial life and asserted that Proctor offered no scientific evidence for rejecting Lockyer's theories on the solar corona or for regarding Lockyer as "impeding the progress of science."[62] In his angry response, published in *Nature* in July 1870, Proctor complained that the reviewer had purposely ignored sections of his book where he presented sound reasons for his criticisms of Lockyer. He denied that he looked upon Lockyer as an obstacle to progress and anticipated "admirable work from him in the future." But Proctor's attempt to focus his assault on the reviewer and to spare Lockyer was not reciprocated. Lockyer exercised his privilege as editor by adding a note to Proctor's letter, declaring that Proctor had misconstrued his theory of the corona and that Proctor had made himself look "ridiculous" by attempting to "evolve the secrets of the universe . . . out of the depths of his moral consciousness."[63]

In 1872 the Council of the Royal Astronomical Society was divided by a bitter debate over whether or not to pass a motion supporting the establishment of a new solar physics observatory in England, separate from Greenwich, with Lockyer in charge. Proctor was one of those who opposed the motion. Just before a special meeting of the Council at the end of June, he tried to convince George Airy to vote against the motion. Proctor warned that it would "set up a certain junior member of our council [i.e., Lockyer] as a sort of 'Astronomer Royal for the Physics of Astronomy,' " and that Lockyer's focus on solar physics and solar photography represented an extremely limited view that excluded the study of stars and nebulae from the newly emerging discipline of astronomical physics.[64] In November of 1872, Lockyer, and others who supported the resolution, resigned from the RAS Council. When asked by Airy to reconsider his resignation, Lockyer refused, and stated that his main reason for quitting the Council was "the offensive manner" in which Proctor was behaving. "Week after week," Lockyer told Airy, "in more or less obscure journals which as Editor of Nature I must see I find myself attacked by one who takes good care to advertise himself as 'Honorary Secretary of the Royal Astronomical Society.' "[65]

Just as the quarrel with Lockyer began to die down, Proctor began to criticize Airy's preparations for gathering scientific information on the upcoming

62. Pritchard 1870, 161–62.

63. Proctor 1870b, 190.

64. Proctor to Airy, June 26, 1872, RGO Archives, Cambridge University, as cited in Meadows 1972, 98.

65. Ibid., 98. Meadows also discusses the political maneuvering that took place when both Proctor and Lockyer were nominated for the RAS Gold Medal in 1872, a situation that could not have helped to smooth over the antagonism between them. See Ibid., 99–102.

transit of Venus in 1874. Proctor had already antagonized Airy on this subject a few years earlier. When he published a paper making fun of the Admiralty in connection with the transit of Venus in a supplementary number of the *Monthly Notices*, a journal he edited, Proctor was accused of abusing his editorial power and was forced to resign as secretary of the RAS in 1873.[66] A few years after the transit of Venus had taken place, Proctor commented on the results of the British transit expeditions. As he saw it, their chief value was to prove conclusively what he had been insisting all along: the Delisle method of estimating the sun's distance defended by Airy was not trustworthy given the current state of scientific instrumentation.

Airy's blunder in this particular case, however, revealed a much deeper systemic problem with government astronomy in Britain, which, Proctor pointed out, was based on a hierarchical, military model. "Many seem to suppose that astronomical matters are in some sense like military or naval (warlike) manoeuvres," Proctor asserted, "to be discussed effectively only by those who 'are under authority, having (also) soldiers under them,' in other words by Government astronomers." Proctor argued that science could not operate in such a system, for "those chiefly responsible for the selection of methods and the supervision of operations would be perfectly free from all possibility of criticism."[67] If Airy's subordinates were discouraged from speaking up when their superiors erred, then it was the duty of those like Proctor who were outside the official ranks to correct the mistake.

Proctor painted a rather unflattering picture of the practical value of the work done by government astronomers. Although he admitted that it was important, especially for navigation and commerce, their systematic observations had "scarcely any closer relation to the real living science of astronomy than land surveying has to such geology as Lyell taught, or the bone-trade to the science of anatomy." Government astronomers really engaged in "celestial surveying," not "astronomy," and "to one who apprehends the true sublimity of astronomy as a science the routine of official astronomy is by no means inviting."[68] The severe disciplinary regime established at Greenwich— the "factory mentality" that dominated the observatory in order to ensure precision—were to Proctor merely signs of the official, unimaginative science practiced by Airy and his government colleagues.[69]

---

66. Ibid., 102–3.

67. Proctor 1903, 77.

68. Ibid., 77–78.

69. For a discussion of Airy's Greenwich program of precision transit measurement and magnetic and meteorological recording, see Schaffer 1988, 115–45.

Proctor's heated controversies with Airy and Lockyer profoundly shook his faith in British practitioners of science and led him to oppose the recommendations of the Devonshire Commission. During the 1860s, Huxley, Tyndall, Thomas Hirst, John Lubbock, Edward Frankland, and Lockyer, among others, had pushed for increased state funding for science.[70] Even Proctor spoke in favor of state aid to science in 1869, and applauded the proposal for establishing national institutions expressly for the practical advancement of scientific research. In 1870 he criticized the British government's refusal to fund transportation for an eclipse expedition to Spain and Sicily. He argued that science was the "greatest power our country possesses," yet "it has been treated for a long while as a troublesome beggar—a few hundreds doled out there and a few thousand there. The country does not yet know its own interest."[71] In 1870 Prime Minister Gladstone agreed to set up the Royal Commission on Scientific Instruction and the Advancement of Science under the seventh Duke of Devonshire to examine the issue. The Commission's final recommendations were an endorsement of state aid for science, and contained a proposal for Lockyer's pet project, the establishment of an observatory for astronomical physics independent of Greenwich.[72]

The dramatic shift in Proctor's attitude toward enhanced state aid for science is glaringly evident in his book *Wages and Wants of Science-Workers* (1876, Smith). In the abstract, the notion of state endowment of science was good, Proctor agreed, because science was a "potent means of culture"—it promoted a belief in universal law and the gradual extinction of superstition.[73] But as soon as "details are considered, and especially when candidates for the nation's money come forward and tell us precisely what they want, the matter assumes a different aspect," Proctor declared.[74] Proctor suggested that he had changed his mind when he perceived the selfish motives of those who put forward detailed proposals for spending any increase in funding:

> Even while as yet they were in their infancy, mischievous tendencies began to show themselves which had certainly not been anticipated by those earnest students of science who first supported the general principle that science deserves

---

70. MacLeod 1996, 201.

71. Proctor 1871, 90, 92, 96.

72. Meadows 1972, 95.

73. Arguments based on science as a "source of individual and national might" or on science as "a means to adding to material wealth" were only of secondary importance in Proctor's mind.

74. Proctor 1970, 18, 22, 43.

the recognition of the State. Greedy hands were stretched out for the promised prizes. Jobbery began its accustomed work; and those who sought to check its progress were abused and vilified. If this happened when schemes for endowment were but mentioned, what evil consequences might not be looked for if those schemes succeeded?[75]

Proctor did not mention Lockyer by name, but he obviously had in mind the proposal for the new state-funded observatory for astronomical physics that had been the cause of the controversy within the Royal Astronomical Society. As the Commission brought its work to a close, Lockyer was hoping to be named director of this new observatory so that he would not have to return to the civil service.

In order to offer an alternative to the Commission's recommendations to enhance state aid to science, Proctor presented a very different vision of how science could be funded, which was based on the principles of laissez-faire individualism and on a more significant role for science writing within the profession. Instead of looking to the state for support, scientists could supplement their income by writing books on their scientific disciplines, rather than on their specific researches, which would have an appeal beyond a small circle of specialists. Examples of books in this genre that attained commercial success were John F. W. Herschel's *Outlines of Astronomy,* Lyell's *Principles of Geology,* Tyndall's *Heat* and *Sound,* and Darwin's *Origin of Species* and *Descent of Man.* There was also the option of writing textbooks, which Huxley, Tyndall, and Clerk Maxwell had found to be "sufficiently remunerative." Proctor presented his own career as an example of what could be done by a hardworking science writer. He referred to the cultivation of science writing as "the natural development of the resources of science as a profession," and affirmed that there was more potential here than in pouring money into a science ministry that would haphazardly fund a variety of scientific schemes. "I believe that men of science can do much more for themselves in this matter than the nation can do for them," Proctor declared, "or, to speak plainly, that it is the duty of men of science to refrain from an appeal to the nation for assistance when the means be ready at hand for making science self-supporting."[76]

In *Wages and Wants of Science-Workers,* Proctor compared the quality of scientific research done by those engaged in science writing and by those who worked at government-funded institutions. His views had obviously

75. Ibid., 89–90.   76. Ibid., 5–8, 9.

been affected by his controversy with Airy. According to Proctor, original research often ceased once an individual secured a government-funded post. He claimed that the situation in state-funded observatories was even worse: it was not likely "to lead to discoveries in science"—in reality it was not "properly speaking, scientific" at all. "The object for which government observatories are erected in fact," Proctor announced, "precludes almost entirely the pursuit of original researches." Subordinates who did not pay strict attention to routine duties, or who did not carefully avoid originality, became involved in "a modified form of transportation—called appointment to colonial observatories."[77]

While original research was not encouraged by conditions in state-funded institutions, the "most fruitful of our scientific workers are also those who have succeeded best in scientific literature." Proctor pointed to Herschel, Lyell, Darwin, Huxley, Tyndall, Spencer, Lubbock, and Wallace as examples of scientists whose original research had not suffered despite writing books for the general public. On the contrary, Proctor remarked, learning to present a theory clearly in such books was excellent training for the practitioner. "I feel sure that certain crude theories which have been maintained by some who pride themselves most on avoiding the popular and rhetorical element," Proctor wrote, "would have been abandoned had they been submitted to this process."[78] Putting a new scientific theory into a form suitable for a popular audience was a kind of "test," in Proctor's opinion, of its cogency. Proctor was arguing that adopting the role of popularizer for the Victorian reading audience was an integral part of being a scientist.

By the end of the 1870s, Proctor's notion of what it meant to be a practicing scientist was in marked contrast to the views of his adversaries Lockyer and Airy. He was offering an alternative vision of how to pursue scientific knowledge. As Stefan Collini has pointed out, a fixed standard for what constituted a professional did not yet exist in Victorian intellectual life in general.[79] At stake in the controversy arising out of the findings of the Devonshire Commission was the definition of the modern scientist. Proctor was not alone in contesting the notion that state aid was essential for the development of a scientific profession.[80] But he may have overestimated support for his position outside of astronomical circles. Throughout *Wages and Wants of Science-Workers*, Proctor identifies, though not always by name, astronomers such as Lockyer and Airy as his opponents. Evolutionists, such as Tyndall,

77. Ibid., 27, 38, 40.

78. Ibid., 30–31.

79. Collini 1991, 203, 220.

80. MacLeod 1996, 220.

Spencer, and Huxley, are praised throughout the book as examples of individuals who combined original research and science writing for the public. Proctor was particularly impressed with Spencer, who, more than the others, supported himself on his writing and had thereby maintained his independence from the state. But Huxley was one of the members of the Commission who presumably stood by the majority of the recommendations.

Proctor somehow managed to convince himself that Lockyer, as secretary, had manipulated Huxley and other members of the Devonshire Commission so that they had come around to his views. The evidence on which the Commission based its recommendations "was in reality carefully sifted before reaching them, so as to represent the views of those only among scientific men who approve of the general principle, that because science in itself is good, it cannot be too lavishly endowed by the nation." On the issue of funding for a new solar physics observatory, Proctor charged that only the minority who supported this defeated recommendation in the Royal Astronomical Society Council, a group including Warren de la Rue and Norman Lockyer, had been interviewed by the Commission.[81] Yet Proctor seems to have totally misread Huxley. Ironically, during the early 1870s Huxley was in the process of establishing at his new "Science Schools" at South Kensington a regime every bit as disciplined as Airy's at Greenwich, described by Adrian Desmond as a "proper top-down training for students by State-paid professionals." In the beginning, Royal Engineers, drafted by the War Office as free labor, surrounded Huxley, who compared himself to a recruiting sergeant who enlisted men into the army of science. According to Desmond, when Huxley ("The General") retired in 1885, he left a "modern command structure" in South Kensington.[82]

As Proctor pondered launching his own journal in the early 1880s, he had foremost in his mind a vision of a scientific community that, unlike government observatories, was not hierarchical. He had an example of a journal that embodied almost everything he hated about Lockyer's definition of professional scientist: *Nature*. Proctor aimed for a unique journal that would retain a place for the activities of the popularizer within the emerging definition of professional science, that would present original research, and that could challenge the place of *Nature* as the leading weekly magazine for a broad reading audience. Drawing on his experience editing the *Monthly Notices of the Royal Astronomical Society*, writing articles for a wide variety of periodicals,

81. Proctor 1970, 10, 68.   82. Desmond 1997, 396–97, 420, 542.

and publishing a series of science books for the public, Proctor came up with the idea for *Knowledge*.

KNOWLEDGE AND NATURE

Throughout the early 1880s Lockyer became increasingly anxious about the financial position of *Nature*. From the beginning in 1869 the journal had been running at a loss. Many science journals had failed on account of the unwillingness of publishers to continue their support for periodicals with annual balances in the red. Messrs. Macmillan had been patient throughout the 1870s, but Lockyer worried that by now the publishing house expected *Nature* to turn a profit. Lockyer's biographers state that this was in fact the "most difficult period of its [*Nature*'s] existence." It was not until 1899 that *Nature* was in the black for the first time.[83] Lockyer's anxiety over the future of *Nature* could only have been increased when he received a letter from his old enemy Proctor in October of 1881, announcing the publication of the first number of *Knowledge* on November 4.[84] It is unclear whether or not Proctor knew about the financial difficulties of *Nature*, but the timing is suspiciously coincidental. Proctor had chosen to launch his new journal just when *Nature* was most vulnerable. It was the perfect time to challenge *Nature* for control of the upscale science periodical market aimed at a popular audience. Proctor set up *Knowledge* as a weekly, just like *Nature*, so that he faced Lockyer on a level playing field. Like *Nature*, and other weeklies, his periodical could combine the speed and timeliness of a newspaper while drawing on the intimacy of a quarterly.[85] A careful examination of *Knowledge*, of its price, title, motto, masthead, aims, format, and contributors, will reveal how carefully Proctor set up his journal as a competitor to *Nature*.

When Lockyer originally founded *Nature*, he purposely set the price at the rather low rate of four pence per issue to attract subscribers. He expected that the advertising revenue would help make up the loss. Most weeklies cost sixpence per issue in the 1860s. Circulation figures for *Nature* are difficult to determine. In 1870 Lockyer boasted that there were nearly five thousand subscribers and fifteen thousand readers, but Roy MacLeod estimates that there were no more than one hundred to two hundred subscribers at that

83. Lockyer and Lockyer 1928, 50, 114, 173.

84. Proctor to Lockyer, October 4, 1881, University of Exeter.

85. MacLeod 1968, 16.

point and a maximum of five hundred subscribers by 1895. However, by 1878 the journal had gained enough subscribers that Lockyer felt he could risk increasing the price per issue from four pence to sixpence.[86]

Proctor thought he could undercut *Nature* by setting the price of his weekly at two pence per issue. In his decision on the appropriate price for *Knowledge*, Proctor may have been acting on the advice of his publisher, Wyman and Sons of Great Queen Street. Charles Wyman had published the short-lived weekly *Scientific Opinion* (1868–70), which at four pence weekly had been in direct competition with *Nature*.[87] Readers suggested in 1882 that the circulation of *Knowledge* could be increased among "the higher and superior educated branches of society" by enlarging it and raising the price to sixpence. Proctor replied that this would put the weekly "beyond the reach of many to whom I wish to be of use . . . No; our plan was to make *Knowledge* as low-priced as possible, and to give as much as we possibly could for the money. To that plan we must adhere."[88] Proctor claimed on June 2, 1882, that the first volume of *Knowledge* had reached more than twenty thousand readers, and about a year later, he triumphantly announced that circulation had been steadily increasing.[89] Proctor was forced to raise the price to three pence per copy in March of 1884, but it was still half the cost of *Nature*.

Proctor, like Lockyer before him, selected the title of his new journal carefully. He realized that part of *Nature*'s success was owing to its simple, concise, and comprehensive name. In October 1869, James Sylvester, the mathematician, wrote to Lockyer congratulating him on his choice of title for the journal:

> What a glorious title, *Nature*—a veritable stroke of genius to have hit upon. It is more than Cosmos, more than Universe. It includes the seen as well as the unseen, the possible as well as the actual, Nature and Nature's God, mind and matter. I am lost in admiration of the effulgent blaze of the idea it calls forth.[90]

Graeme Gooday has pointed to the strategy behind the choice of title; it was "clearly one of appropriating a powerful representation of 'Nature' for its metropolitan audience."[91] Proctor needed something equally simple, concise, comprehensive, and symbolic of the power of his view of science. "Knowledge" was just the thing. As a journal title, "Knowledge" actually

86. Lockyer 1870, 1; MacLeod 1969, 442–44.
87. Barton 1998b, 6.
88. [Proctor] 1882b, 596.

89. [Proctor] 1882c, 13; 1883b, 391.
90. Lockyer and Lockyer 1928, 48.
91. Gooday 1991, 313.

trumped "Nature," because knowledge was what the scientist produced after having actively examined nature.

Proctor also paid close attention to the epigraph that graced the front page of every issue of *Knowledge*. *Nature*'s epigraph, "To the solid ground of Nature trusts the mind which builds for aye," was taken by Lockyer from Wordsworth's sonnet "A Volant Tribe of Bards on Earth Are Found" (1823). Lockyer had to alter the original. He capitalized "nature" and decapitalized "mind" in order to avoid the typical Wordsworthian image of nature as pastoral. The epigraph was meant to serve *Nature*'s interests in promoting laboratory research and institutionalized teaching in science by evoking a notion of nature as accessible in both wilderness and city.[92] Proctor selected a line from Tennyson's "In Memoriam" (1850), "Let Knowledge Grow from More to More" (line 1, stanza 7). This choice had the advantage of being from a better-known poem by a more current and scientifically literate poet.[93] The words stressed the progress of science. Furthermore, Tennyson's notion of knowledge, not to mention his notion of nature, did not require any cleansing of pastoral connotations.

Proctor's attempt to position *Knowledge* as a challenger to *Nature* is perhaps clearest in his choice of masthead. *Nature*'s masthead, a cosmical representation of the Earth, was selected by Lockyer, possibly on the advice of Huxley, to reflect the impact of "*naturphilosophie*, even of Goethe, in British scientific thought, and perhaps also a wishful plan for the journal's distribution" (Fig. 6.5).[94] Shrouded by clouds in the foreground, and enveloped by darkness from behind, despite the twinkling stars in the distance, the scientific secrets of the Earth are partially hidden from view. The letters spelling out the title "Nature"—dark letters that seem to be carved out of some organic woodlike substance into an antique font—obscure the Earth from the gaze of the reader as well. The basic idea for Proctor's masthead seems almost identical (Fig. 6.6). He also uses an orb-shaped astronomical

92. Ibid., 314.

93. A number of scholars have commented on the popularity of "In Memoriam" with Tennyson's contemporaries (Ross 1973, 101, 153). Even T. H. Huxley spoke highly of Tennyson's insight into scientific method in "In Memoriam." Huxley once wrote that Tennyson was the only modern poet "who has taken the trouble to understand the work and tendency of the men of science" (L. Huxley 1902, 2: 359). Literary allusions to Tennyson's poems were used to denote decency or respectability. Proctor was not alone in trying to benefit from establishing a link to Tennyson. Darwin drew from Tennyson's poetry in the *Descent of Man* to distance himself from charges of immorality (Dawson 2005, 52–53).

94. MacLeod 1969, 439.

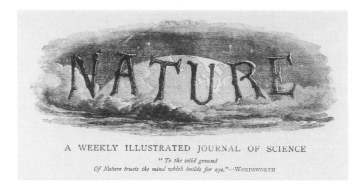

FIGURE 6.5 The masthead of *Nature*, with its murky depiction of the globe. (*Nature* 25 [1881]: 1.)

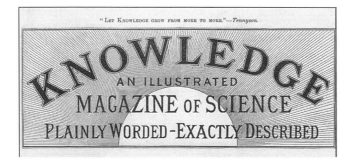

FIGURE 6.6 The title page device, which is the same as the masthead used in the individual issues, depicts a brilliant sun illuminating the sky with its rays. (Title page of *Knowledge* 1 [November 1881–June 1882].)

object as the primary image. He also places this object at the lower center of the page, superimposing text upon it. But the difference between the mastheads is as great as night and day. Proctor's astronomical object is the sun, whose dazzling rays of light dispel all clouds and illuminate the entire sky, just as *Knowledge* will shed its light on the truth. Only the upper third of the Earth is revealed in *Nature*, while half of the sun appears in *Knowledge*. In Proctor's masthead, all is clarity and light, in contrast with the murky darkness of *Nature*'s shrouded Earth. Proctor's letters are more sharply and precisely cut, and his font is far more modern. Proctor surely designed his masthead with an eye toward outshining his rival's. It could only have irked

Lockyer even more that the central image in *Knowledge*'s masthead, the sun, was the chief subject of his own scientific research at the Solar Physics Observatory.

Proctor was careful to distinguish the aims and intentions of *Knowledge* from those of *Nature,* although both journals identified the general public as an important part of their audience. In the circular for *Nature* at the time of its founding, Lockyer asserted that the journal had two objects:

> First, to place before the general public the grand results of Scientific Work and Scientific Discovery, and to urge the claims of Science to a more general recognition in Education and in Daily Life; and Secondly, to aid Scientific men themselves, by giving early information of all advances made in any branch of Natural Knowledge throughout the world, and by affording them an opportunity of discussing the various Scientific questions which arise from time to time.[95]

Lockyer himself is not entirely clear on how he saw these two audiences, the general public and scientists, interacting through the pages of his journal. Most scholars are more certain though; they agree that one of Lockyer's priorities was to gain the support of the general public for the men of science.[96]

The "Prospectus" for *Knowledge,* which Proctor sent out to Lockyer, potential contributors, and possible subscribers, was reproduced in the first issue of *Knowledge* on November 4, 1881. "'Knowledge,'" Proctor declared, "is a weekly magazine intended to bring the truths, discoveries, and inventions of Science before the public in simple but correct terms—to be, in fact, the minister and interpreter of Science for those who have not time to master technicalities (whether of Science generally or of special departments)." In

95. Lockyer and Lockyer 1928, 46-47.

96. Sheets-Pyenson 1976, 219; Desmond 1997, 372, 460. In contrast to the position of Barton, Sheets-Pyenson, and Desmond, Roos argues that *Nature* did not begin as a professional journal for specialists. At least one of the original intentions of *Nature* was to act as a mediator between "increasingly diverse and sometimes antagonistic segments of late Victorian society—between scientists and artists, between professional scientists and interested amateurs, between scientific generalists and specialists, and between specialists in different fields." Roos asserts that the aims and intentions of *Nature* were broader, more generalist and less professional than other journals like the *Natural History Review,* but narrower, more specialist, and more professional than those of *The Reader.* He denies that Lockyer intended to establish a journal that had, as one of its aims, the attempt to bolster the authority of professional scientists. However, he does acknowledge that *Nature* underwent a transformation not totally in keeping with the wishes of its founders as it became more and more like a professional journal during the 1870s and 80s. Even if it were true that the original intentions and aims of *Nature* did not include strengthening the power of professional science, it could be argued that by the early 1880s, when *Knowledge* was founded, *Nature* had strayed far enough from its course that Proctor could make a case that a new journal was needed to mediate between science and the public (Roos 1981, 161, 166-67).

contrast to Lockyer's prospectus, there is no mention of a second objective of "aiding scientific men." Proctor claimed that *Knowledge* would be unique in emphasizing the benefits derived from science "as a means of mental and moral culture," as opposed to its material benefits, and insisted that he had pointed out the necessity almost ten years earlier.[97] But, he explained, he waited until the early 1880s to establish such a journal, as he believed that he should "obtain as wide an experience as possible of the wants of the class of readers for which it is intended." He referred to the experiences he gained over the last ten years as a lecturer and writer, including the letters he had received from members of his audience, as giving him insight into "the nature of the difficulties which commonly perplex scientific students and the readers of scientific treatises."[98]

Proctor believed he knew what the reading public wanted in a science journal—and he also claimed that he knew what they did not want. "The general public do not want Science to be presented to them as if they were of intelligence inferior to their teacher's," he asserted. The public could not be expected to "take interest in statements couched in abstruse or technical terms."[99] Proctor was particularly critical of papers like the *Times,* which included articles on science that merely confused the reader. "Scientific articles in *The Times,*" Proctor declared, "are generally worth reading . . . as awful examples." How, Proctor asked, could the general reader be attracted to science when it was presented "in such garb as this?"[100] As originally conceived by Proctor, *Knowledge* was not a journal bent on furthering the aims of would-be professional scientists. *Knowledge* was geared solely toward interested members of the mass reading public.

Proctor chose a format for *Knowledge* that distinguished it from *Nature.* In Lockyer's original plans for *Nature,* the journal was divided almost in half. Some portions would be directed at the general public, others were solely for scientists. For the public, Lockyer included articles written by men eminent in science, full accounts of scientific discoveries of general interest, records

97. Proctor 1881h, 3. Proctor was not as unique as he claimed in presenting science as a means to culture. Huxley and other practitioners had argued for some time that the greatest benefits of science were moral and intellectual, not technological or practical. They claimed that it trained the faculties of the mind, stimulated the imagination, and formed character just as effectively as the classics. Proctor shared the views of practitioners (Barton 1981, 14, 16–17).

98. Proctor 1881h, 3.

99. Ibid.

100. Proctor 1882p, 342.

of efforts made for the encouragement of science in colleges and schools, and reviews of scientific works. For scientists, Lockyer included abstracts of important papers communicated to British, American, and European scientific societies and periodicals, and reports of the meetings of scientific societies in England and abroad. Columns devoted to correspondence were thought to be of interest to both scientists and the public.[101]

In the prospectus for *Knowledge,* Proctor announced that his paper would contain original articles by the ablest exponents of science, *"Serial Papers* explaining scientific methods and principles; *Scientific News* translated into the language of ordinary life; a *Correspondence Section* (including columns of *Notes and Queries*) for free and full discussion . . . and *Reviews* of all scientific treatises suitable for general reading." In addition, there would be sections on mathematics, chess, and whist, regarded as scientific games.[102] There were no sections or articles included for the practitioner. When in 1882 a reader requested supplements on special subjects, Proctor reminded him "the purpose of *Knowledge* is rather to encourage general than special study of science." The "professed botanist would hardly expect to learn much that was new to him" from botanical articles in *Knowledge.* Nor would "professed students of astronomy," who might find something to interest them in the light essays on astronomy. But, Proctor asserted, "I do not write for them, but to interest . . . those who are not astronomers."[103] Although Proctor altered the format of *Knowledge* from time to time, he stuck to his basic plan through the journal's life as a weekly up until volume eight in 1885.

The format of *Knowledge* reflected Proctor's aversion for the professionalizing, hierarchical vision of science contained in the pages of *Nature.* Proctor's format drew on a republican image of scientific community, which, Susan Sheets-Pyenson notes, had begun to disappear in the new science journals in the 1860s, when periodicals sought support for the ideals of professionalized science.[104] Proctor's experiment in periodical publishing also looked forward to the goals of editors like William Thomas Stead, who promoted the "New Journalism" near the end of the century. Stead attempted to transform periodicals into fully inclusive public forums, and, when it came to science, he was critical of the hierarchical nature of elite science. His campaign against the arrogance of scientific experts, and his lack of deference toward Huxley in particular, led to strained relations between them.[105] Stead warned fellow

101. Lockyer and Lockyer 1928, 47.
102. Proctor 1881h, 3.
103. Proctor 1882d, 301.

104. Sheets-Pyenson 1985, 563.
105. Dawson 2004, 181, 192.

editors never to employ an expert, scientific or otherwise, to write a popular article on his own area of research. He believed that it was wiser to use an ignorant journalist, who could tap the expert's brains to write the piece, and then send the proof to the expert to correct.[106]

Proctor's inspiration for a journal format, which was based on the broad participatory notion of science, came from journals that he had read or to which he had contributed. He had tremendous respect for at least one of the journals from the 1860s, *The Intellectual Observer,* to which he had regularly contributed pieces on astronomy early in his career as science writer. He referred to *The Intellectual Observer* as "one of the very best science magazines ever published."[107] Two other journals were more important in shaping the format of Proctor's *Knowledge,* the *English Mechanic* and *Hardwicke's Science Gossip.* The *English Mechanic* (f. 1865) was a cheap mass-circulation science journal run cooperatively with its largely working-class readers. A rival of *Nature, The English Mechanic* boasted in 1870 to have a circulation larger than all the other English scientific publications put together.[108] One reason for the success of *The English Mechanic* was its loyal readership, which valued the opportunity to exchange views in extensive correspondence columns and to obtain information on a wide range of topics.[109] Proctor offered an upscale version of *The English Mechanic,* which stressed reader participation as a way of cultivating a loyal readership, but with a focus on science rather than its practical applications, which would appeal to a broad audience of members of both the working and middle class. In addition to a large correspondence section where readers replied to other readers' queries, which at times threatened to gobble up the entire issue, Proctor was not adverse to copying other columns from *The English Mechanic.* In 1882 he announced the opening of a new "Exchange Column, similar to that which has for several years

106. Stead 1906, 297. Proctor's vision for *Knowledge* coincided with a number of general features of the new journalism, which, according to Gowan Dawson, became prominent in the number of journals founded during the last two decades of the century. Proctor was not imitating Stead's daily *Pall Mall Gazette,* as it was founded in 1883, a few years after Proctor launched *Knowledge.* It is more likely that Proctor picked up on techniques originally used in North America that Stead and other "New" journalists also drew on. Proctor was frequently in the United States in the 1870s and 80s. Like Stead, Proctor engaged in a crusade against vested interests, but in the case of *Knowledge* it was government observatories; he also rejected the tradition of impersonal journalism and developed a distinct personal style of writing and editing (Dawson 2004, 172–74).

107. [Proctor] 1882e, 332.

108. MacLeod 1996, 224.

109. W. Brock 1980, 111, 113.

formed a feature in our excellent contemporary, *The English Mechanic*."[110] In 1884, a new column was added, "The Inventor's Column," which offered accounts by experts of "all inventions of really popular interest and utility."[111]

Science Gossip (f. 1865), one of the major science periodicals for a popular audience in the 1870s, provided Proctor with an example of a successful journal for more upscale readers. Edited by Mordecai Cubitt Cooke, a mycologist, *Science Gossip* was conceived of as a cheap monthly magazine—priced at 4d—devoted to natural history and microscopy. Its subtitle, "an illustrated medium of interchange and gossip for students and lovers of nature," emphasized its role as a clearinghouse of knowledge for amateurs. Early issues began with an unsigned editorial by Cooke, followed by signed articles on a wide range of topics and some contributions from readers. Then there were regular sections on zoology, entomology, fish, botany, microscopy, and geology, composed of short extracts from recent scientific books and periodicals. Several pages of brief correspondence, queries, replies, and notes for readers came in the final section. Cooke used experts from the British Museum and Kew to answer questions from readers. Proctor avoided the inclusion of extracts from other publications in *Knowledge*, and his focus was on astronomy rather than natural history. But some features of *Science Gossip* were later adopted by Proctor, particularly the accent on correspondence and "gossip" to draw in an audience of nonexperts.[112]

But Proctor emphasized that what distinguished *Knowledge* from any other science journal for the general audience, including *The English Mechanic* and *Science Gossip*, was both the quality and quantity of its "original matter." "Some readers," Proctor wrote in his "Editorial Gossip" column, "have pointed to publications kindred to *Knowledge* in certain respects, which offer more matter to their readers." Proctor replied that he could easily fill his journal with correspondence and articles taken from other publications, but had chosen to emphasize never-before-published pieces by science writers.[113] In contrast, Lockyer sought well-known practitioners to write for *Nature*. When the first issues of *Knowledge* appeared in October

---

110. [Proctor] 1882q, 367.

111. "Inventor's Column" 1884, 329. Proctor's link with the *English Mechanic*, and his lecturing tours of England, may have attracted working-class readers to his books. The library for the mechanics' institute at York obtained his *Light Science for Leisure Hours* and five of his other books (Paylor 2004, 151, 156).

112. English 1987, 107–12.

113. [Proctor] 1883c, 350.

1881, Lockyer's contributors to *Nature* that month included eminent men of science such as Archibald Geikie, E. Ray Lankester, Edward Tylor, Lord Rayleigh, Robert Ball, and Peter Guthrie Tait.[114] Proctor's chief criteria for choosing contributors had nothing to do with their credentials as experts. They had to be able to write light and entertaining, but informative, essays. Often he selected individuals like himself, who had experience writing for a popular audience.

Among his most regular contributors over the years were popularizers Grant Allen (on natural history) and Edward Clodd (on anthropology and evolution). Henry Slack, journalist, former editor of the *Intellectual Observer*, and in 1878 president of the Royal Microscopical Society, wrote a series titled "Hours with the Microscope" throughout Proctor's tenure as editor of *Knowledge*. The astronomy pieces were written by Proctor himself, Thomas William Webb, and Arthur Cowper Ranyard, Proctor's successor as editor of *Knowledge*. W. Jerome Harrison, Chief Science Master of the Birmingham School Board, and a Fellow of the Geological Society, dealt with geological subjects.[115] E. A. Butler, listed as BA and BSc, wrote a series on household insects. William Mattieu Williams, science writer, Fellow of the Chemical Society, phrenologist, and formerly master of the science classes at the Birmingham and Midland Institute, was the author of two series, "The Chemistry of Cookery" and "Philosophy of Clothing."[116] William Jago, who penned a series on the chemistry of the cereals, was a Fellow of the Chemical Society (FCS) chemical consultant, author of textbooks on chemistry, lawyer, and had been trained as a mining engineer at the Royal College of Chemistry and Royal School of Mines.[117]

Andrew Wilson, physician and author of a series on human physiology, was a science journalist and lecturer on zoology and comparative anatomy at the Edinburgh Medical School.[118] John Browning, Fellow of the Royal Astronomical Society (FRAS) and chairman of the London Tricycle Club, wrote on a variety of topics, including the eye, weather forecasts, and a

---

114. MacLeod asserts that in the first decade of *Nature*'s existence a good many contributors were like Lockyer, privately educated and not university men. But afterward men who came from the Scottish universities, the University of London, the Royal College of Science, and Cambridge wrote the majority of the journal's leading articles (MacLeod 1969, 448).

115. "Harrison, William Jerome" 1988, 232.

116. "Williams, William Mattieu" 1921, 468–69.

117. "Jago, William" 1941, 702.

118. "Wilson, Andrew" 1988, 568.

series of articles on bicycles, from tricycles to mudguards. Another prolific contributor on a series of subjects was electrical engineer William Slingo, principal and founder of the Telegraphic School of Science at the General Post Office.[119] In addition to helping with the editing of *Knowledge,* Slingo composed articles on the electromagnet, electroplating, photography, the electric tramcar, scientific industries, and the history of a lightning flash. Proctor was quite happy to accept pieces from female contributors. Amelia Edwards, novelist, journalist, and Egyptologist, wrote a series on Egyptian archaeology; Ada Ballin, journalist, author, editor, and later founder of *The Mother's Magazine* and *Womanhood,* offered a series on thought and language; and the poet Constance Naden submitted a series on the evolution of the sense of beauty.[120] One frequent contributor, Thomas Foster, even composed an article on "Feminine Volubility," which was not out of keeping with Proctor's desire to openly discuss the scientific proof for female inferiority.[121] The mysterious Foster also wrote a series, titled "The Morality of Happiness," dealing with themes raised by Herbert Spencer's evolutionary ethics.

Eminent British scientists were rarely contributors. When their names did appear as authors, it was usually under a piece that was actually a summary of a public address, as in the case of William Huggins's "Photographic Spectrum of Comet (Wells)" and Robert Ball's "The Sun's Distance."[122] Proctor did recruit a few American scientists, E. D. Cope, the biologist and paleontologist, who penned an article on the evolution of human physiognomy, and the Princeton astronomer C. A. Young, whose pieces were on his area of specialty, but there was little overlap between Proctor and Lockyer's lists of contributors.

CONDUCTING KNOWLEDGE

In 1886, Proctor drew the attention of his readers to a piece in another periodical that recommended that the motto of *Knowledge,* "Let knowledge grow for more to more," should be dropped for something more appropriate. "The *Topical Times* says," Proctor reported, "rather neatly, that I ought

---

119. "Slingo, Sir William" 1941, 1246–47.

120. J. S. C. 1917, 601–3; "Ballin, Ada S." 1988, 27; "Naden, Constance Caroline Woodhill" 1917, 18–19. For more on Naden, see Moore 1987, 225–57.

121. Foster 1885, 17–18.

122. Dr. Huggins 1882, 89–90; Ball 1883b.

to take for a motto for *Knowledge* 'Le Savoir; c'est moi.'" Rather than feeling "goaded" by this proposal, Proctor declared there was "some truth in the suggestion that *Knowledge* depends a good deal on me." He admitted that he wrote a lot of the articles for *Knowledge* because he could not find authors "who are ready to work on the lines I have adopted." Few men of science used "plain untechnical words," while few popularizers had sufficient acquaintance with scientific subjects.[123] The pages of *Knowledge* really did bear Proctor's imprint. Not only did he compose articles on a wide variety of topics, contribute a regular scientific gossip column, and respond to correspondence and queries, Proctor also tried to develop a distinctive editorial style. In all of his roles, whether as editor, contributor, or gossip columnist, Proctor criticized practitioners like Lockyer while maintaining the importance of reaching out to the public with entertaining and informative scientific news. The attempt to draw the popular audience into the scientific arena succeeded so well that Proctor was slowly forced to cut back on the number of pages devoted to providing an open forum for his readers.

Proctor left his mark on *Knowledge* through the vast number of articles he contributed. Each issue usually contained at least two original articles by Proctor. Most of them were on astronomical topics, such as meteors, comets, Venus nearing transit, the moon, star clouds, solar eclipses, sunspots, proofs of the earth's rotundity, the harvest moon, the universe of suns, the earth's shape and motions, the sun's heat, and the red spot on Jupiter. Others dealt with scientific issues of a somewhat broader nature. Proctor wrote articles on weather, betting, thought-reading, learning languages, corset-wearing, the comma, strange coincidences, flying and flying machines, our two brains, French balloon experiments, the gambling spirit in America, monster sea serpents, Dickens and Thackeray, and illusions of the senses.

In contrast to other science journals, Proctor did not often report on the activities of scientific societies in the pages of *Knowledge*. At first, he refused to include even the briefest abstract of papers, touching off a debate among his readers. W. Mattieu Williams agreed with Proctor's policy, arguing that brief abstracts would be of no value to the general public.[124] Later, Proctor relented in perverse fashion. In the second volume Proctor wrote an article on the British Association (BAAS) meeting at Southampton in 1882. In opposition to the press's view that the BAAS stirred up enthusiasm for science, he argued that the work of the various sections was for men of science only, and amounted to "the kind of work known as 'mutual admiration,'" while the

---

123. Proctor 1886a, 228.   124. Williams 1881, 143–44.

presidential address, supposedly offered for the general public, was nothing but a bore. How could a recapitulation of scientific progress, necessarily presented in a "crude and condensed form," possibly interest the public? During Dr. Siemens's presidential address that year, "one bald pate after another was bowed in sleep," and "during the last half of the address, the cries of the coachmen were heard by a much larger section of the audience than the voice of the president."[125] To those readers who objected to his remarks as unduly critical, Proctor was unrepentant. Two weeks after his first article a second one appeared on the British Association, where he asked if the work done by the Association during the fifty years of its existence was really "worth the time and labour bestowed on the meetings"? He was especially critical of the jargon used in the presidential addresses, which could not be understood by the vast majority of the audience, and of reports of these addresses in such periodicals as *The Times* and the *English Mechanic*, which merely repeated "this egregious nonsense, as if it had been profound science."[126] Proctor was hard on Arthur Cayley's presidential address on imaginary geometries the following year because no one understood it.[127]

Other elite scientific societies came in for their share of criticism in the pages of *Knowledge*. Proctor related an anecdote about a scientist who, when asked about the meaning of the letters F.R.S. after his name, truthfully responded, "fees raised swiftly."[128] Proctor reserved his choicest barbs for the Royal Astronomical Society, charging that it was dominated by a small clique of professional astronomers who delivered tedious papers at meetings, published their own observations in official society publications, and used the society merely to further their own careers. In 1885 Proctor reported that, according to an informant, the presidential address at the annual meeting of the RAS had been "one of the dullest and dreariest ever listened to." This led Proctor to remark on the despotic rule of the RAS Council.[129] Later that same year Proctor supported the motion to change the existing mode of election of officers and the Council, arguing that the same people were nominated year after year.[130] Although all scientific societies were guilty of abuses in award competitions and election to offices, Proctor singled out the RAS. The process whereby medals were awarded "very readily lends itself to jobbery," while the presidency was seen as a stepping-stone to a salaried post.[131] In light

125. [Proctor] 1882g, 224–45.
126. [Proctor] 1882h, 257.
127. Proctor 1883k, 287.
128. [Proctor] 1885a, 12.

129. [Proctor] 1885c, 155.
130. [Proctor] 1885d, 376.
131. Proctor 1886e, 215.

of the severity of Proctor's criticisms of the RAS, it is understandable that
readers thought he wanted to see the RAS destroyed, a charge Proctor hotly
denied.[132] In general, Proctor was critical of most scientific societies, not just
the RAS. He declared that the influence of scientific societies tended more
"to the discouragement than to the advance of science," and felt confirmed
in this belief "by noting that not one series of scientific researches of any importance has attained success through the influence of any scientific body."[133]

Proctor's scathing comments about elite scientific societies were matched
by his criticisms of the men of science, in particular astronomers. Whereas
Nature and other science journals presented scientists in a positive light in
order to entice readers to appreciate their accomplishments and admire them
as intellectuals, Proctor could be harshly negative. His old enemy, Lockyer,
of course, came in for his share of abuse. Proctor ridiculed Lockyer's publications on several occasions. In comparison to Robert Ball's trustworthy
writings, Proctor declared in 1888 that Lockyer's astronomical works were
full of blunders.[134] Similar criticisms were leveled when Lockyer's *Outlines
of Physiography,* published by Macmillan, was reviewed in *Knowledge* later
that same year. The anonymous reviewer, likely Proctor, did not believe that
Lockyer could "seriously regard this absurd production as a contribution to
exact knowledge," so he ruled out the possibility that Lockyer had intentionally played a "joke on his good friends, the publishers of *Nature.*"[135] The
book was so full of inaccuracies that the reviewer claimed to be puzzled. In
another article, Proctor portrayed Lockyer as an egomaniac who praised his
own work under the cloak of anonymity. He exposed Lockyer as the author
of an unsigned piece in the *Times,* which attributed all of the progress in
astronomy to his own work.[136] In an article titled "The Dignity of Science,"
Proctor returned to the theme of state funding of science, so crucial for the
controversies in the 1870s. It was Lockyer, Proctor claimed, who had raised
this issue first, in the pages of *Nature.* To Proctor, Lockyer's position was
that "your man of science ought to be incapable of anything so degrading as
work, but to beg he should be by no means ashamed."[137] But Lockyer was not
the only professional astronomer to be at the receiving end of Proctor's barbs.
The American astronomer Edward Singleton Holden (1846–1914), who had
dared to question the value of Proctor's work, was castigated for his failure
to "do satisfactory work" despite the large annual salaries he received.[138]

132. [Proctor] 1885m, 278.
133. Proctor 1886e, 216.
134. Proctor 1885n, 322.
135. [Proctor] 1888b, 234.

136. Proctor 1887a, 115.
137. Proctor 1886d, 93.
138. Proctor 1887b, 210.

Proctor still did not think very highly of the work done by astronomers who worked at government-funded observatories.[139] Although Proctor tended to treat scientists with disdain in the pages of *Knowledge,* his attitude toward Huxley and Tyndall was a noticeable exception. Their names were never included in his rants against would-be professional scientists.

Proctor's original articles, reports on scientific societies, and comments on practicing scientists were only part of his contribution to *Knowledge.* Proctor wrote regular columns on scientific news, responded to correspondence, and assumed all of the other important editorial duties. Some readers were confused by Proctor's multiple roles. His habit of signing his contributions with a slightly different name led some correspondents to ask if he was trying to be mystifying. In 1882 Proctor had to explain, "The Editor of *Knowledge* and Mr. Proctor, Mr. R. A. Proctor, and Mr. Richard A. Proctor are, of course, one and the same person. Yet it is not an idle fancy or mere joke which causes me to speak at one time as Editor, and at another in my own name. As Editor I may have to explain that some statement cannot be accepted, which as Richard Proctor I would not take the trouble to contradict."[140] Other readers did not like Proctor's editorial style, and complained that he did not comport himself with proper editorial dignity. Why, they asked, did he talk so familiarly with his readers and why did he note and correct his own mistakes? Proctor acknowledged that he did not behave like other editors. "We choose our own way," Proctor proudly announced, "because we like it much better. We prefer to have no pedestal on which to assume statuesque editorial dignity."[141]

Proctor actually referred to himself as the "conductor," rather than the editor, of *Knowledge.* Alison Winter has discussed how the notion of conducting took on new associations during the Victorian period in light of mesmerism, and not just in musical contexts. For the journal editor, the issue concerned how to provide a central harmonizing influence for a collective reading public, how to be both charismatic and an ordering influence.[142] Proctor was not the first editor of a science periodical aimed at the public to insist on the title "conductor." The American *Popular Science Monthly,* founded previous to *Knowledge* in 1872, bore on its title page "Conducted by E. L. Youmans." But Youmans did not explain how he distinguished between the roles of "editor" and "conductor" in the initial issues of his journal, and he did not stray far from the traditional, dignified editorial persona. For Proctor, the

139. Proctor 1886c, 339.
140. [Proctor] 1882f, 365.

141. [Proctor] 1882o, 489.
142. Winter 1998, 320–22.

term "conductor," when linked to science, could have referred to his role as a communicator of scientific ideas, just as certain materials were excellent conductors of heat. Or, if taken in a musical sense, Proctor may have seen himself as a conductor of a band, as one who brought harmony to the noise produced by clashes between the upholders of science and the defenders of religion. The line from Tennyson's "In Memoriam" that served as the motto for *Knowledge* was followed by a musical analogy resonating with this sense of the term. The full stanza reads:

> Let knowledge grow from more to more
> But more of reverence in us dwell;
> That mind and soul, according well,
> May make one music as before,
> But vaster.

Like Tennyson, who embraced science while rejecting scientific materialism, Proctor saw himself as offering a cosmic, or "vast," perspective on the world that blended science and religion into a harmonious whole.[143]

The term "conductor" likely also included the idea that his readers could be an integral part of the band that produced the music. In the first volume of *Knowledge*, Proctor told his readers that "conducted by" meant "'edited,' or rather includes it and something more."[144] That "something more" was establishing a dynamic relationship with his readers that gave them, and not the men of science, a voice in the making of both knowledge and *Knowledge*. Proctor used extensive correspondence columns with varying formats as a way of allowing his readers to participate in the shaping of the journal, and he wrote columns designed to respond to readers' comments. In giving a large portion of the weekly to correspondence, Proctor was following the lead of egalitarian periodicals like the *English Mechanic*. There was not much of an editorial presence in the *English Mechanic*, whereas Proctor developed a distinctive editorial style. Eventually, Proctor felt compelled to limit the amount of pages devoted to correspondence, as it threatened to overwhelm the other departments of the journal, generated an intolerable drain on his time, and exposed him to irate readers who challenged his authority as editor.

143. I am indebted to Jonathan Smith for this point.
144. [Proctor] 1882b, 595.

In the very first issue of *Knowledge,* Proctor made it clear that correspondence would occupy a central place in the new journal. "I am very anxious that Correspondence should become a distinguishing feature of this magazine," Proctor announced. "I wish all readers to feel that in these columns, including the section for Queries and Replies, they have a means of resolving doubts which may occur to them in scientific study or investigation, when reading articles on science in magazines and journals, and in studying the pages of this magazine itself."[145] At first, Proctor set up two columns devoted to correspondence, "Our Correspondence Columns," which featured letters from readers followed by Proctor's response in square brackets, and "Queries" and "Replies to Queries," which contained questions from readers and the answers to the questions from the previous issues by readers and Proctor. The correspondence column also played a central role in *Nature,* but here a seemingly impartial Lockyer provided a space for different groups of scientists to engage in controversy.[146] By contrast, Proctor did not hesitate to offer his opinions, and the letter writers to *Knowledge* were not eminent scientists. The correspondents to *Knowledge* in 1881 were not a distinguished lot, including Newton Crosland, J. McGrigor Allan, G. F. P. Dyer, and C. J. Shaw, while others preferred to use pseudonyms such as Cerebrum, Tyro, Upsilon, Sun, Crusader, and Speculum. In the same year the correspondence columns in *Nature* contained communications from eminent scientists such as George J. Romanes, Charles Darwin, Lord Rayleigh, William B. Carpenter, and G. G. Stokes.

Readers responded enthusiastically to Proctor's offer to open up the pages of *Knowledge* for discussion of scientific issues. Throughout the first year of *Knowledge,* the high volume of letters forced Proctor to modify continually the correspondence columns of the journal. In the third weekly issue of November 18, he introduced a new column titled "Answers to Correspondents," which gave brief answers to questions but not the questions themselves. A week later, Proctor introduced more changes. "We begin to see more clearly than we did at first the lines on which our Correspondence columns will have to be conducted," Proctor declared, "though we shall always be glad to have hints and suggestions from our readers towards the improvement of this section of KNOWLEDGE." Referring to the "unexpectedly rapid growth of the correspondence," Proctor stated that from hereon

---

145. [Proctor] 1881b, 15.   146. Meadows 1972, 34, 36.

in he would limit his replies to letters.[147] Two weeks later, although thankful for the interest shown by his readers, Proctor asserted that he would have to condense published letters in order to include the bulk of those sent.[148] Just a week later, in response to a letter complaining that too much space was being given to correspondence, queries, and replies to queries, Proctor now declared that there was room for only a third of the letters sent in, and that the most original, interesting, and concise would have the best chance of being included.[149] In the following issue of December 23, 1881, the final number of the year, Proctor acknowledged that he was having difficulty deciding how to handle the avalanche of mail he was receiving. To make matters worse, some of the mail contained conflicting advice on how to resolve the problem, which had implications for the size of the journal and the amount of original material included. Proctor complained that "We are urged—(1) to widen our space for correspondence; (2) to keep our correspondence; . . . (5) to explain simply the principles of every science on the face of the earth, all at the same time; (6) to leave all such explanations to text books; (7) to increase our size; (8) by no means to do so . . .; (19) to have more astronomy; (20) to have less astronomy."[150]

By February of the following year, Proctor decided on a course of action that altered the complexion of *Knowledge*, restricting the participatory quality of the journal. A series of rules governing the submission of letters, queries, and replies was adopted, which demanded brevity and a standard form, and in the cases of those letters that were too long, redundant, or unsuitable for publication, the names of the letter writers would be listed in a new column titled "Letters Received."[151] Then in March, Proctor announced that the "Queries and Replies" columns would be extensively revamped and merged with the "Answers to Correspondents" column. In the future, queries would be sent to experts in departments, such as astronomy, geology, chemistry, and botany, who would deal broadly with specific questions, so that the answers were of interest to the entire readership. Proctor admitted that this was a more expensive arrangement than leaving readers to answer each other's questions, but more useful.[152] It also relieved him

147. Proctor 1881e, 73.

148. Proctor and Proprietors of "Knowledge" 1881, 112.

149. Proctor 1881c, 139.

150. Proctor 1881d, 160.

151. Proctor 1882l, 320; 1882k, 327.

152. [Proctor] 1882r, 434.

of the burden of wading through huge reams of correspondence. Proctor claimed the following month that he was following the readers' wishes in discontinuing "Queries and Replies," because "while taking up much space, they only interested a few." Strikingly, he acknowledged that this change put *Knowledge* closer to more traditional journals. "Hereafter," Proctor affirmed, "*Knowledge* will occupy the same position in this respect as the *Athenaeum, Academy,* and *Nature,* in which there are no columns for queries or replies, and very few answers to correspondents."[153]

When the second volume of *Knowledge* appeared in June of 1882, Proctor had stabilized the correspondence section of the journal, which was now composed of about three pages of "Letters to the Editor" and "Answers to Correspondents." Subsequently, in the third volume the "Answers to Correspondents" section was discontinued. However, soon after the elimination of "Queries and Replies," Proctor experimented with another way to maintain a "give and take" relationship with his readership. A relatively new column, "Science and Art Gossip," was moved to the front page of the journal in August 1882. Proctor invited "notes on matters of interest . . . briefly (as well as plainly) worded" from "contributors, correspondents, and subscribers."[154] He reported the gossip, but added his own witty comments. With the end of "Answers to Correspondents," this column became the chief means for Proctor to speak out on controversial issues, to reflect self-consciously on his role as conductor of *Knowledge,* and to adopt a jaunty, personable persona not in keeping with the traditional, impartial editor. The "Letters to the Editor" had become strictly scientific.

Proctor's use of correspondence and gossip columns to establish a journal that championed a less hierarchical conception of science—which carried on the "Republic of Science" tradition of earlier science journals for general readers—presented him with a series of challenges. On the one hand, he wanted to encourage his readers to participate in the making of *Knowledge.* But on the other, some of his readers insisted on the right to have their communications published whenever and as often as they pleased. Others disliked Proctor's editorial style, while some wanted more space for original material and less for gossip. On May 4, 1883, Proctor proclaimed that the gossip column would no longer appear. Although "many correspondents of *Knowledge* have, from the beginning of its career, indicated a wish that we should adopt something like the chatty style which . . . we have actually employed in our Gossip columns," other correspondents, "few and

153. [Proctor] 1882a, 539.   154. [Proctor] 1882m, 191.

ill-mannered" grumblers, grudge the space and "mar the tone of our converse with our readers."[155] Two weeks later, Proctor reversed his decision. "So many readers express (despite our protest) the request that *Gossip* should be resumed, that in a modified form we resume it," Proctor declared.[156] The column was reduced in size and placed near the end of the journal. It continued for as long as Proctor conducted *Knowledge*.

Some readers also had serious objections to Proctor's editorial style in the gossip and correspondence columns. Readers were not always sure if Proctor's sarcasm was playful or serious. In May 1882 he received a letter protesting his severity and unfairness to correspondents, characterizing his answers as "a weekly dose of sulphuric acid." Proctor denied in "Answers to Correspondents" that he was overly harsh. "About one answer in forty since KNOWLEDGE first appeared has been *really severe;* that is, not meant for banter," Proctor asserted, and in the case of severe replies they were "more than merited."[157] In his "Editorial Gossip" column of February 13, 1885, he again responded to those who thought that he was too sharp with correspondents, this time in his "Letters Received and Short Answers" column. He pointed out that readers only saw the reply, not the communication that elicited it. "No one but the editor of a journal like this can have any conception of the mass of heterogeneous correspondence which is showered on his devoted head," Proctor complained, "in fact, of the hundreds of letters he receives from persons of all social ranks, and of the most diverse intellectual capacities. I need scarcely say that people with fads and crazes make the most strenuous efforts to advertise their views through such a medium as *Knowledge*," including "Anti-Vaccinators, Anti-Vivisectors, Believers in Swedenborg, Vegetarians, Spiritualists, Earth-flatteners, and Immovable Earth men."[158]

Some faddists would not take "no" for an answer and demanded their right to have their communications published in a democratic journal such as *Knowledge*. In May 1885, the hylo-idealist Robert Lewins badgered Proctor for weeks to have his letter appear in print. Proctor relented, but in an editorial comment he said that nothing more on the subject could be admitted.[159]

155. [Proctor] 1883j, 267.

156. [Proctor] 1883a, 298.

157. [Proctor] 1882i, 613.

158. [Proctor] 1885b, 133.

159. [Proctor] 1885q, 421. For an analysis of Lewins and hylo-idealism, see Moore 1987, 225-57.

The following week he claimed "absolute power to reject any communication, without assigning reasons for so doing" and he refused to debate the reasons for his decisions with unhappy authors.[160] Yet Lewins was not so easily put off. In July of 1885 Lewins was still hounding Proctor to publish one of his articles. Proctor responded publicly, in "Letters Received and Short Answers," that Lewins had forced him to "speak plainly." *Knowledge* would not be converted into "propaganda" for Lewins's doctrines because hylo-idealism was unproven metaphysics and atheistic to boot.[161] The tenacious Lewins would not give up. Letters demanding that his previous letters be inserted in *Knowledge* were sent during August, leading Proctor to refuse to publish them on the grounds that they merely reiterated the earlier exposition of hylo-idealism. Then came a special notice stating that answers to correspondents would henceforth be reduced.[162]

Proctor founded *Knowledge* in order to challenge *Nature*'s hold on the science periodical market for the public and to counter its hierarchical vision of a scientific community dominated by would-be professional scientists. He contributed original articles, meant to be models of clearly accessible science, and wrote pieces that ridiculed practicing scientists and their societies. As conductor of *Knowledge,* he experimented with various types of correspondence columns, and adopted an editorial style that, he believed, brought him in closer contact with his readership while encouraging them to participate in the making of knowledge. Ensuring that such an enterprise was financially viable, limiting the amount of time to respond to readers, and dealing with particularly persistent faddists was not easy for Proctor. He struggled to balance editorial control with the right amount of openness. It was a difficult balancing act that often brought him into conflict with his readers. In February of 1885 he informed a reader that he did not resent the expression of any reader's opinion on a question on which they were competent, but he was entering a "mild protest against your quasi-Papal fashion of teaching me how to edit this paper."[163] Proctor's little experiment in conducting an egalitarian science periodical threatened to turn readers into editors.

A NEW EXPERIMENT

In the summer of 1885, in the midst of Lewins's challenge to his editorial authority, Proctor began to drop hints about a change in the journal that would

160. [Proctor] 1885o, 445.
161. [Proctor] 1885j, 36.

162. [Proctor] 1885k, 168.
163. [Proctor] 1885i, 182.

leave little room for correspondence and none for replies.[164] In the September 4 issue, he announced plans for publishing *Knowledge* as a monthly magazine that would cost sixpence. "Like other monthly magazines," Proctor declared, "*Knowledge* in future will not be open to correspondence properly so called. . . . I am obliged to admit that the introduction of the correspondence element, and especially of the replies to correspondents, was a mistake from the beginning." Rather than aiming at the discovery of truth, too many correspondents wrote "argumentative letters," which aired theories originating in "ignorance or misapprehension." Proctor then recalled how he was forced to withdraw from the correspondence columns of the *English Mechanic* because he became associated with paradoxists, flat-earth men, weather prophets, and circle-squarers.[165] "Controversy equalises in the eyes of outsiders the ignorant and the well-informed" and "science suffers by such controversy" as well-established facts are made to appear as if they were subjects for discussion. Proctor had hoped to avoid such unprofitable controversy in the weekly pages of *Knowledge,* but had failed. Argumentative persons, however, "would not be content with so slow a method of conducting their discussions as a monthly magazine would afford." Furthermore, in the monthly *Knowledge* "scientific, literary, and artistic matters will be dealt with by those who know, not discussed between those who know and those who do not know."[166] The following month Proctor explained that the change in form was made "chiefly because I could not longer stand the burden."[167] Unwilling to spend the time needed to enforce the boundary between science and nonsense, Proctor gave up on his vision of an egalitarian weekly that would challenge *Nature*'s dominance of the periodicals market for general audiences.[168]

The new monthly, Proctor admitted, "must be an experiment." Original articles would occupy "a larger relative proportion of our space." More significantly, Proctor planned to allow articles on a controversial topic that had previously been taboo in *Knowledge:* the influence of science on religion.

164. [Proctor] 1885l, 189.

165. Proctor was by no means alone in his struggle to distinguish himself from weather prophets, paradoxists, and flat earthers. Katherine Anderson has examined the similar situation of Victorian meteorologists, while Garwood has explored Wallace's problems when he tried to debate flat earthers (K. Anderson 2005, 41–82; Garwood 2001).

166. Proctor 1885e, 204.

167. [Proctor] 1885n, 322.

168. Proctor did not exclude his readers entirely from active involvement in the journal. He agreed to the request of readers to extend the "Editorial Gossip" section since "Replies to Correspondents" had ceased (Proctor 1885g, 67).

Proctor's aim was to demonstrate that developments in science were not subversive of religion. To dramatically highlight this bold, new direction, Proctor revealed that, under the pseudonym of Thomas Foster, he had been contributing articles to *Knowledge* that subtly touched on the implications of science for religion.[169] The new monthly appeared on November 1, 1885. The title page still listed Proctor as "conductor" of the journal and retained the same logos and Tennyson quote as the journal slogan, but the subtitle had changed to "An Illustrated Magazine of Science, Literature, and Art," and the publisher was now Longman. The leadoff article "The Unknowable; or, The Religion of Science," wherein Proctor rejected the notion of a conflict between science and religion, set the tone for subsequent issues.[170] Articles praising Herbert Spencer, religious agnosticism, and cosmic evolution, many by Proctor, appeared in *Knowledge* in the following months. He explained that *Knowledge* now had "an object far higher . . . than the mere interpretation or explanation of scientific matters," teaching the *"great lesson which the recent progress of science teaches is the universal prevalence of law and the consequent futility of lawlessness, no matter under what high or seemingly sacred names it may be disguised."*[171] In de-emphasizing pieces on astronomy and natural history, and focusing on evolutionary themes, the science and religion debate, agnosticism, and other such controversial issues, Proctor may have decided to imitate aspects of James Knowles's successful monthly, the *Nineteenth Century*.[172] Circulation of *Knowledge* from the time that Longman became the publisher to the time of Proctor's death in 1888 ranged from about 4,500 to

---

169. Proctor 1885e, 205. In this article Proctor remarked that the use of a second name "has served its purpose, in securing for articles outside science, attention." But there had been inconveniences in using a pseudonym. Mr. Foster had been invited by one of the most eminent Spencerians in America to meet other admirers of Spencer, including Mr. Proctor, at a public gathering, "an invitation which he could not readily decline yet could not possibly accept without duplicity." However, some readers must have been appalled by the duplicitous statements issued earlier by Proctor in 1885 when he denied that he contributed pieces to *Knowledge* under a pseudonym. "Those who credit the conductor of *Knowledge* with the authorship of everything that appears in it may note," Proctor announced, "once for all, that *Mr. Proctor signs every single article, letter and paragraph that emanates from his pen.* What is not so distinctly signed in this magazine is none of his" ([Proctor] 1885h, 160).

170. Proctor 1885r, 1–3.

171. Proctor 1886b, 314.

172. Throughout 1884, when Proctor was thinking about the switch from weekly to monthly, *Nineteenth Century* had featured articles by Spencer on "Religion: A Retrospect and Prospect," "Retrogressive Religion," and "Last Words about Agnosticism and the Religion of Humanity"; by Frederic Harrison on "The Ghost of Religion" and "Agnostic Metaphysics"; by Sir James F. Stephen on "The Unknowable and the Unknown"; and by St. George Mivart on "A Limit to Evolution."

7,500 copies. But sales dropped off slowly during this period. In 1886 Longman's initial print run was between 6,500 and 7,500 copies, while in 1887
the range was 5,250 to 5,700 copies. By 1888 the print run fell between 4,600
and 5,000 copies.[173] In 1888 there were sixty-six subscribers from around the
world, including Canada, Britain, New Zealand, Austria, the United States,
and India.[174]

While it had lasted as a weekly, *Knowledge* had offered to the British reading public a conception of science in stark contrast to the one they found in
Lockyer's *Nature*. Initially, Proctor resisted the temptation to attack Lockyer
directly in the pages of *Knowledge*. He held out an olive branch to Lockyer a
month before the first issue of *Knowledge* appeared. On October 4, 1881, Proctor sent Lockyer the prospectus for *Knowledge* and invited him to contribute
a paper on the elementary constitution of matter, adding that a paper by the
American astronomer C. A. Young on the same subject would also appear.[175]
On November 17 of the same year, Proctor wrote again to Lockyer, asking
why he had received no answer to his earlier letter. He assured Lockyer that
pairing an article by him with Young was not intended to raise a dispute. He
claimed that he now avoided controversy. "In the past," Proctor affirmed,
"there were some passages not altogether pleasant between us,—hastily written and (I speak for myself) doubtless foolish letters and public utterances
not altogether well judged—I have forgotten how it happened and what it
was all about." Whatever had happened in the past, Proctor believed that
it should not prevent Lockyer from writing for *Knowledge*. Neither, Proctor
declared, should Lockyer refuse his request on the grounds that *Knowledge*
had been established as a competitor to *Nature*. Disingenuously, Proctor
claimed that he had *not* set up *Knowledge* to compete with *Nature*. "You may
notice that the plan and purpose of Knowledge," Proctor affirmed, "are so
distinct from those of *Nature* that there is and can be no question of rivalry
(we thought of this in putting our price so low)."[176]

Apparently, Lockyer responded to Proctor in a curt letter, declining
to contribute to *Knowledge*. In his "Answers to Correspondents" column
of December 2, 1881, Proctor publicly replied to Lockyer, expressing his
regret that Lockyer could not spare the time.[177] Later that month C. A.

---

173. *Archives of the House of Longman, 1794-1914* 1978, B18, 459, 461-62, 536, 538, 540, 585, 587,
595.

174. Ibid., N112.

175. Proctor to Lockyer, Oct. 4, 1881, University of Exeter.

176. Proctor to Lockyer, November 17, 1881, University of Exeter.

177. Proctor 1881a, 106.

Young's article on "The So-Called Elements" was published. It contained criticisms of Lockyer's use of spectroscopic evidence to support the theory that the chemical elements are actually composed of smaller and simpler substances.[178] There the matter rested until April of 1883, when in his "Science and Art Gossip" Proctor reported that he had learned from a "coarsely-worded post-card communication forwarded to our publishers" that "Mr. Lockyer was offended at our answering in this way what he regarded as a private communication." According to Proctor the postcard was "full of illbred vituperation." Although the postcard was unsigned and not in Lockyer's handwriting, Proctor claimed that it could only have come from him since it contained information known only to Lockyer and Proctor.[179]

In his "Letters Received" column Proctor included a long response in which he explained that he had replied in the pages of *Knowledge* to Lockyer's letter declining to contribute "to indicate publicly that desire to be on friendlier terms." Proctor lauded Lockyer's earlier observational work, but "the Editor (not being the Author) of *Nature* has made a few mistakes; and he does not very readily forgive those who pointed them out.... Give the world some more good work like that of 1868 and the mistakes will be forgotten." Lockyer could hardly have been happy with the implication that he had not produced anything good since 1868. Moreover, a public response to Lockyer's protest over the initial public response could only have angered him more.[180]

The following week Proctor blasted Lockyer in an article titled "Social Dynamite," which compared the "social offense" of sending anonymous letters to one's enemies to the use of dynamite by political terrorists. Now Proctor was not so sure that Lockyer had actually written the postcard, but if not Lockyer, it had to be a friend in whom he had confided the details of his private correspondence with Proctor. If Lockyer were really innocent, Proctor called upon him to help expose those "ready to tamper with social dynamite." Proctor threatened to publish a facsimile of the postcard using photographic means to help "out" the culprit.[181] A few weeks later, in an article titled "Personal," Proctor acknowledged the letter sent by Lockyer to the publishers of *Knowledge* dated May 11. Lockyer asserted that he did not send the postcard and that he was ignorant as to who did. Proctor regretted his earlier accusation that it was Lockyer who had sent the postcard, but, going back to the original issue that set everything off, the public response to Lockyer's letter declining to contribute, he claimed that he answered friends

178. A. Young 1881, 151–52.

179. [Proctor] 1883m, 229.

180. [Proctor] 1883g, 240.

181. Proctor 1883n, 244–45.

like Clodd through the correspondence columns in *Knowledge*. Proctor then apologized to his readers for "giving so much space to the subject," but he hoped it would "help to unearth a mischief-making individual, who, if early detected, or even threatened with imminent detection, may be hereafter innocuous." Proctor went on to describe the postcard in more detail.[182] About four months later, Proctor referred again to the anonymous postcard. Now he claimed to know the identity of the "sneak." But he was not about to let Lockyer off the hook. "I have no doubt either," Proctor declared, "that after reading what I there wrote [in "Social Dynamite"], Lockyer himself could form a tolerably shrewd guess on the subject. He has not however thanked me yet for enabling him to see what a treacherous 'friend' that particular person was."[183]

Proctor's controversy with Lockyer on the propriety of responding publicly to private correspondence, and his stand on the harmful effects of anonymous letters, underlines the original strategy behind *Knowledge*. All correspondence to *Knowledge* was considered by Proctor to be within the public arena since the production of scientific knowledge involved all members of society, not just elite scientific experts. Lockyer would be treated no differently than any other correspondent. Once sent to the conductor of *Knowledge*, the contents of Lockyer's letter were public knowledge. By insisting that the readers of *Knowledge* had no right to know that he had declined to write a piece for Proctor, Lockyer challenged the entire notion of a "republic of science." The anonymous postcard was an even more insidious threat to the public nature of science, the very raison d'être of *Knowledge*. It sabotaged Proctor's attempt to carry on an open, reasoned dialogue with correspondents. That was why he compared it to a terrorist act that undermined the very fabric of society. Science could not be conducted if Lockyer or any of his expert friends were not willing to bring knowledge before the bar of public opinion. The story of Proctor's establishment of a competitor to *Nature* illustrates the existence in the 1880s of two different ways of envisioning the role of the practitioner and the role of the public. Proctor, like Chambers before him, maintained that scientific controversies were resolved by appealing to the public. Elite scientists had no monopoly over the discovery and determination of scientific truth. At stake in the pages of Proctor's journal was the very meaning of science, the issue of who should participate in the making of knowledge, and the boundaries between

182. Proctor 1883l, 287.   183. [Proctor] 1883h, 208.

practitioner and public. The concept of "professional scientist" that drove the agenda of scientific naturalists like Huxley could still be challenged in the 1880s, though, as Proctor discovered, not without exposing him to the wrath of scientists like Lockyer and not without inviting challenges to his own authority as conductor from his readers.

CHAPTER SEVEN

Practitioners Enter the Field

Huxley and Ball as Popularizers

ON NOVEMBER 10, 1862, Thomas Henry Huxley delivered the first of
six weekly lectures to an audience of workingmen. Inelegantly titled "On
Our Knowledge of the Causes of the Phenomena of Organic Nature," Adrian
Desmond has described these lectures as "plebeianizing the *Origin*."[1] Huxley
later recalled that he imagined the lectures "would be of little or no interest
to any but my auditors."[2] He allowed them to be recorded in shorthand and
published by Robert Hardwicke in four-penny weekly parts, and thought-
lessly gave up any claim to payment. Huxley enclosed them in a letter to
Darwin, sent on December 2, 1862. He told Darwin "I have no interest in
them and do not desire or intend that they should be widely circulated." Perhaps, in the future, he wrote, "I may revise and illustrate them, and make
them into a book as a sort of popular exposition of your views." Huxley
downplayed their importance further, claiming that "there really is nothing
new in them nor anything worth your attention," but he asked Darwin to
glance at them and let him know if he objected to anything he found.[3]

But Huxley had seriously underestimated the impact of his lectures on
the public. To his astonishment, the lectures sold well, and not only in England. Huxley regretted that he had treated the publication of his lectures in
such a cavalier manner. He wrote to Hooker that "I had never imagined the
lectures as delivered would be worth bringing out at all," and now "I lament
I did not publish themselves and turn an honest penny by them as I suspect

---

1. Desmond 1997, 310.
2. T. Huxley 1894a, vi.

3. Burkhardt et al. 1997, 579.

Hardwicke is doing." He added that Hardwicke was "advertising them everywhere, confound him."[4] Huxley was also surprised by the enthusiastic reaction of his friends. Lyell wrote to Huxley on January 28, 1862, to tell him that his lectures were valuable as a means of educating the public about science.[5] Darwin was excited by the lectures and sent Huxley a series of letters full of praise. On December 7, 1862, he wrote, "they would do good and spread a taste for the Natural Sciences."[6] Eleven days later he reported to Huxley that he had read lectures four and five. "They are simply perfect," he told Huxley. As he read the fifth lecture, containing the discussion of natural selection, Darwin claimed to have thought, " 'what is the good of my writing a thundering big book, when everything is in this green little book so despicable for its size?' " Jokingly, he declared, "in the name of all that is good and bad I may as well shut shop altogether."[7] Huxley took all of this as encouragement from Darwin to write a more polished book for the public that dealt with evolutionary themes. He pleaded that he did not have the time. Darwin responded on December 28, 1862, that Huxley's little book was "in every way excellent and cannot fail to do good the wider it is circulated. Whether it is worth your while to give up time to it, is another question for you alone to decide; that it will do good for the subject is beyond all question."[8]

Over the next few years Darwin continued to push Huxley to write science books for a popular audience. Huxley was still reluctant to give up any of his precious time. On November 5, 1864, Darwin wrote to Huxley, saying "I want to make a suggestion to you," which came out of a discussion with his wife Emma. She had been reading Huxley's lectures to her son Horace, and remarked to Darwin, "I wish he would write a book." Darwin answered that Huxley had just written an important book on the skull, a reference to *Lectures on the Elements of Comparative Anatomy* (1864). She replied, "I don't call that a Book," and added, "I want something that people can read; he does write so well."[9] Darwin then made his suggestion to Huxley. "Now,"

---

4. L. Huxley 1902, 1: 223.

5. Ibid., 224.

6. Burkhardt et al. 1997, 589.

7. Ibid., 611.

8. Ibid., 633.

9. This remark could easily have applied to Huxley's *Man's Place in Nature* (1863), a compilation of three essays that even Huxley worried would be too technical for his audiences. In a note to the reader he stated that "the readiness with which my audience followed my arguments, on these occasions, encourages me to hope that I have not committed the error, into which working men of

he wrote, "with your ease in writing, and with knowledge at your fingers' ends, do you not think you could write a 'Popular Treatise on Zoology.'" Darwin acknowledged that it would be some "waste of time," but he argued that "a striking Treatise would do real service to Science by educating naturalists." When asked to recommend something for a beginner, Darwin could only think of Carpenter's textbook on zoology.[10] Darwin did not think to recommend any of the works of John George Wood, or any of the other popularizers of science who worked in the area of natural history. He did not consider these works to be adequate introductory texts. On January 1, 1865, Huxley wrote that wished he could act on Darwin's suggestion, but he found writing to be boring and "a very slow process" unless he was interested. Moreover, he was too busy to take on such a project.[11] But Darwin would not give up. He declared on January 4 that Huxley was the "one man" who could write a popular treatise on zoology. Although it would be "almost a sin for you to do it, as it would of course destroy some original work," Darwin insisted on the importance of writing for the public. "I sometimes think," he affirmed, "that general and popular Treatises are almost as important for the progress of science as original work."[12]

Huxley's ambivalence in the 1860s toward writing books for a general audience stands in stark contrast to the conventional picture of him as the most important popularizer of science during the Victorian period. John Carey referred to him as "the greatest Victorian scientific popularizer," while Charles Blinderman described him as "the foremost popularizer of science in the nineteenth century."[13] Subbiah Mahalingam presented Huxley as "the major scientific popularizer of the nineteenth century."[14] J. Vernon Jensen asserted that Huxley "became perhaps the leading expositor and advocate of science, and of Darwinism in particular."[15] Huxley did become more involved in popularizing activities later in his career, but, as he admitted to

---

science so readily fall, of obscuring my meaning by unnecessary technicalities" (T. Huxley 1895b, [xv]).

10. Burkhardt et al. 2001, 399.

11. Burkhardt et al. 2002, 7.

12. Ibid., 13.

13. Carey 1995, 139; Blinderman 1962, 171.

14. Mahalingam 1987, iii.

15. Jensen 1991, 15. Many of the past studies of Huxley's literary and lecturing activities have emphasized his great skills as public communicator (Knight 1996, 129; Blinderman 1962, 171–74; Block 1986, 386).

Allen in 1882, he still found that conjoining "precision with popularity" to be "a very difficult art."[16] Later, in 1894, near the end of his life, when he had the reputation of being a formidable speaker, Huxley confessed that it was difficult for him to write popular lectures. "I found that the task of putting the truths learned in the field, the laboratory and the museum," he wrote in the preface to *Discourses Biological and Geological,* "into language which, without bating a jot of scientific accuracy shall be generally intelligible, taxed such scientific and literary faculty as I possessed to the uttermost."[17] Huxley's attitudes toward popularizing science are too complicated to be captured in a narrative that stresses his prowess as a great speaker and writer or in one that emphasizes his heroic efforts to educate the British public.

This is a chapter about scientific practitioners who tried to adopt the role of popularizer. I will deal with two examples, Huxley, and Robert Ball, who became the foremost popularizer of astronomy after Proctor's death. Unlike Proctor, who virtually gave up his astronomical researches in order to lecture, write, and edit for popular audiences, Huxley and Ball cultivated an identity for themselves as professional scientists, and they maintained their place among the practitioners throughout their entire careers. They also shared a commitment to evolutionary naturalism. An examination of their lectures and written work for popular audiences reveals that they borrowed narrative formats previously established by popularizers who were not practitioners. Huxley frequently made a common object or animal the focus of his work, while Ball was fond of the evolutionary epic. Despite Huxley's reputation as the age's foremost popularizer, it could be argued that Ball is a better example of a successful popularizer of science. Ball was by far the more prolific and had just as much success selling his books to the Victorian reading public. Although Huxley's writing style was widely appreciated, his inability to meet deadlines rendered him ill suited for the world of professional writing that Wood, Proctor, and Allen inhabited.

HUXLEY AS THE ARBITER OF "POPULAR SCIENCE"

If Huxley's writing and lecturing activities are placed in the context of the work of popularizers of science in the second half of the nineteenth century—the Anglican clergymen, women, showmen, and exponents of the evolutionary epic—then some interesting questions arise about why and

16. Clodd 1900, 112.   17. T. Huxley 1897a, v.

when he became involved in work that took him away from his own scientific research. Examining Huxley from this perspective also undermines the claim that Victorian audiences perceived him and his fellow practitioners as the most important spokesmen for science. The success of popularizers such as Wood, Brewer, and others, in selling their books, and their religious vision of nature, to Victorian readers, is a reminder that the public did not necessarily go to the practitioner to learn about the ultimate meaning of modern science. Huxley wrote and lectured about science knowing all too well that the explosion of science writing by popularizers who were not practitioners presented some serious problems for realizing the agenda of scientific naturalism. If the writers and lecturers like Wood had the ear of the Victorian public, then Huxley's goal of eliminating theologies of nature in science while reinforcing the importance of expertise was in danger of being frustrated. Huxley's efforts to communicate with the public are illuminating not because he may possibly have been the foremost popularizer of his age. They are revealing because he was so active in trying to define and control the meaning of science in different settings for different audiences.

However, the popularization of science was not a priority for Huxley in the 1850s and 60s. During this period, Huxley was mainly concerned with establishing a career for himself, a difficult task given the state of British science and his background. Born in 1825 in Ealing, at that time a small village west of London, Huxley was the youngest of six children in a lower-middle-class family. He was apprenticed to general medical practitioners in Coventry and London's East End, took courses at Charing Cross Hospital, and at the age of twenty finished his first examination for a medical degree at London University. Too poor to continue his education, he entered the navy and applied for a position on the survey ship HMS *Rattlesnake*. From 1846 to 1851 he served as assistant surgeon while the ship charted the waters off the eastern coast of Australia. Hoping to make a name for himself in the study of the anatomy of marine invertebrates, he conducted dissections while on board and sent the results back to England. The naturalist Edward Forbes arranged for them to be published in the *Philosophical Transactions of the Royal Society*.

Upon Huxley's return from the *Rattlesnake* voyage in November 1850, he was a man with few prospects. He desperately wanted a scientific post that would allow him to continue his research on marine invertebrates. He also had an added incentive. During the voyage he had met his future wife, Henrietta Heathorn, in Australia. Finding suitable scientific employment was the key to bringing her to England and allowing them to marry. But given the small number of paid scientific positions in England there were few opportunities. His application to obtain funding from the Admiralty to write up the discoveries he had made while on the *Rattlesnake* was turned down. Worse still, Huxley was £100 in debt.[18] For three years he struggled financially on the meager pay he received from the navy while he searched for a paying and permanent scientific job. Since he came from outside the Oxbridge system, Huxley already had a nasty grudge against the Anglican establishment. His search for suitable employment intensified his hostility as he was passed over for positions in favor of less qualified applicants with establishment connections. Although he was elected a Fellow of the Royal Society in 1851 and awarded its Royal Medal the following year, his future looked bleak. Adrian Desmond has appropriately referred to this period as his "season of despair."[19]

An indication of how Huxley survived is contained in a letter to his sister, written in November 1854, shortly after he was offered a position as lecturer at the School of Mines and as paleontologist to the Geological Survey. Here he tells her "how many irons I have in the fire at this present moment." The list is a daunting one: "(1) a manual of Comparative Anatomy for Churchill; (2) my "Grant" book; (3) a book for the British Museum people (half done); (4) an article for Todd's *Cyclopaedia* (half done); (5) sundry memoirs on Science; (6) a regular Quarterly article in the *Westminster;* (7) lectures at Jermyn Street in the School of Mines; (8) lectures at the School of Art, Marlborough House; (9) lectures at the London Institution, and odds and ends."[20] Huxley managed to survive by working on a textbook, a specialist monograph, a museum catalogue, high-profile public lectures and rather lowly technical lectures, and highbrow reviewing. This kind of work spanned popular and specialist domains, but much of it carried the stigma of lowly scientific piecework. During this desperate period he formed a very low opinion of many of these activities, as he felt as though he was forced to undertake them partly under financial duress and partly as stepping-stones to a future career as a practitioner of science. But he could justify undertaking these activities as contributing to the scientific education of the British public.

Huxley's ventures into journalism during this period reveal his ambivalence toward popularizations of science. Late in 1853 the publisher John Chapman asked Huxley to write a regular column on science for the *Westminster Review.* The terms offered, twelve guineas per sheet (for every sixteen

18. Desmond 1997, 152.   20. L. Huxley 1902, 1: 129.
19. Ibid., 172.

pages), were too good to be turned down given Huxley's precarious financial situation.[21] Huxley may also have been interested in communicating his views on science to the new middle-class audience that the *Westminster Review* sought to reach. Although Huxley became friendly with the writers who were part of Chapman's circle, such as Marian Evans, George Lewes, and Herbert Spencer, and though he shared their program for reform, he did not want to become one of them. Huxley saw himself first and foremost as a man of science, and, as James Secord has put it, he "feared exile on Grub Street."[22] Huxley wanted to avoid being stigmatized as a crass journalist, as it could have damaged his chances of establishing a career in science. In his reviews Huxley made sure that the *Westminster Review* readers for the journal would distinguish him from popularizers who merely disseminated knowledge discovered by practitioners. The goal of these authors was amusement and profit, and their works were derivative and superficial, while the true practitioner was motivated by genius and the pursuit of truth. Huxley was favorable toward those popularizers who recognized this distinction and who presented their work as introductions, rather than competitors, to the works of genuine men of science. Huxley assumed the role of critic of the scientific author, as a judge who stood above the economically driven world of Grub Street, rather than presenting himself as a popularizer. His authority for doing so was his status as a practitioner. As the arbiter of scientific works for the public he could protect uncritical readers from accepting unscientific ideas and he could avoid being seen as a literary hack.[23]

Huxley's strategy for preserving his authority as, he hoped, a temporary resident of Grub Street brought him into conflict with the other members of the *Westminster* circle. Lewes and his colleagues conceived of their literary activities as incorporating scientific knowledge and practice. They believed that their criteria for criticism and original fiction were informed by valid scientific principles. But they were not practitioners of science, and, according to Huxley's strategy, they too had to submit their work to the judgment of elite scientists, especially when they produced scientific works.[24] From Huxley's point of view, his critique in the pages of the *Westminster Review* for January 1854 of Lewes's *Comte's Philosophy of the Sciences* was entirely consistent with his strategy. Here Huxley asserted that if Lewes's book was merely a clear and lively exposition of Comte's philosophy, he could review

21. Desmond 1997, 185.
22. J. Secord 2000, 499.

23. White 2003, 71–72.
24. Ibid., 70.

it positively. But Lewes had aimed to bring Comte's principles to bear on the present state of science. "We regret to be obliged to say," Huxley declared, "that neither as regards chemistry nor physiology are these pledges redeemed." Huxley pointed to errors in the book that revealed Lewes's ignorance of recent science. They were "no accidental mistakes," and they demonstrated "how impossible it is for even so acute a thinker as Mr. Lewes to succeed in scientific speculations, without the discipline and knowledge which result from being a worker also."[25] Lewes's literary life was no replacement for the specialist's training and experience needed to deal adequately with scientific issues.[26]

Huxley's attempts during this period to distance himself from the sphere of "popular science" can also be seen in his anonymous attack on the tenth edition of the *Vestiges of the Natural History of Creation* in the *British and Foreign Medico-Chirurgical Review*, published three months after his caustic review of Lewes's book on Comte. Although the development hypothesis, with its emphasis on universal progress, was central to all reform-minded intellectuals, Huxley was unable to condone its unscientific formulation in the *Vestiges*. In his role as judge of hack science, Huxley reluctantly took on the "unpleasant duty" of evaluating this work of "fiction." In part, it was unpleasant because he would have to undertake the "wearisome" task of examining "all the blunders and mis-statements of the 'Vestiges'" that, according to Huxley, "abound in almost every page." Only through such an examination, though, could it be shown that the author of the book received his science second-hand. "We look for evidence of knowledge," Huxley insisted, "and we find—what might be picked up by reading 'Chambers's Journal' or the 'Penny Magazine.' We look for original research, and we find reason to doubt if the author ever performed an experiment or made an observation in any one branch of science." Huxley predicted that if the author of *Vestiges* ever revealed his identity, it would be found that he was "prominent neither in the mechanical nor any other department of even *one* science."[27] Popular works such as the *Vestiges* were not written to further the cause of science; rather, they were written for profit by unqualified hacks. These writers were in danger of turning knowledge into a commodity governed by the questionable ethics of commercial bookmaking.[28]

In his reviews, Huxley could shape the public understanding of the meaning of "popular science," emerging at this time as a new category of

25. [T. Huxley] 1854, 255.   27. T. Huxley 1903, 2, 17–19.
26. White 2003, 74–75; J. Secord 2000, 500.   28. J. Secord 2000, 504.

writing. He constructed criteria for distinguishing between the elite science of the practitioners and the works of the popularizer, and then posed as judge of the value of the latter. He viewed Charles Kingsley's *Glaucus* as the ideal work for the general reader. It contained a true appreciation of "the value of Natural History Science, as a means of mental education." But Kingsley also knew his place. His book was "without the least pretension to scientific lore."[29] He did not present himself as an original investigator. The *Vestiges* and Lewes's book on Comte provided Huxley with models of scientific works that were unfit as vehicles for educating a popular audience. All of his criticisms of them applied to the work of popularizers such as Wood, whose contributions to Routledge's Common Objects series were to appear later in the decade. But Huxley toiled as a review writer for only a short time. In 1855 he was appointed Fullerian Professor of the Royal Institution and lecturer at St. Thomas's Hospital, London. As he was offered more institutional posts, he became busier and more secure financially. In 1858 he wrote to his sister Lizzie about his article on glaciers in the *Westminster Review* for 1857. "I used at one time to write a good deal for that Review," Huxley told her, "principally the Quarterly notice of scientific books. But I never write for the Reviews now, as original work is much more to my taste."[30]

Although undertaking original research, rather than being a critic, was clearly more important to Huxley in 1857, he recognized that there was some value to policing the work of popularizers. Huxley sought to advance the authority of the laboratory-based practitioner and to educate the reading public about science. These aims could be realized through the periodical press. Friends such as Hooker and Tyndall agreed. Early in 1858 they discussed the possibility of starting a scientific review, which would play the same role for science as the *Quarterly Review* or the *Westminster Review* did for literature. It was decided that the plan was not feasible. This was the first of many unsuccessful schemes to establish a periodical under their control.[31] Huxley, though, with the help of Mervyn Herbert Nevil Story-Maskelyne, then Keeper of the Minerals at the British Museum, arranged for the *Saturday Review* to devote a column to science every fortnight and to include reviews of significant books on science. Huxley wrote to Hooker on April 20, 1858, explaining that it would be best if seven or eight practitioners,

29. [T. Huxley] 1855, 246, 248.

30. L. Huxley 1902, 1; 170.

31. *Natural History Review* and the *Reader* both failed in the 1860s.

each representing a different scientific discipline, joined together to supply at least one article every three months. He mentioned James Sylvester for mathematics, Tyndall for physics, Story-Maskelyne and Edward Frankland for chemistry and mineralogy, A. C. Ramsay for geology, Warington W. Smyth for technology, and Hooker and himself for biology.[32]

Huxley's contributions to the *Saturday Review* cannot be determined with certainty since the articles and reviews in this periodical were unsigned. Yet he shared a common intellectual perspective with the group of practitioners that he selected to represent the different sciences. Since Huxley divided up the biology pieces with Hooker, it is very possible that he wrote some of them. John George Wood's *Common Objects of the Country* was discussed in an article on "Science in the Country." Wood was presented as a writer who explored the "practical and workday side of semi scientific studies" rather than its poetical dimension. The anonymous author was critical of the reflections that Wood offered on design. Wood's digressions "either take the shape of small jokes, or of exercises of that perverted ingenuity which loves to hit on unexpected ways of making out creation to be rather better and more wisely designed than at first sight it would seem to be." The reviewer then broadened his criticism to include Wood's peers. "These capricious extensions of the argument from design," he wrote, "are a very favourite pastime with the smaller kind of scientific writers." Wood's emphasis on the practical and his tendency to habitually explore "the marvels around us" resulted in "a *dilettante* trifling, unless it is subordinated to the constant apprehension of poetical truth."[33]

Huxley, or one of his fellow contributors to the *Saturday Review*, wrote about other works for popular audiences. The reviewer in the *Saturday Review* panned David Page's *Handbook of Geological Terms and Geology*, and his explanations were referred to as "often incomplete, unsatisfactory, and even wrong."[34] Gosse was reviewed more favorably since he offered precise descriptions of the objects under study. But the reviewer used the occasion to criticize writers of popular books who did not devote their attention to the clearness of their descriptions, and he recommended that they leave "any vivacities of philosophy they may desire to express to come in as quite secondary." The reviewer was also hostile to those who put too much emphasis

---

32. L. Huxley, 1: 151.

33. "Science in the Country" 1858, 393–94.

34. "Mr. Page's Handbook of Geological Terms" 1859, 713.

on entertaining their audiences. "It is one of the absurdities current among the writers of popular books," he declared, "to pretend that whatever is 'dry' can be rendered 'amusing.' "[35] In general, Huxley and his colleagues were unhappy with the works of Wood, Page, and other popularizers who were writing for the public.

Although Huxley's involvement with reviewing tailed off after he obtained secure employment, his activity as a popular lecturer increased. Huxley lectured at the Royal Institution to fashionable middle-class audiences, including his lectures as Fullerian Professor.[36] While his lecturing at the Royal Institution, as well as his writing for the *Westminster Review* and the *Saturday Review*, allowed him to reach middle-class audiences, he was also interested in communicating his scientific ideas to new working-class constituencies. He had begun to lecture to workingmen as part of his teaching duties in 1855. On February 27, 1855, he told his friend Frederick Dyster, "I want the working classes to understand that Science and her ways are great facts for them—that physical virtue is the base of all other, and that they are to be clean and temperate and all the rest—not because fellows in black and white ties tell them so, but because these are plain and patent laws of nature which they must obey 'under penalties.' "[37] Huxley enjoyed speaking to workers, and in 1857 he gave his first address at the Working Men's College to an audience of about fifty, including the liberal Anglican F. D. Maurice.

During the late 1850s Huxley continued to establish himself as a research scientist. He shifted his research interests from invertebrate morphology to vertebrate paleontology.[38] He gained notoriety in 1859, and thereafter, when he defended Darwin, and became part of a cadre of practitioners who worked to reform metropolitan science. In the 1860s Huxley and his allies began to infiltrate and dominate the key scientific institutions and societies and argued for the need to establish a scientific elite that could function as true professionals. Huxley and his friends perceived that to reform science they would have to become an integral part of the movement that sought to reform British society as a whole. Reform could only occur if the Anglican clergy was dislodged from its powerful position as the ruling intellectual elite. As a result, Huxley became involved in countless controversies with

35. "Evenings at the Microscope" 1859, 570–71.

36. Huxley was a popular lecturer at the Royal Institution, setting new records for attendance in 1876 (1,068) and 1877 (1,104) (Jensen 1991, 60).

37. L. Huxley 1902, 1: 149.

38. Di Gregorio 1984, xviii.

conservative Anglican representatives. Lecturing to popular audiences in the 1860s was part of the effort to educate British society on the importance of science, a key strategy for achieving reform. The lectures to workingmen of 1862, which had so delighted Darwin, reflected Huxley's interest in popular education. Popular writing, however, did not become an integral component of his commitment to popular education until the 1870s.

THE EARLY 1870S AND THE CALL TO PENS

In 1870 Huxley was appointed as president of the British Association. This was not only the crowning triumph of his early career, it also indicated how influential scientific naturalism had become within British science. All of the strategies that Huxley and his allies had adopted since the 1850s to reform scientific institutions had placed the young guard into increasing positions of power. The scientific community had become far more independent and professionally self-defined. But events in the late 1860s and early 1870s led to a new environment that called for a change in strategy. First, the poor showing of British manufacturers at the Paris International Exhibition in 1867 had touched off a fierce debate on the quality of technical instruction in Britain. Scientific lobbyists such as Lyon Playfair argued that Britain needed to keep pace with the rest of Europe by training the manufacturing population in experimental science in order to banish widely used, but hopelessly outmoded, rule-of-thumb practices. This led to a demand for teachers trained in natural science in schools run by the Department of Science and Art to serve the industrial population, which in turn led to a demand for more and better science textbooks. Huxley saw the need for science textbooks firsthand as one of the examiners for the Department of Science and Art.[39] Second, the Reform Act of 1867 had given the working class the vote, and it was clear to many that they had to be educated in order to participate in an informed manner in future elections. From Huxley's point of view it was crucial that they be educated about the importance of science and molded into supporters of the scientific enterprise. Third, in 1870, the government passed the Education Act, a response to the perceived need to educate the workers. In those areas of the country where the churches and other voluntary organizations had not provided enough schools, the Act decreed that locally elected school boards were given the power to establish schools and to levy a rate to pay for them.

39. Gooday 1990, 44–45, 49.

During the early 1870s Huxley and his friends began to realize that their plans for reforming science could only succeed if they put more of their energy into improving the scientific education of the general populace and of the teachers who were going to instruct them about science.[40] It was not enough to exercise control over the practitioners of science. Converting the public into enthusiastic supporters for their vision for science was also imperative, for they voted members of Parliament into power, and Huxley wanted the state to get into the business of funding science. With the extension of the franchise to the urban working classes in the Reform Act of 1867, many of those who were reading the works of popularizers such as Wood were part of the electorate. Moreover, the Education Act of 1870 had the potential to increase the literacy rate and create an even larger reading public. Huxley began to reevaluate his priorities. Original research came to be seen as less important. Gaining control over the organs of education, both formal and informal, moved closer to the top of the list of priorities. Darwin's earlier arguments about the importance of writing scientific books for a popular audience seemed more compelling in light of the changed circumstances. Huxley was now at the peak of his career, and it was no longer necessary for him to devote himself to research in order to establish his reputation.[41] He was in a position to give more time to the pressing issue of scientific education.

Huxley was not unusual in recognizing that education had become a significant matter. Nor was he unique in linking this issue to the work of popularizers of science. Scientific journals sounded the call to arms throughout the early 1870s. In 1870 the author of an article on "Science and the Working Classes" in *Nature* charged that science lecturers had failed to teach

40. Frank Turner has pointed to the 1870s as a time when scientists shifted tactics in their presentation of "public science," by becoming more civic-minded and state-oriented, though he places his emphasis on the mid-decade as the significant turning point due to two events. First, the passing of the Cruelty to Animals Act in 1876, instigated by the antivivisectionists as a means to limit the scope of physiological experimentation; and second, the refusal of the Conservative and Liberal governments to carry out the recommendation of the Devonshire Commission for a ministry of science and a science advisory council (Turner 1993, 205, 208). But Huxley and his allies were already in the early 1870s beginning to see that cultivating the scientific habit of mind was essential to the future prosperity of scientific naturalism. Developments later in the decade would only have reinforced their resolve to push for a reform of educational institutions and to work toward curbing the influence of popularizers of science who possessed neither the expertise nor the allegiance to scientific naturalism.

41. A cursory examination of Huxley's scientific memoirs illustrates how his research output dropped off after 1870. The five volumes of his scientific memoirs include 168 essays in total. Of these, fifty were published after 1870. More than two-thirds of Huxley's output was done before 1870. The bulk of this work was completed during the 1850s and 60s. See T. Huxley 1898–1903.

the workingman how to "make use of his eyes," since they pitched their talks at too high a level. "The error of 'popular' scientific lectures," the critic declared, "of evenings with working men at mechanics' institutes, is that which is so commonly attributed to clergymen, that of speaking over the heads of their audience." Scientific education, *Nature* proclaimed, had to become a national priority, or England would fall behind in the international struggle for existence. If the working classes were taught how to exercise reasoning powers and "thus trained to form the strength of the nation in all fresh advancements in Science and the Arts, England would quickly outdistance all competitors, and assume that position which it would now seem younger rivals are likely to snatch from her grasp."[42]

About half a year later, *Nature* returned to the theme of education and the popularization of science, observing that practitioners had finally begun to take the lead in disseminating knowledge. *Nature* called on the men of science to teach "the whole mass of the community" to become passionate about acquiring knowledge. "Many of its chiefs have now begun to perceive this," the journalist wrote, "and we are glad to record the successes of one of the best organized attempts that have hitherto been made to extend the knowledge and love of science among the working classes." *Nature* praised the delivery of a series of lectures in Manchester by such eminent scientists as the astronomers William Huggins and Norman Lockyer, the chemist Henry Roscoe, and Huxley himself. *Nature* believed that this star-studded lineup indicated a change of attitude among the leading practitioners of science. After this set of lectures, "let it be said no more that the chiefs of science are either unable or unwilling to explain to others the discoveries which they themselves have made. They are at last emerging from their seclusion, and recognize their functions as teachers of truth."[43] *Nature*, that organ of the scientific elite, presented the popularization of science as an integral dimension of scientific education and proposed that a systematic program be established in order to maintain the dominance of England over other nations. Popularizing science could be seen as a national service performed by practitioners rather than a private pursuit for commercial gain.

Huxley made time to serve on two high-profile bodies dealing with the state of British education. He was elected to the London School Board in 1871. Even more important was his role on the Royal Commission on

---

42. "Science and the Working Classes" 1870, 21–22.

43. "Science Lectures for the People" 1871, 81; Riley 2003.

Scientific Instruction and the Advancement of Science, established by Prime Minister Gladstone in 1870 and headed up by the Duke of Devonshire. Not only did this focus national attention on the issue of scientific education, it also put Huxley, appointed as one of the commissioners, at the center of the debate. Huxley and the Duke were joined on the commission by the Treasurer of the Royal Society, W. A. Miller (who died shortly afterward), both of the Secretaries of the Society, the physicist G. G. Stokes, the physiologist William Sharpey, Lubbock, and Lockyer. The Commission sat for six years, interviewed over 150 witnesses, collected four huge volumes of evidence, and published eight reports.[44]

During the course of questioning the Rev. J. Fraser on May 12, 1871, Huxley voiced his concerns about the emphasis on common things as presented in the works of one important popularizer. Fraser recalled that fourteen or fifteen years ago "there was a great talk about knowledge in common things, and there was a book brought out at that time by Doctor Brewer, among others, which probably you may have heard of, I think it was called 'Knowledge of Common Things.'" Fraser affirmed that Brewer's book "made its way into a considerable number of schools," though not approved by the men of science. Huxley's evaluation of Brewer's book was extremely harsh. In Huxley's view, "the 'Knowledge of Common Things' had reference entirely to practical purposes; there was no means taken to lead the mind of a child to what may be called purely scientific considerations." Highly defective, the scheme of education based on Brewer's book provided "pure information" without showing the student how to use it. Huxley recommended teaching topics that could be understood and "so arranged that the child becomes acquainted with natural phenomena in a scientific way, and is led up to a higher step of information afterwards." Huxley was in favor of a much more systematic presentation of scientific facts rather than Brewer's supposed hodge-podge of information.[45]

The debates about the role of science in education in the early 1870s, and the failure of popularizers like Brewer to provide the public with suitable reading material, led Huxley to overcome the distaste he had acquired, while reviewing for the *Westminster Review* and the *Saturday Review* in the 1850s, of writing science books for a broad audience. Instead of seeing this activity as suitable only for literary hacks, or an unacceptable interruption

44. MacLeod 1996, 205.

45. *Royal Commission on Scientific Instruction* 1872, 576.

of his original research, he began to view it as worthy of his time. In his "Autobiography," written near the end of his career, Huxley asserted that he had subordinated his ambition for scientific fame to higher priorities, such as "the popularization of science; to the development of organization of scientific education; to the endless series of battles and skirmishes over evolution; and to untiring opposition to that ecclesiastical spirit."[46] Huxley continued to lecture to popular audiences during the 1870s, and thereafter. In 1876, for example, he undertook an extensive and successful lecture tour of the United States.[47]

During the 1870s he also began to devote considerable energy to the printed word as a means of conveying scientific knowledge. Here he ran up against writers like Wood, Brewer, and Gatty, who had already invested a considerable amount of energy into popularizing science. Huxley's writing and lecturing must therefore be seen against the backdrop of the activities of popularizers who had been at work in the 1850s and 60s. In his article on biology and natural history in late Victorian Yorkshire, Samuel Alberti has given us a wonderfully complex picture of how amateurs and professionals redefined their identities in response to one another.[48] A similar account can be offered of the concomitant refashioning of the identities of scientific practitioners and popularizers of science. Popularizers were constantly positioning themselves in relation to practitioners in the ways that they incorporated theologies of nature into their works, in the literary strategies that they used to write about the natural world, and in the methods they relied on to present themselves as authoritative guides to their audience. But practitioners like Huxley had to position themselves in relation to popularizers if they wanted to be successful in communicating with the public. Huxley had to think carefully about the style he would adopt in his lectures and writings. Like other popularizers, he needed to reflect on how to reach the new audiences for science that had developed in the midcentury period. He had to take into account the fact that these new audiences were reading Wood, Brewer, and others and that their conceptions of science were being shaped by those who did not share his objectives. He needed to formulate a strategy to gain control of public science so that his secularizing agenda would overpower attempts to perpetuate the use of discourses of design. Although the journal Nature had seemed, at first, to be a promising means by which to establish control of the popular sphere, by 1874 Huxley and

46. T. Huxley 1897b, 16.   48. Alberti 2001, 115–47.
47. Jensen 1991, 87–110.

his friends were disillusioned with the journal. Lockyer refused to rein-in critics of scientific naturalism, such as the North British physicist Peter Guthrie Tait.[49] Moreover, *Nature* was in the process of losing its popular audience and becoming a journal for practitioners, though this may not have been completely evident in the mid-1870s.[50] New projects had to be initiated to serve Huxley's interests.

Huxley was a key figure in two ambitious publishing projects that were started in the 1870s, the International Scientific Series and Macmillan's Science Primers. Later that decade he completed his work on the *Physiography* (1877), his first book written with a popular audience in mind. The long delayed introduction to the Macmillan series appeared in 1880, titled *Science Primers: Introductory*, and his contribution to the International Scientific Series, *The Crayfish* was published in the same year. Discussing Huxley's activities as a popularizer will therefore lead us to an analysis of works that are not always included in the Huxley canon.[51] Having spent so much time, up to late 1860s, attempting to professionalize science, Huxley now turned with enthusiasm to science in the public sphere. Establishing connections with publishers, writing, and setting up ambitious monograph series with the goal of controlling the market for science books aimed at the general audience, Huxley now made these projects in print culture a central part of his overall strategy to reform British science and society. He recognized that public support for science was essential to establish the cultural authority of the scientific elite and to free up funding for scientific institutions. Just as Huxley and his X Club colleagues tried to take over scientific societies, dominate major scientific institutions, and manage government commissions, as part of a concerted strategy to reform science, they also tried to gain control of the publications read by the Victorian public.

---

49. Barton 2004.

50. Roos 1981, 176.

51. Scholars have tended to focus on Huxley's periodical articles, written in the heat of controversy, as examples of his efforts to convey scientific knowledge to the Victorian public. But many of these articles are rhetorical exercises, designed primarily to persuade audiences of the truth of evolutionary theory or the falseness of Anglican theology. Moreover, they were published in periodicals with a fairly prescribed audience. Huxley's attempts to present a body of knowledge in a systematic form for public consumption are not to be found in these works, but rather in his *Physiography, Science Primers: Introductory,* and *The Crayfish*. Of course, in the end, there was a rhetorical dimension to these works as well. Understanding Huxley's strategies for reaching a popular audience will also require attention to his work behind the scenes in organizing the International Scientific Series and Macmillan's Science Primers.

THE *PHYSIOGRAPHY*

Published in November 1877, by Macmillan, Huxley's *Physiography* was really his first book-length study written with a popular audience in mind, in particular the young reader. Huxley's friends poured lavish praise upon it. John Morley wrote on December 14 that his stepson, who was not "too bookish," was finding the book so fascinating that he could not put it down. "Your Physiography is worth silver and gold," Morley asserted. "There is nothing at all like it in the way of making nature, as she comes before us everyday, interesting and intelligible to young people." Morley confessed that even a "Master of Arts and an able editor" like himself had learned a great deal from reading the book.[52] Alexander Henry Green, then professor of geology at Yorkshire College in Leeds, and later, in 1888, professor of geology at Oxford, wrote to Huxley on January 22, 1878, that he was reading the *Physiography* "with no less delight than perfect."[53] Morley had "predicted to Macmillan that it must have a great and perpetual sale," and he was right.[54] Although not in the same league with the successes of Wood's *Common Objects of the Country* or Gatty's *Parables of Nature,* the book sold well. Priced at seven shillings and sixpence, 3,386 copies had been purchased within the first six weeks.[55] Huxley received £50 for selling the copyright to Macmillan, and a royalty of one-fifth of the selling price per copy on all books sold after the first 2,500.[56] During the first three years 13,000 copies were sold. Appleton published it in the United States in 1882, and the book was translated and published by Brockhaus in Germany in 1884 and by G. Baillière et cie in France in 1882. Macmillan put the *Physiography* out in a revised edition (by R. A. Gregory) in 1904, over twenty-five years after its first publication and years after Huxley's death in 1895. The final Macmillan edition appeared in 1924.[57]

52. Imperial College, Huxley Collection, 23.36.

53. Ibid., 17.115.

54. Ibid., 23.36.

55. L. Huxley 1902, 1: 510.

56. British Library, Macmillan Archive, Additional MSS 55210, f 62, November 27, 1877. By 1880, Macmillan planned to produce a cheaper edition. On May 12, 1880, they proposed to reduce the price to six shillings, and to pay Huxley a royalty of one shilling per copy sold. Since Macmillan believed that it had great potential as a textbook, they wanted to go to press at once with this new and cheaper edition so that they would be ready with a good supply for the opening of the schools in July and August (Imperial College, Huxley Collection, 22.151).

57. Stoddart 1975, 19.

Although Huxley had finally managed to produce a book for popular audiences, it had undergone a long gestation process before seeing the light of day. The basic idea for the *Physiography* seems to have originated in 1868, when Huxley gave a public lecture on "A Liberal Education; and Where to Find It" at the South London Working Men's College. Here Huxley affirmed that children must be properly prepared to receive Nature's education in order to escape the great evils of disobedience to natural laws. The mind must be stored with "knowledge of the great and fundamental truths of Nature and of the laws of her operations." The teaching of physical geography was one important aspect of a liberal education. Huxley presented it as "a description of the earth, of its place and relation to other bodies; of its general structure, and of its great features—winds, tides, mountains, plains: of the chief forms of the vegetable and animal worlds, of the varieties of man." Physical geography, according to Huxley, was "the peg upon which the greatest quantity of useful and entertaining scientific information can be suspended."[58]

Huxley explored physical geography in a set of lectures, titled "Elementary Physical Geography," delivered a year later at the London Institution. He repeated the lectures, now titled "Elements of Physical Science," to women at the South Kensington Museum in 1870. Huxley intended to publish the course of lectures from the start, as he arranged for verbatim reports to be taken at the London Institution for his own use. He did not want to make the same mistake as he had made in 1862 with his working-class lectures. However, Huxley's good intentions were frustrated. "I am sorry to say that, in this, as in other cases," Huxley admitted in the preface to the *Physiography*, "I have found a great gulf fixed between intention to publish and its realization." Huxley found the process of seeing a book through the press to be "a laborious and time-wasting affair," especially in the case of the *Physiography*, which had maps and figures to attend to. Since he "never could muster up the courage, or find the time, to undertake the business," the manuscript remained untouched for seven years. Huxley was almost perpetually overextended. But early in 1872 he was so overworked that he suffered a serious illness and was forced to go on holiday for several months in order to recover.[59] His health remained poor up until 1873, so he was unable to make much progress on the *Physiography* in the early 1870s. Later, with the help of an assistant, Huxley then rewrote parts of the work, added some new sections, and carefully revised the proofs of every chapter.[60] By

58. T. Huxley 1897c, 86, 109.   60. T. Huxley 1878, viii–ix.
59. Desmond 1997, 410.

waiting to work up the lectures for publication, however, Huxley could take advantage of his experiences on the London School Board. He revised the manuscript keeping in mind the elementary course in physical science proposed by the Board in the new curriculum.[61]

In the *Physiography*, Huxley took a page from the works of such popularizers as Wood by including a large number of illustrations. While his earlier "Six Lectures to Working Men" contained five fairly simple figures, the *Physiology* offered 122 illustrations and five colored plates, including diagrams, maps, scientific instruments, and even some depictions of natural phenomena such as volcanoes, geysers, and gorges. In the *Physiography* Huxley adopted the tone and stance of the natural historian, and he mimicked the strategy adopted by popularizers for reaching a lay audience. Like Wood and Brewer, Huxley focused his *Physiography* on aspects of nature that were familiar. Despite his criticism of Brewer's emphasis on common objects of nature during the Devonshire Commission hearings, Huxley found that he could not dispense with this approach to communicating with a popular audience. His use of common objects was a silent acknowledgment of the success of popularizers like Brewer and Wood. It was a strategy that Huxley had used previously, and that he would use again, when he put such objects as the lobster, the horse, chalk, coal, and yeast at the center of an address.[62] Faraday, who used a common-objects approach in his lecturing, though applied to natural philosophy rather than natural history, may also have influenced Huxley. In Christmas of 1860 Faraday gave his last course of Juvenile Lectures on the chemical history of a candle at the Royal Institution. This was not the first time that he used a candle as the focus of his Christmas Lectures. Faraday, who gave his first course of Juvenile Lectures in 1827, was renowned as a superb public speaker and widely regarded as the finest lecturer in science in London.[63] In 1894 Huxley referred to Faraday as a

61. White 2003, 129.

62. These common objects were the focus in "A Lobster; or, The Study of Zoology" (1861); "On Our Knowledge of the Causes of the Phenomena of Organic Nature" (1862), which used the horse as his common object of study; "On a Piece of Chalk" (1868); "On the Formation of Coal" (1870); and "Yeast" (1871).

63. B. Jones 1870, 2: 429; S. Thompson 1901, 227. One of his biographers described his lecturing strategy as "introducing his subject on its most familiar side," so that he put himself in a close rapport with his audience at the outset, "and then leading on to that which was less familiar. Before the audience became aware of any transition, they were already assimilating new facts which were thus brought within their range" (S. Thompson 1901, 232). Since Faraday's lectures depended heavily on experiments, Huxley could only have derived the basic strategy of focusing on a common object from him. The wider public would have identified the use of the common

"prince of lecturers," whose main guideline for conceptualizing "lectures of a popular character" was to assume that the audience knew nothing about the topic.[64]

Although Huxley recognized that a focus on common objects was an effective way to reach a popular audience, it was also a dangerous strategy to adopt. It was associated in the minds of his audience with Wood, Brewer, and other popularizers who were pushing a theology of nature. It also placed his readers in the world of natural history, and Huxley was arguing in his essay "On the Study of Biology" (1876) that the term "Natural History" was confusing and outdated, and that it should be replaced by the more modern term "Biology." The new term accurately reflected the progress of science since the beginning of the century, and it denoted the emphasis on laboratory experiment that Huxley insisted was an essential component of the discipline.[65] Huxley aimed to sever the connection between common objects, natural history, and a theology of nature by using lobsters, horses, chalk, and coal to teach his audience how to think systematically about nature as secular. Although he began in the world of natural history in his *Physiography,* he ended up in the world of the scientific naturalist by following causal connections to larger and larger visions of the world as a product of natural laws rather than emphasizing divine harmony.[66] In effect, he aimed to subvert the goal of the common objects approach as used by popularizers such as Wood and Brewer.

For the *Physiography* Huxley selected the Thames and its basin as a physical focus for his text, and he placed the reader at the very start of the book on the London Bridge. "No spot in the world is better known than London," he explained, "and no spot in London better known than London Bridge." The reader was invited to imagine that they were standing on this bridge, looking down on the river below. Using the Thames as an example of a typical river basin, Huxley then discussed its flood tide and ebb tide, its source, and its tributaries (Fig. 7.1). He moved on to a discussion of the principles behind cartography, returning to the Thames and how it would be mapped, followed by outlines of such topics as determining direction, compasses, map scales,

---

object in a book or lecture about natural history with J. G. Wood, partly because admission to the Royal Institution was limited to the well to do.

64. T. Huxley 1911, vii.

65. T. Huxley 1897c, 268, 281; Desmond 2001, 27-40.

66. Desmond 1997, 485.

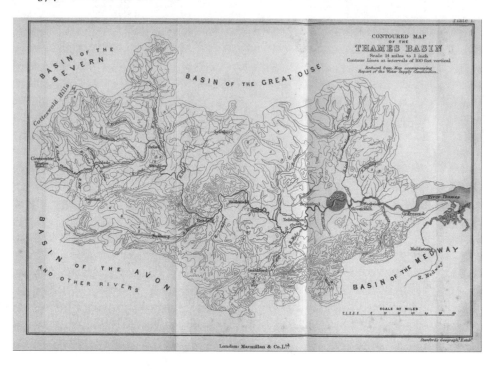

FIGURE 7.1 Huxley's contoured map of the Thames Basin, one of the five colored plates. Huxley brought the reader back to the Thames repeatedly throughout the book, as it provided familiar territory in which abstract scientific concepts could be explained and illustrated. (T. H. Huxley, *Physiography*, 2nd ed. [New York: D. Appleton, 1878], frontispiece.)

contour lines, and river systems.[67] The next set of chapters outlined how the Thames basin was fed with water. This allowed Huxley to introduce his readers to springs, rain and dew, the crystallization of water, evaporation, the atmosphere, and the chemical composition of pure and natural waters. Huxley attempted to make the causal connections between each chapter as clear as possible to his readers. At the conclusion of the chapter on springs, he asserted that these sources of water owed their origin, "directly or indirectly, to the rain which falls upon the collecting ground, and finds its way through the pores and cracks of the rocks beneath." The next chapter therefore focused on the formation of rain. This chapter concluded with the observation

67. T. Huxley 1878, 1, 5.

that atmospheric moisture was precipitated "not only as rain and dew . . . but also occasionally as snow and hoar-frost; the formation of these will form the subject of the next chapter." The following chapter, on evaporation, explained the water cycle and ended by bringing the reader back again to the observer overlooking the Thames, who was instructed by Huxley to remember that the fresh water hurrying toward the sea would not find its final resting place there as it was "distilled afresh" and returned to the earth in showers "which may enter into the stream of Thames again."[68]

Moving on to the atmosphere, Huxley took the opportunity to introduce such topics as the chemistry of gases, the use of scientific instruments and measuring devices like the barometer, and weather charts (Fig. 7.2). He again came back to the Thames basin to apply what had been learned to an understanding of how atmospheric pressure affected the flow of the river.[69] The next suite of chapters discussed the means by which it received its present shape. Here Huxley gave his audience a lesson in uniformitarian geology, emphasizing how the action of natural agents slowly changed the surface of the earth over eons of time. He dealt with the work of rain, rivers, ice (including glaciers), and the sea as natural agents of denudation and destruction. They worked in combination to lower the general level of the land to that of the sea. As a dramatic illustration of the power of water, in this case rivers, to have a significant impact over a long period of time, Huxley pointed to the vast chasms of the Colorado River. Other agents, such as earthquakes and volcanoes, elevated sections of the earth's surface, thereby counterbalancing the work of water and preventing the disappearance of all land upon the face of the earth.[70] The third group of chapters took up the issue of the Thames Basin's past history, as told by the analysis of geological sections and the fossil record. Huxley discussed how living matter, both plants and animals, had played a significant part in the formation of the rock masses that built up the earth's crust. Using a section exposed in 1851 during some improvements made to a street in London, he deduced the past geological ages of the Thames Valley and reflected on the forms of life that had previously inhabited the area (Fig. 7.3).

The final set of chapters took a more global and far-reaching view of nature, though Huxley did not abandon the Thames altogether. Huxley tackled such topics as the distribution of land and water throughout the Earth, and

68. Ibid., 38, 54, 74,
69. Ibid., 78, 91, 99.

70. Ibid., 185-86.

FIG. 22.—*Times* weather chart.

FIGURE 7.2 A reproduction of a weather chart from the *Times*, allowing Huxley to explain how to read such images. (T. H. Huxley, *Physiography*, 2nd ed. [New York: D. Appleton, 1878], 93.)

he provided his readers with an account of global mapping. The last two chapters focused on the movements of the earth, both its rotation and orbit and an account of the nature of the sun and its influence on the earth. He ended by returning to the opening sentence of the first chapter. "The movement of the water of the Thames at London Bridge, in fact, formed the starting-point of those studies," Huxley asserted, "which have gradually

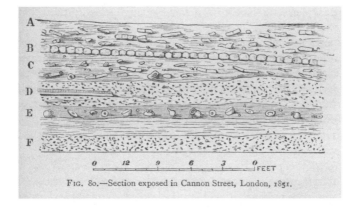

FIG. 80.—Section exposed in Cannon Street, London, 1851.

FIGURE 7.3 Huxley brings the reader back to London to examine a section exposed in Cannon Street in 1851. (T. H. Huxley, *Physiography*, 2nd ed. [New York: D. Appleton, 1878], 275.)

expanded into these one-and-twenty chapters." The simple question, what is the source of the Thames, had led to more questions, and only at the end of the book could Huxley answer the initial query. The Thames was fed, directly or indirectly, by rain, which was condensed from vapor, which "has been raised into the atmosphere by means of solar heat." Without the sun, there could not be rain or rivers, and "hence it is not too much to say that the origin of the Thames is ultimately to be traced to the sun." Through a systematic tracing of interconnected causes and effect, Huxley had brought the reader from the London Bridge to the sun, rather than to an appreciation of the wisdom, goodness, and power of God. He then concluded with a powerful image of the cosmos as subject to universal laws of nature similar to those on earth. "The spectacle of the ebb and flow of the tide, under London Bridge, from which we started," he wrote, "proves to be a symbol of the working of forces which extend from planet to planet, and from star to star, throughout the universe."[71] This was the closest that Huxley ever came to presenting an evolutionary epic. Since he was critical of the speculative dimension of Herbert Spencer's evolutionary thought, he was not an enthusiastic proponent of the notion of cosmic evolution.[72]

71. Ibid., 359, 375–77.

72. The important differences between Allen's and Huxley's use of evolution were obscured when an American publisher, Humboldt Publishing Company, produced a book titled *A Half-Century of Science*, that contained Huxley's "The Progress of Science 1837–1887" (1887), retitled as "The

Huxley's *Physiography* has been seen as presenting a model for post-Darwinian geography during a critical period in the development of this subject in Britain, when it was established in the older universities and incorporated into elementary education.[73] But the transformation of physical geography was not Huxley's primary interest; neither was the dissemination of evolutionism. Huxley managed to bring his readers from the London Bridge to a vision of the universe without relying directly on evolutionary theory. His main goal was to bring naturalism, rather than evolutionism, into the public eye. The *Physiography* was a potent symbol of Huxley's determination to readjust his personal priorities and invest significant time and energy on writing books for a popular audience. It signaled his recognition that education was too important to be left to popularizers without the proper expertise and his realization that the public would play an important role in determining the future of British science. Huxley's interest in these issues was also the motivation behind his involvement in an even larger project, the International Scientific Series.

TARGETING THE INTERNATIONAL MARKET

The International Scientific Series (ISS) was among the most ambitious publishing projects undertaken in the nineteenth century, and it was aimed at disseminating scientific knowledge systematically to a wide reading audience. Between 1871 and 1910 six separate publishers in the United States and in five European countries produced over 120 titles in four languages.

Advance of Science in the Last Half Century," and Allen's "The Progress of Science from 1836 to 1886" (1887). Placing these essays together in one book, and listing Huxley and Allen together as the authors, conferred an artificial unity upon the two pieces and falsely implied that some collaboration had taken place (Allen and Huxley [1888]). This is a good example of how a publisher could change the meaning of texts in the process of reprinting them. It served the purpose of this publisher to add this new volume, Number 96, to his Humboldt Library of Science. But Allen's essay presented a forceful articulation of how all of science had been unified by cosmic evolution. Spencer, of course, emerges as evolutionism's "prophet, its priest, its architect, and its builder" (G. Allen 1887, 876). While Allen covered all of the sciences, including the human sciences, in order to demonstrate that the Spencerian evolutionary synthesis reached throughout all knowledge, in his essay Huxley deals only with the physical sciences. For Huxley, evolution was just one of three closely connected hypotheses that were important for the recent progress of science. In Huxley's view, the chief characteristic of modern science lay in the way it impressed the notion of the natural order on general culture (T. Huxley 1897b, 129). Whereas Huxley's commitment to naturalism is at the center of his essay, the key theme of Allen's piece is his devotion to Spencerian cosmic evolution.

73. Desmond 1997, 484; Stoddart 1975, 17–18, 21.

Edward Youmans, who worked for the New York publishing house of Appleton and Company, conceived the series. Youmans was dissatisfied with existing scientific works for the reading public and viewed the authors as third-rate hacks with only a feeble grasp of their subject. In the preface to the American edition of the first book in the series, Youmans referred to the "tendency of careless and unscrupulous book-makers to cater to public ignorance and love of the marvellous, and to foist their crude productions upon those who are too little instructed to judge of their real quality, has hitherto been so strong as to cast discredit upon the idea of 'popular science.'" Youmans stressed the importance of counteracting "this evil tendency by furnishing the public with popular scientific books of a superior character." Strikingly, Youmans claimed that the ISS was rescuing "popular science" by taking it out of the hands of those who were not qualified to write it.[74]

In 1870 Youmans proposed to Appleton that the series contain books written by the foremost scientists of the world who had the ability to express themselves clearly to educated general readers. He planned to convince the leading men of science to contribute small volumes in their area of expertise by offering ample compensation through an international copyright arrangement. For the British component of the project, Youmans and Appleton entered into a contract with Henry S. King and Company. Youmans had had previous dealings with Huxley, and approached him about playing a leading role in organizing the series. At a dinner given by G. W. Appleton on August 14, 1871, and with King present, Tyndall, Spencer, and Huxley agreed to act as a consulting committee for a series of scientific monographs to be titled "The Anglo-American Series." King consented on behalf of his firm to be the English publisher and to offer authors a royalty of one-fifth of the retail selling price of the first and subsequent editions. Appleton and Company agreed to give all English authors a royalty of 10 percent on U.S. sales in addition to seven and a half percent on sales published on the continent. In effect, English authors were being allowed the practical benefits of an international copyright law.[75]

Youmans sent Huxley a first draft of a printed circular in the same month. It brought to the attention of "English Scientific writers" the plan to "prepare a series of monographs or elaborate essays on selected scientific topics,

74. Howsam 2000, 194; MacLeod 1980, 65; Youmans 1898, v.

75. *Archives of Kegan Paul, Trench, Trübner, and Henry S. King, 1858–1912*, "Memorandum of Results of Conference at a Dinner Given by G. W. Appleton at the St. James' Hotel, Aug. 14, 1871," F, Contracts, Filed under Anglo-American Series; Howsam 2000, 195–97.

and in a form suited for wide circulation." The general aim of the series was described as giving "authentic, popular expression to the latest advances of thought on the leading subjects of progressive inquiry." The circular spelled out the requirements for works "designed to address the unscientific public." They had to be "thoroughly explanatory and expository in character; yet, as they will appeal to classes who have received a certain amount of literary cultivation, some closeness of statement will be admissible." The books were to adopt "the utmost simplicity of style" and to be free of technicalities. The circular explained the international character of the series and how this would be achieved. A London publisher was to produce each book initially. Then duplicate stereotype plates were to be sent to the United States for reprinting, and the book would be simultaneously issued in the other countries. The circular also mentioned that authors would be paid at "the most liberal rates" and for royalties in other countries. Finally, there was a long list of provisional titles and authors, including John Lubbock, William Carpenter, Henry Bastian, Alexander Bain, Archibald Geikie, Balfour Stewart, William Clifford, Norman Lockyer, and even Huxley, who was slotted to contribute a book on "The Races of Mankind."[76]

On December 28, 1871, Henry S. King issued a revised circular address to "British Scientific Authors." By this point the title of the series had become the "International Scientific Series." The general aim and requirement for stylistic simplicity were phrased in similar language to the Youmans document. King described the royalty arrangement and named the French and German publishers. The English and American editions were to appear in crown octavo editions of 250 to 350 pages. They were to sell in England at 3s. 6d. and 5s. The target audience was considerably upscale in comparison to those who had been able to afford the price of one shilling for a book in Routledge's Common Objects series. A tentative list of authors included some new names, such as Walter Bagehot and Stanley Jevons, as well as European scientists such as Claude Bernard, Lambert Adolphe Quetelet, and Jean Louis de Quatrefages. At this point, Huxley was down for a book titled "Bodily Motion and Consciousness." But a new theme emerged in King's circular, the role of scientists in managing the series. The circular stated that "at the last Meeting of the British Association in Edinburgh, a Committee of eminent scientific men was formed, who will decide on the Works to be introduced into the Series in this Country and the United States, upon the

76. Imperial College, Huxley Collection, 29.261.

order of their publication, and upon all questions which may arise affected the character of the enterprise."[77]

Huxley, along with his friends Tyndall and Spencer, were now in full control of the British component of this new scientific series. After seeing the success of other book series with a focus on scientific themes, such as Routledge's Common Object series, they hoped that this series would revolutionize the dissemination of science and create a new group of scientific readers. Moreover, unlike Routledge's series, they were involved in a grand experiment in international publishing. From Huxley's point of view it appeared to be the ideal situation. He could exercise his power as editor to spread the principles of scientific naturalism without having to do all of the writing himself. He was given the resources to attract other scientific practitioners with promises of prestige, a large audience, and a hefty royalty fee. Scientists could be compensated fairly for their publications. Finally, through the series, Huxley's goal of educating the reading public about the latest developments in the natural and social sciences could be achieved.[78] The strong sales of the first five books of the series would seem to indicate that Huxley and his allies attained many of their goals. Tyndall's book *The Forms of Water* (1872) eventually sold a total of 14,750 copies, and Bagehot's on politics, also published in 1872, 12,500.[79] Edward Smith's book on food (6,500) and Bain's on education (9,250) in 1873 had smaller, but nevertheless respectable, sales. Of all of the books in the series, Spencer's *Study of Sociology* set the series record.[80] The number of copies printed in Britain for the first decade totaled 12,500, and by the end of the century it had reached 23,830 copies. In total, 26,330 copies of Spencer's book were printed in Britain when the last edition came out in 1910.[81] Leslie Howsam estimates that the first five years of the British series were the most successful.[82] Although fewer new titles were introduced in the second and third decades of

77. Ibid., 30.98.

78. Howsam 2000, 192–97.

79. Ten thousand copies of Tyndall's *Forms of Water* were printed by the time it reached the eighth edition ten years after its original appearance. By 1899 the book had reached a twelfth edition and 14,250 copies had been printed (*Archives of Kegan Paul, Trench, Trübner, and Henry S. King, 1858–1912* 1973, A1, 177, 473; B1, 60, 61; C1, 1, 3, 234, 235; C6, 234).

80. Howsam 2000, 198.

81. *Archives of Kegan Paul, Trench, Trübner, and Henry S. King, 1858–1912* 1973, A1, 301–2; B1, 68–69; C1, 13–16; C26, 77–79.

82. Howsam 2000, 198.

the series' existence, it survived in England until 1911. The series only lasted until 1889 in Germany, and up to 1910 in the United States.

Even though the series lasted in Britain for almost forty years, scholars disagree on its success as a vehicle for Huxley's goals. Whereas Roy MacLeod asserts that the series illuminated a "sense of a unified, comprehensive evolutionary dynamic," Leslie Howsam states that the series "disseminated a collective image of science that was complex and sometimes contradictory." Their apparent stability, she argues, is as illusory as their claim to represent authoritative science.[83] Katy Ring has questioned the effectiveness of the series in reaching a popular audience. She points out that at a price of five shillings, the books in the series were affordable only to the professional middle classes and the wealthy. In addition, she argues that the majority of the texts were so specialized that they would have been unintelligible to the general reader.[84] Howsam and Ring make persuasive arguments.

Huxley and his allies originally became involved as they believed that under their control the series could become a huge success, but the potential for conflict between the British committee and the publisher existed from the start. King was unhappy with the first book in the series, authored by one of the members of the committee. Tyndall had offered King a lightly revised version of his juvenile lectures to the Royal Institution.[85] On April 12, 1872, King wrote to Spencer, expressing fears that Tyndall's book would give the wrong impression that the series was aimed at boys.[86] King understood that the powerful members of the British committee wanted to maintain tight control of editorial decisions, but he was right to be concerned about Tyndall's book. Both *Nature* and the *Westminster Review* expressed disappointment with it.[87] The issue of editorial control arose again a few years later, and King was obliged on November 25, 1876, to send a contract between his publishing company and the committee for "deciding what works shall be introduced into this Series." King agreed to print all the works that the committee declared suitable to be introduced into the series.[88]

---

83. MacLeod 1980, 76; Howsam 2000, 188, 193.

84. Ring 1988, 72–73.

85. Howsam 2000, 203.

86. Imperial College, Huxley Collection, 19.145.

87. Howsam 2000, 203.

88. Imperial College, Huxley Collection, 30.98.

In October of 1877, Charles Kegan Paul, manager and publisher's reader for King for several years, purchased H. S. King and Company. King had become seriously ill and died the following year. Although King and the British committee had clashed on occasion, he had been in on the project from the start, and he had no strong objections to scientific naturalism. But Paul was not the right publisher for a science series controlled by Huxley and his friends.[89] He had been an Anglican minister and a master at Eton. Educated in the classics, he had little knowledge of science and not much sympathy for it. He was a Broad Churchman, but then abandoned his living in 1874 since he no longer could adhere to the teachings of the Church of England. He associated himself with Comtist Positivism, attracted to the ritual, but in 1888 began to attend mass and in 1890 he converted to Catholicism.[90] When Paul took over he did not feel bound by the contract that Tyndall, Spencer, and Huxley had signed with King. The inevitable showdown took place in 1883. Paul sent Huxley seven new volumes for the series on January 20. Two days later Huxley wrote to Paul, saying that he was unaware that any of the volumes had been accepted. He demanded an explanation as to how it had come to pass that volumes had been included in the series without any consultation with the committee.[91]

Paul replied on January 26, and offered a "short recapitulation of the facts" in order to remove Huxley's misapprehensions. Up to 1876 Spencer had represented the committee and was in frequent communication with Henry S. King and Company in regard to books to be admitted into the series. But in 1876, Spencer said he had done enough, and Paul reminded Huxley of a long conversation in which Huxley had agreed that "the Publishers should take upon themselves a larger share of responsibility for the arrangement of the Series, but that they should consult you at any time when they felt any difficulty whatever."[92] On March 3, Huxley wrote to Paul, and withdrew from the editorship since the reasons for which the editorship was undertaken no longer seemed to exist.[93] Huxley's participation in the International Scientific Series therefore lasted about a dozen years. After he left,

89. I am indebted to Michael Collie for pointing this out. My thinking on this aspect of Huxley's activities has been profoundly influenced by Collie's willingness to share with me some of his work on the International Scientific Series.

90. Howsam 1991, 239; 2004, 136–37.

91. Imperial College, Huxley Collection, 24.76, 24.77.

92. Ibid., 24.78.

93. Ibid., 24.81.

the series, guided by Paul, took a different direction and no longer acted as a forum for the dissemination of scientific naturalism. Henslow's *Origin of Floral Structures* and his *Origin of Plant Structures,* for example, with their emphasis on a revised natural theology and neo-Lamarckianism, were published in the series after Huxley's departure.

In addition to his differences with Paul over editorial control, Huxley had another reason for severing his ties to the International Scientific Series: the dismal situation with his own contribution. It must have been a source of tension between Huxley and Paul. When *The Crayfish* was published in 1880, the twenty-eighth book in the series, it was long overdue. Huxley had been paid an advance of £100 when the series was first announced in 1871.[94] As a member of the British committee he would have been expected to set an example for the other potential contributors by completing his book in a timely manner. Moreover, *The Crayfish* was not the book that he had originally promised for the series. It was entirely different from "Bodily Motion and Consciousness" or "The Races of Mankind," the two proposals originally floated when the series was first being discussed. Although Huxley had switched topics, and though he had taken far too long to complete his book, Paul would have expected it to sell well. But compared to the *Physiography,* and to many of the other books in the series, the sales of *The Crayfish* were disappointing, perhaps even embarrassing. Within a decade of appearing, Paul had printed only 5,275 copies. By the end of the century the total print run amounted to 5,775 copies. In the end, a total of 6,275 copies were printed. Adrian Desmond reports that the *Crayfish* reached a seventh edition, yet the number of editions is misleading in this case.[95] The new editions were reissues and the print runs for each were relatively small.[96]

94. In the midst of Huxley's discussions in 1883 with Paul about the acceptance of seven new volumes without any consultation, he also asked for a statement of the sales of the *Crayfish.* Huxley was wondering why he had not received any royalties. In the Kegan Paul Publication Books there is a note, "In consideration of Prof. Huxley having been paid £100 at the announcement of this Series, he is to receive no Royalty on the first two Editions of this Book" (*Archives of Kegan Paul, Trench, Trübner, and Henry S. King, 1858–1912,* A3 167). In his letter to Huxley, dated January 26, 1883, Paul enclosed a sales statement and reminded Huxley that the terms agreed to at the establishment of the series stipulated £50 for the first edition of 1,250 copies and £55 for each succeeding edition, payable six months after the issue of the edition. Appleton and the continental publishers paid authors directly. Huxley's advance almost covered the first two editions. A note on the same page of the Publication Book records payment to Huxley of the £5 owing to him on the second edition, and another £55 for the third edition.

95. Desmond 1997, 497.

96. *Archives of Kegan Paul, Trench, Trübner, and Henry S. King, 1858 1912,* Publication Book, vol. 1, fol. 114-15 (reel 4, index B); Publication Account Book, vol. 1, fol. 83-85 (reel 6, index C);

Huxley started work on *The Crayfish* in the summer of 1878, after having made it the focus of his Davis Lectures at the Zoological Society and a five-week workingmen's course. At first, Huxley intended to use it for his own series on the invertebrates, vertebrates, and humans. In his diary for 1878 he listed eight volumes, all of an introductory nature, starting with a primer, then books on physiography, biology, and physiology. Then came the crayfish, which was intended to serve as an introduction to zoology, a book on the dog to introduce the mammals, followed by one on "man" to deal with anthropology, and ending with a text on psychology.[97] Later, he decided to use his book on the crayfish to fulfill his obligations to Paul and the International Scientific Series. *The Crayfish* is a key work for understanding how Huxley distinguished his work from that of other popularizers. In *The Crayfish*, Huxley attempted to spell out, as clearly as possible, how his emphasis on common objects differed from those who drew on the natural history tradition to make their writing accessible to a popular audience. In his preface, Huxley explained that his main aim was to "show how the careful study of one of the commonest and most insignificant of animals, leads us, step by step, from every-day knowledge to the widest generalizations and the most difficult problems of zoology; and, indeed, of biological science in general." In this sense, the book was, as the subtitle indicated, an introduction to zoology, the kind of work that Darwin had begged Huxley to undertake in the mid-1860s.[98] Like the *Physiography*, the *Crayfish* was well illustrated, including eighty-two illustrations.

Huxley's first chapter, on the "The Natural History of the Common Crayfish," was designed to set his audience at ease. "Many persons seem to believe that what is termed Science is of a widely different nature from ordinary knowledge," Huxley declared, "and that the methods by which scientific truths are ascertained involve mental operations of a recondite and mysterious nature, comprehensible only by the initiated, and as distinct in their character as in their subject matter, from the processes by which we discriminate between fact and fancy in ordinary life." Huxley reassured his readers that this was not the case. The realm of science was not "shut off

---

Publication Account Book, vol. 26, fol. 122–24 (reel 14, index C). There was also a "large paper" edition published in 1880 of which 250 copies were printed in 1879 (*Archives of Kegan Paul, Trench, Trübner, and Henry S. King, 1858–1912*, Publication Account Book, vol. 1, fol. 58 (reel 4, index B). I am indebted to Leslie Howsam for supplying some of the figures on the *Crayfish* and confirming others.

97. Imperial College, Huxley Collection, Huxley's 1878 Diary, HP 70.21; Desmond 1997, 496–97.

98. Huxley 1880, xix.

from that of common sense," nor did its mode of investigation differ from that used for the commonest purposes of everyday existence. Science was "simply common sense at its best; that is, rigidly accurate in observation, and merciless to fallacy in logic." Huxley then launched into a discussion of the history of biology that imitated Auguste Comte's law that all scientific disciplines must pass through three stages: theological, metaphysical, and a positive. Each of the branches of biology, zoology, and botany, "has passed through the three stages of development, which are common to all the sciences; and at the present time, each is in these different stages in different minds." The country boy possessed common knowledge of the plants and animals that came under his notice. "A good many persons have acquired more or less of that accurate, but necessarily incomplete and unmethodised knowledge," Huxley declared, "which is understood by Natural History," while the few had reached "the purely scientific stage," and were working toward "the perfection of biology as a branch of Physical Science." Huxley maintained that the attempt to construct a complete science of biology began only in the early nineteenth century, with Lamarck, and that it received "its strongest impulse, in our own day, from Darwin." The purpose of the *Crayfish,* Huxley wrote, was to "exemplify the general truths respecting the development of zoological science" by presenting a case study of one animal, the common crayfish.[99]

Huxley then provided a description of the crayfish, its eating habits, how they were caught, and where it was used as an important food source. But all of this, Huxley insisted, was merely "common knowledge," akin to the information possessed by the country boy. Huxley suggested that we "now try to push our acquaintance with what is to be learned about the animal a little further, so as to be able to give an account of its Natural History." He considered the origin of the term "crayfish," saying that this inquiry suggested itself naturally enough at the outset of a natural history even though it "does not strictly lie within the province of physical science." He moved on to a discussion of the crayfish's anatomy, its mode of reproduction, and its young. But these were all "points which an observant naturalist, who did not care to go far beyond the surface of things, would find to notice in the animal itself" (Fig. 7.4). Even the most "observant" naturalist, Huxley implied, dwelled only on the surface of nature.[100]

The *Crayfish* was not just about crustaceans. It also taught the reader that scientific biology was the logical and superior successor to natural history.

99. Ibid., 1–2, 4. 100. Ibid., 10, 16.

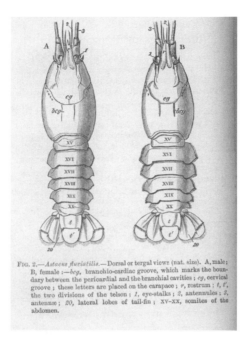

FIG. 2.—*Astacus fluriatilis.*—Dorsal or tergal views (nat. size). A, male; B, female :—*bcg*, branchio-cardiac groove, which marks the boundary between the pericardial and the branchial cavities ; *cg*, cervical groove ; these letters are placed on the carapace ; *r*, rostrum ; *t*, *t'*, the two divisions of the telson ; *1*, eye-stalks ; *2*, antennules ; *3*, antennæ ; *20*, lateral lobes of tail-fin ; XV-XX, somites of the abdomen.

FIGURE 7.4 An illustration of the body of the crayfish shows what an observant naturalist would notice, that it is "naturally marked out into several distinct regions" (T. Huxley 1880, 17). (T. H. Huxley, *The Crayfish* [New York: D. Appleton, 1880], 18.)

In the second chapter Huxley began to move the reader from natural history toward the threshold of biology and sketched out the rest of the book. The natural history of the crayfish provided in the first chapter gave brief and general answers to three questions: What is the form and structure of the animal? What are the various actions of which it is capable? Where is it found? Huxley observed that carrying the investigation into each question further would lead to knowledge of morphology, physiology, and distribution or chorology. It would also raise a fourth question, which could only be dealt with once an advance beyond the natural history stage had been reached. This question was: "how all these facts comprised under Morphology, Physiology, and Chorology have come to be what they are; and the attempt to solve this problem leads us to the crown of Biological effort, *Aetiology*." The subsequent chapters focused on the physiology, morphology, comparative morphology, distribution, and etiology of the crayfish.

In the chapter on comparative morphology, where he examines the common English crayfish in comparison to other crayfishes in different parts of the world, Huxley begins to introduce evolutionary themes. All crayfish may be regarded as exhibiting modifications of a common plan. Even though lobsters and prawns have different external forms and habits, they share a common plan of organization with the crayfish. Huxley uses comparative morphology to push the reader to see ever-greater affinities in plan of organization in living things, between Crustacea and the whole of Arthropoda, then between Arthropoda and the rest of all living things. Comparative morphology leads to the conclusion that there is a common plan, a unity of organization of plants and animals. This sets the reader up for the final chapter on distribution and etiology. Crayfish are geographically distributed around the world in a pattern, according to morphological difference, which suggests that either they originated through divine intervention or due to the usual course of nature, which entailed the evolutionary process. Huxley stated that no informed scientist could accept the first possibility, as it was "an admission that the problem is not susceptible of solution." Huxley was now ready to put forward a hypothesis as to the genealogy of the crayfish. "All the known facts," he wrote, "are in harmony with the requirements of the hypothesis that they have been gradually evolved in the course of the Mesozoic and subsequent epochs of the world's history from a primitive Astacomorphous form" (Fig. 7.5). In the *Crayfish*, Huxley capitalized, as he did in many of his other popular works, on the strategy of starting with a common or familiar object in nature, but also managed at the same time to distance himself from the many popularizers who drew their inspiration from natural history. Whereas they provided only a superficial examination of nature, he claimed to offer genuine understanding of things. Where they appealed to their audience's sense of wonder in order to reveal the divine order in nature, Huxley spoke to their common sense in order to allow them to perceive how the evolutionary process accounted for the unity of plan among living beings.[101]

PRIMING THE PUBLIC FOR SCIENCE

While Huxley was working on the *Physiography* and on the International Scientific Series, he was also deeply involved in yet a third project aimed at disseminating science to the general reading public. Macmillan's Science Primers were designed primarily for British readers, and although not as

101. Ibid., 46, 254, 278, 286, 308, 318, 346.

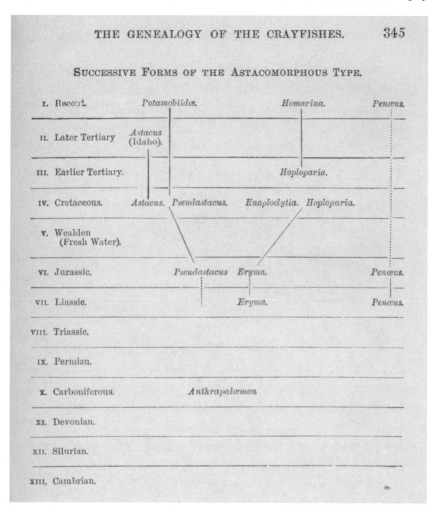

THE GENEALOGY OF THE CRAYFISHES. 345

SUCCESSIVE FORMS OF THE ASTACOMORPHOUS TYPE.

		Potamobiidæ.	Homarina.	Penæus.
I.	Recent.			
II.	Later Tertiary	Astacus (Idaho).		
III.	Earlier Tertiary.		Hoploparia.	
IV.	Cretaceous.	Astacus. Pseudastacus. Enoploclytia. Hoploparia.		
V.	Wealden (Fresh Water).			
VI.	Jurassic.	Pseudastacus Eryma.		Penæus.
VII.	Liassic.		Eryma.	Penæus.
VIII.	Triassic.			
IX.	Permian.			
X.	Carboniferous.	Anthrapalæmon		
XI.	Devonian.			
XII.	Silurian.			
XIII.	Cambrian.			

FIGURE 7.5 Huxley's diagram of the evolution of the crayfish from one form. (T. H. Huxley, *The Crayfish* [New York: D. Appleton, 1880], 345.)

grandiose in scope as the International Scientific Series, Huxley was just as enthusiastic about them. Perhaps he relished the opportunity to work with Alexander Macmillan, who, between 1860 and 1872, had established Macmillan and Company as one of the leading publishing houses in London. The Macmillan family had strong connections to liberal Anglican F. D. Maurice and the Christian Socialists and was receptive to liberal ideas. Alexander attracted prominent literary figures to social gatherings in London where the leading topics of the day were discussed. In addition to Huxley, Alfred Tennyson, Herbert Spencer, Francis Turner Palgrave, Coventry Patmore, Charles Kingsley, Thomas Hughes, and Maurice frequently attended.[102] Over the course of the 1860s and 70s, Macmillan became more interested in science and began to produce more publications in the area. It was Macmillan who had founded the illustrated scientific weekly *Nature* in 1869. Whereas in the early 1860s scientific books constituted one in every ten published by Macmillan, by the latter half of the 1870s this had risen to almost one in every four. By this time Macmillan had become a leading publisher of scientific books for the general reader, along with H. S. King, Longman, and John Murray.[103]

The introduction of the Science Primers was one of Macmillan's innovations in science publishing. In the wake of the Education Act of 1870, Macmillan saw the potential for publishing a series of science textbooks that dealt with subjects at a fairly elementary level. Katy Ring has described the Science Primers as a good example of how publishers began marketing "the more mainstream sciences for lower class consumption."[104] Friendly with Huxley and other scientists, Macmillan decided to turn to the acknowledged masters in each branch of science. He recruited Huxley, and two faculty members of Owen's College, Balfour Stewart, a physicist, and Henry Roscoe, a chemist, to act as joint editors, as well as to contribute volumes to the series. Contentious subject matters were ruled out. Anticipating a large institutional demand as a result of the Education Act, each volume of the series was to have a large initial run of 10,000 copies, and each copy was to be priced at one shilling. Sales for the first three years were good, but by 1878 there was a noticeable decline, in part because the market was saturated with a glut of elementary science books designed for class use. The Society for the Promotion of Christian Knowledge (SPCK) had begun their "Manuals of Elementary Science" in 1873. They also sold for a shilling per volume. In 1875 Chambers brought out their "Chambers Elementary Science Manual," selling them at a similar price. Other presses set up their own series later in the decade.[105]

For Huxley, becoming involved in Macmillan's Science Primers offered another opportunity to undermine the influence of popularizers steeped in natural history and theologies of nature. According to Adrian Desmond, the

---

102. VanArsdel 1991, 179, 182.
103. Meadows 1980, 55–56.

104. Ring 1988, 80.
105. Ibid., 81–83.

primers were written "to control the dissemination of elite knowledge, [and] to usurp women's traditional 'mentorial' role as kindergarten teachers."[106] Huxley agreed to write the introductory volume. On June 29, 1871, he wrote to Roscoe, "your chemical primer appears to me to be admirable—just what is wanted." In his letter, Huxley enclosed a sketch for his primer, explaining that it would function as a preparation for all of the other books in the series. "When it touches upon chemical matters," Huxley affirmed, "it would deal with them in a more rudimentary fashion than yours does, and only prepare the minds of the fledglings for you."[107]

Consumed with other projects, Huxley could make no real headway on his primer, while Roscoe made steady progress. When the primers by Roscoe and Stewart were published, Huxley wrote to Roscoe on April 8, 1872, and expressed his satisfaction with the result. "I like the look of the 'Primers' (of which Macmillan has sent me copies to-day) very much," he declared, "and shall buckle to at mine as soon as possible. I am very glad you did not wait for me."[108] *Nature* reviewed the two primers enthusiastically. While "scientific class-books hitherto have been either too difficult or too easy," Roscoe and Stewart had avoided both extremes. The reviewer praised the emphasis on experiment, stating that from "first to last the student finds himself in immediate contact with Nature." He also commended the way the primers presented a systematic account of their subjects. The student's "empirical knowledge of external things is systematized; simple every-day phenomena reveal to him their principles and *rationale*." All in all, the writer concluded, the primers marked "a stage in the advance of scientific education."[109]

The recruitment of authors for the series continued, and still Huxley's introductory volume had not been written. Alexander Macmillan wrote to Huxley on May 31, 1872, that he had engaged Archibald Geikie to write the primer in geology.[110] Geikie's *Geology* appeared in that same year (1872). Three years later, Huxley still could not find the time to work on his contribution. Exasperated by Huxley's tardiness, Roscoe continued to prod his coeditor. On January 24, 1875, he reported to Huxley on the splendid sales of the primers. Stewart's book on physics had increased in sales from 5,000 in

106. Desmond 2001, 23.

107. L. Huxley 1902, 1: 387.

108. Ibid., 1: 410.

109. Tuckwell 1872, 3-4.

110. Imperial College, Huxley Collection, 22.144.

the previous financial year to over 7,000 during the last six months. "Now we really want your Introductory! And I think you will not find it difficult to put the thing together," Roscoe asserted. When the idea for the series had been in its infancy, and authors had not yet been recruited, writing an introduction would have been based on anticipating the final shape of each volume. By now many of the volumes had come out, so Huxley had an exact notion of the scope and content of the series. To Roscoe, all Huxley had to do was "give a sketch of the different special divisions of physical science beginning (as you did) from the simplest concepts and ending in life phenomena."

To motivate Huxley, Roscoe pointed out that other publishers, including the religious firms, were stealing the idea behind Macmillan's Primer Series, and that to compete they desperately needed the introductory volume that tied everything together. Roscoe reminded Huxley that they now had volumes on physics (by Balfour Stewart), chemistry (by Henry Roscoe), physical geography (by Archibald Geikie), geology (also by Archibald Geikie), astronomy (by Norman Lockyer), botany (by Joseph Hooker), and physiology (by Michael Foster), "and your Introductory is now necessary for us to meet the coming struggle with the Christian Knowledge Society and other sinners who are pirating our ideas!" The public saw the Christian Knowledge Society publications as copies of their series, but, Roscoe wrote, "our plan differs from all the wretched imitations in making each book an experimental inquiry." Hoping to rouse Huxley to action, he asserted that the Christian Knowledge Society publications were "in reality whitened sepulchers for they are hideously bad (at least Chemistry is) inside!" Roscoe believed that Huxley's introductory volume was crucial, as it would give their series a distinctive identity.[111]

Huxley's inability to fulfill his commitments to the series became more and more distressing as the years rolled by. He was upset when contributors to the series produced volumes that seemed to compromise the need for an introductory volume. On September 1, 1875, he told Macmillan that "I am greatly bothered over the 'Introductory Primer'—Geikie's book takes the wind out of my sails."[112] In 1877 he admitted to a correspondent that the introductory Primer "weighs heavily on my mind." He also recognized that it "will look rather absurd that a Primer on Zoology in a series into which I

111. Ibid., 25.277.

112. British Library, Macmillan Archives, ADD 55210, f30.

have anything to do should be written by anybody but myself," but he recommended that Macmillan consult Roscoe and Stewart to see if the series
would be best served if someone else did the volume.[113] He was still interested
in the success of the series and wrote to Macmillan on January 23, 1877,
that the mathematician William Kingdon Clifford should do a geometry
primer.[114]

Huxley finally found the time and energy to work on the long neglected
project in 1879. On July 19, 1879, he wrote to Macmillan, "the spirit is
moving me to polish off that 'Introductory Primer.' Are you prepared to
get it done at once and come down with something handsome if I sent you
the MS?"[115] Macmillan's response was to send Huxley £200 as an advance
payment.[116] By September 10, 1879, he told Macmillan that the "Primer is
substantially done and I am rather pleased with it than otherwise. I am going
to send it to be mauled by Roscoe and Stewart."[117] Two days later he sent the
"long-promised Primer" to Roscoe and Stewart. He told Roscoe that "it is
quite different from my first sketch, Geikie's primer having cut me out of that
line—but I think it much better." The basic idea, Huxley wrote, "is to develop
Science out of common observation, and to lead up to Physics, Chemistry,
biology, and Psychology."[118] By October 4, Stewart was supposed to send
the primer back to Huxley, at which point Huxley was going to consider all
of the suggestions he had received and then finish it off.[119]

Huxley's *Science Primers: Introductory* came out early in 1880, priced at
one shilling. It was even more elementary than either the *Physiography* or the
Crayfish. It contained no illustrations. The book was broken up into three
main sections, titled "Nature and Science," "Material Objects," and "Immaterial Objects." Each of these sections was further subdivided into short,
numbered subsections that could be as short as a page or as long as several
pages. In the first section, Huxley began with the senses, and explained basic
terms, such as "cause," "effect," "natural order," and the "laws of nature."
He made it clear that the goal of science was to obtain knowledge of the

113. Ibid., ADD 55210, f44.

114. Ibid., ADD 55210, f35.

115. Ibid., ADD 55210, f82.

116. Ibid., ADD 55210, f84.

117. Ibid., ADD 55210, f85

118. L. Huxley, 1902, 2: 2.

119. British Library, Macmillan Archive, ADD 55210, f87.

laws of nature by using observation, experiment, and reasoning. Echoing comments in his other works, he affirmed that it was important to discover as many laws of nature as possible in order to guide our conduct by them, and he stated, "science is perfected common sense."[120]

In section two, on material objects, Huxley drew the reader's attention to a specific natural object: water. Just as common objects such as a piece of chalk, or the River Thames, or a crayfish, had provided a means for Huxley, and other popularizers, to begin at the reader's level of understanding, water was to serve that purpose in this book. Water was actually the supreme common object. "One of the commonest of common natural objects is *water*," he remarked, "everybody uses it in one way or another every day; and consequently everybody possesses a store of loose information—of common knowledge—about it." But a great deal of this knowledge "has never been attended to by its possessor," so Huxley proposed to "make a beginning of science by studying water." Huxley dealt with water in relation to both mineral and living bodies. He discussed water's weight, which allowed him to raise the issues of gravity, attraction, force, and the measurement of weight. He then examined the motion of water, the affect of heat on it (which takes him to gases in general), the affect of cold, and the structure of it (opening up a discussion of chemistry). Moving on to living bodies, Huxley switched from water to various common plants and animals, such as wheat and the ordinary fowl. He emphasized that certain constituents of the body are very similar in both, and then investigated the issue of what was meant by the word "living." Section two on material objects ends with an explanation of the division of the study of living bodies, biology, into botany and zoology. Huxley managed in this second section to introduce the basics of physics, chemistry, and biology in order to prepare readers for the series primers in each of these areas.[121]

Section three on immaterial objects focused on mental phenomena, bringing them within the range of natural science. Huxley provided a simple description of this area of science, asserting, "sensations, emotions, and thoughts, thus constitute a peculiar group of natural phenomena, which are termed *mental*." He insisted that the study of mental phenomena had to be conducted along the same lines as the study of material objects. Falling within the province of the science of psychology, the scientist investigated the order in which mental phenomena succeed one another and "the relations

120. T. Huxley 1881, 18.   121. Ibid., 19.

of cause and effect which obtain between them and material phenomena." Huxley concluded by drawing a very strict boundary around the domain of science. "All the phenomena of nature are either material or immaterial, physical or mental," he argued, "and there is no science, except such as consists in the knowledge of one or other of these groups of natural objects, and of the relations which obtain between them." Mental phenomena could be studied only insofar as they were natural objects, and he excluded from science everything that fell outside of the world of nature, including theological topics. This made scientific naturalism constitutive of natural science. Every school child that read this introduction to science would be trained to reject the very premises of theologies of nature. Huxley's economical use of words in this book, his matter-of-fact tone, the lack of any illustrations, and his extreme clarity, was in stark contrast to the natural theologians' use of colorful language to evoke a sense of wonder in their readers. The Huxley of his controversial essays is nowhere to be seen.[122]

From the point of view of sales, the introductory primer was a success, as Roscoe had predicted. Far more copies of it were sold than of the *Crayfish*. Macmillan told Huxley on April 23, 1880, that eighty copies had been purchased during the previous week, bringing the grand total to 19,000 copies sold.[123] In the agreement that Huxley had made with Macmillan, Huxley sold the press the copyright and the proceeds of the first 10,000 copies for £250. After the first 10,000 had been sold, Huxley was to receive a royalty of two pence per copy.[124] Huxley's positive experience working with Macmillan on the Science Primer series, and the success of his book, may have contributed to his decision to resign from the committee overseeing the International Scientific Series. By the early 1880s, Huxley clearly saw that his loyalties to Macmillan ran far deeper than his allegiance to Charles Kegan Paul. Huxley had more in common intellectually with Macmillan than he did with Paul. Moreover, Huxley's publishing projects with Macmillan had become quite lucrative. In 1873 Huxley signed an agreement with Macmillan to publish *Critiques and Addresses,* receiving £50 for the copyright and a royalty of two shillings on each copy after the sale of the first 500 copies. The copyright

122. Ibid., 93–94.

123. Imperial College, Huxley Collection, 22.154.

124. British Library, Macmillan Archive, ADD 55210, f90. If the selling price was altered after the first 10,000 copies, the royalty was to bear the same proportion to the selling price as two pence bears to one shilling.

for *American Addresses* garnered him £50 from Macmillan in 1877, plus one-sixth of the selling price per copy after the first 1,500 sold. For *Hume* (1878), Macmillan paid him £150 for the copyright, and a royalty of 10 percent of the selling price after the sale of the first 10,000 copies. In 1882, Huxley's *Science and Culture and Other Essays* fetched £50 for the copyright, and two shillings per copy of all sales over the 500 copies mark. By this point, Macmillan had become his chief publishing partner. Huxley chose to work with Macmillan on his nine volumes of *Collected Essays* (1893–94). In 1892 he signed an agreement with Macmillan for the first five volumes, for which he was to receive one-fifth of the price of every copy sold.[125]

Huxley's views on the importance of conveying science to a popular audience underwent a remarkable change over the course of his life. During the 1850s he distanced himself from scientific hacks like Chambers and Lewes by adopting the role of critic. In the 1860s, he resisted Darwin's pleas to write a book on zoology aimed at the public. Although he recognized the need to introduce more science into education, he saw it as a battle that required more energy and time than he was willing to offer. Huxley's priorities in the 1850s and 60s were his own original research, the reform of science from within, and, after 1859, the fight to ensure a fair hearing for Darwin's theory of evolution by practitioners of science. Huxley's attitude about communicating science to the public near the end of his career was strikingly different. In his "Preface" (1894) to *Discourses Biological and Geological* he criticized those who regarded "a popular lecture as a mere *hors d'oeuvre,* unworthy of being ranked among the serious efforts of a philosopher."[126] Here Huxley, in part, was defending the role of the man of science as a public figure and educator against practitioners who were putting too much emphasis on professionalization, with its exclusive valuation of specialist research. Huxley's change of heart on the importance of writing for a general audience took place at some point in the late 1860s and early 1870s,

---

125. Ibid., ADD 55210, f26, f39, f78, f99, f126.

126. Huxley 1897a, v. In addition to being an appropriate activity for a philosopher, Huxley also asserted, in an undated note, that the "work of the popular expositor" was a dimension of scientific work that deserved to be seen as "literature." Huxley emphasized that this work could only be undertaken by a man who was a "well qualified interpreter of nature" and his role was to translate "that interpretation out of the hieratic language of the experts into the demotic vulgar tongue of all the world." This form of writing deserved to be classed as literary since it embodied "great emotions and great thoughts in such form that they touch the hearts and reach the apprehensions not merely of the select few but of all mankind" (T. Huxley, n.d., "On Literary Style," Imperial College, Huxley Collection, HP 49.55).

a period when the number of legal voters was dramatically expanded and the Education Act of 1870 was passed. Seeing the popularization of science as an integral part of the pressing need to educate the public about science, he became involved in three major projects. If Huxley is to be granted the title of the foremost popularizer of science of the nineteenth century, then he must be judged on the basis of the success of these projects.

ROBERT BALL: SCIENTIFIC NATURALIST AND LECTURER

Whereas Huxley's reputation as one of the premiere popularizers of the age was based, in part, on his prowess in controversy, Robert Ball tried to avoid being seen as a contentious figure, though he shared Huxley's commitment to evolutionary naturalism. Ball became involved in scientific writing and lecturing because he enjoyed it, and not because it was a financial necessity or because it was part of a suite of activities undertaken to carve out a career in science. He put far more time and energy into his activities as popularizer than Huxley and most other practitioners. As Robert Smith has observed, "what is different is the remarkable extent to which Ball, an eminent professional, also took on so successfully and extensively the role of popularizer with his articles, books and lectures." Author of thirteen astronomy books aimed at a popular audience, which appeared between 1877 and 1906, Ball became the most widely known astronomical author and lecturer in the English-speaking world after Proctor's death in 1888.[127]

The circumstances leading to Ball's career as practitioner and popularizer included his upbringing in a home where science was often the focus of family life, his remarkable success as a university undergraduate, and his fateful posting as a tutor at an aristocratic estate that housed the largest telescope in the world. Robert Stawell Ball (1840–1913) was born in Dublin to Dr. Robert Ball, a clerk in Dublin Castle, and Amelia Gresley Hellicar (Fig. 7.6). Ball senior was an avid naturalist, active in the Royal Irish Academy and at one time honorary secretary of the Royal Zoological Society of Ireland. Robert junior recalled that the Ball household was often the scene of scientific activity. There were occasions when animals purchased for the Dublin Zoo found a temporary home in the Ball house, including a sloth from Brazil that was hung on the back of a chair in front of a fire in the dining room. Ball junior also remembered meeting eminent scientists who were friends

127. R. Smith 2004, 5, 14.

SIR ROBERT STAWELL BALL.

FIGURE 7.6 A picture of Robert Stawell Ball that emphasizes his relationship to books rather than his work as practicing astronomer. (A. A. Rambaut, "Sir Robert Stawell Ball," *The Observatory* 37 [January 1914]: facing p. 35.)

of his father, such as Edward Forbes and Richard Owen. He vividly recollected sitting on Owen's lap as a child while the famous anatomist drew pictures of a Bengal tiger pursuing his prey. Educated at first by a series of governesses, he was sent to a preparatory school in Dublin, and then in 1857 to Trinity College, Dublin. The death of his father in the same year, and the resulting financial difficulties, forced him to obtain a scholarship in his first year, and to become a disciplined student. He distinguished himself in 1861 by winning the University Studentship, a scholarship worth £100 a year for seven years. The scholarship provided Ball with financial independence and the opportunity to determine his own future. At one point he considered a career in the Church and started attending divinity lectures, but the study of heresy in early Christian history did not interest him. Ball's forte as an undergraduate was mathematics, though he had also attended William Henry Harvey's lectures on botany. He became interested in astronomy after reading O. M. Mitchell's *Orbs of Heaven* in the same year that he entered Trinity College. While at the College he studied John Brinkley's *Elements of Plane Astronomy* and other classics of mathematical astronomy in a course required of all Trinity students.[128]

Ball's future as an astronomer was determined when he accepted a position in 1865 as tutor to Lord Rosse's sons at Birr Castle, Parsonstown, a small town about eighty miles from Dublin. Attracted by the prospect of having access to the giant six-foot reflector, then the largest telescope in the world, Ball stayed for two years, viewing distant nebulae. Rosse also introduced him to many distinguished men of science, including Charles Babbage, William Huggins, Charles Wheatstone, and Warren de la Rue. In 1867 he accepted a position as Professor of Applied Mathematics and Mechanism at the newly founded Royal College of Science in Dublin. Elected a Fellow of the Royal Society in 1873, the following year, at the age of thirty-four, he was appointed Astronomer Royal of Ireland, Director of Dunsink Observatory, and Andrews Professor of Astronomy at the University of Dublin. At Dunsink he investigated stellar parallax, but his activity as observer declined in the early 1880s when he began to have problems with his right eye, which, by 1885, was totally blind.[129] Ball succeeded John Tyndall in 1882 as Scientific Advisor to the Commission of Irish Lights and was knighted in 1886. But the crowning glory of his career was his appointment in 1892 as Lowndean Professor of Astronomy and Geometry at Cambridge University, succeeding John Couch Adams, and as Director of the University Observatory a short time later. To his sister, Ball jokingly wrote on February 20, 1892, that the post was "perhaps the highest scientific chair in England, if not in Europe, the Solar System, the Milky Way, or the Universe!" More honors awaited him, however, as he later served as president of both the Royal Astronomical Society (1897–99) and the Mathematical Association (1899–1900).[130]

One aspect of Ball's life, important for understanding how he handled the larger meaning of scientific ideas in his books and lectures, is largely

---

128. W. Ball 1915, 7, 9, 12–13, 15, 28–29, 31–33; Baum 2004b, 106–7.

129. Ball told David Gill in 1897 that he had been totally blind in his right eye for twelve years (Ball to Gill, July 15, 1897, Cambridge University Library, Royal Greenwich Observatory Archives, RGO 15/128 187). In 1897 Ball had his useless eye removed and replaced with a glass one. In a letter in 1897 he joked about his disability. "You will remember that Tycho Brahe, having lost his nose in a duel, made a copper nose which both his friends and enemies alike agreed in declaring was as good as the original!" He reminded his correspondent that Tycho had predicted that "a future 'Tychonides' would arise who might be worthy of comparison with him," and Ball dared to aspire to this role in his "sanguine moments" (W. Ball 1915, 125, 175–76).

130. Baum 2004a, 106–7; W. Ball 1915, 73, 137.

absent from most biographical accounts: his loss of faith when he was in his thirties. His serious consideration of a clerical career at Cambridge suggests that when he was in his late teens and early twenties he was a committed Anglican. While at Parsonstown Ball helped Rosse in his quest to refute the nebular hypothesis, with its radical social and religious associations. The supporters of the nebular hypothesis argued that the Orion nebula was composed of gaseous material, and that this was a confirmation of Laplace's theory and of the existence of natural laws that exemplified universal progress. Since the Leviathan telescope had become operational in the mid-1840s, Rosse had embarked on a campaign to undermine the validity of the nebular hypothesis. He claimed that the Leviathan provided evidence that the Orion nebula was resolvable into separate stellar objects. But in the mid-1860s William Huggins announced that spectroscopic evidence showed that objects like Orion were gaseous and not stellar. Ball sided with Rosse, confirming the authenticity of the images of the nebula drawn from observations made at Parsonstown.[131] Allying himself with Rosse and those who rejected the nebular hypothesis, Ball also repudiated the heterodoxy associated with it. Up until 1867, then, Ball's faith seems to have remained intact.

On August 5, 1868, after having taken up the Professorship of Applied Mathematics at the Royal College of Science in Dublin, he married Frances Elizabeth Steele. Lady Ball, who was a devout Christian when they were married, described in her diary the pain she experienced when she realized that her husband was drifting away from his early religious beliefs. On October 10, 1868, she wrote, "Am sorry to feel that since my marriage, religion is declining in my soul." Although she blamed herself, she also grumbled about the lack of discussion with her husband on religious matters. On January 16, 1870, she expressed her wish that she and Ball had more "heart to heart communion" on the subject of religion. By January 3, 1873, she was feeling depressed, and wished that "R and I had more sympathy in common particularly on Sundays." When they were first married, she recalled, Sundays were spent "profitably," but "now the day is not observed at all as it ought to be."[132] In a revealing letter to the physicist Oliver Lodge, dated February 28, 1908, Ball discussed his loss of faith, telling him "to you old friend I have written what I have never written before." As he grew up, he told Lodge, his evangelical mother and her devout relations had smothered him and his siblings with religion. "I must say what between

---

131. Schaffer 1998a, 220; 1998b, 465–67.
132. Cambridge County Record Office, Lady Ball's Diaries.

family prayers morning and evening and Sunday Schools and Pray meetings and long services on Sunday we were all rather overdosed." As a result, Ball and his two brothers gradually dropped "from nearly all their exercises and the only one I have retained is a Sunday morning attendance at Kings Chapel." Ball recalled that he still believed in a future life when he was near thirty, but that he had "renounced that belief by the time my children could run about." Ball's first child, a son, was born on December 17, 1869, followed by four more children by 1876, and the last in 1881. This places Ball's rejection of a future life in the first half of the 1870s, the same time that his wife began to sense him pulling away from her on religious issues. He told Lodge that due to his loss of faith he "let things slide" with his children. "The result," Ball believed, "has been that I think my sons are much the same as I am while the very sweetest of daughters (now Mrs Carcroft) is as pious as even her grandmother or aunts could have wished."[133]

Although not a member of Huxley's inner circle, Ball became an evolutionary naturalist. He was friendly with Tyndall. In an undated letter to his wife written when he was at Dunsink, he told her that he had lunched with Tyndall and his wife again while visiting London.[134] During his life Ball was discreet about his loss of faith, and unlike Tyndall or Huxley he refrained from public attacks on organized religion. Perhaps he did not want to damage his hopes for advancing his career, and, later, when he began to build a reputation as a lecturer and writer, he did not want to lose speaking engagements and publishing opportunities. The strategy seems to have worked at least for a time, as a religiously oriented publisher, the Society for the Promotion of Christian Knowledge, put out his *Time and Tide* (1888), and he wrote a four-part article on the sun for the periodical *Good Words* in 1890.[135] But the Catholic clergy compared his book *In Starry Realms* (1892, Isbister), with its chapter on Darwinism, to Tyndall's notorious "Belfast Address."[136]

By the 1890s Ball was already a well-established scientific lecturer and author. In this decade Ball became the star of the astronomy lecture circuit.[137]

133. Ball to Lodge, February 28, 1908, University College London Library Services, The Lodge Papers, MS ADD 89.

134. Ball to his wife, n.d., Cambridge County Record Office, Letters, Chiefly of R. S. Ball to His Wife. The letter would have been written after Tyndall's marriage in 1876 but before Ball took up the Cambridge professorship in 1893.

135. R. Ball 1890b.

136. G. Jones 2001, 192.

137. R. Smith 2004, 7.

Although still based in Ireland, Ball came to England throughout the second half of the 1870s and throughout the 80s to give lectures to a variety of audiences. By 1884 he had already given over seven hundred lectures. He gave his first public lectures in 1874 at the Midland Institute in Birmingham. He spoke about the transit of Venus and his experiences at Parsonstown with the Leviathan. In 1880 he was asked to replace Proctor, who was on one of his lecture tours outside Britain, as a Gilchrist lecturer. The Gilchrist Trust funded an annual series of public lectures in British industrial centers. Ball's lectures took place mostly in Yorkshire and Lancashire, and they were presented largely to working-class audiences. In his first set of lectures in 1880, he went to Rochdale, Accrington, Huddersfield, Preston, and Bury. In less than two weeks during the month of January 1882, he lectured on "The Telescope and Its Uses," delivering the same talk in ten different towns. Ball maintained a connection with the Gilchrist Trust for twenty years. He also lectured to more upscale audiences. In 1881 he gave his first lecture at the Royal Institution on "The Distance of the Stars." Ball went to the United States to deliver lectures three times, in 1884, 1887, and 1901. In the final tour, he delivered forty-eight lectures in nine weeks. He apparently left a lasting impression. The *Boston Evening Transcript* obituary asserted, "since the days of the great and only Huxley, no one has put more of natural science into the minds of men through the medium of the tongue."[138]

Ball had several reasons for becoming active as a public lecturer. First, it was a lucrative source of income. He cleared £165 when he gave the prestigious Lowell Lectures in 1884 in Boston and was convinced that he could make £100 a week if he stayed. Ball's normal charge per lecture was between £25 and £40. In 1892, when he was weighing whether or not he should accept the Lowndean Chair, he drew up a budgetary scheme for living in Cambridge. He estimated that he could earn £600 annually from his lecturing, which was equal to the salary he was to receive as Lowndean Chair. In a letter to a friend in 1897 he pointed out that he could make far more lecturing than he could writing books. "Lecturing is a more permanent source of income than writing," he declared, "for the same lecture will be available scores of times, while there is (or ought to be) a limit to the number of times the same thing can be written." Ball later discovered a second reason

---

138. W. Ball 1915, 191–92, 204, 217, 223-24; R. Smith 2004, 6; Cambridge County Record Office, Volumes of Notes for the Gilchrist Lectures on the Telescope 1882; Butterworth 2005, 167; Cambridge County Record Office, Boston Evening Transcript, Saturday April 10, 1915, Obituaries and Letters of Sympathy to Lady Ball, 1913.

for giving public addresses: it improved the lectures he gave in a university setting. Writing to his wife in 1892 about the great success of his introductory Cambridge lectures, he remarked "now that I have understood things better I see that my hold on the public generally is a vast advantage to me even in purely university matters." Ball also had a third reason for undertaking extensive public lecture tours. "Lecturing," he once told a correspondent, "is an amusing occupation, a rest, and a change."[139]

Ball spoke on a number of different astronomical topics, but he had a set repertoire of lectures that could be used over and over again. When asked in 1892 what subjects he could lecture on, Ball responded in a lighthearted manner:

> I can congeal you with the "Ice Age" or burst you up with the thunders of Krakatoa. I can tell you awful whoppers about "Time and Tide," or petrify you with a burst of eloquence about "Invisible Stars." I usually put the greatest rot into a lecture called "Other Worlds"! There is a faint (very faint) *soupçon* of theology in "An Evening with the Telescope."[140]

After 1890 Ball lectured on such topics as earthquakes, the moon, Venus, and the stars.[141] Ball also did not hesitate to exploit the theme of extraterrestrial life, as had Proctor. In one lecture, titled "Other Worlds Than Ours," he stole the title from Proctor's most famous book. But Ball did not assert that life definitely existed on other planets. In his lecture he speculated that of all the planets in the solar system, Venus was the only most likely to be inhabited by beings with affinities to earthly organisms. "If there be life upon Venus," Ball declared, "we may be assured that the inhabitants, whatever they are like, will dwell on the sunny side, and night will be to them unknown." He thought that there could be tropical forests and luxuriant vegetation in the region always facing the sun, but he acknowledged that the atmosphere could also be fatal to life.[142] Ball gave a variation of this lecture at Goole

139. Wayman 1986, 194-95; 1987, 124; Cambridge County Record Office, Memoranda of Merits and Disadvantages of Accepting the Lowndean Chair, 1892, R83/61; W. Ball 1915, 221; Ball to his wife, October 21, 1892, Cambridge County Record Office, Letters, Chiefly of R. S. Ball to His Wife.

140. W. Ball 1915, 225.

141. The texts of a number of Ball's lectures have survived and are located at the Cambridge County Record Office, Shire Hall. See: "Abstracts and Texts of Lectures: 'The Earth's Note'; 'A Discovery about Venus'; 'The Eternal Stars'; 'The Moon's Story'; 'Other Worlds Than Ours' c. 1890-c.1910."

142. "Other Worlds Than Ours," Cambridge County Record Office, Abstracts and Texts of Lectures.

in January of 1890 as part of the Gilchrist series. According to the report in the *Goole Weekly Times* "every inch of available space" was occupied three-quarters of an hour before the lecture was scheduled to begin, "whilst outside the building large numbers vainly sought admission." Over 1,100 people had paid for admission and filled the largest building in Goole.[143]

In addition to lecturing on the "hot" astronomical topics of the day, Ball was a colorful showman, like Wood, Pepper, and other speakers who were in demand on the lecture circuit. Ball frequently heard Tyndall lecture in the Royal Institution, acknowledged that he had a special genius, and tried to learn how to "Tyndallise" his imagination "up to the point of being able to devise picturesque phraseology." He also recognized the need for visual illustrations, especially for young audiences, rather than experimental demonstrations. He went to great lengths to collect the best magic lantern slides of observatories, instruments, sunspots, and lunar craters. In the 1870s, early in his lecturing career, good photographs were hard to come by. The dry plate had not yet "taken its place in the astronomical observatory for the purpose of obtaining pictorial representations" and some of the best astronomical photographers, like Isaac Roberts and E. E. Barnard, were not yet on the scene. By 1881, Ball was using photographs, by 1884 charts and the stereopticon, and in 1890 the oxy-hydrogen lantern was brought into play. Ball liked to sprinkle his lectures with jokes and humor. He often began with a witty anecdote, such as the time he was to meet a man he did not know at the train station in a town where he was speaking. When he arrived, the station was full of people and he noticed one individual carefully scrutinizing those who had departed from the train. When Ball asked him if he was there to meet him, the man replied negatively. Gradually the station emptied out and the only ones left were Ball and the man, who then approached him and said that it was very strange as he was supposed to meet Sir Robert Ball. Ball said, "that's me," whereupon the man replied that he had expected to see "a careworn creature in blue spectacles; but as to you—why, you look like a fellow that could enjoy himself!" Ball dropped the humor by the end of the lecture and often wound things up with a quotation or a poem.[144]

143. "Gilchrist Lectures at Goole. Mr. Robert Ball on 'Other Worlds'" 1890, 3.

144. W. Ball 1915, 203, 206–7, 220, 226, 228–29; "Lecture at the Midland Institute on Tides" 1881, 6; "Modern Astronomy 1884, 4; "The Gilchrist Lectures at Goole" 1890, 3. Ball's notebooks at Cambridge, Mark Butterworth argues, make it clear that he had a set lecture structure that alternated between lantern slides and interludes composed of separate though related subjects of interest. His lectures began with a humorous story, then proceeded to an introduction, followed

According to Ball, his most widely known lecture was "A Glimpse through the Corridors of Time," which he first gave in Birmingham in 1881 to inaugurate a new hall built by the Midland Institute. Ball selected a topic, the long-term effects of the tidal interactions of the moon and the Earth, which would be unknown to the audience. The subject existed only in scientific papers (largely by G. H. Darwin) and had not yet "been divested of its mathematical garb" in a form "intelligible to the general public."[145] As an evolutionary naturalist, Ball rarely appealed to the religious sensibilities of his listeners. But, like Allen, he did try to elicit a sense of awe and wonder in his readers. This was his concession to the discourse of design favored by many popularizers, just as Huxley had imitated the emphasis on common objects. Ball began by noting that he had chosen a poetic title since he believed that the subject "appeals powerfully to the imagination as well as to the reason." Ball asked his audience to try to imagine a time when the earth was a hot, inorganic mass that gave birth to the moon. He announced that he would try to explain the processes that had brought this about with the aid of mathematical reasoning, the chief tool of the mathematical astronomer whose task it was to sit at his desk and interpret the facts provided by observational astronomers.[146]

Ball showed how the tides, caused by the moon, slowly increased the length of day while the Earth pushed the moon farther away, so that its distance slowly increased. Run things back in time a million years, Ball told his audience. The length of day was only three hours long and the moon was much closer to the earth than at present. Go further back still, and at some point the two bodies would have been so close that they were almost touching. Ball then painted a dramatic picture of the fiery earth and the nearby moon that was rapidly orbiting around it. "Such is the picture," Ball announced, "which I wish to present to you as a glimpse through the Corridors of Time." When the moon was so close to the earth, the action of the tides was much more powerful. Higher tides would have submerged the city of Birmingham, even though it is five hundred feet above sea level and a good many miles from the sea. After giving his audience a terrifying glimpse of the past, Ball now gave them a vivid peek into the future, when the system had reached its ultimate end. The moon had retreated still farther and the earth revolved

---

by a first group of lantern slides, an interlude, a second set of slides, another interlude, a third set of slides, a conclusion, and then a piece of poetry (Butterworth 2005, 171).

145. W. Ball 1915, 193.

146. R. Ball 1881, 80.

more and more slowly. Ball concluded with a cosmic comment. Tides were
not confined to the moon and the earth; every body in the universe was
capable of producing tides in every other body. "We know that tides
have wrought our solar system into its present form," he asserted, "and are
we to say that the wondrous powers of the tide have no grander scope for
their exercise?"[147] In "The Corridors of Time," and in his other lectures, Ball
elicited a sense of wonder in his listeners through his use of dramatic verbal
pictures.

BALL'S STORIES OF THE HEAVENS

An energetic lecturer, Ball was also a prolific author. For him, the two
activities were closely related. Ball's repertoire of lectures often shaped his
books.[148] He found that it was relatively simple to turn lectures into written
publications. Both *The Earth's Beginning* (1901, Cassell) and *Star-Land* (1889,
Cassell) were originally delivered as lectures at the Royal Institution, while
Time and Tide (1892, SPCK) was based on lectures given at the London
Institution. Between his lectures and his books, Ball earned a personal
fortune.[149] In some cases, Ball sold the copyright to his book outright to
his publisher. He received £200 for *The Cause of an Ice Age* (1891) from
Kegan Paul.[150] He commanded this large sum by virtue of his reputation
as an eminent man of science and as a proven author of works for public
consumption. He was also receiving royalties from other works on an annual
basis. When he was considering the move to Cambridge, he estimated that
he was earning £300 per year from his books. Although it was half of what
he was earning as a lecturer, it was still a substantial amount and equal to
half of the salary that accompanied the Lowndean Chair.[151] Ball continued to
earn royalties from his books right up until the end of his life. In his notes for
executors, written a few years before his death, he left detailed instructions
on the royalties to which he and his heirs were entitled. In this document
he stated that his books with Cassell and Company had been the most
lucrative, including *Star-Land*, *Story of the Sun* (1893, Cassell), and *The Earth's*

147. Ibid., 81–82, 104–5, 107.

148. R. Smith 2004, 7.

149. Wayman 1986, 187.

150. *Archives of Kegan Paul, Trench, Trübner, and Henry S. King, 1858–1912* 1973, F.

151. Cambridge County Record Office, Memoranda of Merits and Disadvantages of Accepting the
Lowndean Chair, 1892, R83/61.

Beginning.[152] Ball was one of those popularizers who used the diversification strategy when it came to publishers. Besides Cassell and Company and Kegan Paul, Longman, G. Philip and Son, Isbister and Company, Cambridge University Press, George Bell and Sons, and the Society for the Promotion of Christian Knowledge also published his books.

Ball wrote books for both children and adults, seeing them as two separate audiences with different intellectual abilities. Mixed audiences could create problems, whether he was lecturing or writing. When he was delivering a course of juvenile lectures at the Royal Institution in 1891, he found that only one-quarter of the audience of eight hundred were youths and children. A child of eight, the president of the British Association, and the Lord Chancellor attended one lecture. "Composed as it was of people possessing an almost infinite variety of intellectual attainment," Ball recalled, "it was calculated to put the lecturer in a position of some difficulty."[153] Ball wrote a series of books for children and beginners. Longman published two school textbooks. *Astronomy* (1877, Longman), Ball's first published book, was priced at one shilling and sixpence, and by 1890 had sold close to 4,000 copies.[154] The book went through several editions up to 1916 and was published by H. Holt and Company in the United States. An elementary introduction that covered all of the basics, it was part of the London Science Class-Books series. *Elements of Astronomy* (1880, Longman) was "mainly intended for beginners" and provided an elementary survey of key terms and ideas in astronomy.[155] Priced at six shillings, it passed through eight editions, the last appearing in 1917, and was published by Appleton in America. A number of Ball's other books were aimed at novices. *A Primer of Astronomy* (1900, Cambridge University Press, 5th ed. 1955), which sold for one shilling and sixpence, started with the sun, moved out planet by planet, then went on to discuss comets, stars, and nebulae. The goal of Ball's *A Popular Guide to the Heavens* (1905, George Philip, 5th ed. 1955) was to "provide a popular guide to the study of the sky by furnishing a summary of our present knowledge of the Solar System" and a guide to the positions of the planets. Priced at a rather expensive fifteen shillings, as it contained eighty-two plates, including star maps and examples "of the finest achievements in the art of drawing and photographing celestial objects," the illustrations dominated

152. Ibid., Probate and Executor's Papers 1913–14, Notes for Executors, 1.

153. W. Ball 1915, 209.

154. *Archives of the House of Longman, 1794–1914* 1978, A12, 140, 358; E2, 55–56.

155. R. Ball 1889, v.

the book.[156] All of these books were straightforward introductions that used clear and concise language to convey the elementary concepts of astronomy. None of them dwelled on religious themes or made use of poetic language.

Ball's *Star-Land* (1889, Cassell) is an interesting exception to his usual approach to writing for children and beginners. Selling for six shillings, in print up until 1904, and available in an American edition, 13,000 copies had been bought by 1891. The book was based on a set of lectures for children that Ball had given in 1881 and 1887 during Christmas at the Royal Institution (Fig. 7.7). More of Ball's public speaking style is preserved in this book than in others, as the publisher sent a reporter to record his lectures verbatim. At first, Ball was not willing to give his permission to be recorded, saying that he sometimes said things in his lectures that he did not want to appear in one of his books. According to Ball's brother, Charles, the publisher replied, "These are precisely what we want to catch!" The shorthand notes ended up forming the basis of *Star-Land*, and the book provides an intriguing glimpse of how playful Ball could be as both author and lecturer.[157] Ball attempted to be entertaining and instructive. He presented his lectures as an excursion "to the country of Star-Land," an enchanted world filled with wonders and populated mainly by astronomers.[158]

Ball illustrated the marvels of this world in dramatic, yet sometimes humorous, fashion. One technique he relied on was telling stories about scientific heroes like Newton or about great discoveries such as the detection of Uranus by Frederick William Herschel. The purpose of these stories was to demonstrate the almost wizardlike predictive power that astronomers possessed. After spinning a yarn about Urbain Le Verrier and the discovery of Neptune, Ball asked, "have we not shown you how entitled the calculations of astronomers are to our respect, when we find that they actually discovered the existence of a majestic planet before the telescope had revealed it?" (Fig. 7.8). Ball also liked to evoke his audience's sense of wonder when confronted by the immense size of the heavens. To impress his readers with the huge distance between the earth and the sun, Ball asked them to consider how long an express train would take to make the trip. Going forty miles an hour, the train would take a staggering 265 years to cover the entire distance. Ball then pointed out that even if King Charles I had been present when the train began to move, the journey would not yet have finished. Anyone who

156. R. Ball 1905, v.   158. R. Ball 1890a, 69.
157. W. Ball 1915, 212.

A JUVENILE LECTURE AT THE ROYAL INSTITUTION.

FIGURE 7.7 Ball delivering his juvenile lectures at the Royal Institution. (Robert Stawell Ball, *Star-Land* [London, Paris, and Melbourne: Cassell and Company, 1890], frontispiece.)

boarded the train today could not expect to reach the end of the trip; only their great-great-grandchildren would see that day. To emphasize the consequences of life on a world like the moon where gravity was significantly reduced, Ball asked his readers to imagine how that would transform the most familiar games. "In cricket, for instance," he joked, "I don't think the

Fig. 41.—The Moon always turns the same face to the Earth.

FIGURE 7.8 For the illustrations in the book, a bearded and vaguely wizardlike figure becomes the guide to the wonders of the heavens rather than the clean-shaven Ball. (Robert Stawell Ball, *Star-Land* [London, Paris, and Melbourne: Cassell and Company, 1890], 111.)

bowling would be so much affected, but the hits on the moon would be truly terrific."[159]

In his books for adults, Ball became bolder as the years passed in his support for evolutionary naturalism. Cassell and Company sold his *The Story of the Heavens* (1885), among his most successful works, for a whopping thirty-one shillings and sixpence. The high cost was due to the inclusion of twenty-four colored plates, as well as numerous illustrations. However, the exorbitant price did not stifle strong sales. By 1891 18,000 copies had sold. The book continued to sell into the twentieth century, the final edition appearing in 1910. Ball explained at the outset that the title of the book referred to the "wondrous story" he was about to tell his readers. If told "adequately it would prove of boundless interest and of exquisite beauty." In this book Ball often appealed to the reader's sense of awe in the face

159. Ibid., 18, 123, 168, 224.

of the beauty of the heavens, but another theme was just as important: the tremendous accomplishments of the great astronomers. Ball announced that the "wondrous story" in the book "leads to the contemplation of grand phenomena in nature and great achievements of human genius."[160] Unlike Proctor, who put religious themes at the heart of his astronomical writings, Ball presented a secularized cosmic story that had more in common with the approach of Allen and Clodd. However, Ball also avoided open hostility toward religion. He offered a story focused on the glory of man's works in disclosing the natural wonders of the heavens.

Ball first dealt with the sun and the moon, since they were the two heavenly objects most familiar to humans. He argued that to "unfold the story of the heavens in the most natural manner" he was obliged to "follow as far as possible the order of distance of the different bodies," which led him to discuss the planets in the solar system, then the comets, stars, and other distant astronomical phenomena. Ball often pointed to the beauty of heavenly objects. "A lover of the picturesque," he remarked, "cannot behold Saturn in a telescope without feelings of the liveliest emotion." Each heavenly object was frequently coupled with an important astronomer. In the chapter on Mars, the astronomical hero is Asaph Hall, who discovered the Martian moons in 1877, while the story of Uranus, "in its earlier stages at all events, is the story of the early career of William Herschel." Ball declared that it "would be alike impossible and undesirable to attempt to separate them." Although Herschel had an important place in the history of astronomy, which Ball insisted was "the record of brilliant discoveries," one of the most dramatic moments in the book focused on Le Verrier's discovery of Neptune. Here Ball celebrates the triumphs of the mathematical, rather than the observational, astronomer. "We picture to ourselves the great astronomer buried in profound meditation for many months," Ball wrote, "his eyes are bent, not on the stars, but on his calculations and his formulae." Le Verrier needs no telescope, "the human intellect is the instrument he alone uses." Guided by his "consummate mathematical artifice" Le Verrier eventually "discerns with a certainty little short of actual vision the planet glittering in the far depths of space." He gives his calculations to the observational astronomer "and lo! There is the planet in the indicated spot." Ball claimed, "the annals of science present no such achievement as this." But Le Verrier's triumph was confirmation of the law of universal gravitation, and the great hero of the book is actually Newton. An earlier chapter on "The Law of Gravitation"

160. R. Ball 1885, 1.

established this universal law as providing the basis for astronomy as a science and offered a tribute to the "mighty genius of Newton." By the time Ball arrives at the chapter on "Comets," he has demonstrated that the entire solar system operates according to strict natural law, even the comets, formerly a superstitious sign of impending calamity. For Ball, the tools of the old mathematical astronomy confirm scientific naturalism, in comparison to Proctor, whose spectroscope justified the idea of teleology in nature. Ironically, Ball is more modern than Proctor in his secularization of astronomy, though he makes little use of new technology like the spectroscope or camera.[161]

Reviews were largely favorable in the general periodical press, and some remarked upon Ball's attempt to elicit wonder in his readers. Ball's scientific naturalism was not generally noticed. The *British Quarterly Review* announced, "this is certainly one of the most masterly and successful attempts to popularize science which has been made in our day." The reviewer praised Ball's ability to be exhaustively systematic and yet accessible to a popular audience. Moreover, the tone was "fitted to excite wonder, reverence, gratitude, and admiration."[162] The *Athenaeum* applauded the book's "combination of scientific accuracy with lucid arrangement and attractive style," and the reviewer anticipated that it would be a "brilliant success."[163] In the *Academy*, the reviewer predicted that a "certain class of critics" would probably sneer at Ball's book and label it "sensational." Yet as far as the reviewer was concerned, insofar as Ball had presented the facts in such a way "as to excite astonishment on the part of the reader," his sensationalism was "just what every popular scientific book ought to be." Scientific facts should "awaken wonder when first presented." The reviewer "unreservedly recommended" Ball's book to those who wanted to "obtain trustworthy information, conveyed in an attractive manner, respecting the most striking phenomena of the heavens."[164] The scientific periodicals were not so enthusiastic. *Nature* commended Ball for providing "the fullest and most complete exposition of the leading facts and principles of astronomy which has yet been laid before the entirely unscientific public" and for devoting special attention to the most recent astronomical discoveries. But there was some criticism of Ball's occasional tendency to be too condescending toward his

161. Ibid., 82–83, 96, 120, 192, 251, 257-58, 265, 289–90, 318, 329.

162. *"Story of the Heavens,"* British Quarterly Review 1886, 202-3.

163. "Science. Astronomical Literature. *The Story of the Heavens*" 1885, 703.

164. "Some Astronomical Books" 1886, 191.

audience, addressing them almost as if they were children.[165] The reviewer in *Knowledge* went even further in condemning the level of discussion in Ball's book, complaining that the inexperienced student would not realize that they were reading page after page of "nothings."[166]

Ball's *In Starry Realms*, which appeared seven years after *The Story of the Heavens*, was bolder in presenting a vision of evolutionary naturalism. Published by Isbister and Company, and on sale for seven shillings and sixpence, this book went through several editions, the last one appearing in 1912. It was composed of articles from *Good Words*, *Girl's Own Paper*, *Macmillan's Magazine*, *Contemporary Review*, *Longman's Magazine*, and various newspapers. Among the chapters, the most notable is "Darwinism and Its Relation to Other Branches of Science." Ball recalled reading the *Origin of Species* as a student in college when it first appeared. He claimed that he was an "instantaneous convert to the new doctrines, and I have felt their influence during all my subsequent life." Ball did not deal directly with the religious dimension of Darwin's thought. He was more interested in Darwin's impact on astronomy. Ball stated that most astronomers now agreed that the history of the solar system was evolutionary as described by the nebular hypothesis. "Astronomers were thus the first evolutionists," he exclaimed, in that "they had sketched out a majestic scheme of evolution for the whole solar system and now they are rejoiced to find that the great doctrine of Evolution has received an extension to the whole domain of organic life by the splendid genius of Darwin." The astronomer traced in outline the manufacture of the earth from the primeval nebula, and then he handed over the rest of the earth's history, with its organic evolution, to the Darwinian biologist. Ball referred to Darwin as the "Newton of natural history," whose "immortal work has revolutionized knowledge."[167] The loyal assistant of Lord Rosse, whose Leviathan was aimed at the heavens in order to undermine the nebular hypothesis, had evolved into an ardent supporter of Laplace's theory. Ball may have expressed his desire in his *In Starry Realms* to provide his readers with "a fresh appreciation of the sublimity of that scheme of creation in which our earth plays a small though dignified part."[168] But creation was merely sublime, not divine.

165. "Ball's 'Story of the Heavens'" 1885, 124.

166. "Story of the Heavens" *Knowledge*, 1886, 98.

167. R. Ball 1892, 344, 349-50, 363-64.

168. Ibid., 140.

The chapter on Darwinism elicited the wrath of at least one Catholic critic. An angry reviewer in the *Irish Ecclesiastical Record* insisted that it was his duty to demonstrate the unreliability of Ball's views on evolution, particularly since his books were so popular. "As a writer of science for the million," the reviewer declared, "he has few rivals; and it may be safely said that he has taught the general public more astronomy than any other man who has written on the subject." Ball's critic wanted to prevent his Darwinism from profiting "by the popularity of his astronomy." He pointed out that it was not generally known that the controversial chapter on Darwinism had been published much earlier in a periodical.[169] It did not attract much attention at the time, and if it had been "left at rest" in the pages of that periodical "it might have been spared adverse criticism." Now it had been brought forth from its "comparative obscurity, and given anew to the public as the closing chapter of a book called *In Starry Realms*." Since Ball's fame had increased since the article had originally appeared fourteen years earlier, and since "any opinion from one so renowned and accomplished a teacher must necessarily make a deep impression on the public mind," Ball was going to receive the verbal abuse that he so richly deserved. The author of this review then presented a close analysis of Ball's handling of the nebular hypothesis, biological evolution, and the origin of life in his chapter on Darwinism. His aim was to show that Ball did not speak with authority when he left his field of expertise. "However eminent as an astronomer," the reviewer insisted, "as a biologist Sir R. Ball is, we are sorry to have to say, absolutely unreliable."[170]

Ball later wrote an entire book on the nebular hypothesis in which he tried to clarify where Rosse fit into the history of the study of nebulae. *The Earth's Beginning* (1901), published by Cassell and Company in Britain and priced at seven shillings and sixpence, was also published by Appleton in the United States. The last British edition appeared in 1909. Ball began by drawing the reader's attention to the "sublime magnificence" of his subject. All human affairs, past, present, and to come, "shrink into utter insignificance when we are to consider the majestic subject of the evolution of that solar system of which our earth forms a part." It was telling, Ball pointed out, that three great men of science, Kant, Lamarck, and William Herschel, were all led to the same conclusion. Although no human eye had witnessed the stages "in the grand evolution of our solar system," there was evidence as to its nature in the study of the parallel stages of currently existing nebulae. Whereas

---

169. It had originally appeared in 1883 in *Longman's Review* and had been delivered as a lecture at the Midland Institute in Birmingham on November 20, 1882 (R. Ball 1883a, 76–92).

170. Gaynor 1897, 243–44.

Ball had earlier supported Rosse in his controversy with Huggins, now he acknowledged that Huggins had removed the first great objection to the nebular theory by establishing the existence of gaseous nebulae through the use of the spectroscope. Ball presented photographic proof, an illustration of the Great Spiral Nebula produced by the Lick Observatory (Fig. 7.9). "If an artist thoroughly versed in the great facts of astronomy had been commissioned to represent the nebular origin of our system as perfectly as a highly cultivated yet disciplined imagination would permit," Ball maintained, "I do not think he could have designed anything which could answer the purpose more perfectly than does that picture which is now before us." As Ball saw it, the picture exhibited a system "in a state of formation," in actual evolution. In addition to presenting evidence based on the study of nebulae currently in the heavens, Ball offered other forms of proof. There were the remarkable coincidences observed in the movements of the bodies of the solar system. The nebular origin of the system explained the extraordinary concordance in the positions of the planes of the orbits of the planets and the agreement in the directions of the planets' rotation.[171]

Although Ball presented himself as an enthusiastic supporter of the nebular hypothesis, he also attempted to give Lord Rosse credit for his contributions to astronomy. Rosse's announcement of the spiral construction of the Great Spiral Nebula, believed by some at the time to exist only in his imagination, had been vindicated by recent astronomical photography. The Lick Observatory had announced, on the basis of the "unchallenged evidence of the photographic plate itself," that there were at least "60,000 spiral nebulae in the heavens." In Ball's hands, Rosse emerged as part of a pioneering tradition in astronomy that had established the existence of spiral nebulae. Since the spiral nebula was now "of extraordinary importance as a celestial object," Ball argued that Rosse's work was extremely valuable, no matter what his views were on the resolvability of nebulae or on the implications of their nature for religious orthodoxy.[172]

Ball wrote an entire book lauding the achievements of astronomers, including Rosse. Ball's *Great Astronomers* (1895, Isbister) developed at greater length the theme of astronomical heroes used in *The Story of the Heavens*.[173] Ball's lineup of heroic astronomers included eighteen figures from Ptolemy

171. R. Ball 1909, 1, 5, 50, 74, 194, 308, 332.

172. Ibid., 197-98, 199-200.

173. *Great Astronomers* was in print up to at least 1907. An American edition by Lippincott was published in 1895. Another Isbister edition appeared in 1901. G. Philip and Son produced a cheap edition ca. 1906. Sir Isaac Pitman and Sons, Ltd., published editions in 1906 and 1907.

Fig. 28.—THE GREAT SPIRAL NEBULA (Lick Observatory).
(*From the Royal Astronomical Series.*)

FIGURE 7.9 Ball maintained that this photograph of the Great Spiral Nebula was a "marvellous natural illustration" of the views of Kant, Laplace, and Herschel (R. Ball 1909, 194). (Robert Ball, *The Earth's Beginning* [London, New York, Toronto, and Melbourne: Cassell and Company, 1909], figure 28.)

to Adams. In his introduction he presented the book as an account of the evolution or growth of astronomy. "From the days of Hipparchus down to the present hour," he wrote, "the science of astronomy has steadily grown." The history of astronomy involved the interplay between observation and theory, as "one great observer after another has appeared from time to time, to reveal some new phenomenon," followed by "one commanding intellect after another" that explained the "true import of the facts of observations." Working together, the stargazers (such as William Herschel and John Herschel), who watched the heavens, and the mathematical astronomers (such as Newton and Le Verrier), who worked at their desks, slowly comprehended the secrets of this vast universe. Ball also celebrated the accomplishments of the instrument makers, such as the Earl of Rosse. The astronomical heroes were at the center of the story. He declared, "the history of astronomy thus becomes inseparable from the history of the great men to whose labours its development is due."[174]

Ball's emphasis was on British astronomical heroes, particularly those of the modern period, and his Irish origins led him to include William Rowan Hamilton, Rosse, and John Brinkley. He had tremendous respect for Le Verrier, though he insisted that the discovery of Neptune had to be shared with the British astronomer J. C. Adams. Ball again paid homage to Le Verrier, as he had in *The Story of the Heavens*. To Ball, Le Verrier embodied all of the qualities of the first-rate mathematical astronomer. The task of the mathematical astronomer was to elicit from the observations of the practical astronomer "the true laws which govern the movements of the heavenly bodies." This was a task "in which the highest powers of the human intellect may be worthily employed." Ball himself had a mathematical bent, and he tended to favor the mathematical over the observational astronomer. Le Verrier held a "highly honoured place" among the mathematical astronomers, as he provided a "superb illustration of the success with which the mind of man can penetrate the deep things of Nature."[175] Sitting at his desk, Le Verrier had studied the observed disturbances of the movements of Uranus, determined the existence of another planet, and even calculated its position before a telescope was used to confirm the discovery. This episode, Ball remarked, will be "celebrated so long as science shall endure."[176] This was a triumphant confirmation of the powers of the human intellect and of the importance of the astronomical hero. Portraying a single scientist as a hero was not an innovation. But presenting a series of heroic scientists to illustrate the making of a discipline was unusual. Edward Clodd's *Pioneers of Evolution* appeared two years later. Clodd's innovation was to emphasize the role of scientific heroes in the making of the evolutionary epic, rather than in the construction of a specific discipline. But for Clodd, and other popularizers who would follow, Ball had established a secular narrative for discussing the evolution of science based on the heroic figure of the brilliant scientist.

ESTABLISHING A ROLE FOR PRACTITIONERS

Both Huxley and Ball became increasingly involved in the popularization of science during their careers, after they had obtained their credentials as

---

174. R. Ball 1895, 5.
175. Ibid., 336.
176. Ibid., 346.

reputable practitioners. By becoming deeply involved in the popularization of science, Huxley and Ball attempted to convince their colleagues that it was important for practitioners to take the lead in communicating the results of science to the public. Huxley retained his belief, expressed in the 1850s when he played the scientific critic, that only practitioners of science possessed the knowledge to write good books for the reading public. Since the time that he had attacked the *Vestiges* and Lewes in the early 1850s, a whole new cadre of popularizers had been writing books aimed at the public. In their emphasis on a discourse of design, the Woods of the world presented an agenda at odds with Huxley's and Ball's scientific naturalism. The public needed to be educated so that they could determine who spoke with authority on behalf of science. In his debate with the Duke of Argyll in the pages of the *Nineteenth Century* in 1887, Huxley claimed that he was exposing a process of mystification "based upon the use of scientific language by writers who exhibit no sign of scientific training, of accurate scientific knowledge, or of clear ideas respecting the philosophy of science, which is doing very serious harm to the public. Naturally enough, they take the lion's skin of scientific phraseology for evidence that the voice which issues from beneath it is the voice of science, and I desire to relieve them from the consequences of their error."[177] Huxley's point applied as much to Wood and his ilk as it did to the Duke of Argyll. The public needed to understand that only practitioners had the authority to speak for science. Huxley and his allies may have consolidated their control over the institutions of science by the end of the 1860s, but Huxley believed they now needed to mount a campaign to establish their power in the public sphere. Popularizers drawn from the ranks of the Anglican clergy or women who were perpetuating theologies of nature stood in their way, as did publishers who produced their books.

Huxley and Ball had to fight against the stigma that some contemporaries attached to popularization activities. In his reflections on whether or not to take up the Lowndean Chair, Ball listed a series of drawbacks if he moved to Cambridge. The second on the list was "the fact that there is opposition to me as not being a Cambridge man and as a 'Popular Lecturer.'"[178] Huxley's close friend, Tyndall, was attacked by the physicist P. G. Tait in a letter to the editor of *Nature* in 1873. Tait cautioned readers against accepting "as correct

177. T. Huxley 1894b, 117.

178. Cambridge County Record Office, Memoranda of Merits and Disadvantages of Accepting the Lowndean Chair, 1892, R83/61.

great parts of what he has written." According to Tait, only a few men could become involved in science popularization "without thereby losing their claim to scientific authority." Tyndall, he believed, was not one of them. "Dr. Tyndall has," Tait insisted, "in fact, martyred his scientific authority by deservedly winning distinction in the popular field." Tyndall was unable to "'make the best of both worlds.'"[179]

By the early twentieth century, the situation had changed. The scientific community had expanded and become more thoroughly professionalized. Eminent scientists could retain the respect of their peers when they took on nonspecialist writing as long as they had made a substantial contribution to research at the same time. Practitioners such as J. Arthur Thomson, J. B. S. Haldane, James Jeans, and Arthur Eddington could engage in popularizing activities without facing the same degree of disapproval from their colleagues as had Huxley and Ball. As Peter Bowler has asserted, in contrast to the late Victorian era, a large proportion of the nonspecialist writing on science in the early twentieth century was done by working scientists in their spare time.[180] Today, there are many eminent practitioners who devote a great deal of time to writing for a popular audience. It could be said that they dominate the field.[181] The names of Brian Greene, E. O. Wilson, Richard Lewontin, Richard Dawkins, Stephen Hawking, and the late Stephen J. Gould and Carl Sagan, are familiar to the contemporary reading public. These men all belong to the modern tradition of the practitioner-popularizer, a tradition that Huxley and Ball played a key role in establishing.

In the latter part of the nineteenth century, Huxley and Ball encountered a number of obstacles in their efforts to establish a role for practitioners in the popularization of science in addition to the negative attitude of contemporaries toward this kind of activity. First, there were the personal factors. For Huxley, these include his early reluctance to adopt the role of popularizer, his health problems, and his inability to complete his books in a timely fashion. For Ball, it was his wife's opposition to his lecturing

179. Tait 1873, 382.

180. Bowler 2005, 232–34.

181. Despite the active participation of so many eminent practitioners in contemporary popularizing projects, writing for nonscientists is still accorded a lowly status, at least according to Gould. He objected to the deprecation of science writing for a general audience, stating, "I deeply deplore the equation of popular writing with pap and distortion." Gould called for a recovery of "accessible science as an honorable intellectual tradition," and traces this tradition back to Galileo (S. Gould 1991, 11–12).

tours.[182] Second, there were the inevitable problems in dealing with publishers, though Huxley seems to have had more difficulties than Ball. Huxley, for example, had to deal with publishers like Kegan Paul, who did not necessarily share his goals and who frustrated his attempts to exercise control over the projects he was involved in. When Paul published the works of individuals like Henslow after Huxley left, it could only have undermined the basic thrust of the series that Huxley had worked to establish.

Third, and finally, the success of popularizers led Huxley and Ball to mimic their approach. Because of Wood, Brewer, Page, Buckley, and others, Victorian readers expected a book or a lecture designed for a general audience to make science accessible through a focus on a common object or by telling the story of the evolutionary epic. Ball adopted the evolutionary epic pioneered in its modern form by Chambers, and later developed by Page, Buckley, Clodd, Allen, and Proctor. His naturalistic interpretation exposed him to criticism from religious readers and posed awkward questions about his relationship to his first patron, Lord Rosse. The success of those like Wood who reached their audience through an emphasis on common objects, forced Huxley to start in the world of natural history. But such a strategy was risky for Huxley, since his aim was to undermine the authority of Wood and other popularizers who were not practitioners. Moreover, he wanted to replace natural history with the study of biology, as this was more conducive to his vision of a professionalized and secularized science. Starting with common objects, like other popularizers, but offering a more systematic analysis of them in order to lead readers away from a theology of nature, required a readership to have the ability to discern the subtle distinction between Huxley's approach and that of Wood's. But if Huxley's audience had been reading him properly, their enthusiasm for theologies of nature should have abated, and they should have recognized that popularizers like Wood did not speak with the voice of science. Yet, during the 1880s, British audiences continued to buy and read the works of popularizers with no scientific training.

The transmission of knowledge in Britain was fundamentally altered by the communications revolution during the middle of the century. This revo-

182. When Ball was about to leave Dunsink, he told his wife that he had had enough of lecturing and that he had done it mainly for financial reasons. Since Ball was receiving a higher salary at Cambridge she expected that he would cease lecturing. But she complained in her diary after Ball had taken on the Cambridge professorship that he continued to lecture (October 15, 1893, Lady Ball's diaries, County Record Office, Shire Hall, Cambridge).

lution led to the growth of new audiences for science, and it led publishers, who were motivated in part by commercial considerations, to seek out popularizers who could communicate effectively with these audiences. Practitioners of science were not always their first choice as writers. Roy MacLeod has argued that scientific publishing was decisive in establishing the hegemony of scientific naturalism, "and in supplying its mediations with the reading public."[183] Although scientific naturalists may have been successful in gaining control over key scientific institutions, Huxley and his allies had only limited success in gaining control of the market for books designed to circulate in the public sphere, and in their efforts to drive the popularizers working in the clerical and maternal traditions out of the field.

183. MacLeod 1980, 64.

—— * ——

CHAPTER EIGHT

Science Writing on New Grub Street

"'POPULAR SCIENCE,'" H. G. Wells once declared, "it is to be feared, is a phrase that conveys a certain flavour of contempt to many a scientific worker." Wells made this observation in a provocative article on "Popularising Science" for the journal *Nature* in 1894, ten years after he had been a student of Huxley's at the Normal School of Science, and only a year before the publication of his scientific romance *The Time Machine*. He acknowledged that "this contempt is not altogether undeserved," since "a considerable proportion of the science of our magazines, school text-books, and books for the general reader, is the mere obvious tinctured by inaccurate compilation." In his opinion, this did not justify a "sweeping condemnation" of all "popular science." In fact, Wells insisted, good "popular science" was needed to educate the public, whose interest in science was vitally important to its continued success "in an age when the endowment of research is rapidly passing out of the hands of private or quasi-private organizations into those of the State." Wells therefore advised "even the youngest and most promising of specialists" to be reconciled "to the serious consideration of popular science," as had Darwin and the scientific leaders of his generation, who "addressed themselves in many cases to the general reader, rather than to their colleagues." But Wells was critical of practitioners who courageously braved the contempt of their colleagues for "popular science" by writing books for the general public, though he refrained from singling out any individuals by name. Those works "ostensibly addressed to the public by distinguished investigators, succeed in no notable degree, or fail to meet with appreciation altogether."[1]

Wells then offered an analysis of "the defects of very much of what is proffered to the public as scientific literature," and the main target seems to have been books written by practitioners. He admitted that he might appear to be "impertinent" in his criticisms, but he spoke as both a "mere general reader" and as a "reviewer" who had "given some special attention to the matter." One group of practitioners ignored the general principle that "a book should be written in the language of its readers" and wrote in the "dialect of their science." They also made the error of dealing only with the technical side of science, dwelling on morphological, mathematical, and classification issues. Another group went to the opposite extreme and underestimated their audience. They avoided scientific terminology and technical issues and ended up with vague, ambiguous, or misleading language. Rather than challenging their readers with the philosophical dimensions of science, they injected their books with too much humor. "I may testify," Wells declared, "that to the common man who opens a book or attends a lecture, this clowning is either very irritating or very depressing." Wells believed that scientific exponents had to be "absolutely serious in their style." He also condemned those in this group who merely wandered around their subject. He argued that "intelligent common people" did not come to scientific books for "vulgar wonders of the 'millions and millions and millions type'"; rather, they wanted books with "an orderly progression and development." Satirizing the tendency to write books with no discernible structure, Wells gave as an example an author who selected the title "Badgers and Bats." "It is alliterative, and an unhappy public is supposed to be singularly amenable to alliteration," Wells observed. The author begins with Badger A. "'We come now,' he says, 'to Badger B'; then 'another interesting species is Badger C'; paragraphs on Badger D follow." The segue to the next topic, Wells implies, is crude, and reveals the lack of structure in the book. "'Let us now turn to the Bats,' he says. It would not matter a bit if you cut any section of his book or paper out, or shuffled the sections."[2] Of course, though Wells did not make this point, many of his criticisms of this group of writers could have applied equally to nonpractitioners.

Wells concluded by expressing his fear that "awful examples" of the "populariser of science" seem "to be increasing." He called for the development of a "critical literature dealing generally with the literary merits of popular scientific books" and for the establishment of "canons for such criticism." In

1. H. Wells 1894, 300.   2. Ibid., 300.

his opinion, it was "a matter that is worthy of more attention from scientific men."[3] Wells was still concerned about the inability of practitioners to write proper books for the public forty years later. In his autobiography he stated, "it is not always the professors, experts and researchers in a field of human interest who are the best and most trustworthy teachers of that subject to the common man." This point was "excessively ignored by men of science" who "do not realize their specialized limitations."[4] Wells agreed with W. T. Stead that many scientific experts did not have the skill to communicate with the reading public.

Wells voiced his disapproval of the current state of science for the public during a time when the world of publishing and journalism was undergoing, as Nigel Cross puts it, "a radical transformation." In his survey of patterns in British publishing, Simon Eliot echoes this point, referring to the last two decades of the nineteenth century as marking "a new beginning." During this period the combination of a series of innovations significantly changed the size and nature of British print-culture, including the development of the new journalism in newspapers and magazines, the expansion of syndication, the establishment of the Society of Authors, the increase in new public libraries, the emergence of the literary agent, the adoption of a more permissive attitude toward the themes dealt with in fiction, the collapse in 1894 of the three-decker novel, the development of cheap, paper-covered reprints, the huge expansion in printing capacity as a result of the use of large rotaries, and the advent in the last decade of the hot metal typesetting machines such as the Linotype and Monotype. As Cross notes, all of these developments led to the climate of change and controversy pervading Gissing's *New Grub Street*, which depicted both struggling writers, who could not adjust to the new conditions, and successful authors who figured out how to make a good living in changed circumstances. In reference to the latter group, Eliot has asserted that the concept of the "man of letters" as a workaday professional, comparable in training, status, and income with a lawyer or a doctor, was the creation of this period. Promoted by the Society of Authors, this image was born of a time when the demand for light and entertaining works for syndication and serialization seemed to be almost endless.[5] Scientists, then, were not the only ones who could speak of, or who desired, the professionalization of their discipline.

---

3. Ibid., 301.
4. H. Wells 1934, 227–28.

5. Cross 1985, 204, 209; Eliot 1994, 13–14.

A number of prominent popularizers were beneficiaries of the changed environment, unlike Grant Allen, who was unable, in the long run, to focus his activity primarily on writing about scientific topics. Some embodied traditions of the past, such as Agnes Giberne, Eliza Brightwen, and Henry Hutchinson. In this group we can discern the continuing power of the clerical and maternal traditions right up until the end of the century. In the decades following the publication of his introductory primer for the Macmillan series, and long after Huxley had relinquished control of the International Scientific Serics, the clerical and maternal traditions, with their emphasis on a theology of nature, continued to flourish. A second group, represented by Agnes Clerke and Alice Bodington, offered a justification for popularization that harkened back to Mary Somerville by drawing on the need for a synthetic approach to scientific disciplines in an era of increasing specialization. Prolific and dynamic, these popularizers of science contributed to the outpouring of publications that marked that last two decades of the century.

CONTINUITIES: GIBERNE, BRIGHTWEN, AND THE MATERNAL TRADITION

Although practitioners became more active as popularizers during the 1870s, representatives of older traditions continued to write for a popular audience. Prolific and successful, Agnes Giberne and Eliza Brightwen were important popularizers at the end of the century who can be placed within the revised maternal tradition. Their success with the late-Victorian reading audience is testimony to the continuing power of religious themes in science. During the span of her forty-year career as a popularizer, Agnes Giberne (1845-1939) wrote a series of books for the public that dealt with the larger meaning of contemporary science (Fig. 8.1). Educated at home by her parents and by tutors, her interest in science was inspired by her father, Maj. Charles Giberne of the Indian Army, who was descended from an old and noble French family with Huguenot blood. Her mother, Lydia Mary, was the daughter of William Wilson, vicar of Walthamstow. Giberne was a devout Anglican for her entire life. Born in India, Agnes returned to England with her family when her father retired from the army. In her *The Wonderful Universe* (1896), she acknowledged that it was her father "who first awoke my interest in such subjects; he who made Astronomy a living force in my imagination." Giberne believed that the teaching began very early, as she vividly recalled standing by his side one winter day when she was seven or

AGNES GIBERNE.
Photo by Russell & Co

FIGURE 8.1 Agnes Giberne, writer and popularizer of astronomy. (*Girl's Own Paper* 20, no. 1,000 [February 25, 1899]: between 344 and 345. Shelfmark P.P.5993.w.(1.). By Permission of the British Library.)

eight asking how it could be colder when they were nearer to the sun at this time of year. Giberne asserted that these early lessons in science familiarized her "as a child with scientific modes of thought and expression; laying a firm foundation, upon which a superstructure of further study could so easily be reared."[6]

Giberne also developed a love of inventing stories as a child, which grew into a passion for writing. At the age of seventeen her first story was published by the Society for Promoting Christian Knowledge (SPCK). It was followed by more religious tales for children that appeared in books or as

6. Giberne 1920, 13.

serials in magazines such as *Fireside, Sunday at Home, SPCK Magazine,* and *Treasury.* Giberne later wrote fiction for adolescent girls, as well as historical fiction. Beginning in the 1880s she turned her attention to writing books on scientific subjects, many of them for beginners and children. Working at her home in Eastbourne, she wrote more than ten books that ranged across an impressive number of scientific disciplines, from astronomy to meteorology, oceanography, geology, and even botany.[7] According to the *Times* obituary, she was best known for her astronomy books for beginners. The *Times* attributed the "wide appeal of her books on science" to the "grace and clearness of her writing," and referred to her as a "pioneer" in "providing easy and popularly written books on scientific subjects."[8] Giberne established a close working relationship with several presses. Seeley and Company, with whom she published six books, priced them at five shillings, with the exception of *Starry Skies* (1894), which sold for two shillings and sixpence. The SPCK published her *Wonderful Universe,* charging one shilling for it, and much later, *The Garden of Earth* (1921), which cost six shillings and sixpence. The only science book of Giberne's published by Pearson, *The Mighty Deep* (1901), went on sale for five shillings.

Yet she found herself in desperate financial straits by 1905, when she applied to the Royal Literary Fund for assistance. Now aged sixty, she was single (she never married), and had, according to one supporter, given up the best years of her life to providing for her ailing father. The cause of her distress was failing eyesight, due to cataracts in both eyes, and a weak heart. The loss of her eyesight, as another supporter pointed out, "is a dreadful prospect for one who writes to live." The application listed her main sources of income as an annuity, the royalties from her books, and the £100 a year she received from a pension due to being an officer's daughter. Although then Prime Minister Arthur Balfour had looked at the case and decided that it was not strong enough to grant a pension from the Civil List, as he did not consider Giberne's work to be of "the highest class," he saw justification for providing a grant from the Royal Bounty Fund. According to M. G. Ramsay of H.M. Treasury, Balfour judged her books to be "eminently sound and wholesome," and "of service in spreading a knowledge of elementary science among a wide class, not exclusively among children." Giberne was awarded £200 from the Royal Literary Fund, and £273 from the Royal Bounty Fund, to be put toward the purchase of a Post Office annuity. But Giberne applied again to the Royal Literary Fund in 1917, as her books were not bringing in

---

7. Lightman 2004c, 777-78.   8. "Miss Agnes Giberne" 1939, 12.

as much in royalties as they had previously. Her nominal income of £170 per annum was insufficient because of the rising cost of living, and she had been forced to move into smaller rooms and to sell some of her furniture and all of her silver. This time she received a grant of £50.[9]

Giberne wrote books for a variety of audiences, including children, beginners, and more advanced adults. In her children's books, she adopted a storytelling format. *The Starry Skies*, for example, opened with the traditional fairy tale phrase. She wrote, "Once upon a time—thus runs a certain tale—there was a man who wanted to see what could be found at the other end of the world." The man kept walking for weeks, months, and even years in a straight line. What do you think he found, Giberne asked her youthful audience? Was it a world of giants or a land of fairies? The man, of course, finds himself in the same spot from which he had started, and learns that the earth is not flat. But Giberne also intended this discovery, and the others to be found in the book, to have a magical quality, like Arabella Buckley in her *Fairy-Land of Science*. Giberne offered lucid and simple explanations of gravity, the seasons, the rotation of the earth, the moon, the sun, the planets of the solar system, comets, meteors, stars, and nebulae. In her chapter on the solar system, for example, she tried to convey a sense of the proportional sizes of the planets by letting a crab apple stand for Mercury, large apples for Venus and the Earth, a small apple for Mars, and large globes for Jupiter and Saturn.[10]

Among the Stars (1885, Seeley), another book for children, was presented in a fully fictional form. This book sold over 2,000 copies within a year, and was also published in several editions in the United States. Here Giberne told the story of a young boy named Ikon, an only child whose mother had died. Lonely, he takes pleasure in the companionship of the stars and is eager to learn all he can about them. A German professor, Herr Lehrer, informally teaches him about astronomy, taking him to an observatory that belongs to his friend Fritz, and showing Ikon the stars through his own telescope. When Lehrer must leave to go on a trip, he promises to write. At this point, Giberne adopts the familiar format, as the two exchange letters. But the letters come quickly to an end when Ikon falls ill and has a dream that the professor has come to take him on a long journey to the moon, which they almost reach using their wings. After Ikon awakens, he is visited by Fritz, who completes Ikon's dream by using the "wings of imagination." Guided

9. British Library, Royal Literary Fund, File no. 2702, documents 3, 5, 7, 13, 16, 18–21.

10. Giberne 1894, 1, 190.

by a girl with wings, named Stella, Ikon visits the moon. This provides the pretext for conveying information about the moon's surface. Fritz returns several times to tell Ikon more stories about the heavenly bodies in the solar system. But imagination is constrained by the limits of scientific knowledge. On a trip to the sun Fritz explains sunspots and the corona, and prepares to end the journey. But Ikon protests, and asks why they cannot get any closer to the sun. Fritz replies that scientists know little about the sun. "You don't want a nonsense-story," Fritz declared, "and yet you are vexed because I don't attempt to describe more than I know."[11] Fritz's imaginative stories, and therefore Giberne's, are based on current scientific knowledge, and do not indulge in fruitless speculation. Although *Among the Stars* is a novel, the use of extensive dialogue and, in the one chapter, letters, is reminiscent of the familiar format of the early nineteenth century. Lehrer and Fritz are fatherlike figures, and, at the end of the book, the professor's daughter, also named Stella, guides Ikon on a journey to the stars.

Giberne stated that *Among the Stars* was intended to be a "little volume for children" that was "much easier" than her *Sun, Moon, and Stars* (1880, Seeley).[12] An anonymous reviewer in the *Review of Reviews* agreed with Giberne's assessment, and placed *Sun, Moon, and Stars* in the same category as Kingsley's *Madam How and Lady Why* and Ball's *Story of the Heavens,* as suitable for girls as they grew older.[13] The Rev. C. Pritchard, Savilian Professor of Astronomy at Oxford, wrote the preface for *Sun, Moon, and Stars* and praised it as one of the few astronomy books suited to beginners.[14] The book sold well, and by 1885 it was advertised as having reached its eleventh 1,000 copies printing.[15] At least two more editions appeared in Britain after 1885 and before 1900, so that by the end of the century at least 13,000 copies were printed. In her preface to the book, Giberne asserted that she had written it with two aims in mind. First, it was intended to be suitable for beginners of all kinds, "whether children, working-men, or even grown people of the educated classes." Her second goal was to help readers "look

11. Giberne 1885, 215,

12. Ibid., v.

13. "What Shall School Girls Read" 1892, 159.

14. Giberne n.d., v. Pritchard, a self-avowed supporter of women's entrance to higher education, also endorsed Giberne's *Ocean of Air* in a preface. Giberne needed Pritchard's endorsement to bolster her own scientific credibility (P. Gould 1998, 166).

15. Giberne 1885, 311.

upward through Nature unto Nature's God" by placing the "Book of Nature, side by side with the Book of Revelation."[16] Religious themes played an
important role in Giberne's books for children. In *The Starry Skies*, Giberne
referred frequently to God as the creator of the heavenly bodies and the
beautiful, as well as orderly, system that governed their movements.[17] Similarly, *Among the Stars*, Lehrer, and the other mentor figures, try to impress
Ikon with the "wonderful power of God" as demonstrated in His works.
There are frequent references to the Bible to help Ikon understand the
larger significance of the order in the heavens.[18]

But in *Sun, Moon, and Stars* religious themes were emphasized even more.
Each chapter began with a biblical quote. The immensity of heavenly objects
like the sun offered Giberne the opportunity to discuss divine omnipotence.
"It is impossible to learn much about the sun," she wrote, "without feeling
how mighty must be the power of our God, who could create such a vast
and dazzling globe of fire." Later, the size of the universe and the almost
unimaginable number of stars, revealed the "unutterable might of God's
power!" A section on how suns and worlds could have been built up by
tiny meteorites leads Giberne to declare that whatever the "order of God's
working...still He throughout was the Master-builder." Giberne also did
not hesitate to support the pluralist position. Like Proctor, Giberne believed
that God has created distant stars for a purpose. "Surely around many of
them, as around our sun, must journey worlds; and not only worlds, but
worlds containing life." She referred to Proctor's map of Mars, with its
continents and oceans, as support for the notion that the red planet could
support life. Having stressed God's power throughout the book, as the
creator of a vast universe, near the end she began to emphasize His infinite
love by reminding her readers of the historical fact of Christ's birth and
death on the Earth. "Let us look into the matter as we will—let us weigh
measure, calculate—let us find our earth to be but as a grain of fine dust, lost
amid myriads of worlds and suns," and yet, Giberne affirmed, "at the close

16. Giberne n.d., xi–xii.

17. When discussing the sky, Giberne asserts that "He made the Sky, the Sun and the Moon and
the Earth, the Planets and the Stars; and HE is everywhere, around and amidst in them all." In
her examination of the sun, Giberne reminds her readers that heat, light, and life on Earth comes
from the sun, and "one step farther brings us to the thought of OUR FATHER IN HEAVEN, who
created the Sun, and who appointed it to be our storehouse of Heat and Light." Giberne constantly
brings her youthful readers back to a recognition of God's impressive power as creator (Giberne
1894, 46, 99).

18. Giberne 1885, 96, 120–23, 237, 300.

of all, we stand face to face with the simple historical fact that the King of heaven, the Creator of the Universe, Himself lived as a Man for thirty-three years upon earth."[19]

Giberne's more demanding books for beginners also placed science within a religious framework. Her *Radiant Suns* (1895, Seeley), which discussed the revolutionary impact of the spectroscope and camera in astronomy, was no exception. But in the closing pages she tackled the controversial issue of evolution. Here she signaled her willingness to accept evolutionary theory and denied that it necessarily undermined revealed religion. "Divine Truth is strong enough to stand," she believed, "and our earnest desire should be to find what is actually true in Nature as well as to possess it in Revelation." To Giberne, the question was whether God created the stars instantaneously by the mere fiat of His word, or "did He create them by His Word no less, but through the processes of evolution and development?" If the latter explanation were found to be true, Giberne asserted, then it would not shake her faith, though some "private and pet notions on religious questions" would have to be given up. Although not proven as fact, the nebular hypothesis explained a great deal and was accepted by those who are scientifically best fitted to judge. Giberne found Isaac Roberts's photograph of the Andromeda Nebula, and William Huggins's interpretation of it, to be persuasive evidence of a stellar object in a state of cosmic evolution on a gigantic scale (Fig. 8.2).[20]

Giberne explored the religious significance of nature for her reading audience across a variety of scientific disciplines, not just in astronomy, whether she was dealing with meteorology, geology, or even oceanography. In her *Ocean of Air: Meteorology for Beginners* (1890, Seeley), which was published in the United States as well as Great Britain, Giberne analyzed the gases, water, forms of life, movement, disturbances, and forces within air. In this book Giberne frequently plays the "what if" game. "If water could be warmed and cooled with the ease and rapidity of other substances," she declared, "the climates of many parts of Earth would be in consequence so changed, that lands now more or less densely inhabited would become almost uninhabitable." If there were no heat, "the Ocean of Air would be a frozen mass of solid substance." If there were no light, "the dwellers in that Ocean would be plunged into the depths of perpetual midnight." The consequences of things being otherwise are unimaginable or too horrible to conceive of. To Giberne, this implied a design in the arrangement of things.

19. Giberne n.d., 22, 116, 125, 181, 230, 295.   20. Giberne 1895, 298–99, 310.

THE NEBULA IN ANDROMEDA.
FROM A PHOTOGRAPH BY DR. ISAAC ROBERTS, F R.A.S.

FIGURE 8.2 The photograph, Giberne wrote, has "a whirlpool-like aspect, suggesting a rotary motion; and around the larger central mass are distinct rings of shining matter, separated from the main body by dark rifts or spaces." Although it was not possible to "dogmatize" on the meaning of the photograph, Giberne acknowledged that it lent "no little support to the theory of development" (Giberne 1895, 299). (Agnes Giberne, *Radiant Suns: A Sequel to "Sun, Moon and Stars"* [London: Seeley and Company, 1895], opposite p. 298.)

She therefore asserted "one cannot look too earnestly into that world of Nature, which is the Handiwork of our Father in heaven, the outward expression of His Thoughts."[21]

When she discussed the religious significance of geology, Giberne adopted a different approach. Her popular *The World's Foundations* (1882, Seeley) reached a second edition in 1883, and by 1888 5,000 copies had been

21. Giberne 1890, 142, 270, 295.

printed. The last edition appeared in 1908. The book was divided into three parts. "How to Read the Record" taught the reader to learn the "ABC of the Rock-Alphabet." The second part, "A Story of Olden Days," takes the readers through the different geological epochs. Finally, part three on "The Past in the Light of Present" turns to the work of geological agents like water and fire. Although some considered geology to be a "dangerous subject," Giberne affirmed that it "speaks to us, as surely as the Bible itself speaks to us, of the Creator and His ways, albeit in terms more ambiguous, in language more easily misunderstood." The two could not contradict each other, as the Bible was God's word while geology described "His Handiwork." The reconciliation of geological science and biblical knowledge is a major theme in the book. Each chapter begins with a biblical quote. In part one, Giberne compares the earth's crust to a chopped-up book written in a strange language that was only just beginning to be understood. This allowed her to draw on the doctrine of the two books, the Book of Nature and the Book of God. Referred to by James Moore as the "Baconian compromise," naturalists and interpreters of the Bible had since the early seventeenth century agreed to a convention that preserved the peace. Naturalists could instruct interpreters of the Bible if they used their knowledge, gained through the study of things in their own right, to better understand divine power and wisdom.[22] Although this compromise had been disrupted in the middle of the century by evolutionary theory and biblical criticism, Giberne tried to rehabilitate it in light of recent geological knowledge. Giberne reminded her readers that God was the author of many books, including the Bible and the "great many-volumed Book of Nature, to which belongs the torn and battered Volume of Geology." She acknowledged that a full reconciliation between geology and the Bible was beyond "the utmost stretch of human intellect," yet nevertheless the "grand Truth of either record remains still unshaken." Giberne recommended that in cases in which reconciliation seemed impossible, as in the stories told by Genesis and in the Earth's crust, that "the one Record may and should be used as a help to the understanding of the other," and she believed that the reader should consider a metaphorical reading of the account of creation.[23]

Giberne maintained that a reading of the geological record revealed a powerful and wonderful God at work. "We shall find prospects of wonder and power," she declared, "of mystery and beauty, unfolding before our eyes.

22. Moore 1986.   23. Giberne 1888, iv, 9, 86, 91, 204.

THE MER DE GLACE.

FIGURE 8.3 Two human figures at the right are dwarfed by the majestic mountains, their strength insignificant compared to the power of ice, one of the geological agents that God used to shape the earth's crust over time. (Agnes Giberne, *The World's Foundations; or, Geology for Beginners* [London: Seeley and Company, 1888], frontispiece.)

Wrapped up in the crust of this earth are marvellous tokens of the goodness and greatness of God, and strange histories of olden days are written in her stories." Giberne's examination of the forces shaping the earth's crust over the eons is actually a study of how God slowly "moulded and fashioned this earth of ours—fire and water, sea and river, tiny polyps and rhizopods, one and carrying out His will" so that this world would become a suitable home for humanity. Giberne made it clear that when she discussed "powers of Nature," she meant "simply powers used by God in nature," as "Nature" was "but the handiwork of God, the Divine Architect." Just as Giberne emphasized in her astronomy books the immensity of the heavens to impress upon the reader God's omnipotence, in this work she stressed the might of natural forces as a reflection of divine power (Fig. 8.3). The story told by the earth's crust was also one of wonder, not just unimaginable power. The past was filled with strange creatures and eerie landscapes. "It is no

FOREST OF THE CARBONIFEROUS AGE.

FIGURE 8.4 A forest of the carboniferous period looks more like a scene from an alien planet. (Agnes Giberne, *World's Foundations; or, Geology for Beginners* [London: Seeley and Company, 1888], opposite p. 128.)

fairy-tale which I have to relate," Giberne wrote, "but the wildest fairy-tale of heroes and dragons, griffins and monsters, never surpassed the wilder reality of the earth's inhabitants in those days." To help the reader imagine what the earth looked like during the "Age of Coal," Giberne invited them on a ramble. "Come with me, then, in fancy," she requested, "leaving the present far behind us, back to those early ages, and stand with me upon some low ancient hill, which overlooks the flat and swampy lands of the American continent." Dominated by vast, strange forests, the American continent is nevertheless a scene of divine activity (Fig. 8.4). As Giberne and the reader stand together on the hill, she moves time forward in order to paint a panoramic picture of how generation after generation of forest trees live, wither, and die, then become submerged under water, eventually to be transformed by a caring God into coal for the future benefit of humanity. "How wonderful the tale of olden days told to us by these buried forests," Giberne exclaimed. Although she acknowledged that *The World's Foundations* was not "a religious work," she maintained, "no volume on the subject of Geology can be fairly and honestly written without frequent reference to the Divine Architect of the great Earth-crust Building."[24]

In her *The Mighty Deep*, Giberne drew a theology of nature out of the study of oceanography, aided by the discoveries of the *Challenger*, which had dispelled human ignorance about the vast depths of the seas. Here she began by discussing storms and hurricanes; the role of ice, icebergs, and glaciers; the river-ocean-rain cycle; the erosion of land by the sea; and geological information on movement within the earth's crust. One of her main themes in this section was the power of nature, aptly captured in the frontispiece, a photograph of a huge wave about to pound the shore. Once she reached the topic of chalk and how it formed from the shells of tiny sea creatures, she told her readers that they had passed the boundary line between inanimate and animate nature. The shells of these creatures, though so small, are delicate and elaborately constructed. There is a "different design" for each species and "each shows the carrying out of a definite and beautiful plan." Giberne declared, "such Design we must ascribe to a MIND lying beyond that which we see; not to the jelly-speck itself, which acts as an unconscious architect, working automatically." Turning to other miniscule forms of lower animals, Giberne also discussed the diatom plant. Too minute to be seen by the unaided human eye, they were nevertheless so "marvellous in their make" that Giberne pronounced, "perhaps in all the world no greater marvels are to be seen than these extraordinary minute vegetable cases" (Fig. 8.5). The following discussion of coral as formed from the dead remains of once-living jelly specks again emphasized the divine design behind them. Giberne then moved on to the crustaceans, the fishes, and the behemoths of the ocean. "When we reach Whales," she declared, "we leave lower animals behind us; we step upon the uppermost rungs of the Ladder of Life; we are in the society of Mammals." Having started with the inanimate world, stressing its raw power and the partial success of scientists in finding lawfulness when penetrating its mysteries, Giberne's subsequent account of animate nature puts design center stage as she moved from the most primitive to the most advanced animals. The final section of the book discussed the British command of the seas, which provided the basis of the "God-given" Empire put into the "hands of the Anglo-Saxon

24. Ibid., 5, 36, 87–88, 129, 134, 142, 203.

FIGURE 8.5 The complexity, loveliness, and wonderful minuteness of diatoms provided "unmistakable evidence of a MIND, hidden, out of sight, inventing and designing" (Giberne 1902, 156). (Agnes Giberne, *The Mighty Deep and What We Know of It*
[Philadelphia: J. B. Lippincott; London: C. Arthur Pearson, 1902], opposite p. 154.)

Race, to be used for God and for the good of Man, to be governed in the
Name and in the Spirit of Christ."[25]

Eliza Brightwen (1830–1906), another important female popularizer during this period, shared much in common with Giberne (Fig. 8.6). Both incorporated a theology of nature in their works and both can be placed with
those women from earlier in the century who revised the maternal tradition
in order to address a mass reading audience. Starting in 1890, and rather late

25. Giberne 1902, 142, 154–55, 253, 289. Giberne's conclusion presents the existence of the
Empire as a means for humanity to transcend the struggle for existence. Whereas in the lower
creation, under ocean waters, a "fierce perpetual struggle for existence goes on," humans are
capable of forgetting the self for the sake of their country. "He may lose sight of ease and gain,"
Giberne declared, "in thought for the poorer, the weaker, the darker tribes of Earth—under that
protecting Flag" (Ibid., 290). Written shortly after Queen Victoria's death, the book is dedicated
to her as the "Queen of the Ocean and Mother of Her People."

FIGURE 8.6 Eliza Brightwen, natural historian and animal biographer. (Eliza Brightwen, *Eliza Brightwen: The Life and Thoughts of a Naturalist*, edited by W. H. Chesson [London: T. Fisher Unwin, 1909], frontispiece.)

in life, Brightwen wrote seven books on natural history. Brightwen came to her fourteen-year career as scientific author in a rather circuitous fashion. She was adopted at the age of four by her father's brother, Alexander Elder, who was one of the founders of Smith, Elder and Company, the London publishing firm. The family moved to a ten-acre estate at Stamford Hill, Essex, shortly after Eliza's adoption, where she observed the animals on the grounds and had access to her uncle's impressive library of natural history books, fiction, and poetry. In her autobiography, Brightwen described her early life as being "extremely lonely and quiet." Other than spending a few unhappy months at a nearby boarding school when she was twelve, and the lessons she had in music and drawing, she had no formal education. Soon after her twelfth birthday, Brightwen began to experience powerful religious feelings, especially a "deep sense of sin, a great dread of death and judgment." She felt as though she were leading "the life of a heathen," since her uncle, a Presbyterian who disliked the English Church, scarcely ever took her to church and never led the family in prayer.[26]

In 1855 she married George Brightwen, at that time a senior clerk at a London bill-broker firm, but later co-manager of the London Discount Company. The marriage was a happy one, but she frequently suffered from ill health and bouts of nervous depression. Largely through her husband, she came into contact with a number of prominent scientific figures. She was related by marriage to an eminent popularizer when George's sister became Philip Henry Gosse's second wife in 1860. In 1856 and 1864 she stayed with William and Lady Hooker at Kew. During the summer of 1865 she collected flower specimens for George Henslow, who was lecturing weekly on botany at St. Bartholomew's Hospital Medical School. In 1871 she heard anatomist Richard Owen and the popularizer Frank Buckland lecture, and was introduced to the latter. Brightwen became increasingly religious after her marriage, and at some point, according to the *Times* obituary, became an evangelical Anglican. She claimed that in 1865 she experienced conversion after hearing a sermon at her parish church at Harrow Weald. She wrote in her diary that "my whole soul was stirred, the time had come, and the Holy Spirit brought home the Word with such power that I then and there accepted Christ as my Saviour, thanked and praised Him with tears of joy that he was willing to save even me who for thirty-four years had known no real peace on account of sin." Six years later she wrote *Practical Thoughts on Bible Study* (1871, Adams), where she asserted that a "careful study of the Word of God" was needed more than ever "in these days of error and unbelief."[27]

In 1872 George Brightwen bought The Grove, a magnificent estate, nearly 170 acres in size, near Stanmore on the Hertford-Middlesex boundary. But Eliza suffered a serious nervous breakdown at about the same time that lasted for ten years. According to her nephew, the author Edmund Gosse, her death appeared to be imminent on several occasions. However, when her husband experienced heart problems in 1882, she recovered sufficiently to care for him. After his death in 1883, though still occasionally plagued by depression and pain, she was well enough to pursue her revived interest in natural history. She began to observe animals, some of which she acquired as pets, and collected enough specimens to transform a billiard room into a

26. Creese 2004c, 1: 273–74; Brightwen 1909, 2, 12–13.

27. Creese 2004c, 1: 274; Brightwen 1909, 56, 68, 77–78, 93; Brightwen n.d., 1; "Obituary. Mrs. Brightwen" 1906, 6.

crowded museum. Based on the notes of her observations she wrote a book, offered it to the publisher Fisher Unwin, who accepted it for publication. It was a tremendous surprise to Brightwen that *Wild Nature Won by Kindness* (1890), which appeared when she was sixty years old, sold well and received positive reviews. Priced at three shillings and sixpence, it reached a fifth edition in 1893, and a seventh edition by 1909.[28] According to Edmund Gosse the success of this first natural history book "revolutionized" her life by focusing her interests, spurring her on to fresh investigations, and introducing her to new friends. In addition to writing six more books, she engaged in a number of activities stemming from her study of natural history. From 1890 she served as vice president of the Selbourne Society, a natural history group founded in 1885. She presented public lectures to local groups, many on the protection of animals. She was also a Fellow of the Entomological Society and of the Zoological Society.[29]

Brightwen was fortunate in hooking up with Thomas Fisher Unwin, with whom she worked on the majority of her books. He had started his own publishing business in London in 1882 and was part of a new generation of publishers who were willing to take on young writers at the start of their careers, despite the financial risks. Brightwen was not young, but she was unknown when she first approached Unwin. He lived up to his reputation for being generous to his authors. In one letter Brightwen thanked Unwin for a "lovely and most kind gift" he had sent her. "It is a touching proof of our friendly relations as publisher and author," she wrote to him. In the same letter she reflected on the joys of being an author and of working with Unwin in particular. "If authorship always brought as many pleasant things as I have received," she declared, "I think everybody would try and write but I think I have been specially favoured by a kind publisher and a generous public and am grateful accordingly." In 1891 Brightwen invited Unwin to visit her at her estate and to stay overnight. Her relations with Unwin seem to have been friendly right up until her death. In 1904 he wrote to her, asking if she was considering any new book-length projects, but she replied, "I fear there is no hope of any more books from me." She explained that she had had a year of "continual suffering" and was currently in bed "just slowly recovering from a time of agonizing pain which recurs from time to time." Two ladies who were caring for her did not want to look after her pets, and since her books were based on her observations of her pets, she had no new material

28. Gates 2004b, viii–ix.

29. Creese 2004c, 274–76; Brightwen 1909, xi, xiii, xv.

to draw on. "It is very kind of you to wish for another book," she wrote, "and would gladly send you one if I could but at 75 I suppose I must rest on such laurels as I have now and be content." Brightwen had been suffering from cancer and her health continued to deteriorate. On the Sunday before she died during the spring of 1906, Edmund Gosse, who had been a regular visitor throughout her lengthy illness, read to her from the Bible.[30]

Wild Nature Won by Kindness was composed of a series of short life histories of various animals—birds, water shrews, squirrels, moles, mice, snails, earwigs, spiders, butterflies that Brightwen had observed and studied on the grounds of her estate. She was by no means the only popularizer to work in the genre. Her book appeared at a time when the animal biography reached its zenith in the 1890s. She established a warm rapport with her readers by embedding her observations in a conversational tone. In her introduction she told her readers, "in the following chapters I shall try to have quiet talks with my readers and tell them in a simple way about the many pleasant friendships I have had with animals, birds, and insects." Most chapters focused on a specific animal. In "Richard the Second," for example, she recounted the story of her pet starling Richard. She discussed the mischief he created, his close brush with death when he joined a group of wild birds, and his ability to pronounce some words. He emerges as a personality and as a friend to the author over the course of their five-year friendship, which became "part of my home-life." Birdie, a Virginian nightingale, became her daily companion for fourteen years. The bird became so attached to Brightwen that he adopted her as "a kind of mate," constructed a nest for her, and tried to feed her flies. One chapter gave readers some tips on how to domesticate a wild animal. Her advice was simple. "The only way to enjoy friendships with full-grown birds is to tame them by good and kindness," she believed, "till such a tie of love is found that they will come into our houses and give us their sweet company willingly."[31]

If such a bond could be established, then a privileged vantage point for observing the animal was created and its genuine nature was revealed. "A pet creature can only show its true nature," she later declared in her *Inmates of My House and Garden* (1895, Unwin), "when it is brought up so kindly as to be without fear."[32] Brightwen's expertise in taming animals

---

30. Codell 1991, 304; Brightwen to Unwin, November 9, West Sussex Record Office, Letters f285, f370, and f372; Thwaite 1984, 423.

31. Gates 1998, 220; Brightwen 1890, 12, 42, 73, 81–83.

32. Brightwen 1895, 93.

allowed her to speak with authority on their habits, diet, and physiology. While scientific naturalists advocated the experimental model for knowing nature, with its emphasis on questioning nature so as to force it to reveal its secrets, Brightwen's experiential knowledge came through a personal encounter based on love. Brightwen's anthropomorphizing of animals, her treatment of them as individuals rather than as members of a species, and the fact that her work was done in her secluded country estate and not a laboratory, flew in the face of the scientific naturalists' conception of proper science. While Brightwen acknowledged that her "little volumes" lacked "scientific importance," she nevertheless maintained that, "so far as they go, [they are] original." Rather than borrow from "other and cleverer writers," she had "set down as plainly as I could what I have myself observed and experienced." By doing so, Brightwen claimed to have made a real, though perhaps modest, contribution to science. "Unpretending as are the chronicles of the inmates of my house and garden," she affirmed, "they are scrupulously true, and every fact that a veracious observer records is a contribution, however small, to our general sum of knowledge."[33]

Brightwen's loving relationship to nature not only led to scientific knowledge, it also led to moral enlightenment and knowledge of God's existence and wisdom. Like Giberne and the other popularizers of the second half of the century, her works presented a theology of nature rather than a demonstrative argument characteristic of natural theology. The cleanliness of shrews reminded her of "how many a lesson we may learn from the small as well as the great creations of God's hands." The story of her friendship with wild robins leads to a lesson on the nature of true companionship, which must be given "freely." Her animal biographies were also framed in the introduction and the concluding chapter by religious themes. In the introduction she expressed her hope that the book would "tend to lead the young to see how this beautiful world is full of wonders of every kind, full of evidences of the Great Creator's wisdom and skill in adapting each created thing to its special purpose, and from the whole realm of nature may be taught lessons in parables, and their hearts be led upward to God Himself." The concluding chapter on "How to Observe Nature" reiterated that nature was full of wonders and evidences of divine wisdom. Like Giberne, she insisted that God had given us two great books for instruction, the "book of nature" and "the written Word of God." Although it was acknowledged that

33. Ibid., 9–10.

the Scriptures revealed the will of God and "His wondrous works for the welfare of mankind," many failed to "give any time or thought to reading the book of nature!" Brightwen shared with Gatty the firm belief that nature is designed by God to teach us moral and religious lessons. "The whole realm of nature is meant, I believe," Brightwen announced, "to *speak to us,* to teach us lessons in parables—to lead our hearts upward to God who made us and fitted us also for our special place in creation." Gatty's fictional short stories *Parables in Nature* are no more didactic than Brightwen's "lessons in parables" in *Wild Nature Won by Kindness.*[34]

Encouraged by the success of *Wild Nature Won by Kindness,* Brightwen subsequently wrote a series of natural history books. Intended as a sequel to her first book, *More about Wild Nature* (1892, Unwin) reached a third edition by 1897. Here Brightwen tried to establish a close rapport with her audience by inviting her readers into her home for a visit (Fig. 8.7). The first section, titled "Indoor Pets," drew on the same format as *Wild Nature.* Brightwen offered animal biographies of Katie the shrew, two parrots named Polly and Ruby, a mischievous mongoose named Mungo, the short-lived Impey the bat, the high-spirited Joey the kestrel, Sylvia, the graceful wood mouse, and the jilted starling Pixie. In the second section of the book, titled "Inmates of the Grove," she moved out of the house and related her experiences with a variety of animals on her estate, including curious and impetuous cattle, foxes endangered by the hunt, squirrels, and hungry birds. She also reminisced about the donkey she had as a pet when she was child.

Brightwen departed from her recipe for success in the rest of the book and served notice to her readers that she would not limit herself to animal biography. In the next section of the book, "Recording Impressions," she outlined her philosophy on how to capture the richness of nature in the written word and compared it to painting. Her aim was to produce "word-paintings" or "word-pictures" that captured the "pleasure of such intangible things as sunsets, wide-stretching views, the thousand odours of wild flowers and scented leaves, the joyous songs of birds, the music of falling water, and the wondrous beauty of distant mountains or snow-clad peaks." By writing down what had been observed, the naturalist would become more aware of

---

34. Brightwen 1890, 17, 125, 175, 204–5. Katy Ring perceptively noticed the link between Gatty and Brightwen back in 1988. She pointed out that both refer to the lessons in parables, both concentrate on anthropomorphized animals, and both challenge the "ostensibly detached and neutral attitudes towards nature of writers like Tyndall and Huxley." Ring asserts that Brightwen felt alienated by the new professionals who threatened to destroy the revelatory experience of Nature (Ring 1988, 139–42).

THE GROVE.

FIGURE 8.7 Brightwen's house at The Grove. Including an image of The Grove near the beginning of the book was Brightwen's way of welcoming her readers into her home. (Eliza Brightwen, *More about Wild Nature,* 3rd ed. [London: T. Fisher Unwin, 1897], 2.)

"how much we have missed of the real beauty of the delicate things that are given for our enjoyment." Writing therefore strengthened the power of appreciation, invigorated the thought processes, and disciplined the senses. What followed were a series of poetic snapshots of familiar sites in nature, though Brightwen was describing specific locations on her estate, such as a forest glade in spring or a peaceful spot by a lake. The meditations on nature invariably led to moral revelations, just as had her animal biographies. They elicited thoughts of "the harmonious power of beneficent Nature," a feeling of gratitude for Nature's "silent work," and a desire to imitate her "tender loving charity" in the face of the "harsh and rugged things that meet us in our daily life." Nature became a source of personal happiness and contentment. In the final section of the book, "Home Recreations," Brightwen returned indoors. But here she offered suggestions on how the reader could bring the "treasures" of nature to others, especially those who were confined inside as a result of illness. Displaying nature became an exercise in charity and unselfishness. She discussed the pleasure she received from her own home museum, gave tips on how readers could set up their own, and delivered instructions on how to study birds and insects.[35]

A number of Brightwen's other books also avoided a focus on animal biographies. Her final book, *Quiet Hours with Nature* ([1904], Unwin), which was dedicated to Edmund Gosse, was composed largely of articles that had appeared in *The Girl's Own Paper*. Although the book contains some essays on birds and insects, the bulk of the volume was devoted to trees, "to their forms and their development." Trees were treated as distinct individuals, just like animals, and each one elicited a different reaction. Brightwen asked her readers to regard the essays on trees as "monograph-portraits of certain individuals, which possess, either from age or size, some special interest for me." Again, Brightwen invited the reader to visit her. The frontispiece, depicting a "View at the Grove," enticed readers to visit the beautiful trees that grew on her estate. The "noble specimen of the Scotch fir" that stood on a rising slope of her lawn called to mind "a series of charming compositions" that had given her pleasure for over thirty years. A specimen of the Wellingtonia stood on her lawn and challenged her to guess its age. The unfolding of the buds of a tulip tree in the spring showed "creative wisdom and design for the protection of the fragile leaves in their early stage."[36]

In her *Glimpses into Plant Life* (1897, Unwin), Brightwen concentrated once more on botany rather than zoology. Intended as an introduction to the elementary facts of botany for the young, this book was unlike any of Brightwen's other works in many respects. It had a more systematic structure, offering chapters on adaptation, roots, tree stems, leaves, buds, flowers, pollination, fertilization, fruit, dispersion of seeds, germination, plant physiology, insectivorous plants, and plant growth. Brightwen's estate did not provide the site for discussion or supply the bulk of the examples. For this book, Brightwen adopted elements of the more traditional introduction to natural history. However, she retained a first-person narrative in order to establish a bond with her reader. She also positioned this work within a religious framework. "In the hope that even so humble an effort as this may not be without a use in enlarging and quickening a sense of that infinite harmony which runs through every part of the Creator's marvellous

---

35. Brightwen 1897b, 129–33, 176, 180–81.   36. Brightwen [1904], xi, 84, 99–100, 110.

plan of nature," she declared in the preface, "I put forth, not without a full sense of its inadequacy, this little volume." Although there is one chapter on adaptation, this theme plays an important role throughout the book. In her chapter on roots and in the one on tree stems, she attempted to demonstrate "how wonderfully plants are adapted to their wants and environment." The chapter on flowers emphasized the contrivances for ensuring fertilization. The "various means and adaptations by which it is attained," she declared, "are a continual source of admiration and wonder to the reverent student of nature." Brightwen also constantly drew the reader's attention to the wonders of nature. She enthusiastically proclaimed, "nature is an inexhaustible storehouse of wonders." Clearly, the glimpse that Brightwen provides into plant life reveals the patterns of design in their life and structure.[37]

In her *Rambles with Nature Students* (1899), published by the Religious Tract Society rather than Unwin, Brightwen offered herself as a guide to her readers, who would "help them to understand the thousand and one things that they may see in a country walk. The curious objects in hedges, trees, and fields all have a purpose and a meaning, but very often those need interpretation for those who never have had the opportunity of acquiring facts in natural history." Brightwen also explained in the preface why a monthly record had scientific merit. Making notes on the first appearances of birds and insects, or the flowering of trees and plants, "will result in the course of a few months in a record possessing a certain value." The record could be used to "compare one year with another, and note the differences in each, and the effect of temperatures in hastening or retarding the appearance of flowers and insects, and the arrival of migratory birds."[38]

What follows are a series of observations, arranged by month, that are taken from Brightwen's own experiences of interacting with nature on her estate. Attempting to convey to her readers the excitement of going on a ramble, she moves quickly from insects to flowers, birds, and seeds—in short, through an extensive range of flora and fauna. "Nature keeps us breathless in the attempt to overtake her marvellous energy," she wrote. "Every day something fresh appears, wild flowers are springing up, buds are opening, even early horse-chestnuts are to be met with in full leaf." When she pauses to reflect, more often than not her thoughts dwell on the purpose of what has been observed. The wasp, foolishly dreaded by most, has been

37. Brightwen 1897a, 11, 95, 146, 173, 190.   38. Brightwen 1899, 8.

"endowed by the Creator" with a "wonderful instinct," and carries out "the various useful purposes for which they were created," including its role of clearing away pollutants. She insisted that bats were "perfectly harmless and extremely useful," as they cleared the air of "millions of flies, gnats and moths, which would otherwise be a torment to us and very injurious to the farmer and gardener." When Brightwen is confronted by parasitic plants, whose purpose she cannot fathom, she leaves it "as a problem for my readers to solve," secure in her faith that a solution does exist.[39]

Brightwen's adoption of the seasonal genre in *Rambles with Nature Students,* used earlier by Loudon, Pratt, and Roberts; her use of the fiction of a ramble, as did Johns and Houghton; and her discussion of religious themes is a reminder of what she shared in common with earlier representatives of the maternal and clergyman-naturalist traditions. Like them, she stressed that everyone could participate in the discovery of scientific truth and that important scientific work did not need to be done at a laboratory or some other privileged scientific site. Useful scientific activity could be undertaken anywhere, in a rural location, like her secluded country estate, or even in the most unlikely locations close to a bustling metropolis. "It need not be thought that one must be far away from cities in order to learn about nature," she maintained. Once, while she was waiting for a train at the Bedford station, she found several rare fossils, pieces of jasper, a shell impression, and other treasures. She had heard of "as many as fifty species of wild flowers being found in a single field, and a well-known scientist discovered an equal number of wild plants in a piece of waste ground in the outskirts of a large town." Brightwen believed that "there exists all around us curious hidden lives of creatures unknown to us," that were only revealed when some trace of them came to our attention. "Then, indeed," she wrote, "investigation may often lead to interesting discoveries."[40] Rejecting the notion that only trained experts could practice science at particular sites, Brightwen also referred positively to many popularizers who preceded her. She recommended Johns's *Forest Trees of Britain* to young readers interested in the study of trees and in another book directed readers to it in a discussion of seedling trees.[41] In a section on the ivy-leaved toadflax, she quoted from Anne Pratt's *Flowering Plants of Great Britain.*[42] References to Wood's *Insects at Home* appear in two different books.[43] Since Brightwen did not frequently

39. Ibid., 46, 90, 151, 162.
41. Brightwen 1895, 143; [1904], 230.
42. Brightwen 1899, 94.

40. Ibid., 9, 195.
43. Brightwen [1904], 138; 1897b, 249.

mention other sources, as she tended to rely on her own observations and insights, these references are that much more significant.

Brightwen's works contained few references to would-be professionalizers of science, and when she did cite them or their work, they were often appropriated for her own ends. In a discussion of leaves, for example, she launched into a discussion of the "wonderful substance called protoplasm," but did not mention Huxley by name. When Huxley had used the term in the late 1860s he was accused of materialism. But for Brightwen, Huxley's term could be reinterpreted in such a way that it had religious connotations. She likened protoplasm to the lump of clay that a potter used. While the potter molded his lump "according to his purpose, into a rough pot, or a lovely vase," protoplasm seemed to be "just such a foundation material from which the Divine Creator causes animal and vegetable forms to proceed." According to John Odell, the steward for Brightwen's estate, and a knowledgeable botanist, Brightwen had great respect for Darwin's work on orchids. When she referred to Darwin in her *Glimpses into Plant Life,* he became a symbol for the importance of examining even the smallest natural objects as containing revelatory wonders. "I stand as it were only on the threshold of scientific research," she stated, "and look with wonder at the work of such a student as Darwin, who gave twenty long years to observation of the common earth-worm before he wrote his deeply interesting book upon it." Darwin's example showed "that the minutest objects in nature are worthy of reverent attention." Brightwen did not discuss evolutionary theory in her natural history books, but in her autobiography she acknowledged that the "preying of animals upon each other, the constant wail and pain of death going up from the innocent creatures," presented a "difficult problem." She saw "but one explanation." The struggle for existence was the result of original sin. "Before man's fall it is evident that all animals ate vegetable food [Gen. I. 30]," she affirmed, "there was *then* no preying one upon another, all lived in peace and harmony, as will be the case again when man is fully restored [Isaiah xi. 6-9]." Because of the "ruin wrought by man's sin," all of creation was "out of order for a time." Brightwen never attacked practitioners directly, not even scientific naturalists. A number of her books were dedicated to eminent practitioners, *Glimpses into Plant Life* to Joseph Dalton Hooker, *More about Wild Nature* to William Henry Flowers, and *Wild Nature Won by Kindness* to James Paget.[44]

44. Brightwen 1897a, 30, 111; 1909, xx, 130-31.

CONTINUITIES: HUTCHINSON AND
THE CLERICAL TRADITION

Just as Brightwen and Giberne belonged to an earlier tradition of popularization, the career of Henry Neville Hutchinson (1856–1927) calls to mind the
work of clergymen-naturalists from earlier in the century. During the 1890s,
Hutchinson was one of the most prolific writers on geological topics for a
popular audience, writing five books over the course of the decade. He was
critical in introducing new fossil discoveries to the British public, especially
those of Othniel Charles Marsh, an American paleontologist who led expeditions to the western United States starting in the early 1870s. Hutchinson
worked with a number of publishers, including E. Stanford, Chapman and
Hall, and Smith, Elder and Company. The son of the Rev. T. Neville Hutchinson, science master at Rugby, he received his education at Rugby, and then
at St. John's College, Cambridge, where he graduated with a BA in 1879. Assistant master at Clifton College from 1879 to 1880, in 1883 he was ordained
deacon and appointed curate to St. Saviour's in Bristol. Hutchinson was ordained priest in 1885, and a year later took up the position of private tutor
to the sons of the Earl of Morley, at Plymouth, until 1887. Suffering from ill
health, he went abroad, but then returned to London in 1890 to undertake
literary work. Besides his books on geology, Hutchinson wrote several works
on anthropology. He became active in several scientific societies, becoming
a Fellow of the Royal Geographic Society, the Zoological Society, and the
Geological Society. In 1914 he received an MA degree from Cambridge.[45]

Hutchinson saw himself as following in the footsteps of previous popularizers who had written for a popular audience. In the preface to his *Prehistoric
Man and Beast* (1896, Smith, Elder) he asserted that his book was "intended
for everybody, not for the specialist either in Geology or Archaeology." The
purpose of the book was to communicate, "in a simple style some of the
most interesting results arrived at of late years by the two diligent armies
of workers" who were laboring in the fields of geology and archaeology.
Hutchinson compared his role to that of an interpreter who translated the
difficult and strange language of the elite scientist for the benefit of his
audience. "The writer has endeavoured to place himself in the position of
an interpreter," he declared, "not of a Brahmin speaking a language unknown to the people." He praised the work of popularizers who had taken
on the same role in the past. Mentioning Ball's "A Glimpse through the

---

45. Lightman 2004d, 1040–41; Venn 1947, 503.

Corridors of Time," which he characterized as "an interesting lecture" that gave his audience "a graphic and eloquent *resumé* of the new and interesting results," Hutchinson presented Ball as a fellow interpreter of elite science. In his lecture Ball had taken George Darwin's papers in the *Philosophical Transactions* on the birth of the moon, which were "not intelligible to the ordinary reader," and translated them. "It is well," Hutchinson remarked, "therefore, when an interpreter comes between learned scientific men and the general public who is willing to undertake the task of giving a simple and popular account of their work and the methods of reasoning employed." Although he did not mention Wood by name, he referred to some of Wood's most famous works while stressing how important it was for the geologist to have a firm grasp of natural history in general. Hutchinson maintained that "the would-be geologist should be encouraged to collect and study living forms—the 'common objects' of the country or the sea-shore—at the same time that he collects fossils from the rocks, and to place living types side by side with those of the past which fill his cabinet of treasures."[46]

Hutchinson also felt a bond with previous popularizers who had made geology the focus of their works. In his opinion, "no living geologist" had the ability to "entrance his readers" as had Hugh Miller. He referred to Miller's *Old Red Sandstone* as a "delightful work." Hutchinson felt particularly indebted to Buckley. In the preface to one work he expressed his "best thanks to Miss Buckley (Mrs. Fisher) for permission to use some of the illustrations given in her well-known popular books on science, and for her valuable suggestions while the book was passing through the press." He judged Buckley's *Winners in Life's Race* to be "one of the best popular books on Natural History ever written."[47] He also fondly recollected an account of the ichthyosaurus in a Peter Parley book that "impressed itself on our youthful imagination."[48] But he added that this "inestimable instructor of youth" would have been surprised if he were alive to see the still more wonderful remains of the great fish-lizards discovered in various locations around the world. Just as Miller's works were now "somewhat behind the times," the Peter Parley book was sadly out of date. Hutchinson saw himself as filling a void. "The scarcity of popular works on Geology at the present

46. Hutchinson 1896, ix; 1890, ix, 11–12.

47. Hutchinson 1890, viii, x; 1894, 32, 36. In his *Creatures of Other Days*, Hutchinson took an account of the crocodile's habits from "one of Mrs. Fisher's (Miss Buckley) delightful works on Natural History" (Hutchinson 1894, 99).

48. Hutchinson 1892, 34. This is likely a reference to the section on "The Icthyosaurus [*sic*]" in *Peter Parley's Wonders of the Earth, Sea, and Sky* (1837) ([Clark] 1837, 6–14).

time," he wrote in 1890, "has induced the writer to put together this little book in the hope that it may supply a want."[49]

Hutchinson's handling of religious themes was extremely subtle. Whereas many of the clergymen-naturalists, as well as Giberne and Brightwen, tended to return over and over again to religious issues in their books, Hutchinson was more circumspect in his presentation of a theology of nature. In *Creatures of Other Days* (1894, Chapman and Hall) he made it clear at the beginning of the book that he believed that throughout history God had been immanent in the laws of nature, which implied that his subsequent discussion of paleontology was an account of divine activity in the past. "However many ages ago it was," Hutchinson insisted, "whether millions or billions of years ago, that these primaeval inhabitants of the world enjoyed their existence, the same unbroken laws of nature—the visible expressions of a Divine and All-powerful Will—were at work, fulfilling His purpose, as now." Hutchinson rarely touches on religious themes during the rest of the book, except to scold those who ignorantly rejected geological science on the grounds that it conflicted with the Bible. He discussed the destruction of Waterhouse Hawkins's dinosaur models in New York more than thirty years earlier and blamed politicians who did not see the results of geology as harmonizing with revealed religion. He hoped that "by this time people have learned the truth so well expressed by the great naturalist Agassiz, who said, speaking of living animals: 'These are but the thoughts of the Almighty uttered in material forms.'"[50] Similarly, in *Prehistoric Man and Beast*, he told his readers in the preface that humanity was "the final product of countless ages of Evolution, the 'beauty of the world, the paragon of animals,' and the highest manifestation of Creative Power that has yet appeared thereon." A few pages later, he rejected the notion that a belief in human evolution was "contrary to true theology," or that there was anything "degrading in the idea." Anticipating the objections of biblical literalists, Hutchinson asserted, "in the opinion of some students of ancient systems of thought, the account of the creation in the opening chapters of Genesis *implies* evolution."[51] Having placed the study of human evolution firmly into a religious framework, he then outlined what geology and archaeology had revealed about the men and women of the Stone and Bronze Ages.

49. Hutchinson 1890, viii.
50. Hutchinson 1894, 2, 142.
51. Hutchinson 1896, ix, 5.

Like some of the other clergymen cum popularizers, such as Kingsley and Henslow, Hutchinson tackled the issue of evolution head on, but his final position resembled Page's. Although Hutchinson accepted evolution as a fact established by modern science, his attitude toward Darwin's theory of natural selection seemed to change over time. In his *Extinct Monsters* (1892, Chapman and Hall), his most successful book in terms of sales, Hutchinson drew on the concept of natural selection to explain how snakes evolved from lizards and how huge and heavy sloths became the smaller and agile types of the present day.[52] In his discussion of the triumph of mammals over reptiles, he stated that this development was "in accordance" with the "biological truth" that "in the great onward and upward struggle for existence, higher types have supplanted lower ones."[53] Two years later Hutchinson was celebrating Darwin's discovery while, at the same time, undermining natural selection in his support for Lamarckian notions. Evolution had been placed on a "truly scientific basis by the illustrious Charles Darwin, and is now generally accepted by naturalists and palaeontologists," Hutchinson affirmed. "Indeed, it is hard to be a palaeontologist in these days without being also an evolutionist—so abundant is the evidence derived from a study of extinct animals." The evidence was accumulating year by year with the discoveries of long-lost types around the world. Later in the book, though, Hutchinson makes the case for allowing will into the evolutionary process. When the earliest ancestors of the horse desired to run faster, "Nature would therefore encourage such a wish, and give success to the swiftest." Hutchinson quickly reassured his readers that he had no doubts "as to the wonderful results which Natural Selection can produce; but we cannot help thinking that the animals themselves may have made a *determined effort in the direction of attaining speed.*" Hutchinson saw no reason why they should not have "worked out their own salvation" with the cooperation of Nature.[54]

By 1896, Hutchinson distanced himself even more from Darwin. He declared that he accepted evolution, but not the theory of natural selection as the sole factor in the process. He pointed out that "many even of the followers of Darwin are beginning to perceive that this cause, important

52. *Extinct Monsters* reached the 3,000-copies-printed mark by 1893, five editions by 1897, and a sixth, and final, edition, in 1910.

53. Hutchinson 1892, 122, 133, 171.

54. Hutchinson 1894, 198, 214–15.

as it certainly is, has been given too prominent a place in modern specu-lation." Hutchinson noted that Darwin himself had felt that he had gone too far. Although Darwin's theory "marked an important epoch in human knowledge," Hutchinson left open the possibility that "another discovery may, ere long, be made which shall be received with even greater joy."[55] Hutchinson's doubts about natural selection continued for many years. To Clodd, with whom he corresponded, he wrote in 1911, "with regard to the state of recent thought on Evolution I should much like to have a chat with you some day." Hutchinson told Clodd that he had been reading Samuel Butler, Hugo De Vries, T. H. Morgan, and other evolutionary theorists. "I don't see that Nat'l Selection can ever be so important again," he asserted, "as it was in time of Darwin and Huxley."[56]

Hutchinson's treatment of the development of humans and animals in his books can be classified as a type of evolutionary epic. "Everywhere there has been evolution: in society, in art, in morals, in religion, as well as in the veg-etable and animal kingdoms," he declared. But he took it one step further. "The world itself," he affirmed, "nay more, the whole solar system, is the result of gradual unfolding or development." However, Spencer was not the inspiration behind his vision of evolution, even though he gradually moved away from Darwin during the mid-1890s. Of the cosmic evolutionists, it was Buckley with whom he shared the most, though he was not attracted to spir-itualism. Like her, he used an innovative form of storytelling to bring home to the reader the imaginative dimensions of the evolutionary process. In the introduction to his first book, *The Autobiography of the Earth* (1890, Edward Stanford), Hutchinson presented his version of the evolutionary epic as if it were a story written by the Earth itself, and, ultimately by God. He began his introduction by telling his readers, "the story which we are about to read has not been written by man, but . . . by the Creator himself; and therefore we can trust absolutely the truthfulness of the record." Hutchinson claimed to be reading the story along with his audience. In stating that the story "is a plain unvarnished account in which man has no hand," Hutchinson implied that even he, the supposed author of the book, also had had no hand in writ-ing it. Written by God, the story was indelibly recorded in the Earth's strata. The Earth's crust was analogous to a series of passages in a text, which, Hutchinson maintained, were free from error, even though the earlier

<hr />

55. Hutchinson 1896, 4.

56. Hutchinson to Clodd, November 20, 1911, Leeds University Library, Clodd Correspondence.

chapters of the record were incomplete and difficult to read. "The symbols and the strata, or pages, on which they are written are absolutely free from interpolations or misrepresentations," he asserted, whereas human history was not "absolutely impartial and truthful." The testimony of the rocks could be trusted because it recorded God's story mechanically, without the taint of human intervention. Hutchinson declared, "*the Earth is its own Biographer, and keeps its diary with the impartiality of a recording machine.*"[57]

Having set up in the introduction the fiction of reading the story of the Earth's autobiography together with his audience, Hutchinson then presents an outline of geological history, continually drawing on his elaborate version of the parallels between the Book of God and the Book of Nature. He must defer to the chemists, astronomers, and physicists in his discussion of the period before the earth had cooled since "land and water are the paper and ink with which the geological record was written." Like a good cosmic evolutionist, Hutchinson therefore provides the current facts supporting the nebular hypothesis, which he refers to as "cosmoglyphics," and invites the reader to judge whether or not the theory has been confirmed. "If not," he asked, "would the reader try and see whether any other story can be spelt out from these cosmoglyphics"? Hutchinson then calls on Ball for aid, referring to his lecture "A Glimpse through the Corridors of Time," to explain the birth of the moon. Before moving on to the earliest chapters in the recorded history of the Earth, Hutchinson discussed how the story was read from rocks and fossils. "No one can read the language of the geological record," he declared, "unless he first acquires some knowledge of the grammar of that language." This grammar was learned through knowledge of the present processes that shaped the earth's crust. The "doctrine of 'uniformity'" was the key to understanding the symbols recorded in rocks.[58]

Hutchinson then began to move through the various geological eras, starting with the earliest. The story that he and the reader read in the rocks was an evolutionary one. The "successive introduction of higher and higher types" was one of the "most interesting features of the geological record, and is justly appealed to by evolutionists." As Hutchinson moved into more recent geological epochs, he observed that it was becoming easier to read the story. By the time he arrived at the "Bath Oolites," the pages on which the story is written "are perhaps less mutilated and torn away by denudation than those of any previous chapter." In this section Hutchinson extended his analogy between the Earth and a text by comparing nature to a printing press. He

57. Hutchinson 1896, 10; 1890, 1–2.   58. Hutchinson 1890, 2–3, 9, 11–12, 17–19.

asserted that Mother Nature kept her records mainly in the sea. "When these have been quietly accumulating in her submarine printing-press," he wrote, "where she stamps her pages with many strange characters—ripple-marks, worm-tracks, sea-weeds, and such like,—then after a time they are heaved up and left to dry and harden while the records of another period similarly grow." In this way, Hutchinson playfully suggested, "she 'periodically' issues a chapter."[59]

Hutchinson's imaginative form of storytelling, akin to Buckley's, also emphasized the magical dimension of the natural world. He repeatedly drew on the idea of nature as a fairyland, so vividly developed in Buckley's *Fairy-Land of Science*. In *The Autobiography of the Earth*, Hutchinson asserted that the carboniferous forests of the coal period "must have been a veritable fairyland, where 'mosses' (so-called) grew to the height of an oak or an ark!" He repeatedly drew the reader's attention to the resemblance between dinosaurs and the dragons of ancient mythology throughout his *Extinct Monsters*. In a chapter on "The Dragons of Old Time—Dinosaurs," he declared that "geology reveals to us that there once lived upon this earth reptiles so great and uncouth that we can think of no other but the time-honoured word 'dragon' to convey briefly the slightest idea of their monstrous forms and characters." Hutchinson submitted, "there is some truth in dragons, after all." To those who did not believe that flying dragons once existed, Hutchinson presented the scientific evidence for the pterodactyl. "The notion of a flying reptile may perhaps seem strange, or even impossible to some persons," he wrote, "but no one has a right to say such and such a thing 'cannot be,' or is 'contrary to Nature,' for the world is full of wonderful things such as we should have considered impossible had we not seen them with our eyes." The "antique world" of extinct monsters was "quite as strange as the fairy-land of Grimm or Lewis Carroll." Hutchinson stressed to his readers, "all these monsters once lived" and everything they would see in the book was "quite true." "Truth is stranger than fiction," he declared, "and perhaps we shall enjoy our visit to this fairy-land all the more for that reason."[60]

In highlighting the fairyland of science theme in his works, Hutchinson was appealing to the imagination of his readers. Like Buckley and Page, he offered vivid literary images that compared the dramatic sweep of geological history to a panorama or some other feature of mass visual culture. In the preface to *Extinct Monsters* he stated that most natural history books dealt only with creatures that were alive in the present. "Few popular writers have

---

59. Ibid., 153, 177.   60. Hutchinson 1890, 116; 1892, 1, 60, 110.

attempted to depict," he insisted, "as on a canvas, the great earth-drama that has, from age to age, been enacted on the terrestrial stage, of which we behold the latest, but probably not the closing scenes." The story of cosmic evolution was best told as if it were a stunning play, or a stupendous spectacle on a canvas. The antique world, "this panorama of scenes that have for ever passed away—is a veritable fairy-land," Hutchinson exclaimed. Four pages later Hutchinson was explaining how the testimony of the rocks beneath our feet contained clues as to the existence of ancient creatures and their natural environment. It only required "the use of a little imagination to conjure up scenes of the past, and paint them as on a moving diorama." But Hutchinson offered his readers more than just literary images. He offered them fabulous visual images as well that were intended as restorations of extinct monsters that roamed the earth long ago. *Extinct Monsters* contained twenty-four full-page illustrations, and thirty-eight additional figures (Fig. 8.8).[61] Hutchinson expressed surprise that little had been done to expose the public to the strange creatures of the past even though there was a "wealth of material for reconstructing the past." He was critical of the few past efforts. Guillaume Louis Figuier's *World before the Deluge* was dated and could not be considered a "trustworthy book." Other books contained poor illustrations.[62]

Although Hutchinson appealed to the readers' imaginations, their taste for vivid illustrations, and their desire to be entertained, he nevertheless argued that the images and information contained in his books were based on scientific fact. He preferred to think of the restored dinosaurs in his books as "living in the eye of imagination rather than in a common menagerie, to be teased and prodded by the vulgar crowd." He was critical of readers who suggested that the dinosaurs would have made a highly interesting addition to a modern wild-beast show. If they lived today "some enterprising showman" might have been tempted to put them on public display merely to cater to "the amusement of an idle public." But Hutchinson did not want to be seen as an "enterprising showman" who pandered to the vulgar crowd. He put the dinosaurs on display to exercise the readers' imagination and to stimulate their rational faculties. He was aiming at the elusive

61. Many of Hutchinson's other books contained illustrations of ancient geological epochs, primitive human beings, and restorations of extinct creatures similar to those in *Extinct Monsters*. *The Autobiography of the Earth* offered twenty-seven illustrations; *Prehistoric Man and Beast* included ten plates, and *Creatures of Other Days* boasted twenty-four full-page illustrations and seventy-nine figures.

62. Hutchinson 1892, x, 1, 5.

A GIGANTIC HORNED DINOSAUR, TRICERATOPS PRORSUS.
PLATE X.   Length about 25 feet.

FIGURE 8.8 The frontispiece to Hutchinson's *Extinct Monsters,* a recreation of *Tricer-atops prorsus.* (Rev. H. N. Hutchinson, *Extinct Monsters* [London: Chapman and Hall, 1892], frontispiece.)

balance between amusement and instruction. Hutchinson intended to bring the "long-neglected 'lost creations' of the world" out of their obscurity so that they "may be made to tell to the passer-by their wondrous story." This could only be accomplished if "reason and imagination will, if we give them proper play, provide us eyes wherewith to see the world's lost creatures." The dry bones of dinosaurs came alive for such men as "Cuvier, Owen, Huxley, and others," and it was to their work that Hutchinson looked to guide him in his task of resurrecting them for the reader. Scattered throughout his works are his expressions of gratitude to the men of science, past and present, for providing valuable assistance in his books and for supplying the scientific information necessary for him to put flesh on the dry bones of his extinct creatures.[63] Even when Hutchinson produced a more lighthearted

63. In his *Autobiography of the Earth,* Hutchinson asserted that British scientists such as William Smith, Hugh Miller, Lyell, Hutton, and Playfair had laid the foundations of geology. He listed the leading textbooks he had consulted in order to write the book, including works by Archibald Geikie and others. In the preface to *Extinct Monsters* he thanked Henry Woodward and Smith Woodward of the Natural History Museum for their help, and then in an appendix listed the

FIGURE 8.9 A primeval scene titled "Dining in the Open," which depicted an unwelcome intrusion to a picnic in the Older Stone Age. (Rev. H. N. Hutchinson, *Primeval Scenes: Being Some Comic Aspects of Life in Prehistoric Times* [London: Lamley, 1899], 13.)

book, *Primeval Scenes* (1899, Lamley), he insisted that the creatures depicted had been carefully drawn with due regard to scientific accuracy (Fig. 8.9). Writing in the age of the "new journalism," Hutchinson once remarked that there existed a "large class of persons who unfortunately read magazines and newspapers rather than books." Through a combination of scientific instruction and entertainment—drawing on vivid literary and visual images—Hutchinson hoped to entice the reading public to the study of geology.[64]

If the reviews are any indication, Hutchinson did not satisfy all of his readers that he had successfully combined instruction and entertainment. At least one reader denied that his work was scientifically sound. *Nature's*

---

works by Buckley, Miller, Darwin, Owen, Othniel Marsh, and Edward Cope that he had read to prepare the book. For his *Prehistoric Man and Beast*, he told his readers that he had submitted the rough sketches and drawings to several archaeologists and scientists, including William Flower, Henry Woodward, and Arthur Smith Woodward. In *Creatures of Other Days* Hutchinson even referred to Huxley as a great "authority." Hutchinson deferred to the authority of practitioners and at the same time was able to bolster his own credibility through his associations with them. Several eminent practitioners wrote prefaces for his books, such W. H. Flower for his *Creatures of Other Days*, and Henry Woodward for his *Extinct Monsters* (Hutchinson 1890, vii; 1892, xiii, 245; 1896, xiv; 1894, 130).

64. Hutchinson 1894, xii; 1892, 2; 1899, 8; 1896, 1.

appraisal of *Extinct Monsters* was negative. Although the book was novel in presenting a series of restorations, and although it was "clearly and simply written," the reviewer damned Hutchinson for straying "into less safe matter," such as in his unsupported assertions about the eye of the ichthyosaurus. Hutchinson appeared to go beyond the restricted role that *Nature* had set for popularizers who were not recognized practitioners. The reviewer complained that it was "undesirable that a popular work, whose main merit is that it does not pretend to teach the facts of science, should appear to enunciate judgments on scientific problems."[65] The *Saturday Review,* however, viewed Hutchinson's book positively. "His book, in short," the reviewer declared, "is both attractive and useful and will add to his reputation as a popular, but accurate, writer on geological subjects." Hutchinson was praised for avoiding the use of long words, as well as for steering clear of the other extreme, writing in a sloppy style merely to gain popularity.[66]

LOOKING BACK TO SOMERVILLE'S SYNOPTIC OVERVIEWS: BODINGTON

While Giberne, Brightwen, and Hutchinson appropriated aspects of the clerical and maternal traditions, some popularizers were prepared to adopt the writing formula devised by Mary Somerville. In the preface to her *On the Connexion of the Physical Sciences,* Somerville declared, "the progress of modern science, especially within the last few years, has been remarkable for a tendency to simplify the laws of nature, and to unite detached branches by general principles."[67] Somerville's goal in her *On the Connexion* was to make the connections between the branches of science as explicit as possible. By focusing on the unity of science, she earned William Whewell's praise in 1834 in the *Quarterly Review.* Whewell was worried about the impact of specialization, which seemed to be leading to the fragmentation of science. "Physical science itself is endlessly subdivided," he complained, "and the subdivisions insulated." Kathryn Neeley has argued that *On the Connexion* established a new literary form: the extended synthetic literature review. But even Somerville recognized that writing such a book made tremendous demands on the popularizer, who had to assimilate and synthesize a huge amount of scientific information in a variety of fields. In her *Personal Recollections* she remarked that "no one has attempted to copy my 'Connexion of

65. H. G. S. 1893, 250-51.
66. "Dragons of the Prime" 1893, 20.

67. Mary Somerville 1834, [ii].

the Physical Sciences,' the subjects are too difficult."[68] This is exactly why Buckley hesitated to undertake an updated and revised version of *On the Connexion* for Murray in 1875, and it also explains why women looking for an alternative to the maternal tradition in the middle of the century could not emulate Somerville.

Near the end of the century the growing pace of specialization seemed to offer the opportunity for popularizers to redefine their roles. In a discussion of the possible uses of spectroscopic analysis in both astronomy and chemistry, Proctor remarked on how specialization had made it difficult for new methods of research to be used across scientific disciplines. "The workers in one field have been too busily employed to note either on the one hand the value of their results for other fields than their own," Proctor wrote, "or on the other, the advantage they might obtain from employing in their own field other workers' results." Proctor believed that important discoveries would be missed or delayed as a result. "Hence the necessity," Proctor argued, "which Spencer and others have pointed out, for scientific overseers, who, not working specially in any field, but having a general knowledge of what is going on in all, may note what the workers themselves would be apt to overlook—the general rather than the special significance of the results obtained in the various fields of scientific labor."[69] At least two popularizers, Alice Bodington and Agnes Clerke, saw themselves taking on the role of "scientific overseer" as spelled out by Proctor and as embodied in Somerville's objectives in *On the Connexion,* at least in their own disciplines. For both of these women Somerville's emphasis on synthesis had provided a rationale for the existence of popularizers in an era of increasing specialization. Unlike Page, Buckley, Clodd, or Allen, neither of them used evolution to fashion a synthesis of knowledge in epic form. Both Bodington and Clerke wrote primarily for adults, and considered practitioners to be among their audiences, which was highly unusual for popularizers in the second half of the century. Whewell's vision of unity in science was advanced by popularizers at the end of the century, but not by practitioners. The extended

---

68. Ibid., [vii]; Margaret Somerville 1874, 294; [Whewell] 1834, 59; Neeley 2001.

69. Proctor 1887c, 10. Spencer perceived specialization as part of the evolutionary process from indefinite homogeneity to definite heterogeneity. The scientific-philosophical class had evolved out of the clerical class. Then there had been a series of further differentiations between the man of science and the philosopher, between those scientists who dealt with the organic and the inorganic, and between those biologists who dealt with botany and those who focused on zoology. "Nowadays," Spencer wrote, "men who occupy themselves with mathematical, physical and chemical investigations are generally ignorant of biology" (Spencer 1895-96, 746-47).

synoptic scientific review was later to become a staple of twentieth-century popular science.

Alice Bodington (1840–1897) undertook science writing during a particularly turbulent period in her life after leaving England. Raised by her paternal grandmother, she developed wide intellectual interests, particularly in the sciences. She married, and then divorced, General Bell, with whom she had one son. Later she became the second wife of George Fowler Bodington, a London doctor who specialized in mental disorders. In 1887 they immigrated to Canada with their children, Helena and Winifred. According to the children, who wrote an account of their first six years in British Columbia, Bodington and her husband had restless, enterprising minds. They were enticed by pamphlets put out by the Canadian Emigration Department to come to Canada to build up the family fortune. Helena and Winifred described their mother as an extremely well educated woman who was used to good society. They recalled the "veritable storehouse of her mind, from which knowledge of the most diversified and entertaining variety could be extracted with the least possible effort." Her training and tastes made her "a brilliant conversationalist and a fascinating companion." But in British Columbia, Bodington felt intellectually isolated. Helena and Winifred remembered that their parents discussed politics and the news of the day endlessly. "When the English mail arrived bringing 'The Illustrated London News,' 'Truth,' 'Punch,' and the 'British Medical Journal,'" they recollected, "we were like hungry wolves falling upon their prey—for we were all starving for literature."[70]

The Bodingtons spent a year in Vancouver and then took up farming in the Fraser River valley. Winifred and Helena wrote that their parents were entirely unfitted for the life of hardy pioneers. George was sixty when they arrived in British Columbia, and he had poor business sense. Alice had been a semi-invalid for years and spent the greater part of her days lying on a sofa. An intellectual, her talents were "utterly inadequate" to undertake farming, or even to manage a household. According to her daughters, "she was entirely ignorant of the first rudiments of housekeeping knowledge, except as lay in directing well-trained and competent servants." Unhappy with her life in the wilds of British Columbia, she slipped into a lifelong depression. Her daughters recorded that the hardships she encountered, "combined with a temperament always slightly inclined to melancholy, made her own

---

70. Creese 2004a, 229–30; Irvine and Meiklejon n.d., 1–2, 32.

life nothing more nor less than a bitter and long drawn-out tragedy." In desperate shape financially, the Bodingtons decided to abandon farming in 1895. George was appointed as Medical Superintendent of the hospital at the Provincial Asylum for the Insane at New Westminster. Alice died two years later.[71] Somehow, before her death, Bodington was able to write under these trying conditions. Perhaps she wrote to counter her sense of isolation. Through her publications she could maintain contact with the intellectual scene. Her only book, *Studies in Evolution and Biology* (1890), was published by E. Stock and sold for five shillings. A compilation of some of her earlier pieces, the book covered the evolution of the eye, mammals, past flora, interesting facts in evolution, paleontology, and neo-Lamarckism. She wrote articles for a number of periodicals, including *Westminster Review, Journal of Microscopy and Natural Science, Open Court,* and *American Naturalist.* In her writings she tackled a broad range of issues, including the bearing of recent scientific research on marriage, race, agnosticism, prehistoric man, the existence of ghosts, parasites, mental evolution, insanity in royal families, and mental action during sleep. She also produced a series of natural history studies on such topics as water insects, variation in plants and animals, and the courtship rituals of spiders.

Although *Studies in Evolution and Biology* did not sell well enough to reach a second edition, it is important as an example of a book that presented a rationale for the work of popularizers in language reminiscent of Somerville's preface to *On the Connexion.* In the note to the reader that opened her book she declared, "the fields of physical science are now of such vast extent, the workers therein so numerous and so busy, that no specialist has time to give a general view of what is going on. Yet all branches of physical science are so inter-dependent that it is difficult to understand one branch thoroughly without some knowledge of the rest." Like Somerville and Whewell, Bodington claimed that a fundamental unity lay at the bottom of all science. Real understanding of any scientific discipline was impossible without some grasp of the others and how they were connected. But practitioners were too busy themselves to work up the knowledge of disciplines outside their own. The growth of specialization, Bodington therefore argued, provided an important niche for popularizers of science. Yet, despite the urgent need for popularizers who could follow in Somerville's footsteps, "a stigma is supposed to rest, for some mysterious reason, upon the person who ventures

71. Creese 2004a, 230; Irvine and Meiklejon n.d., 2–3, 38.

to write upon any branch of science without being an original discoverer." Bodington could not understand why it was "considered almost wrong to write about physical science without having made original experiments." This was not the case in other fields of knowledge. "A historian is not required to have fought in the battles he describes," Bodington pointed out, "nor a geographer to have personally traversed the wilds of Africa." A popularizer did not have to be a practitioner to produce useful books. "Why cannot a wide view be taken by some competent person of the results of the labours of hundreds of scientists," Bodington asked, "so that we may more clearly see what manner of fabric is being reared?" Having justified the role of popularizer in the age of specialization, Bodington then offered her list of the labors of eighteen scientists she had consulted in order to write *Studies in Evolution and Biology,* including those by Cope, Ernst Haeckel, Darwin, and Wallace.[72]

Bodington again presented arguments validating the role of popularizer in her chapter on "Micro-Organisms as Parasites." This time she called upon two eminent scientists to back her up. "In venturing to write upon a subject of which I have no experimental knowledge, I will endeavour to take shelter behind two great names," she wrote. "It is often assumed that no one has a right to criticise or to write about, or even give an opinion upon, scientific subjects without being a professed zoologist or biologist, as the case may be." Bodington managed to get both Darwin and Huxley on her side. Darwin, she maintained, had once said that he received the most valuable criticism of his theories from a professor of engineering rather than a professed naturalist. He believed that it was important that his theories were read by intelligent men accustomed to scientific argument, though not naturalists. Bodington cleverly quoted from Darwin's letter of January 4, 1865, to Huxley, where he begged his bulldog to write a "popular treatise," as they were "almost as important for the progress of science as original work." She had found all of this material in Francis Darwin's *Life and Letters of Charles Darwin,* which had been published three years earlier. Bodington also had an appropriate quote from Huxley on the dangers of specialization. Huxley had stated that "'the man who works away at one corner of nature, shutting his eyes to all the rest, diminishes his chance of seeing what is to be seen in that corner.'"[73] Bodington enlisted Darwin and Huxley in her bid to defend the position that she and other similar popularizers had something important to contribute to

72. Bodington 1890c, ix–xi.   73. Ibid., 141–42.

the discussion between practitioners. Bodington claimed the right to be able to enter freely into scientific debates, particularly on evolutionary theory, without inviting the condemnation of male practitioners.[74]

But Bodington added a second point. As well as helping scientists grasp the larger picture in their field, popularizers were needed to ensure that discoveries were communicated clearly and plainly to the masses. Only then could real progress be made "in eradicating ignorance and superstition." Darwin had the unusual ability to make "original observations" and to "make the result of his investigations clear and plain to any person of ordinary intelligence." But "many original observers are unable to do this," Bodington asserted, in part because "each works at his speciality." Ignorant of what their fellow workers are discovering, "they are hardly aware of the magnitude of the coral reef they have been building." So work done by practitioners never reached the ears of the public. "Hundreds and thousands of original workers are carrying on their investigations," Bodington wrote, "yet one may search the leading periodicals in vain for any account of the results they have accomplished, probably because some stigma is supposed to attach to the fact of writing upon science without original investigation." Bodington believed that it was time "that Natural Science should have her historians" so that the work of practitioners was not "buried in the pages of scientific journals, to be read only by specialists."[75] Bodington's *Studies in Evolution and Biology* contained a series of essays that synthesized recent biological research.

Studies in Evolution and Biology, with its forceful rationale for popularization and its collection of synthetic essays, drew a mixed reaction from the critics. The *Saturday Review* was impressed by "the author's plea for her writing," referring to it as "ingenious." "The historian, she urges, is not required to have fought in the battles he describes," the reviewer wrote, "or the geographer to have personally traversed the wilds of Africa; why then should she, though not an original discoverer, refrain from writing on science?" But in the *Athenaeum,* Bodington's justification was held up to ridicule. The reviewer contended that it was easy to understand why it was considered wrong to write about physical science without having made original experiments. In fact, the reviewer was going to explain to her "by the help of her book why it is vexatious to the expert." What followed was a long list of errors that would not have been committed by anyone with

---

74. Gates 1998, 62.   75. Bodington 1890c, 142.

"knowledge of other naturalists' work." "These are the kind of errors," the unrelenting critic declared, "that make it not 'almost wrong,' but dangerous for unskilled persons to publish excerpts from their note-books." The reviewer even corrected her grammar, and then damned her with faint praise. On the whole she had "not done her work badly," and her book, flawed as it was, might induce readers to take a real interest in evolution and biology. But "her facts are second-hand, her style is not good, and her selection of authorities is not always judicious."[76]

Despite the hostility of some reviewers, Bodington identified with many of the goals of practitioners, especially those influenced by scientific naturalism. In *Studies in Evolution and Biology,* she emphatically rejected the design argument in the course of discussing the danger of cystic disease to female mammals as a result of the atrophy of useless kidneys. Bodington pointed out that the survival of useless organs demonstrated how "the argument for design is utterly put out of court by the awkwardness of the whole plan." She was disdainful of popularizers like Frank Buckland, who opposed evolutionary theory because it undercut Paley's design argument. In one essay she referred to Buckland's "airy contempt for strict scientific investigation."[77] Bodington also espoused views on religion that she shared with a number of scientific naturalists. Her daughters describe her as an "agnostic" during the time that they were living in British Columbia. While they were there, Bodington wrote an essay on "Religion, Reason and Agnosticism," published in the *Westminster Review,* which made it clear that she had no desire to present herself as a religious teacher, the role often adopted by women looking to the maternal tradition for their authority. In this piece she was critical of Christian theologians, who could offer no satisfactory answer to the question of why evil existed, and she recommended a rational religious position based on the recognition of the "Unknowable, or, as some philosophers would rather say, the Unknown." She asserted that she had earnestly tried to believe in the existence of the Christian God and the divinity of Christ, but she felt "one belief after another give way in my

76. "New Books and Reprints" 1890, 653; "Science" 1890, 803-4.

77. The reference to Buckland was in an essay by Bodington on the evolution of the eye, a subject at the center of discussions of design in both Paley's *Natural Theology* and Darwin's *Origin of Species.* What led to Bodington's criticism of Buckland's method of scientific investigation was his way of resolving the question of whether or not the common mole is blind. According to Bodington, it was likely Buckland who said, "the mole is not blind, because if you part its fur you can see its eyes" (Bodington 1887, 83).

grasp, as a falling man feels boughs and twigs and tufts of grass yield as his whole weight hands upon them."[78]

Having explained her position as the result of examining traditional religion by the light of reason, Bodington turned to the relationship between science and religion. She predicted that theologians would always fail when they tried to find God through science. Since they had never been trained to reason in the way demanded by physical science, they started with God, rather than with induction. But if induction became the starting point, Bodington argued, then the seeker of truth would conclude that the "Universe, as we know it, seems the product of impersonal, unvarying law." Bodington acknowledged that this was a sad thought, but it was far more satisfactory than the "hideous conception of a personal God who can condemn His sentient creatures to infinite punishment for finite offences," or the notion of a capricious being who is responsible for the suffering caused by natural disasters. As for "the evidence of design, once regarded as the corner-stone of natural religion," Bodington believed that it had "become a thorn in the side of those who relied upon it." Now supporters of natural theology had to explain the design in vestiges of human descent from ancestors who tended to become seats of disease. "The man who should now cite the example of a watch as proving the existence of the watchmaker," Bodington argued, "would have to cite the example of a watch full of awkward unnecessary structures, tending to throw it utterly out of gear, or to stop its motions altogether." The human body did not suggest "the nice adjustment of means to ends which characterizes a machine, but rather that rough-and-ready adjustment to chance circumstances which characterizes organic growth." Instead of providing evidences of design, the organic world contained evidences of imperfection that pointed to the operation of "blind evolutionary forces." Bodington closed by referring to herself as a "scientific agnostic."[79]

Bodington also accepted the position of leading scientific naturalists on gender and race. Her opinion of women in general was quite negative. Her daughters wrote that "she disliked most women who, she frankly owned, bored her by their silly conversation and she much preferred talking to men on subjects in which she was interested which included amongst others, astronomy, biology, poetry, literature and art."[80] In her essay on "The Marriage Question from a Scientific Standpoint" (1890), she argued that

78. Bodington 1890c, 109; 1887, 83; Irvine and Meiklejon n.d., 3; Bodington 1893, 369–71.

79. Bodington 1893, 373–76, 378.

80. Irvine and Meiklejon n.d., 2.

women were inferior to men, just as had Darwin in his *Descent of Man*. "The inferiority of women to men" was number one on her list of "four factors" that she considered as "immutable where the great majority of men and women are concerned."[81] Bodington sounded the same theme in her essay on "Importance of Race and Its Bearing on the Negro Question." Although the influence of women was beneficial "and conducive to the ethical advance of humanity," she maintained, "women are far inferior to men." In response to those who, like Becker, were supporters of the women's suffrage movement, she argued, "man has granted already as much as women can safely demand; and if she still has grievances, she must find fault with Nature herself." Bodington then turned to the issue of race, bringing to bear the results of science. "The same line of reasoning," she declared, "which leads those who adopt it to the conclusion that the vote of women would be practically null and void, applies with incomparably greater force to the Negro." The color of a race's skin, from the scientific point of view, was not the issue here, Bodington insisted. This was a matter of utter indifference. But an analysis of the brain underneath the skin of the black man demonstrated he could not become the equal of the white man. Bodington affirmed that the white race still had the duty to protect the weak and helpless, and to guide them to a higher level of civilization. "In a word," she declared, "for the weaker races paternal government is the best and kindest form of rule."[82]

But Bodington was more independent when it came to her views on the status of current evolutionary theory. She was by no means an opponent of evolution. On the contrary, she affirmed that "every branch of biological research, whether it deals with the animal or vegetable kingdom, leads more and more to the conviction that evolution is one of the most important laws of organic nature." Where she departed from many of the scientific naturalists was in her attitude toward Darwin and his theory of natural selection. She revered Darwin for opening "to our view new paths leading towards the discovery of the great laws government evolution," but she maintained, "the great master himself had only grasped one form of the law governing evolution." In his earlier works, Darwin had sought "to account for all changes in animals and plants by natural selection; whereas we now see that the infinite, delicate variations in the world of organic beings are owing to the intense irritability and susceptibility to molecular changes of protoplasm, and the consequent action of the environment upon it." Bodington thought

81. Bodington 1890b, 174.   82. Bodington 1890a, 422–24.

that modern biology had passed Darwin by, and she insisted on her right to supplement his theories with Lamarckianism. She argued that Darwin had himself later recognized that he had overemphasized natural selection and that he needed to allow more weight to the direct action of the environment.[83]

In her chapter in *Studies in Evolution and Biology* on "Neo-Lamarckism," a label she seemed to be willing to accept for her position, she claimed that Lamarck had "showed the road by which alone we could reach the true knowledge of the origin of diverse animal forms, and pointed out that these forms were perpetually altering, under our eyes, from the influence of the circumstances in which they were placed." She reproduced some extracts from chapter seven in volume one of the *Philosophie Zoologique*, to illustrate how Lamarck had anticipated modern developments in biological research. In reading this chapter "we might readily imagine in many passages we were studying the work of some great zoologist of to-day." Lamarckian ideas also provided the connecting theme in her entire book. "Neo-Lamarckism," she wrote, "supplies the 'motif' which runs through almost every study in this little book." Every advance in the physical sciences that she had attempted to chronicle added "a fresh laurel to the fame of this most unjustly decried genius." To those who saw this as disloyalty to Darwin, she suggested that it would not detract from his fame if he were viewed as the Newton of evolution, and "Lamarck as its Galileo." Lamarck echoed Galileo when he said, in opposition to the powerful defenders of the idea of the changelessness of animal organisms, "still they move, either forwards or downwards, in the race for life." By preferring Lamarck to Darwin, Bodington was promoting the role of will in the evolutionary process, like Buckley and Butler.[84]

WRITING FOR THE PRACTITIONER: CLERKE'S ASTRONOMICAL SYNTHESIS

When, in her *Studies in Evolution and Biology,* Bodington made her case for the importance of the person "who assiduously collects and arranges the facts discovered by others," one of her examples was the "writer of the 'History of Astronomy in the Nineteenth' century." Although this writer had "probably made no astronomical discoveries herself," Bodington argued that she had made a significant contribution "in her clear account of the enormous

---

83. Bodington 1890c, 1, 22–23, 144; Gates 1998, 62.
84. Bodington 1890c, 174, 184, 186; Gates 1998, 63.

progress of astronomical science."[85] Here Bodington was referring to none other than Agnes Clerke. Both Bodington and Clerke presented themselves as writing for practitioners as well as for the public, and both offered their readers synoptic overviews of the important scientific literature. While both of these popularizers attempted to carve out a place for themselves in an era of specialization by emulating Somerville, Clerke was more successful in gaining recognition that she had assumed a new role for the popularizer. Not only was she far more prolific than Bodington, she also established herself as an authority on the "new astronomy," the field at the cutting edge in her discipline. Whereas Bodington was writing in isolation in the wilds of British Columbia, Clerke lived in London at the center of the British scientific scene where she was able to build up an international correspondence network of practitioners at the forefront of astronomical research.

Raised as a devout Catholic, and educated by her cultured parents, Clerke had no formal training in astronomy when she embarked upon a career as popularizer. Agnes Mary Clerke (1842–1907) was born at Skibbereen, a small town on the south coast of Ireland in County Cork (Fig. 8.10). The second of three children, her father, John William Clerke, was manager of the Skibbereen Provincial Bank. He had been a classical scholar at Trinity College, Dublin. While her father was an Anglican, Clerke's mother, Catherine Mary (née Deasy) came from a family of affluent and devout and Catholics. Two of Agnes's aunts were nuns. Agnes, her elder sister Ellen, and her younger brother Aubrey, were all brought up in their mother's faith and remained committed Catholics for their entire lives. Educated by her parents, Agnes was taught piano by her mother and Latin, Greek, mathematics, and the sciences by her father. Agnes became fascinated by astronomy as a child, encouraged by her father who had set up a telescope in the garden. She claimed to have mastered John Herschel's *Outlines of Astronomy* by the age of eleven. Eight years later the family moved to Dublin when their father took up a new position as registrar at the court of his brother-in-law Richard Deasy, the newly appointed Baron of the Exchequer. Deasy was one of the country's first Roman Catholic high court judges. In 1867 the family began a ten-year residence in Italy, living in Florence from 1870. The training that Agnes had received from her father had prepared her well enough to undertake advanced research in a number of fields. While in Italy, Agnes frequently took advantage of the Florence city libraries to study Italian history

---

85. Bodington 1890c, 142.

FIGURE 8.10 Agnes Mary Clerke, emulator of Somerville's model of scientific authorship. (Margaret Huggins, *Agnes Mary Clerke and Ellen Mary Clerke: An Appreciation* [Printed for private circulation, 1907], frontispiece.)

and literature, the history of science, contemporary affairs, European languages, and the classics. The philosophy and science of the Renaissance became her special area of expertise. She also developed into an accomplished pianist, once playing for the composer Franz Liszt when he was in Rome.[86]

When the family returned to England in late 1877, or early 1878, and settled in London, Agnes and her sister Ellen both embarked on literary careers. Agnes first established herself as an essayist by publishing a series of articles with the *Edinburgh Review* on the science and philosophy of the Renaissance. She started to write what was to become her *Popular History* in 1881. Most of the research was done in the British Museum library. Since all of the *Edinburgh Review* articles were unsigned, she seemed to burst upon the scene out of nowhere when her first book, *A Popular History of*

86. Brück 2002, 3, 8–9, 14, 16–17, 20, 23, 28–29.

Astronomy in the Nineteenth Century, appeared in 1885. Published by Adam and Charles Black, and priced at twelve shillings and sixpence, the book was reprinted within two months and Macmillan published an American edition. It eventually reached a third edition in 1893, and a fourth edition in 1902. During the course of her career she wrote five more books, all on astronomy. She worked primarily with Black, though she also undertook projects for Longman, Hutchison, and Cassell. Clerke was selective about the projects she undertook. In 1890 she received a visit from Adam Black, who hoped that she would agree to write a book on elementary astronomy for a series that he intended to publish. But she decided to refuse the offer, in part due to her brother's advice. She told one correspondent that the "time was too short, and the profits too small, and I breathe more freely for being out of it." Black wanted the manuscript ready in eight months and she was not willing to endure the "grind" it would have involved to complete it in time while also working on her other projects. Clerke could afford to be selective. Although she never married, she lived with her family and did not subsist solely on the money she earned as a popularizer. In her opinion, she did not earn enough to really make a living on her books. Having just received Black's annual account, she wrote to a friend in 1889 that publisher's accounts are seldom "cheering documents," and that this one was "no exception." The sale of the second edition of her *Popular History* was "going on very languidly"—nearly half of it was still on hand. Clerke estimated that she had earned a total of £93 for both editions. "I do not mean to complain," she wrote, "I think the Blacks quite fair-dealing people; but it shows that one need not expect to make a livelihood of writing books."[87]

For her second book, *The System of the Stars* (1890), she decided to work with Longman, who targeted the upper end of the book-buying public by pricing it at twenty shillings. Having published one book already, she asked Longman for a royalty payment before the manuscript was completed. Longman refused and offered half-profits. In a letter to Longman on May 10, 1887, she agreed to this arrangement for the first edition, opting for the security of a contract. "On the whole, however," she wrote to Longman, "I think it will be more satisfactory to us both to come to a definite agreement at once, and I have therefore made up my mind to accept your alternative offer of half-profits, on the understanding that the agreement is to be limited

87. Ibid., 43-44, 46; Clerke to David Gill, [April 6, 1890], Cambridge University Library, Royal Greenwich Observatory Archives, RGO 15/126 ff. 175r-177r; Clerke to [Gill], January 3, 1889, Cambridge University Library, Royal Greenwich Observatory Archives, RGO 15/126 ff. 114r-v.

to the first edition of 1,000 copies." She reserved the option of asking for royalties if the book ran to a second edition. Clerke received her first check for slightly more than £35 from Longman at the end of 1891. The book sold slowly. By this point about 700 copies had been bought. Two years later she was paid slightly more than £24. She continued to receive between four and ten pounds per year over the next eight years. In 1902 she received her last share of the profits, a check for three shillings and sixpence. Longman was not interested in publishing a second edition, so Clerke arranged for Black to take it on in 1905.[88]

Clerke also earned money by writing periodical articles, encyclopedia essays, and biographical dictionary entries, mostly on astronomy. She wrote a small number of articles for some journals, such as the *Dublin Review, Popular Astronomy, Astronomy and Astrophysics, Fraser's Magazine, Quarterly Review,* and the *Contemporary Review.* After the publication of her *Popular History* she became a regular contributor to the *Observatory* and to *Nature.* When her contributions to the latter came to an abrupt end in 1893, she began to write for *Knowledge.* She contributed fifty-four major articles to the *Edinburgh Review* over the course of thirty years, beginning in 1877 and ending upon her death. Three-quarters of them dealt with scientific subjects. Clerke authored over one hundred articles over the course of her career. Through her link with the *Edinburgh Review* she became a close friend of its editor, Henry Reeve, who likely provided her with an introduction to Adam and Charles Black. When Clerke first arrived in London, this publishing house was in the process of bringing out the ninth edition of the *Encyclopaedia Britannica.* Reeve was already a prolific contributor. Clerke was invited to write essays on matters connected with the history of science. Proctor had been writing most of the astronomy pieces up until the publication of volume ten in 1879, when Clerke replaced him.[89] Her major contributions

---

88. P. Gould 1998, 170-71; *Archives of the House of Longman, 1794-1914* 1978, A15, 9-10, A16, 84, 139, N233, 25; Brück 2002, 189.

89. Proctor did not have a very high opinion of Clerke's *Popular History.* To Clodd, he wrote in 1887 that "Miss A. M. Clerke's book is occasionally useful, but she herself evidently knows nothing and can only quote other folks' opinions, putting them into her own words, and sometimes, nay, often, showing that she has misunderstood them" (Clodd 1926, 61). Clerke would have had several other strikes against her, from Proctor's point of view. She was friendly with his nemesis, Lockyer, and pluralism was not a particularly important theme in her works, though she did not rule out the possibility of life on other planets in the universe (Brück 2002, 158). Clerke had good reason to be annoyed with Proctor. In 1887 he was critical of her friend Holden, of the planned Lick telescope, and of the craze for large telescopes in general (Ibid., 52). Nevertheless, Clerke's review of Proctor's *Old and New Astronomy,* and her evaluation of his entire career, were largely

included pieces on Galileo, Alexander von Humboldt, Huygens, Kepler, Le Verrier, Lavoisier, Lagrange, and Laplace. Clerke was also one of the regular contributors to the *Dictionary of National Biography,* and she dealt with over 150 subjects, including the Astronomers Royal, the three Herschels, and many of the other astronomers. In addition, she wrote the entries on Charles Babbage and John Dalton.[90]

Clerke's prodigious output and her accomplishments earned her recognition and several honors. In 1889 she was offered one of the supernumerary computer positions at Greenwich that had been set up for women. Although tempted, she declined the position. It would have required her to move closer to Greenwich and it would have taken her away from her writing. Likely, she also did not relish the possibility, at age forty-seven, of working along-side women half of her age who had just finished at a Cambridge women's college. She was among the four women elected to the forty-eight-member council of the newly established British Astronomical Association in 1890. In 1893 she was awarded the Royal Institution's Actonian Prize, and in 1902 she was elected a member of the Royal Institution. But perhaps her greatest honor was to be made an honorary member of the Royal Astronomical Society in 1903, along with her close friend Margaret Huggins. Only three women had previously been accorded such a mark of distinction, Caroline Herschel, Anne Sheepshanks, and Mary Somerville. Placed in this select group of women merely reinforced a comparison that had been made between Clerke and Somerville for some time. In an enthusiastic assessment of Clerke's *Popular History* in the pages of the *Edinburgh Review,* the reviewer declared that the book had been written by a lady "on whom the mantle of Mrs. Somerville seems to have descended." Critics had noticed the similarities between Somerville and Clerke as early as 1882 because of the appearance of her *Encyclopaedia Britannica* essays on Lagrange and Laplace.[91]

The comparison to Somerville was made even though Clerke was not nearly as knowledgeable about mathematics, and despite her focus, for the most part, on one discipline within science.[92] Like Somerville, Clerke was

---

positive. She was critical, however, of his "acrid" manner in controversy and the inclusion of a few "offensive allusions to the Scriptures" in his *Old and New Astronomy* ([Clerke] 1893, 546–48).

90. Brück 2002, 32–34, 37, 153–54; P. Gould 1998, 174.

91. Brück 2002, 38, 74–75, 99, 108, 153, 174; [Mann] 1886, 372.

92. It could be argued that dealing with the connections within one discipline was the most that could be expected in an age of specialization, even of a well-informed practitioner. Whewell's notion of finding unity across all scientific disciplines was no longer possible. The time of the

one of the few women who concentrated on the physical sciences, who could condense current research into a synthesis, and who wrote primarily for a more mature and informed audience, including practitioners. For Margaret Huggins, the appearance of Clerke on the scene had no precedent, and it indicated that a new stage in the evolution of science had been reached. She declared, "the progress of Science and the growth of its literature during the last quarter of a century has been so enormous, that a new order of worker is imperatively called for." Clerke was "an admirable example of such a worker" whose "mission is to collect, collate, correlate, and digest the mass of observations and papers—to chronicle, in short, on one hand; and on the other, to discuss and suggest, and to expound: that is, to prepare material for experts, and at the same time to inform and interest the general public."[93]

Clerke started off more as a synthesizer in her first book, *A Popular History of Astronomy during the Nineteenth Century*. In the preface to *Popular History*, Clerke stated that her goal was to introduce the public to spectroscopy, the most important feature of the "new astronomy." "It embodies an attempt to enable the ordinary reader to follow," she declared, "with intelligent interest, the course of modern astronomical inquiries, and to realize . . . the full effect of the comprehensive change in the whole aspect, purposes, and methods of celestial science introduced by the momentous discovery of spectrum analysis." One striking result of the use of the spectroscope by astronomers was that celestial and terrestrial science had become unified. Whereas astronomers had previously depended largely on telescopes and the calculus, and had "looked with indifference on the rest of the sciences," now the chemist, the electrician, the geologist, the meteorologist, and even the biologist supplied the materials for induction. "The astronomer has become," Clerke declared, "in the highest sense of the term, a physicist; while the physicist is bound to be something of an astronomer." Here were Clerke's parallels to Somerville's "connexions" between the sciences. In his positive review in *Nature*, Ball stated that Clerke was too modest when she described her book as a "popular work." He believed that "few men of science who use this book will think that it ought to be classed as a popular work in the

---

polymath was over. Of course there were some attempts in the latter half of the nineteenth century to codify all knowledge through evolution or positivism, but practitioners never accepted such attempts across the board. We have seen how Allen and Clodd's forays into the physical sciences were rejected. Herbert Spencer's synthetic philosophy placed far more emphasis on the life sciences than on the physical sciences.

93. Creese 1998, 238; M. Huggins 1907, 15, 17; Lightman 1997a, 61–75.

ordinary acceptation." He described it as a "masterly exposition of the results of modern astronomy in those departments now usually characterized
as physical."[94] Clerke was already writing for both practitioners and general
readers.[95]

As Clerke became more knowledgeable about astronomy over the years
her books became more demanding for the general reader, and, drawing on
her grasp of the connections between areas of research in the field, she began to offer suggestions to practitioners on which lines of inquiry to pursue.
In her second major book, *The System of the Stars*, Clerke tackled the topic of
sidereal astronomy, which examined the "nature, origin, and relationships
of 30,000,000 stars and of 120,000 nebulae—to inquire into their movements among themselves, and that of our sun among them." The findings of
recent research in sidereal astronomy led Clerke to conclude that all of these
celestial bodies belonged to one system—humanity lived in a one-galaxy universe. Whereas the spectroscope took center stage in her first book, the camera was the star of this second book. Mary Brück has remarked that although
Clerke did not set out to imitate Somerville, her *System* was in many ways
a "new astronomy" version of the *Connexion* in that both books presented
pictures of the physical universe based on a synthesis of scientific research in
their respective eras. In the preface, Clerke again declared that her intention
was to make astronomy accessible to a general audience. "To bring it within
the reach of many is the object aimed at in the publication of the present volume," she wrote. "Astronomy is essentially a popular science." But Clerke's
second book was even more sophisticated than her first. In her review of *System*, Margaret Huggins agreed with Clerke that astronomy was "essentially
a popular science," since of all the sciences it was the one "which appeals
most readily and powerfully to the imagination." But she maintained that
Clerke's book had raised the bar for those writing books for the public. "The
common idea of popular writing is too low a one," she asserted. "There is
too much going down to the popular level, too little effort to bring it up."
Huggins believed that Clerke's book was of value to astronomers. She recommended that practical scientific workers look to Clerke's book to help
them formulate new generalizations from the mass of accumulated facts in

94. Clerke and Ball seem to have been on good terms and respected each other's work. Ball asked
Clerke to read the proofs of his *Story of the Sun* (1893), and thanked her in the preface. In his
Popular Guide to the Heavens (1905) he referred his reader to Clerke's *Problems in Astrophysics* for
the results of the most modern work upon stars, star clusters, and nebulae (R. Ball 1905, 62).

95. Brück 2002, 43–45; Clerke 1885, v, 182–83; R. Ball 1886, 313.

astronomy. Indeed, the American astronomer, E. S. Holden, wrote to Clerke that her *System* was "not only an adequate account of our present knowledge but it is full of pregnant suggestion for our future guidance."[96]

In her third and final major book, *Problems in Astrophysics* (1903, A. and C. Black), Clerke's object was "not so much to instruct as to suggest." Rather than recount the past discoveries of eminent scientists, she examined cutting-edge issues in astronomy and laid out paths for future investigation. A typical chapter in *Problems in Astrophysics,* which dealt in great detail with the physics of the sun, stars, and nebulae, offered a synthesis of the current state of knowledge in a specific area of research, outlined the many questions that remained to be answered, and suggested work to be done by astronomers to begin answering these questions. Clerke dictated to the practitioners what lines of research to pursue. This was also the most technically scientific of her works and therefore the least accessible to a popular audience. It was not a big seller and did not reach a second edition. Reviewers recognized that it did not fit any of the usual categories for science writing. In *Academy and Literature,* the reviewer stated that although it was "written by a popular and learned writer on astronomical subjects, the present excellent book is yet not of a popular nature." Clerke's writing was "clear in the extreme, and would almost render her book popular, were popularity possible to subject-matter so recondite." The reviewer in the *Nation* voiced a similar judgment. "Yet it is not a book of popular science," the journalist affirmed, "it is a popular book on professional science—a thing seldomer to be found and quite otherwise enlightening." Clerke's work challenged the demarcation between books written for a general reading audience and those aimed at practitioners.[97]

Clerke's presentation of herself as an astronomical synthesizer, who could set the agenda for future research, was possible owing to the extensive international network she had built up over the years. Correspondence and personal friendships with key British and American practitioners of astronomy kept her informed about the latest discoveries and boosted her career as popularizer. Margaret Huggins noted that she "cultivated personal relations with a wide circle of astronomical workers, in person, or by correspondence." In her judgment, "these relations had much to do with the success

96. Clerke 1905b, ix, 10; Brück 2002, 97, 194; M. Huggins 1890, 382–383; Holden to Clerke, January 21, 1891, University of California, Santa Cruz, Mary Lea Shane Archives.

97. Clerke 1903, vii; Brück 2002, 160, 163; [Thompson] 1903, 173; "Clerke's Astrophysics" 1903, 98.

of her work." Clerke established friendships with some of the key British astronomers engaged in research with the spectroscope and the camera. Norman Lockyer was the first astronomer that Clerke encountered in person after moving to London. He was a key player in the development of the field of solar physics. She met him through Richard Garnett, superintendent of the reading room at the British Museum. Clerke visited Lockyer's South Kensington observatory sometime near the beginning of her career as popularizer. Lockyer asked her to read the proofs of his forthcoming book, *Chemistry of the Sun and Movements of the Earth* (1887), publicized the second edition of Clerke's *Popular History* in *Nature*, and invited her to contribute frequently to *Nature*. Lockyer played an important role in her early success and helped to make others aware of her talents. Lockyer and Clerke remained friends even when she supported William Huggins's criticism in 1889 of Lockyer's theory that all astronomical objects were composed of the same material as meteorites and comets.[98]

Clerke met William and Margaret Huggins after she had become acquainted with Lockyer, and after the publication of her *Popular History*. Clerke established an intimate friendship with them. The Hugginses pioneered the use of the spectroscope in astronomical research. It was Margaret who penned the appreciative review of *System of the Stars* and who wrote the biography of Clerke and her sister Ellen shortly after their deaths. Clerke was also close friends with David Gill, director of the Royal Observatory at the Cape of Good Hope in South Africa. Gill was a pioneer in the use of the camera to map the heavens. At the Cape, he worked on producing a catalogue of the southern stars. In 1887 he became one of the prime movers behind an international charting project, known as the *Carte du Ciel*, to photograph the entire sky down to the 14th magnitude. She met Gill just after he had been in Paris at the Astrographic Congress, where the *Carte du Ciel* had been launched, when he delivered a lecture at the Royal Institution on the applications of photography to astronomy. At Gill's invitation, Clerke spent two months at the Cape in 1888 to have the experience of doing practical work in an observatory. Gill assigned her a project that involved the spectroscopic study of stars in the southern heavens, which, he told her, was "absolutely virgin soil." The results were later published in two papers in the *Observatory*. Gill became Clerke's closest and most trusted friend. He read over each chapter of *System of the Stars* and recommended refinements

98. M. Huggins 1907, 13-14; Brück 2002, 53, 56, 83-84; Lockyer and Lockyer 1928, 117.

to Clerke. On August 2, 1889, she thanked Gill for his corrections on chapter eighteen, and told him that she would not have had the "courage" to publish some chapters "without submitting them to you." Clerke later dedicated her *Problems in Astrophysics* to Gill.[99]

Clerke's relationship with three key figures in the development of the "new astronomy" in Britain was complemented by her friendship with important American astronomers, many of whom worked in observatories producing spectacular photographs of distant heavenly objects. Hearing about Clerke through Richard Garnett, Edward S. Holden initiated a friendly correspondence in 1884 and offered his help. On January 15, 1884, Clerke replied, "I am deeply touched by such a mark of sympathy with me in my arduous work," arduous to her, she explained, since she had only her "own limited resources." Just after her *Popular History* was published, Holden was appointed president of the University of California and director-designate of the new Lick Observatory on Mount Hamilton in California. He became full-time director in 1888 when the observatory was completed. Holden sent her stunning photographs taken with the Lick telescope, including E. E. Barnard's Milky Way panoramas. After he left the Lick, she corresponded regularly with his successors J. E. Keeler and W. W. Campbell. Keeler, who succeeded Holden in 1898 as director, considered Clerke to be an important molder of opinion in the astronomical world. He kept her updated on his research, including his discovery that a large fraction of the known nebulae were spirals. On September 8, 1899, she thanked Keeler for sending her a superb set of nebula photographs that vividly illustrated "what may be called the 'law of spirality.'" After Campbell had succeeded Keeler in 1900, she wrote to him several times to suggest experiments on the variable star Mira Ceti and on Nova Persei. Campbell wrote a positive review on *Problems in Astrophysics* in 1903, where he stated, "by way of suggestions for future lines of research this book is the richest one known to me."[100]

Clerke did not limit her American network to the Lick Observatory. Beginning in 1885 she corresponded with Edward E. Pickering of the Harvard

99. Brück 2002, 59, 63–69, 72, 79; M. Huggins 1890; 1907; Forbes 1916, 201; Clerke to Gill, August 2, 1889, Cambridge University Library, Royal Greenwich Observatory Archives, RGO 15/126 ff. 138–41v.

100. Brück 2002, 48–50, 60, 86–87, 142, 144–45; Clerke to Holden, January 15, 1884, University of Wisconsin, Madison, Archives of the Washburn Observatory, Series 7/4/2 (Department of Astronomy) Box 1 Folder C; Osterbrock 1984, 315; Clerke to Keeler, September 8, 1899, University of California, Santa Cruz, Mary Lea Shane Archives; Campbell 1903.

College Observatory, which was to become a center for astronomical spectroscopy. Pickering also sent her photographs. On February 20, 1897, Clerke wrote to him to "express my admiration of the splendid photographs of the Argo Nebulae." Clerke also met George Ellery Hale, the future director of the new Yerkes Observatory at the University of Chicago, in 1891 at the British Association for the Advancement of Science meeting at Cardiff. Hale was only twenty-three at the time, and had just invented the spectroheliograph, an instrument used to observe solar prominences and the upper layers of the sun. His largely positive review of Clerke's *System of the Stars* had been published a few months before they met. They corresponded until Clerke's death. Clerke's extensive correspondence network in the United States, together with her circle of astronomical friends in England, provided her with the latest news on developments in the "new astronomy," with a pool of positive reviewers for her books, and, most importantly, with a source of stupendous photographs.[101]

Clerke not only established herself among practitioners as a prolific writer who could synthesize huge amounts of information, she also became accepted as an expert in the interpretation of astrophotography. This was a valuable skill to have when images of strange heavenly objects were circulating among astronomers around the world. Holden, for example, sought her opinion on how to use images to understand the structure of nebulae, star clusters, and other astronomical phenomena. In 1889 he sent Clerke one of the first of Barnard's famous photographs of Milky Way fields and asked for her reaction. Clerke noticed an interesting connection between this photo and one by Isaac Roberts of the great nebula in Andromeda. If you hold Barnard's picture back from the eye, Clerke told Holden, the lines of Robert's picture appear. "In fact," she declared, "this picture might be called the key to the other," for Barnard's photograph revealed what the Andromeda nebula would look like after the lapse of ages. "Not a mere solar system," Clerke wrote, "but just a stupendous stellar collection is forming from it. If this be so, the revelation is of far-reaching significance." Drawing on her knowledge of recently produced astronomical photographs, Clerke's ability to cross-reference provided her with a method for interpreting complex images. Thanks to her analytic skills, she was often the first in England to see the most spectacular triumphs of the Lick Observatory's photographic telescope. She had her pick of the latest, most interesting images for her books,

101. Clerke to Pickering, February 20, 1897, Harvard College Observatory, Archives, Director's Correspondence; Brück 2002, 92, 102, 104–5; Hale 1891.

and the association of her work with the respected names of those who contributed photographs could only help to establish her as an important astronomical authority.[102]

Clerke's synoptic overviews of the "new astronomy," which placed the camera at the center of the recent developments, offered more than just a synthesis of scientific facts. Her synthesis contained a theological dimension. Like those in the revised maternal tradition, she saw religious lessons contained in nature, even in distant heavenly objects to be found in the depths of space. Clerke made the link between astronomy and religious faith explicit throughout her works. In the preface to the first edition of *Popular History* she identified as one of her goals the attempt to help readers "towards a fuller understanding of the manifold works which have in all ages irresistibly spoken to man of the glory of God." Wherever she turns Clerke sees God throughout the entire gamut of astronomical phenomena. Whether it be the evolution of the planets whose growth is guided "from the beginning by Omnipotent Wisdom"; or the "sequence of Divinely decreed changes" by which nebulae are transformed into star clusters; or even gigantic galactic rifts of starless space, wherein "Supreme Power is at work in dispersing or refashioning" star clouds, Clerke sees the hand of God.[103]

However, Clerke recognized that natural theology could not continue unaltered, especially after the revelations of the "new astronomy." Whereas astronomers at the end of the eighteenth century had conceived of the universe as a machine characterized by simplicity, harmony, and order, the picture of the cosmos emerging from recent discoveries emphasized complexity and inexhaustible variety. "The heavens are full of surprises," Clerke exclaimed, after puzzling over changes of brightness in variable nebulae. At times, Clerke could point to discoveries of the "new astronomy" that disclosed orderly patterns in the heavens hitherto concealed by apparent chaos. The "seeming confusion" of asteroids when compared with the "harmoniously ordered and rhythmically separated orbits of the larger planets" proved to be "not without a plan." The distribution of stars and nebulae is "easily seen to be the outcome of design." The "new astronomy," which

---

102. Clerke to Holden, August 15, 1889, University of California, Santa Cruz, Mary Lea Shane Archives.

103. Clerke 1885, vi, 348; 1902, 207; 1903, 541. Mary Brück has presented an interesting, though not completely convincing, argument that Clerke did not write in the English natural theology tradition. She points to the influence of Catholic theologians, in particular St. Thomas Aquinas (Brück 2002, 205–8).

FIGURE 8.11 Photograph of a spiral nebula, the great nebula in Andromeda, by Dr. Max Wolf. Clerke pointed to the symmetrical shape of this nebula as evidence for divine design. (Agnes M. Clerke, *The System of the Stars*, 2nd ed. [London: Adam and Charles Black, 1905], 260.)

was slowly bringing about a unification of science, revealed and reflected the unity of design behind nature. In breaking down the barrier between physics and astronomy, terrestrial and celestial science, the earth and the stars, the "new astronomy" was a science that "aims at being one and universal, even as Nature—the visible reflection of the invisible highest Unity—is one and universal."[104]

Clerke's discourse of design in her theology of nature played an important role in her interpretation of photographs. In *System of the Stars* she included a photo of the great nebula in Andromeda and suggested that the image helped astronomers make sense of what had hitherto been an enigma (Fig. 8.11). "The view given by this magnificent picture of the Andromeda nebula as a symmetrical, though still inchoate structure," she observed, "ploughed up by tremendous, yet not undisciplined forces, working harmoniously towards the fulfillment of some majestic design of the Master Builder of the universe to modify profoundly our notions as to how such designs obtain their definitive embodiment." Clerke was able to reinscribe the concept of design into astronomy through the use of the most up-to-date technology.

104. Clerke 1903, 522; 1885, 183, 328–29, 1905a, 147.

In her review of *System of the Stars,* Margaret Huggins defended her friend's prerogative to discuss the religious implications of astronomy. "Miss Clerke has not been ashamed to show again and again in her book her deep reverence for the Deity—her faith in a divine order in which and towards which all things move," Huggins wrote. "It is interesting to notice this courage of conviction in such a work, for it has become much a fashion to seem really afraid to even mention the word god when science is concerned." Although Huggins agreed that science must not be trammeled by theological dogma, she thought that Clerke had succeeded in keeping science free while not repressing religious instinct. Three years later, Clerke was honored to receive the Royal Institution's Actonian Prize. Every seven years a prize of £105 was awarded for the best essay upon "the Beneficence of the Almighty" as illustrated by discoveries in science.[105]

Clerke had no difficulties accepting the idea that the evolutionary process was at work in the divinely ordered heavens. She had grave reservations, though, about organic evolution as conceived of by Huxley and other evolutionary naturalists. In a chapter in her *Problems in Astrophysics,* titled "The Evolution of the Stars," she presented stellar evolution as part of creative process. She asserted that in the celestial regions, "by a wonderful course of development, the designs of the Maker are being unfolded, but with such majestic leisureliness that every step represents the lapse of millions of years." However, in her *Modern Cosmogonies* (1905, A. and C. Black) she was critical of Laplace and his nebular hypothesis. Laplace arrogantly believed that he knew the secrets of the universe, but his simplistic theory had "ceased to be satisfactory" because the "effects it was designed to elicit" were so intricate. Laplace's theory had "crumbled before the storms of adverse criticism," and it survived "only as a wreck." Laplace had been right about the "unity of the solar world," that the planets "once made an integral part of the substance of the sun." But there was no consensus of opinion on "the mode and manner of cosmic change." Clerke also tackled the issue of organic evolution in a chapter on "Life as the Outcome" in her *Modern Cosmogonies.* While she was willing to accept the notion of an ascending sequence of animal forms, she criticized both Huxley and Spencer for their belief that spontaneous generation took place at some point in the distant past. She referred to Huxley's statement that life is a property of protoplasm, the inevitable outcome of the nature and disposition of its molecules, as an "absurdity."[106]

105. Clerke 1905b, 259; Lightman 2000, 671–79; M. Huggins 1890, 386; Brück 2002, 108.
106. Clerke 1903, 271; 1905a, 58–59, 80–81, 269–72; Brück 2002, 212.

Clerke's critique of Huxley and Spencer did not draw fire from supporters of evolutionary naturalism. But her adoption of the role of astronomical synthesizer and critic was not well received by some scientists, and it did not endear her to feminists. In a series of reviews in *Nature*, Richard A. Gregory became increasingly critical of Clerke's works. Gregory, who had worked in Lockyer's spectroscopic laboratory, had been writing for *Nature* since 1890 and was a regular book reviewer. Likely the anonymous reviewer of the third edition of *Popular History*, he may have been offended by Clerke's support for Huggins in his controversy with Lockyer in 1889. Here the reviewer scolded Clerke for her tendency to cover Huggins's work in depth while ignoring the contributions of others. "If Miss Clerke were more a historian and less a partisan," the reviewer declared, "her work would be of higher value." It seems an unlikely coincidence that Clerke's contributions to *Nature* came to an abrupt end in 1893, the same year that this hostile review appeared and that Gregory became assistant editor of *Nature*.[107] Gregory's signed review of *Problems in Astrophysics* was the first of two pieces that launched a vicious attack on Clerke's scientific credentials that explicitly raised the issue of gender and denied her the right to adopt the role of synoptic critic. Gregory maintained that her function as "historian is to assimilate and describe." To him, this meant that she surveyed the work of practitioners from the point of view of a spectator who "should describe fairly and clearly what she sees, without irritating the men who are doing the work by expressing her opinion upon it or suggesting what course they ought to take next." Gregory reminder her that, "Passengers are respectfully requested not to speak to the man at the wheel." Then Gregory cast aspersions on Clerke's ability to limit her work to the realm of reasonable assertion. Gregory declared that "it is a characteristic of women to make rash assertions," and that "Miss Clerke is apparently not free from this weakness of her sex." Still loyal to Lockyer, Gregory could not resist chastising Clerke for ignoring his theories on novae. She was "not an infallible guide" because she had neglected her duty as a historian to mention that the association of nebulae with new starts was first put forward in the meteoritic hypothesis by Lockyer.[108]

107. Clerke had written six articles for *Nature* in 1886, and up until 1891 contributed between two to five pieces per year. In 1892 she wrote only one article, which was the last to appear. She then began to contribute regularly to *Knowledge*.

108. Brück 2002, 120, 197; "Astronomy of the Nineteenth Century" 1893, 2; Gregory 1903, 339–41; Lightman 1997a, 72–73.

In 1906 Gregory reviewed a new edition of *System of the Stars*. He began by asserting, "the intuitive instinct of a woman is a safer guide to follow than her reasoning faculties." An attractive instinct when applied to the ordinary affairs of life, it was "derogatory when it influences the historiographic consideration of contributions to natural knowledge." Gregory argued that Clerke's neglect of Lockyer's school of spectroscopy was a result of her reliance on intuition over reason. "It is scarcely too much to say," he insisted, "that the evidence brought forward by Sir Norman Lockyer in connection with his meteoritic hypothesis of celestial evolution is chiefly responsible for the change of view that has taken place." Gregory implied that she, like all women, was incapable of offering a true synthesis of the state of affairs in astrophysics. He also criticized those who meekly accepted Clerke's views. "There are many students of science who follow the trend of a writer like Miss Clerke with lamb-like sequacity," he wrote, "and consider it almost a presumption to express any dissatisfaction with her presentment or interpretation of scientific fact." Gregory charged that it was "a sign of weakness to occupy a position of this kind," implying that Clerke's allies were effeminate, and that it was particularly reprehensible in a case "when the author whose views are accepted is not actively engaged in the investigation of the field surveyed." Clerke, Gregory reminded his readers, was not a practitioner. Gregory concluded by emphasizing that Clerke had not provided a complete synoptic overview of the field. By neglecting Lockyer's contributions and privileging Huggins's views, "an incomplete story is presented of the meaning and mysteries of sidereal development revealed by spectroscopic research."[109] Gregory would have rejected the notion that Clerke had successfully emulated Somerville.

While Gregory denied that Clerke presented a genuine synoptic overview of astronomy in her works due to the failings she shared with all women, feminists did not celebrate her accomplishments in the same way that Somerville became a scientific heroine upon her death. Paula Gould has pointed out that at first glance they followed similar career paths. They did not receive a formal education, reached middle age before their first book was published, wrote a similar number of books that were well received, and were both made honorary members of the Royal Astronomical Society. Despite these similarities, Gould argues that they were judged according to different criteria. It was in Whewell's review of Somerville's *Connexion* that

109. Gregory 1906, 505, 507–8.

he first introduced the term "scientist." He coined the term to counter the fragmentation of the sciences. When he praised Somerville for rendering "a most important service to science" in her attempt to illustrate the general principles unifying the detached branches of science, he implied that the new term applied to her. Somerville managed to retain the reputation for being a scientist up until her death in 1872. By the time of Clerke's death, the definition of what constituted a scientist had hardened significantly, and the emphasis on institutionalized research work in laboratories or observatories defined her as a writer. Clerke never became the subject of biographical literature written for women. While Somerville, the Whig icon, was a supporter of the suffrage movement, Clerke, the conservative Catholic, seemed to have had little to offer proponents of the women's movement.[110]

Although Clerke depicted Caroline Herschel as her brother's dronelike assistant in *The Herschels and Modern Astronomy* (1895, Cassell), she often pointed to the accomplishments of women in astronomy in her writings. In her discussion of the work of William Huggins in the *Edinburgh Review,* she made sure that his wife received her due. She asserted that "since 1875 she has been, on equal terms, her husband's coadjutor, and while content to merge her initiative in his, she has known how to make its effect and influence tell as essential factors in the joint product of their labours." She also pointed to the work of Mrs. Willamina Fleming and the team of women under her direction at the Harvard College Observatory in the compilation of the *Draper Memorial Catalogue* of stellar spectra published in 1890. In 1898 Clerke enthusiastically reviewed Alphonse Rebière's *Les femmes dans la science,* which, she declared, "bear[s] remarkable witness to the actual activity of women-workers in science." In the past, the contributions of women to knowledge were valuable, but "unsystematic." Now they were becoming "so serious and habitual as to be admitted without surprise, and appropriated without compliment." Clerke predicted that the increased contribution of women to science would produce "momentous" results. "Intellectual history must be profoundly modified by so large an accession to the forces making for progress," she affirmed.[111] Although by no means an outspoken proponent of feminism, Clerke was supportive of the contributions of women to science.

Although Clerke and Bodington adopted a strategy for science writing articulated by Somerville in the early nineteenth century, in many ways they

110. P. Gould 1998, 190–94.

111. Clerke 1895, 125, 139–41; Brück 2002, 176–77; [Clerke] 1900, 458; Clerke 1902, 385; 1898a, 132.

were more forward looking than Giberne, Brightwen, and Hutchinson. The synoptic overview has become a standard format in popularizations of science. By presenting their work as a response to increased specialization, Clerke and Bodington justified a new role for the popularizer who was not a practitioner. This new role took into account the future of scientific progress. As synthesizers of knowledge they conveyed the results of the research conducted by practitioners to the general audience, as had previous popularizers throughout the nineteenth century. But whether it was a synthesis of astronomy, as in Clerke's case, or of biology, Bodington's forte, popularizers could also serve a second useful role. They presented a synoptic overview of knowledge in one discipline to practitioners who were so specialized that they could no longer keep up with the research outside their narrow areas of expertise. The role of synthesizer enabled Bodington to enter current scientific controversies over the value of Darwin's theory of natural selection and to express her preference for Lamarckianism. Clerke could attempt to set the agenda for future research in astronomy. Adopting this new role helped Bodington and Clerke to survive on new Grub Street, though the latter was more successful in establishing herself in the world of the practitioner through her expertise in interpreting astronomical photographs. But Bodington and Clerke disagreed on the status of scientific naturalism. While Clerke's theology of nature was an integral part of her astronomical synthesis, Bodington identified more with the agnosticism of Huxley and his allies.

Giberne, Brightwen, and Hutchinson looked back to the clerical and maternal traditions. All of them set science in a religious framework. Giberne drew on the old familiar format and continued the tradition of storytelling, while Brightwen adopted the seasonal literary format and took the reader on rambles like Johns. Hutchinson was indebted to Hugh Miller as well as to Buckley and the evolutionary epic. Yet the traditions that inspired them, with the exception of the evolutionary epic, did not fare well after they were gone. The Anglican-clergyman tradition continued into the twentieth century, though considerably weakened. Early twentieth-century Anglican modernists such as Charles Raven, Bishop E. W. Barnes, and Dean W. R. Inge hoped to revive public interest by drawing on a renewed natural theology to demonstrate the compatibility of Christian faith with science. But they were unable to convince the majority of Anglicans that their approach to reconciliation could halt the Church's decline.[112] The power of the

---

112. Bowler 1998, 65–67; idem 2001.

neomaternal tradition seems to have suffered the most. Women were "edged out" of scientific authorship by the end of the century, even though women like Bodington and Clerke were better informed than their predecessors. As soon as science for the masses came to be seen as important in the 1870s, scientific practitioners could not leave it to the care of women. By the end of the century, as scientific endeavors, including the popularization of science, increasingly came to be seen as being linked to masculine intellectual qualities, women were slowly losing their status as interpreters of nature.[113] The decline of the maternal tradition at the end of the century might also have been a result of the increasing access that women had to scientific training in institutions of higher education. Women who were interested in science had more opportunities in the twentieth century to pursue careers. The financially uncertain road of science writing would have seemed less appealing by comparison. The golden age of female popularization of science in the second half of the nineteenth century was, in part, a by-product of the lack of options for women who were fascinated by the world of nature.

113. Gates 1998, 37, 64; Gates and Shteir 1997, 16. Gates and Shteir point out that educational reform directed young women away from science. Formal education in natural history for girls was lessened in the latter half of the century, and more attention was given to the classics, supposedly to make women's study equivalent to men's. Young women in secondary schools were discouraged from taking classes in scientific subjects by the time of the Bryce Commission report in 1894–95 (Gates and Shteir 1997, 17).

Remapping the Terrain

IN THIS book I have tried to establish that a large group of popularizers of science existed during the second half of the nineteenth century who commanded a significant readership and whose agenda was at odds with the aims of many practitioners, especially the scientific naturalists. One of my aims, then, has been to secure a place for these figures on the map used by scholars to depict the topography of nineteenth-century British science. This group's influence can be demonstrated by pulling together the print run data I have supplied throughout this book and comparing it to the information we have for the publications of eminent scientific naturalists. If we list the gross number of books printed by the end of the nineteenth century, concentrating on the single most successful books by popularizers, and restricting ourselves to the most successful writers, it becomes clear that the figures are comparable to those for the major works by Darwin, Huxley, Tyndall, and Spencer (Fig. 9.1). Spencer's *Study of Sociology,* the International Scientific Series sales record holder, is in the middle of the group in between 20,000 and 40,000, along with Darwin's *Descent of Man* and works by Chambers, Pepper, Proctor, Wright, Clodd, and Loudon. From the vantage point of scientific reader at the end of the century, Darwin's *Origin of Species* is in the same company as extraordinarily successful books by Wood and Brewer. It is well below Wood's *Common Objects of the Country,* though, and pales in comparison to Brewer's *Guide.* If we examine the figures for best sellers, Darwin's *Origin* is less remarkable (Fig. 9.2). Books by Brewer, Wood, Chambers, Clodd, Kirby, Wright, Pepper, Giberne, and Somerville all surpassed Darwin's *Origin of Species* in printings in the ten-year period

FIGURE 9.1

Print Runs of Steady Sellers: Number of Books or Editions Printed

Brewer's *Guide to the Scientific Knowledge of Things Familiar* (1847, Jarrold)
(195,000 up to 1892) [66]
Wood's *Common Objects of the Country* (1858, Routledge) (86,000 by 1889) [175]
Darwin's *Origin of Species* (1859, Murray) (56,000) [34]

40,000———

Chambers's *Vestiges of the Natural History of Creation* (1844, John Churchill)
(39,000 by 1890) [33]
Darwin's *Descent of Man* (1871, Murray) (35,000) [34]
Pepper's *Playbook of Science* (1859, Routledge) (34,000) [209]
Proctor's *Other Worlds Than Ours* (1870, Longman) (29 printings by 1909) [305]
Spencer's *Study of Sociology* (1873, Henry S. King) (23,830) [381]
Wright's *Observing Eye* (1850, Jarrold) (20,100) [107]
Clodd's *Childhood of the World* (1873, Macmillan) (20,000 by 1879) [256]
Loudon's *Ladies' Companion to the Flower Garden* (1841, W. Smith) (9 editions of
20,000 copies) [111]

20,000———

Kirby's *Stories about Birds of Land and Water* ([1873], Cassell) (18,000 by 1873) [109]
Ball's *Story of the Heavens* (1885, Cassell) (18,000 by 1891) [410]
Gatty's *Parables of Nature* (1855, Bell and Daldy) (18 editions of first series by 1882) [107]
Somerville's *On the Connexion of the Physical Sciences* (1834, John Murray) (17,500) [22]
Tyndall's *Forms of Water* (1872, Henry S. King) (14,250) [381]
Johns's *Flowers of the Field* (1853, SPCK) (13 editions by 1878) [49]
Page's *Introductory Text-Book of Geology* (1854, Blackwood) (12 editions by 1888) [225]
Giberne's *Sun, Moon, and Stars* ([1880], Seeley, Jackson and Halliday) (13,000) [430]
Kingsley's *Glaucus; or, The Wonders of the Shore* (1855, Macmillan) (10 printings) [75]

10,000———

Morris's *History of British Butterflies* (1853, Groombridge) (8 editions by 1895) [45]
Roberts's *Domesticated Animals* (1833, J. W. Parker) (7 editions) [110]
Brightwen's *Wild Nature Won by Kindness* (1890, T. Fisher Unwin) (7 editions by
1909) [441]
Houghton's *Country Walks* (1869, Groombridge) (6 editions) [83]
Huxley's *Crayfish* (1880, Kegan Paul) (5,775) [384]
Webb's *Celestial Objects* (1859, Longman) (5,500) [59]
Hutchinson's *Extinct Monsters* (1892, Chapman and Hall) (5 editions) [453]

FIGURE 9.1 (continued)

Ward's *Microscope Teachings* (1864, Groombridge) (5 editions) [104]
Zornlin's *Recreations in Physical Geography* (1840, John Parker) (4 editions) [109]
Clerke's *Popular History of Astronomy during the Nineteenth Century* (1885, A. and C.
   Black) (4 editions by 1902) [472]
Henslow's *Botany for Children* (1880, Stanford) (3 editions) [89]
Pratt's *The Field, Garden, and the Woodland* (1838, Charles Knight) (3 editions) [104]
Bowdich Lee's *Anecdotes of the Habits and Instincts of Birds, Reptiles, and Fishes*
   (1853, Grant and Griffith) (3 editions) [102]
Twining's *The Plant World* (1866, Nelson) (2 editions) [111]
Allen's *Evolutionist at Large* (1881, Chatto and Windus) (2000) [274]

Note: The books listed here are mostly from the second half of the nineteenth century, when the
size of print runs for an edition were roughly similar. In the first set of parentheses are the dates
and publishers of the first edition. The second set of parentheses lists the number of books published or editions up to end the century in Britain unless otherwise noted. The numbers in the
square set of brackets are the page numbers in this book where the print run information is discussed.

subsequent to publication. From the perspective of scientific readers, the
Origin was less successful as a best seller than a series of books by well-
known popularizers. Huxley's *Science Primers* can be found in the second
highest group, between 20,000 and 40,000 copies, though undoubtedly
the figures would be higher if information on print runs past 1880 were
available. Contemporaries recognized that writers like Wood were popular
among Victorian audiences. A reviewer in the *Saturday Review,* who was
by no means well disposed toward Wood's work, acknowledged that he
"had a thousand readers where Darwin had but one and Professor Huxley
not more than a dozen."[1] When the Victorian reading public thought about
science, they were as likely to call to mind books by Wood and Brewer as
they were to refer to Darwin's *Origin of Species.* When they considered the
issue of evolution, more often than not it was Darwin's ideas as reflected
through the prism of the work of a popularizer of science like Chambers,
Proctor, Clodd, Page, or Allen.

   In addition to securing a space for popularizers in the topography of
Victorian science, I also have aimed to make it a distinctive one. James
Secord's *Victorian Sensation* located one important midcentury popularizer

1. "The Rev. J. G. Wood" 1890, 479.

FIGURE 9.2

Print Runs of Best-Selling Books: Ten Years from Initial Date of Publication

Brewer's *Guide to the Scientific Knowledge of Things Familiar* (1847, Jarrold)
   (75,000) [66]
Wood's *Common Objects of the Country* (1858, Routledge) (64,000) [175]

40,000_____

Chamber's *Vestiges of the Natural History of Creation* (1844, John Churchill)
   (21,250) [34]
Huxley's *Science Primers: Introduction* (1880, Macmillan) (19,000 sold by April
   1880) [395]

20,000_____

Clodd's *Childhood of the World* (1873, Macmillan) (20,000 by 1879) [256]
Kirby's *Stories about Birds of Land and Water* ([1873], Cassell) (18,000 by 1873) [109]
Ball's *Story of the Heavens* (1885, Cassell) (18,000 by 1891) [410]
Wright's *Observing Eye* (1850, Jarrold) (17,600) [107]
Pepper's *Playbook of Science* (1859, Routledge) (16,000) [209]
Darwin's *Descent of Man* (1871, Murray) (14,000) [34]
Huxley's *Physiography* (1877, Macmillan) (13,000 sold in first 3 years) [370]
Spencer's *Study of Sociology* (1873, Henry S. King) (12,500) [381]
Giberne's *Sun, Moon, and Stars* ([1880], Seeley, Jackson and Halliday) (11,000 by
   1885) [430]
Somerville's *On the Connexion of the Physical Sciences* (1834, Murray) (6 editions of
   10,500 by 1842) [22]
Darwin's *Origin of Species* (1858, Murray) (10,000) [34]
Tyndall's *Forms of Water* (1872, Henry S. King) (10,000) [381]

10,000_____

Page's *Introductory Text-Book of Geology* (1854, Blackwood) (6 editions) [225]
Gatty's *Parables of Nature* (1855, Bell and Daldy) (6 editions of first series by
   1858) [107]
Huxley's *Crayfish* (1880, Kegan Paul) (5,275) [384]
Brightwen's *Wild Nature Won by Kindness* (1890, T. Fisher Unwin) (5 editions) [441]
Houghton's *Country Walks* (1869, Groombridge) (5 editions) [83]
Hutchinson's *Extinct Monsters* (1892, Chapman and Hall) (5 editions) [453]
Loudon's *Ladies' Companion to the Flower Garden* (1841, W. Smith) (5 editions) [111]
Proctor's *Other Worlds Than Ours* (1871, Longman) (4 editions of 4,500 by 1878)
   [305]
Johns's *Flowers of the Field* (1853, SPCK) (4 editions by 1860) [49]

FIGURE 9.2 (continued)

Robert's *Domesticated Animals* (1833, Parker) (4 editions by 1837) [110]
Kingsley's *Glaucus; or, The Wonders of the Shore* (1855, Macmillan) (4 editions) [75]
Ward's *Microscope Teachings* (1864, Groombridge) (3 editions) [104]
Henslow's *Botany for Children* (1880, Stanford) (3 editions) [89]
Zornlin's *Recreations in Physical Geography* (1840, John Parker) (3 editions) [109]
Pratt's *The Field, Garden, and the Woodland* (1838, Charles Knight) (3 editions) [104]
Clerke's *Popular History of Astronomy During the Nineteenth Century* (1885, A. and
     C. Black) (3 editions) [472]
Morris's *History of British Butterflies* (1853, Groombridge) (3 printings) [45]
Allen's *The Evolutionist at Large* (1881, Chatto and Windus) (2,000) [274]
Webb's *Celestial Objects* (1859, Longmans) (2 editions of 2,000 copies total by
     1868) [59]
Twining's *The Plant World* (1866, Nelson) (2 editions) [111]
Bowdich Lee's *Anecdotes of the Habits and Instincts of Birds, Reptiles, and Fishes*
     (1853, Grant and Griffith) (2 editions) [102]

Note: When the information was available the same books referred to in Fig. 1 are listed in Fig. 2. In the first set of parentheses are the dates and publishers of the first edition. The second set of parentheses lists number of editions or books printed in Britain during the ten-year period after initial publication unless otherwise noted. The numbers in the square brackets are the page numbers in this book where the print run information is discussed.

of science, Robert Chambers, his publisher, and his readers on the map as never before. Although the sensational success of the *Vestiges of the Natural History of Creation* demonstrates just how influential popularizers of science could be, Chambers cannot be considered as typical of the figures of the second half of the century. While Chambers was responsible for writing one best-selling book on science, the important popularizers of the second half of the century churned out a series of books and periodical pieces. Some, like Wood and Pepper, were active lecturers as well. Whereas Chambers published his book anonymously in order to present a controversial theory, many of the popularizers of the latter half of the century were interested in building a public reputation that would provide them with more publishing opportunities. The appearance of this group of popularizers in the middle of the century was, in many respects, unprecedented, and intimately tied to the changing conditions in the world of publishing and the emergence of a mass reading audience. For the first time it was possible to construct a career as a popularizer of science, though it was a precarious vocation at best. The role of the popularizer changed significantly in this period. Publishers, editors, and the growing reading audience looked to them, and not necessarily to the practitioner, to interpret the larger metaphysical meaning of scientific theories in terms they could understand and in ways they could appreciate.

Creating a distinctive place for popularizers in the topography of nineteenth-century British science has not merely been a matter of adding new territory to the old map. It transforms the entire landscape by throwing several important issues into sharp relief. First, it is clear that science writing provided an important access route to science for women during this period. The large numbers that took advantage of this avenue ensured the continued participation of women in science, despite the hostility of would-be professionalizers. Second, popularizers offered a way of writing about nature that was in stark contrast to the narrative developed by practitioners to communicate with one another. Since they were writing for the general readership during an age when fiction was the big seller in the literary marketplace, they conceived of themselves as storytellers. It was an era of innovation, as popularizers created new genres, such as the evolutionary epic or the synoptic overview, and experimented with a variety of literary techniques. Third, since they conceived of themselves as offering a combination of instruction and amusement, many of them looked to other forms of mass entertainment for inspiration. Striking verbal and visual images, reflecting the popularity of spectacles, exhibitions, and other aspects of the emerging mass visual culture, found their way into their books and lectures. Fourth, giving popularizers a space on our map has allowed us to see how pervasive religious themes continued to be in science right up to the end of the century. Paley's natural theology was no longer adequate for authors and readers of the second half of the century, nonetheless, many books provided a religious framework as the key to interpreting the larger meaning of scientific developments and offered a discourse of design. However, Clodd, Allen, and Bodington followed Huxley, Ball, and other scientific naturalists in repudiating a religious framework for science. By exploring these four issues, it is has been possible to chart portions of the terrain on a revised map with more precision.

Perhaps the most important changes to the old map indicated by this study of popularizers have to do with the status of "professional" science in this period. By and large, historians of science have considered the second half of the nineteenth century as a time when science became professionalized, due to the efforts of Huxley and other scientific naturalists, who used evolutionary theory as a weapon to reform scientific institutions and ideas.

Yet if it was the age of the (would-be) professional scientist, it was also the age of the popularizer of science. The new conditions in the world of publishing created by the communications revolution of the early nineteenth century produced a counterbalance to the professionalization of science. It created a new polity of readers who became the audience for the works written by popularizers of science and whose patronage made it possible for some to eke out careers as writers and journalists. By midcentury, popularizers were finding that a market existed for their work—at the same time that Huxley and his allies were establishing themselves in positions of power and pushing their agenda of obtaining autonomy for science. The fact that popularizers aiming to connect with the public, and practitioners with aspirations to professionalize, were at work in the same period naturally resulted in a situation in which their activities affected one another in a number of complicated ways. As we have seen, the boundaries between the two groups were somewhat porous. Some scientists, like Huxley and Ball, assumed the role of popularizer, without fully relinquishing their position as practitioners. Proctor established his credentials as a practitioner, but later turned more and more toward his work as editor, journalist, and book writer. Some, like Page and Henslow, began as popularizers and later became practitioners.

The development of "professional" and "popular" science in adjacent, and sometimes even overlapping, spaces led to at least two important zones of mutual influence. First, the professionalization of science created a space for popularizers. As practitioners pursued more highly specialized research, the need arose for writers and lecturers who could convey in understandable language the broader significance of many new discoveries to the rapidly growing Victorian mass readership. Some practitioners absolved themselves of the responsibility of writing for a popular audience, seeing their research as their primary task. They elected to abandon the tradition established by Lyell and Darwin to write books that could be understood and appreciated by members of the public. Others, who were willing to pen works for the public, had become too specialized to communicate effectively with those outside their own field. By the end of the century, Clerke and Bodington were able to justify their roles as popularizers on the basis of their ability to synthesize disconnected fragments of knowledge—a task that could not be undertaken by narrow specialists. Second, the success of popularizers like Brewer and Wood was partly responsible for forcing practitioners, like Huxley, to become move involved in popularizing activities, especially after the late 1860s, and to imitate aspects of their style when they wrote for a popular audience. Since an emphasis on common natural objects had proven to be so well received by Victorian reading audiences, and since it had become a staple of scientific books written for the public, Huxley had no choice but to build his lectures and books around the crayfish, the Thames River, and water.

In more fundamental ways, the very existence of a cadre of popularizers separate from the swelling ranks of the practitioner of science raised a series of crucial questions for the public: Who speaks for nature? Who has the authority to write about scientific issues? What are the necessary qualifications for studying the natural world? Or even more basic, what, exactly, is a scientist? For Huxley and his professionalizing colleagues, only the practitioner could interpret the meaning of the natural world and communicate it to a popular audience. The existence of a group of writers who claimed scientific authority could only confuse the public from Huxley's point of view, especially since their agenda differed so profoundly from that of the scientific naturalist. Besides maintaining a religious framework for science, many popularizers also drew on a more "republican" image of the scientific community. Popularizers therefore represented a real obstacle to Huxley's attempt to establish the ideal of the professional scientist, based on a notion of expertise and special training. For the historian, the inclusion of popularizers on our map of British science is an important reminder that the definition of scientist had not yet hardened into the modern concept of the professional. In the future we must pay careful attention to how the identity of the professional scientist was constructed during the latter half of the nineteenth century in reference to the development of an identity for the popularizer of science.

Popularizers offered effective resistance to the attempts of scientific naturalists to control the face of public science partly because they too were part of a process of professionalization. In this contest, Huxley and his allies came up against a cadre of would-be professional writers who were, in many cases, more skilled in communicating the meaning of scientific ideas to the Victorian reader, and who had the confidence of publishers and editors. The numerous popularizers who wrote books and supplied newspapers and journals with endless copy on scientific topics saw themselves as professional writers. They could draw strength from their link to the profession as a whole. Popularizers did not appear on previous maps of the topography of Victorian science because they did not belong to the scientific elite. They had nothing equivalent to the X Club. They did not have a strong presence in scientific societies. Indeed, their power base was not centered in scientific institutions at all. For some, authority originated in their links to the Anglican Church, or previous traditions in women's writing, but for most the chief source of power came from the institutions of publishing. But, as we have seen, establishing a career as a popularizer of science was a tremendously difficult and precarious undertaking—perhaps riskier than pursuing a career as a practitioner of science. Although the growing market for science books made it possible for Wood and others to begin to forge a career in the midcentury period, even the most prolific authors found themselves to be in financial difficulty at the end of their lives. Despite the backing of Darwin and other eminent scientific naturalists, Grant Allen was forced to support his science writing by penning fictional potboilers. Prospects for building a lucrative career as a popularizer of science were as unlikely at the end of the century as they had been in the middle of the century. Ball was able to make a small fortune from his books and lecturing, but only because his status as an eminent practitioner provided endless opportunities for work, and it enabled him to charge more for his services. Since a plum Cambridge professorship was not in the cards for most, popularizers combined their science writing with other paying activities. For some, such as Allen, Gatty, Loudon, and Giberne, it was more literary work, often fiction, journalism, or journal editing. For others, such as Kingsley and Webb, it was a position in the Anglican Church. Still others, like Clodd, never left their regular jobs in the business world.

British science in the nineteenth century has for some time provided historians with exciting territory for exploration. Up to the 1990s, the towering peaks of elite science, especially the ranges that represented scientific naturalism, dominated our maps of the geography of British science by casting a shadow across everything else. Our attention was inevitably drawn to sites such as laboratories, scientific societies, elite universities, and all those places that figured into the contest for cultural authority. Since then we have charted new territory, adding women, the working class, and others, to the map. At first, the space occupied by popularizers was rendered invisible, or at least terra incognita. Now that we have set foot inside this new terrain, we have found that it is almost a world unto itself, filled with a multiplicity of sites. The sites of print culture, including books, magazines, periodicals, textbooks, children's literature, encyclopedias, and newspapers, were the first to attract surveyors. Sites where nature was put on display, or that involved oral culture, have only recently received serious attention in studies of popular lectures, coffee houses, pubs, museums, gardens, fairs, zoos, spectacles, and exhibitions. We have barely begun to map out the territory occupied by science in the public sphere. Exhilarating vistas continue to beckon to us. Our map of British science in general has become more crowded, more detailed, and more complex. We spend more time tracking the circulation of ideas and practices between territories, understanding how they undergo translation and transformation. Although elite science no longer towers over the rest of the territories on the map, it has become even more interesting, as it takes its new place within this dynamic conception of the world of science. As for popularizers of Victorian science, their location in a key area of the map, adjacent to their publishers and their audiences, is no longer in doubt.

Popularizers of science also inhabit an important region on any map drawn of the terrain of our modern scientific world. The outpouring of popularization in print media and television after World War II has been followed by a boom in the 1980s.[2] Today newsstand magazine racks routinely include an entire section of science periodicals geared toward the mass reading audience. Popularizations of science are prominently displayed on the shelves of the science sections of major bookstores. Carl Sagan's television series *Cosmos* inspired a host of science documentaries that testify to the public fascination with the wonders of cutting-edge scientific discoveries. How has the nature of popularizing science changed since the nineteenth century? Fewer women are now involved in popularizing activities, and the maternal tradition, even in a revised form, has all but disappeared. More scientists of stature, such as Lewis Thomas, Edward O. Wilson, Stephen Hawking, and Ilya Prigogine, are willing to assume the role of popularizer of science. Before his recent death, the biologist Stephen Jay Gould was one of the most prolific popularizers of science.

In many respects, traditions established in the nineteenth century continue to shape the way science is popularized and the way that current audiences consume it. The number of periodicals dedicated to communicating scientific news and information to a general reading audience has increased dramatically since the time of Proctor's *Knowledge*. Dorothy Nelkin has tracked the growth of the "popular science press" in the United States especially after both world wars. She has christened our time as "an age of science journalism." But current scientists are ambivalent about the press and are critical of science journalists. Nelkin asserted in the late 1980s that scientists "complain about inaccurate, sensational, and biased reporting" by those

2. Broks 2006, 88–90.

who report on science in newspapers and magazines. These complaints are reminiscent of the concerns of the author of the article on "Sensational Science" in the *Saturday Review* in 1875. Just as Huxley and his colleagues attempted to control the popularization of science in their involvement with *Nature* and the International Scientific Series, contemporary scientists try to manage the media through public relations or communication controls. There is still disagreement as to the importance of formal training in science for those who become science journalists.[3] Christian clergymen with affinities to Wood, Brewer, Kingsley, and their Anglican colleagues still refuse to cede science to our modern incarnations of scientific naturalism. Ordained scientists such as John Polkinghorne, George Coyne, Alister McGrath, and Arthur Peacocke, as well as ordained Christian theologians, such as John Haught, Philip Hefner, Nancey Murphy, and Ted Peters persist in offering the public a vision of science encompassed in a religious framework. We are still fascinated by the spectacle of science, whether it is the launch of a space shuttle or the demonstration of electricity by a Van de Graaff generator at a science museum. Pepper's Royal Polytechnic Institute, like other galleries of practical science, was among the first to forge a powerful link between entertainment and scientific instruction when nature was on display for a general audience.

The debt of contemporary popularizers to past traditions is particularly noticeable when considering their nineteenth-century counterparts. Among popularizers of astronomy, the late Carl Sagan's sense of awe in the face of the immensity of the universe—encapsulated in his trademark phrase "billions and billions of stars"—is reminiscent of Ball, Giberne, and Clerke. As a popularizer of biology, Richard Dawkins shares much in common with the pugnacious Huxley. Dawkins is an aggressive defender of evolutionary naturalism and a fierce critic of the design argument in his *The Blind Watchmaker* (1987). Or perhaps Dawkins's *Unweaving the Rainbow* (1998) recalls Allen's determination to salvage some sense of wonder when confronted by the workings of organic nature bereft of divine purpose. Dawkins insists that though science reveals an orderly universe indifferent to human preoccupations, nevertheless it gives us a "feeling of awed wonder" that is "one of the highest experiences of which the human psyche is capable."[4] But Allen's modern-day counterpart may actually be the late Stephen Jay Gould, whose books were composed of pithy natural history essays on evolutionary issues

3. Nelkin 1987, 7–8, 87, 102, 135, 169.   4. Dawkins 1998, x.

with cosmic implications. James Secord has already drawn attention to similarities between Pepper and theatrical science demonstrators in television from Mr. Wizard in the 1950s and 1960s to more recent incarnations like Bill Nye, the Science Guy (1993–2002).[5]

Perhaps the largest debt that current popularizers owe to their predecessor concerns the narrative form they use to convey huge masses of information to their audiences. The evolutionary epic presented by Chambers, Page, Buckley, Clodd, Allen, and other nineteenth-century popularizers has became a standard literary genre used by twentieth-century authors, though they take the big bang as their starting point, rather than the nebular hypothesis. In his discussion of the "new" epic of evolution in the twentieth century, Martin Eger acknowledges that the tradition began with Chambers and Spencer. Examples of epics from the 1970s and 80s include Jacques Monod's *Chance and Necessity* (1971), Prigogine and Stenger's *Order out of Chaos* (1984), Steven Rose's *The Conscious Brain* (1976), and Douglas Hofstadter's *Gödel, Escher, Bach* (1979). These books all tell the story of a cosmic or universal evolution that extends beyond biology. They anticipate the construction of a seamless, convincing, and all-inclusive science of development.[6] A long list of recent books and videos containing this story of evolution are listed on the Web site "Epic of Evolution," which is dedicated to discussing the "13.7 billion year-old, scientific story of evolution in a meaningful and inspiring way."[7] The authors include Lynn Margulis, Richard Dawkins, and Connie Barlow.

Just as Page and Buckley disagreed with Clodd and Allen as to the ultimate significance of the cosmic evolutionary process, more modern popularizers have offered a variety of metaphysical interpretations. The evolutionary epic has enthralled both those who champion a materialistic agenda and those who seek religious meaning in the contemporary scientific worldview. The sociobiologist E. O. Wilson found in the evolutionary epic the ideal inspiration for a secular world. "The core of scientific materialism," he asserted in *On Human Nature*, "is the evolutionary epic." The "minimum claims" of this epic were that the "laws of the physical sciences are consistent with those of the biological and social sciences and can be linked in chairs of causal explanation; that life and mind have a physical basis"; that the world

5. J. Secord 2002, 1649.

6. Eger 1993, 187, 191–92

7. Available at http://www.epicofevolution.com/index.html, accessed September 5, 2006.

has "evolved from earlier worlds obedient to the same laws; and that the visible universe today is everywhere subject to these materialistic explanations." Wilson acknowledged that the "most sweeping assertions" of the epic could not be proved with finality, but it was "the best myth we will ever have" from a scientific point of view, and it satisfied the human sense of wonder as well as the older religious myths. The origin of the universe in the big bang, Wilson affirmed, "is far more awesome than the first chapter of Genesis."[8]

William Grassie, a scholar working on the nexus between science and religion, agreed that the evolutionary epic can provide a mythic narrative for our times, but, in contrast to Wilson, he has attempted to build religious meaning into it. The "modern scientific account of physical, biological, and cultural evolution," Grassie declared, "is an extraordinary discovery of our times." It has become "a story that serves to define the fundamental world view of a culture by explaining aspects of the natural world and delineating the psychological and social practices and ideals of a society." Although scientists built the epic of evolution, Grassie argued, "religionists have something important to teach the scientists in how to interpret this marvelous new story."[9] Many twentieth-century figures have read a religious message into cosmic evolution, from the Jesuit theologian and paleontologist Teilhard de Chardin to the cultural historian Thomas Berry and the Christian theologian John Haught. In his book *The Universe Story: From the Primordial Flaring Forth to the Ecozoic Era* (1992), coauthored with the mathematical cosmologist Brian Swimme, Berry called for a "new type of narrative" that gave meaning to the story of the universe. The "new" story he provided begins with the big bang and ends with the age of humanity.[10] Haught, who explicitly rejects scientific materialism in his *The Cosmic Story* (1984), argued that the evolutionary process, from the big bang to the present, is a purposeful one, and therefore contains the key to understanding religion "in a new and adventurous way."[11] Berry and Haught were both speakers at a conference on "The Epic of Evolution" held in Chicago in November 1997. Co-sponsored by the American Association for the Advancement of Science Program of Dialogue between Science and Religion and the Field Museum of Natural History, the conference brought together scientists, philosophers, historians, anthropologists, and theologians to discuss the narrative story of

8. E. Wilson 1978, 201–2.
9. Grassie 1998, 8–9.

10. Swimme and Berry 1992, 2.
11. Haught 1984, 1–3, 23–24.

the coming into existence of the universe, earthly life, and human culture.[12] The evolutionary epic continues to fascinate and to inspire intellectuals, scientists, and popularizers.

Popularizers of science were attracted to the evolutionary epic during the twentieth century, in part, because of the work of scientific practitioners. With the success of the big bang, the idea of cosmic evolution became the leading overarching principle of modern astronomy. The theory of cosmic evolution played a crucial role in the formation of exobiology as a discipline, beginning in the mid-twentieth century, and, strikingly, in the development of the American space program. NASA became the chief patron of cosmic evolution, a role that it continues to play today.[13] James Secord has asserted, "over the longer term, *Vestiges* exercised its most important influence by providing a template for the evolutionary epic-book-length works that covered all the sciences in a progressive synthesis." He refers to the ambitious science surveys of the latter half of the nineteenth century, mentioning Buckley's work in particular, that offered the same cosmic range as the *Vestiges*.[14] But Secord could easily have claimed for Chambers an influence far beyond the end of the nineteenth century. Chambers's *Vestiges* supplied a template for the ages. Similarly, the group of British popularizers who pioneered new traditions in conveying knowledge to the public in the second half of the nineteenth century had a lasting impact on the development of science long after they ceased to be active. They forever changed the topography of western science.

12. Barlow 1998, 12–13.
13. Dick and Strick 2004, 10–18.

14. J. Secord 2000, 461.

Bibliography

LIST OF ARCHIVES

Archives of George Routledge and Co., 1853–1902, 1973. Part 1 of *British Publishers' Archives on Microfilm.* Bishops Stortford: Chadwyck-Healey.

Archives of Jarrold and Sons Ltd., Norwich.

Archives of John Murray (Publishers) Ltd., London.

Archives of Kegan Paul, Trench, Trübner, and Henry S. King, 1858–1912. 1973. Part 1 of *British Publishers' Archives on Microfilm.* Cambridge: Chadwyck-Healey.

Archives of Richard Bentley and Son, 1829–1898. 1976. Part 2 of *British Publishers' Archives on Microfilm.* Cambridge: Chadwyck-Healey.

Archives of the House of Longman, 1794–1914. 1978. Part 3 of *British Publishers' Archives on Microfilm.* Cambridge: Chadwyck-Healey.

British Library. Department of Manuscripts. Alfred Russel Wallace Papers.

British Library. Department of Manuscripts. Macmillan Archive.

British Library. Peel Papers.

British Library. Royal Literary Fund Manuscripts, File no. 2294, Mrs. Sallie Duffield Proctor, Widow of Richard Anthony Proctor.

British Library. Royal Literary Fund Manuscripts, File no. 2702 (Microfilm M1077/11), Agnes Giberne.

Cambridge County Record Office, Shire Hall, Lady Ball's (née Steele) Diaries.

Cambridge County Record Office, Shire Hall, Letters, Chiefly of R. S. Ball to His Wife, 1872–1907.

Cambridge County Record Office, Shire Hall, Obituaries and Letters of Sympathy to Lady Ball, 1913.

Cambridge County Record Office, Shire Hall, Robert Ball's Abstracts and Texts of Lectures.

Cambridge County Record Office, Shire Hall, Robert Ball's Memoranda of Merits and Disadvantages of Accepting the Lowndean Chair, 1892, R83/61.

Cambridge County Record Office, Shire Hall, Robert Ball's Probate and Executors' Papers 1913–14.

Cambridge County Record Office, Shire Hall, Robert Ball's Volumes of Notes for the Gilchrist Lectures on the Telescope, 1882.

Cambridge University Library, Cape Archives in the Royal Greenwich Observatory Archives, RGO 15/128, 180-203, Correspondence of Sir Robert Ball.

Cambridge University Library, Manuscripts, Cape Archives, Royal Greenwich Observatory Archives.

Cambridge University Library, Manuscripts, Charles Darwin Papers.

Cambridge University Library, Royal Literary Fund, File no. 1982 John George Wood.

Cambridge University Library, Royal Literary Fund, File nos. 648 and 1101 Jane Webb and Mrs. Jane Loudon.

Darwin, Francis. "Reminiscences of My Father's Everyday Life." Cambridge University Library, Cambridge, Charles Darwin Papers, DAR.140:3, 63–70.

Dittrick Museum, Allen Memorial Medical Library, Cleveland Medical Library Association.

Harvard College Observatory, Archives, Director's Correspondence, Clerke-Pickering correspondence.

Imperial College, London, Archives, Huxley Collection.

Irvine, Winifred Brooke, and Helena Brooke Meiklejon. n.d. "A Family Arrives in British Columbia 1887." Information and Research Centre. Vancouver Public Library.

John Murray [Publishers] Ltd., Archives, London.

Lambeth Palace Library, Tait Papers. Correspondence and papers of John Henry Wood concerning his officiating at Erith, Kent, 1864. Vol. 162, ff. 262–97.

Leeds University Library. Special Collections, Clodd Correspondence.

Lists of the Publications of Richard Bentley and Son, 1829–1898. 1975. *British Publishers' Archives on Microfilm.* Bishops-Stortford, Herts, England: Chadwyck-Healey Ltd.

National Library of Scotland. Blackwood Archive.

Natural History Museum. London. Entomology Library. Stainton Correspondence.

Natural History Museum. London. General Library. Eight Letters from Sara Lee to Richard/Mrs. Owen. OC.17/280–302.

Natural History Museum. London. Wallace Papers.

Pennsylvania State University Libraries. Rare Books and Manuscripts Division, Special Collections Library, Mortlake Collection.

Reading University Library. Archives, The George Bell Uncatalogued Series.

Royal Astronomical Society. Archives, London.

Royal Institution of Great Britain. London, Tyndall Papers.

Sheffield Archives. Sheffield, Hunter Archaeological Society Collection, Correspondence of Miss Margaret Gatty.

St. John's College. Cambridge University, Samuel Butler's Notes, Vol. 2: October 1883–April 1887 Unpublished MS, Copy C, 152–54.

University College, London. Library Services. Special Collections, The Lodge Papers.

University College, London. The Archives of Routledge and Kegan Paul Ltd., Routledge Contracts.

University Library of Manchester. Loudon to Miss Gaskell, October 20, 1849, Rylands English MS 731/112.

University of California, Santa Cruz. University Library, Mary Lea Shane Archives of the Lick Observatory.

University of Exeter. University Library, Norman Lockyer Collection.

University of London Library. Manuscripts Collection, Spencer Papers.

University of Newcastle. Robinson Library, Special Collections, Trevelyan Papers.

University of Reading Library, Archives. Chatto and Windus Ledgers.

University of Reading Library, Archives. The George Bell Uncatalogued Series, Bell Collection, MS 1640.

University of Westminster, Archives. Press Cuttings, Book of Press Cuttings Relating to RPI. R82.

University of Westminster, Archives. 1861 Programme for Christmas time entertainments and lectures. Dated December 26, 1861. R84.

University of Wisconsin, Madison. Memorial Library, Archives of the Washburn Observatory, Clerke-Holden Correspondence.

West Sussex Record Office. Cobden Papers 982, Brightwen-Unwin Correspondence.

Wilkie, Edmund H. n.d. "Professor Pepper—A Memoir." *The Optical Magic Lantern Journal and Photographic Enlarger*, 72–74, from the University of Westminster, Archive, R66/4i–iii.

PRINTED PRIMARY SOURCES

Allen, Grant. 1875. "Miscellany: To Herbert Spencer." *Popular Science Monthly* 7 (September): 628.

———. 1877. *Physiological Aesthetics*. London: Henry S. King.

———. 1879a. "Evolution, Old and New." *Academy* 15 (May 17): 426–27.

[———]. 1879b. "Evolution, Old and New: From One Standpoint." *Examiner* (May 17): 646–47.

[———]. 1879c. "Pleased with a Feather." *Cornhill Magazine* 39 (June): 712–22.

———. 1880. "The Ethics of Copyright." *Macmillan's Magazine* 43 (December): 153–60.

———. 1881a. *The Evolutionist at Large*. London: Chatto and Windus.

———. 1881b. *Vignettes from Nature*. London: Chatto and Windus.

———. 1883. "The Shapes of Leaves." *Nature* 27: 439–42, 464–66, 492–95, 511–14.

———. 1884. *Flowers and Their Pedigrees*. New York: D. Appleton.

———. 1885. *Charles Darwin*. London: Longmans, Green.

———. 1886. "Science." *Academy* 30 (December 18): 413–14.

———. 1887. "The Progress of Science from 1836 to 1886." *Fortnightly Review* 47: 868–84.

———. 1888a. "Evolution." *Cornhill Magazine* 57 (January): 34–47.

———. 1888b. *Force and Energy: A Theory of Dynamics*. London: Longmans, Green.

———. 1888c. "The Gospel According to Darwin—I." *Pall Mall Gazette* 47 (January 5): 1–2.

———. 1888d. "Obituary. Richard Proctor." *Academy* 34 (September 22): 193.

———. 1889a. *Falling in Love: With Other Essays on More Exact Branches of Science*. London: Smith, Elder.

[———]. 1889b. "The Trade of Author." *Fortnightly Review* 51: 261–74.

———. 1890. "Our Scientific Causerie." *Review of Reviews* 1 (June): 537–38.

———. 1891. *Dumaresq's Daughter: A Novel*, 3 vols. London: Chatto and Windus.

———. 1894a. "Character Sketch: Professor Tyndall." *Review of Reviews* 9: 21–26.

———. 1894b. *The Lower Slopes: Reminiscences of Excursions Round the Base of Helicon, Undertaken for the Most Part in Early Manhood*. London: Elkin Mathews and John Lane.

———. 1895. "The Amateur in Science." *New Science Review* 1: 301–8.

———. 1897. "Spencer and Darwin." *Fortnightly Review* 67 (February): 251–62.

———. 1899. *Flashlights on Nature*. London: George Newnes.

———. 1901. *In Nature's Workshop*. Toronto: William Briggs.

———. 1904. "Personal Reminiscences of Herbert Spencer." *Forum* (New York) 35 (April): 610–28.

———, ed. [1899]. "Introduction." *The Natural History of Selbourne*, by Gilbert White, xxvii–xl. London: John Lane/The Bodley Head.

———, and T. H. Huxley. [1888]. *A Half-Century of Science*. [New York]: Humboldt Publishing.

"Astronomy of the Nineteenth Century." 1893. *Nature* 49 (November 2): 2.

Ball, Robert. 1881. "A Glimpse through the Corridors of Time." *Nature* 25 (November 24): 79–82, 103–7.

———. 1883a. "The Relation of Darwinism to Other Branches of Science." *Longman's Review* 2 (November): 76–92.

———. 1883b. "The Sun's Distance." *Knowledge* 4 (September 28, October 12, October 26, November 9, November 16): 197–99, 226–28, 257, 284, 301–2.

————. 1885. *The Story of the Heavens*. London, Paris, New York, and Melbourne: Cassell and Company.

————. 1886. "Astronomy during the Nineteenth Century." *Nature* 33 (February 4): 313-14.

————. 1889. *Elements of Astronomy*. London: Longmans, Green.

————. 1890a. *Star-Land: Being Talks with Young People about the Wonders of the Heavens*. London, Paris, and Melbourne: Cassell and Company.

————. 1890b. "The Sun." *Good Words* 31: 244-47, 467-70, 553-57, 626-29.

————. 1892. *In Starry Realms*. Philadelphia: J. B. Lippincott.

————. 1895. *Great Astronomers*. London: Isbister and Company.

————. 1905. *A Popular Guide to the Heavens: A Series of Eighty-Three Plates with Explanatory Text and Index*. London: The Geographical Institute; Liverpool: Philip, Son, and Nephew; George Philip and Son.

————. 1909. *The Earth's Beginning*. London, New York, Toronto, and Melbourne: Cassell and Company.

"Ball's 'Story of the Heavens.'" 1885. *Nature* 33 (December 10): 124-26.

[Becker, Lydia]. 1864. *Botany for Novices: A Short Outline of the Natural System of Classification of Plants*. London: Whittaker.

————. 1867. "Female Suffrage." *Contemporary Review* 4: 307-16.

————. 1868. "Is There Any Specific Distinction between Male and Female Intellect?" *Englishwoman's Review of Social and Industrial Questions* 3: 483-491.

————. 1869. "On the Study of Science by Women." *Contemporary Review* 10: 386-404.

Bodington, Alice. 1887. "On Some Curious Facts Connected with the Evolution of the Eye." *Journal of Microscopy* 6: 79-88.

————. 1890a. "Importance of Race and Its Bearing on the Negro Question." *Westminster Review* 134: 415-27.

————. 1890b. "The Marriage Question from a Scientific Standpoint." *Westminster Review* 133: 172-80.

————. 1890c. *Studies in Evolution and Biology*. London: Elliot Stock.

————. 1893. "Religion, Reason, and Agnosticism." *Westminster Review* 139: 369-80.

Bowdich Lee, Sarah. 1853. *Anecdotes of the Habits and Instincts of Birds, Reptiles, and Fishes*. London: Grant and Griffith.

————. 1854. *Trees, Plants, and Flowers: Their Beauties, Uses, and Influences*. London: Grant and Griffith.

Bower, F. O. 1883. "Mr. Grant Allen's Article on 'The Shapes of Leaves.'" *Nature* (April 12): 552.

Brewer, Ebenezer. 1870. *Theology in Science; or, The Testimony of Science to the Wisdom and Goodness of God*. London: Jarrold and Sons.

————. 1874. *A Guide to the Scientific Knowledge of Things Familiar*. London: Jarrold and Sons.

Brightwen, Eliza. 1890. *Wild Nature Won by Kindness*. London: T. Fisher Unwin.

————. 1895. *Inmates of My House and Garden*. London: T. Fisher Unwin.

————. 1897a. *Glimpses into Plant Life: An Easy Guide to the Study of Botany*. London: T. Fisher Unwin.

————. 1897b. *More about Wild Nature*, 3rd ed. London: T. Fisher Unwin.

————. 1899. *Rambles with Nature Students*. London: Religious Tract Society.

————. [1904]. *Quiet Hours with Nature*. London: T. Fisher Unwin; New York: James Pott.

————. 1909. *Eliza Brightwen: The Life and Thoughts of a Naturalist*, edited by W. H. Chesson. London: T. Fisher Unwin.

————. n.d. *Practical Thoughts on Bible Study*. London: Hamilton, Adams.

Britten, James. 1894. "Anne Pratt." *Journal of Botany, British and Foreign* 32: 205-7.

[Buckley, Arabella]. 1871. "Darwinism and Religion." *Macmillan's Magazine* 24: 45-51.

—————. 1876. *A Short History of Natural Science and of the Progress of Discovery from the Time of the Greeks to the Present Day*. London: John Murray.

—————. 1879a. *The Fairy-Land of Science*. London: Edward Stanford.

[—————]. 1879b. "Soul, and the Theory of Evolution." *University Magazine* 93 (January): 1-10.

—————. 1881. *Life and Her Children: Glimpses of Animal Life from the Amoeba to the Insects*. London: Edward Stanford.

—————. 1882. *Winners in Life's Race; or, The Great Backboned Family*. London: Edward Stanford.

[—————]. 1890. "Lyell, Sir Charles." *Encyclopaedia Britannica*, 9th ed., 15: 101-3. New York: Henry G. Allen.

—————. 1891. *Moral Teachings of Science*. London: Edward Stanford.

Butler, Samuel. 1924. *Luck, or Cunning?* London: Jonathan Cape; New York: E. P. Dutton.

Carey, Annie. n.d. *The Wonders of Common Things*. New York: Cassell, Petter, and Galpin.

[Clark, Samuel]. 1837. *Peter Parley's Wonders of the Earth, Sea, and Sky*, edited by Rev. T. Wilson. London: Darton and Clark.

Clerke, Agnes M. 1885. *A Popular History of Astronomy during the Nineteenth Century*. Edinburgh: Adam and Charles Black.

[—————]. 1893. "Proctor's Old and New Astronomy." *Edinburgh Review* 177 (April): 544-64.

—————. 1895. *The Herschels and Modern Astronomy*. London, Paris, and Melbourne: Cassell and Company.

—————. 1898a. "Among My Books." *Literature* 3 (August 13): 131-32.

—————. 1898b. "Section I.—History." In *Astronomy*, by Agnes M. Clerke, A. Fowler, and J. Ellard Gore, 3-38. New York: D. Appleton.

[—————]. 1900. "The Evolution of the Stars." *Edinburgh Review* 191 (April): 455-77.

—————. 1902. *A Popular History of Astronomy during the Nineteenth Century*, 4th ed. London: Adam and Charles Black.

—————. 1903. *Problems in Astrophysics*. London: Adam and Charles Black.

—————. 1905a. *Modern Cosmogonies*. London: Adam and Charles Black.

—————. 1905b. *The System of the Stars*, 2nd ed. London: Adam and Charles Black.

"Clerke's Astrophysics." 1903. *Nation* 77 (July 30): 98-99.

Clodd, Edward. 1888. "In Memoriam. Richard Anthony Proctor." *Knowledge* 11 (October 1): 265.

—————. 1890. *The Story of Creation: A Plain Account of Evolution*. London: Longmans, Green.

—————. 1900. *Grant Allen: A Memoir*. London: Grant Richards.

—————. 1907. *Pioneers of Evolution: From Thales to Huxley*. London: Cassell and Company.

—————. 1926. *Memories*. London: Watts.

"Contemporary Literature." 1880. *Westminster Review* 113: 543-625.

Crosland, Mrs. Newton. 1893. *Landmarks of a Literary Life, 1820-1892*. London: Sampson Low, Marston.

Crosland, Newton. 1898. *Rambles Round My Life: An Autobiography (1819-1896)*. London: E. W. Allen.

[Dallas, W. S.]. 1857. "Contemporary Literature. Science." *Westminster Review* 67: 270-88.

[—————]. 1860. "Contemporary Literature. Science." *Westminster Review* 73: 295-303.

[—————]. 1861. "Contemporary Review. Science." *Westminster Review* 76: 253-63.

[—————]. 1866. "Contemporary Literature. Science." *Westminster Review* 86: 240-46.

Darwin, Charles. 1959. *The Origin of Species by Charles Darwin: A Variorum Text*, edited by Morse Peckham. Philadelphia: University of Pennsylvania Press.

—————. 1985. *The Origin of Species*, edited by J. W. Burrow. Harmondsworth, England: Penguin.

Darwin, Francis, ed. 1887. *The Life and Letters of Charles Darwin*, 3 vols., 3rd ed. London: John Murray.

—————. 1903. *More Letters of Charles Darwin*, 2 vols. New York: D. Appleton.

"David Page." 1879. *Nature* 19 (March 13): 444.

"Diving Bell." 1839. *Literary World* 7 (May 11): 98.

"Dragons of the Prime." 1893. *Saturday Review* 75: 20.

E. H. [1861]. *A Brief Memorial of Mrs. Wright, Late of Buxton, Norfolk*. London: Jarrold and Sons.

[Elwin, Whitwell]. 1849. "Popular Science." *Quarterly Review* 84: 307–44.

"Evenings at the Microscope." 1859. *Saturday Review* 7 (May 7): 570–71.

"Evolutionist at Large." 1881. *British Quarterly Review* 73: 496–97.

"Fatal Accident at the Polytechnic Institution." 1859. *Times*, January 5, 12.

Foster, Thomas [pseudonym for Richard Proctor]. 1885. "Feminine Volubility." *Knowledge* 8 (July 10): 17–18.

Gatty, Margaret. 1855. *Parables from Nature*, 2nd ed. London: Bell and Daldy.

———. 1861. *Parables from Nature*. London: Bell and Daldy.

———. 1872. *British Sea-Weeds*. London: Bell and Daldy.

———. 1954. *Parables from Nature*. London: J. M. Dent; New York: E. P. Dutton.

Gaynor, E. 1897. "Sir Robert S. Ball on Evolution." *Irish Ecclesiastical Record*, 4th series, 1: 243–60.

Giberne, Agnes. 1885. *Among the Stars; or, Wonderful Things in the Sky*. London: Seeley and Company.

———. 1888. *The World's Foundations; or, Geology for Beginners*. London: Seeley and Company.

———. 1890. *The Ocean of Air: Meteorology for Beginners*. London: Seeley and Company.

———. 1894. *The Starry Skies; or, First Lessons on the Sun, Moon and Skies*. New York: American Tract Society.

———. 1895. *Radiant Suns: A Sequel to "Sun, Moon, and Stars."* London: Seeley and Company.

———. 1902. *The Mighty Deep and What We Know of It*. Philadelphia: J. B. Lippincott; London: C. Arthur Pearson.

———. 1920. *This Wonderful Universe: A Little Book about Suns and Worlds, Moons and Meteors, Comets and Nebulae*. London: Society for Promoting Christian Knowledge.

———. n.d. *Sun, Moon, and Stars: A Book for Beginners*. New York: Robert Carter and Brothers.

"Gilchrist Lectures at Goole. Mr. Robert Ball on 'Other Worlds.'" 1890. *Goole Weekly Times*, January 10, 3.

Gordon, Mrs. 1869. *The Home Life of Sir David Brewster*. Edinburgh: Edmonston and Douglas.

"Great Induction Coil at the Polytechnic Institution." 1869. *Illustrated London News* 54 (April 17): 401–2.

Gregory, R. A. 1903. "The Spectroscope in Astronomy." *Nature* 68 (August 13): 338–41.

———. 1906. "Stars and Nebulae." *Nature* 73 (March 29): 505–8.

Guy, F. Barlow. 1880. "Forest School, Walthamstow." [Wood]. *Walthamstow and Leyton Guardian*, September 11, 4.

H. G. S. 1893. "Extinct Monsters." *Nature* 47 (January 12): 250–52.

Hale, George R. 1891. "The System of the Stars." *Publications of the Astronomical Society of the Pacific* 3: 180–94.

Hallett, Lilias Ashworth. 1890. "Lydia Ernestine Becker." *Women's Suffrage Journal* 21 (August): 4–5.

Harrison, Frederic. 1886. *The Choice of Books and Other Literary Pieces*. London: Macmillan.

Henslow, Rev. George. 1873. *The Theory of Evolution of Living Things and the Application of the Principles of Evolution to Religion Considered as Illustrative of the "Wisdom and Beneficence of the Almighty."* London: Macmillan.

———. 1881. *Botany for Children: An Illustrated Elementary Text-Book for Junior Classes and Young Children*, 3rd ed. London: Edward Stanford.

———. 1888. *The Origin of Floral Structures through Insect and Other Agencies*. New York: D. Appleton.

———. 1908. *How to Study Wild Flowers*, 2nd ed. London: Religious Tract Society.

————. n.d. *Plants of the Bible*. London: Religious Tract Society.

Houghton, Rev. W. 1869. *Country Walks of a Naturalist with His Children*. London: Groombridge and Sons.

————. 1870. *Sea-Side Walks of a Naturalist with His Children*. London: Groombridge and Sons.

————. 1875. *Sketches of British Insects: A Handbook for Beginners in the Study of Entomology*. London: Groombridge and Sons.

————. [1879]. *British Fresh-Water Fishes*. London: William Mackenzie.

————. n.d. *The Microscope and Some of the Wonders It Reveals*. London and New York: Cassell, Petter, and Galpin.

Huggins, Doctor W. 1882. "Photographic Spectrum of Comet (Wells)." *Knowledge* 2 (July 2): 89–90.

Huggins, Margaret. 1890. "[Review of] *The System of the Stars*." *Observatory* 13 (December): 382–86.

————. 1907. *Agnes Mary Clerke and Ellen Mary Clerke: An Appreciation*. Printed for private circulation.

Hutchinson, Rev. H. N. 1890. *The Autobiography of the Earth: A Popular Account of Geological History*. London: Edward Stanford.

————. 1892. *Extinct Monsters: A Popular Account of Some of the Larger Forms of Ancient Animal Life*. London: Chapman and Hall.

————. 1894. *Creatures of Other Days*. London: Chapman and Hall.

————. 1896. *Prehistoric Man and Beast*. London: Smith, Elder.

Huxley, Leonard. 1902. *Life and Letters of Thomas Henry Huxley*, 2 vols. New York: D. Appleton.

[Huxley, Thomas H.]. 1854. "Science." *Westminster Review* 61 (January): 254–70.

[————]. 1855. "Science." *Westminster Review* 64: 240–55.

————. 1878. *Physiography: An Introduction to the Study of Nature*, 2nd ed. New York: D. Appleton.

————. 1880. *The Crayfish: An Introduction to the Study of Zoology*. New York: D. Appleton. Reprint: Cambridge, MA: MIT Press, 1974.

————. 1881. *Science Primers: Introductory*. Toronto: Canada Publishing.

————. 1894a. *Darwiniana*. London: Macmillan.

————. 1894b. *Science and Christian Tradition*. New York: D. Appleton.

————. 1895a. *Lay Sermons, Addresses, and Reviews*. New York: D. Appleton.

————. 1895b. *Man's Place in Nature and Other Anthropological Essays*. London: Macmillan.

————. 1897a. *Discourses Biological and Geological*. New York: D. Appleton.

————. 1897b. *Method and Results*. New York: D. Appleton.

————. 1897c. *Science and Education*. New York: D. Appleton.

————. 1898–1903. *The Scientific Memoirs of Thomas Henry Huxley*, 5 vols., edited by Michael Foster and E. Ray Lankester. London: Macmillan.

————. 1903. "Vestiges of the Natural History of Creation. Tenth Edition. London, 1853." In *The Scientific Memoirs of Thomas Henry Huxley. Supplementary Volume*, edited by Professor Sir Michael Foster and Professor E. Ray Lankester, 1–19. London: Macmillan.

————. 1911. *Evolution and Ethics and Other Essays*. London: Macmillan.

"Interesting Lecture at Altrincham." [Wood]. 1882. *Advertiser* (January 14): 4.

"Inventor's Column." 1884. *Knowledge* 6 (October 17): 329.

Johns, Rev. C. A. [1846]. *Botanical Rambles*. London: Society for Promoting Christian Knowledge.

————. 1853. *First Steps to Botany*. London: National Society.

[————]. [1854]. *Birds' Nests*. London: Society for Promoting Christian Knowledge.

————. [1859]. *Picture Books for Children: Animals*. London: Society for Promoting Christian Knowledge.

————. 1860. *Flowers of the Field*, 4th ed. London: Society for Promoting Christian Knowledge.

[————]. [1860]. *Sea-Weeds*. London: Society for Promoting Christian Knowledge.

————. 1862. *British Birds in Their Haunts*. London: Society for Promoting Christian Knowledge.

————. [1869]. *The Forest Trees of Britain*, 2 vols. London: Society for Promoting Christian Knowledge.

Jones, Bence. 1870. *The Life and Letters of Faraday*, 2 vols., 2nd ed. London: Longmans, Green.

"July Reviewed by September." 1860. *Atlantic Monthly* 6 (September): 378–83.

"Just Published, A Guide to the Scientific Knowledge of Things Familiar." 1847. *Publishers' Circular* 10 (December 15): 432.

[Kingsley, Charles]. 1863. "British Sea-Weeds." *Reader* 2 (August 15): 162–63.

————. 1870. *Madam How and Lady Why; or, First Lessons in Earth Lore for Children*. London: Bell and Daldy.

————. 1874. "Preface." *Westminster Sermons*. London: Macmillan, v–xxxii.

————. 1890. *Scientific Lectures and Essays*. London: Macmillan.

————. 1908. *The Water-Babies and Glaucus*. London and Toronto: J. M Dent.

Kingsley, Fanny. 1877. *Charles Kingsley: His Letters and Memories of His Life*, 2 vols. London: Henry S. King.

Kirby, Mary. 1888. *"Leaflets from My Life," A Narrative Autobiography*, 2nd ed. London: Simpkin, Marshall.

Kirby, Mary, and Elizabeth Kirby. [1861]. *Caterpillars, Butterflies, and Moths: An Account of Their Habits, Manners, and Transformations*. London: Jarrold and Sons.

————. 1862. *Things in the Forest*. London: T. Nelson and Sons.

————. 1871. *The Sea and Its Wonders: A Companion Volume to "The World at Home."* London: T. Nelson and Sons.

————. 1872. *Beautiful Birds in Far-Off Lands: Their Haunts and Homes*. London: T. Nelson and Sons.

————. [1873]. *Chapters on Trees: A Popular Account of Their Nature and Uses*. London, Paris, and New York: Cassell, Petter, and Galpin.

————. [1874]. *Sketches of Insect Life*. London: Religious Tract Society.

————. 1875. *Aunt Martha's Corner Cupboard: A Story for Little Boys and Girls*. London: T. Nelson and Sons.

————. n.d. *Stories about Birds of Land and Water*. Boston, New York, Chicago, and San Francisco: Educational Publishing Company.

Lankester, Phebe. 1861. "For the Young of the Household in Cozy Nook: Eyes and No Eyes." *St. James's Magazine* 2 (August–November): 121–27.

————. 1903. *British Ferns: Their Classification, Structure, and Functions Together with the Best Methods for Their Cultivation*. London: Gibbings and Company.

————. 1905. *Wild Flowers Worth Notice*. London: Routledge.

————, ed. n.d. *A Plain and Easy Account of the British Ferns*. London: Robert Hardwicke.

"Late Miss Becker." 1890. *Manchester Examiner and Times*, July 21, 5.

"Late Mr. Richard A. Proctor." 1888. *Times*, September 14, 5.

"Lecture at the Midland Institute on Tides." [Ball]. 1881. *Birmingham Daily Gazette*, October 25, 6.

"Lecture Last Night on 'Unappreciated Insects.'" [Wood]. 1881. *Bolton Chronicle*, December 10, 8.

"Lecture on Ants." [Wood]. 1881. *Leek Times*, November 26, 4.

"Lecture on Jelly Fish." [Wood]. 1881. *Weymouth and Dorset Guardian*, November 30, 5.

"Lectures for Altrincham and Bowdon." [Wood]. 1881. *Altrincham and Bowdon Guardian*, October 8, 5.

"Literary Women of the Nineteenth Century." 1858–59. *Englishwoman's Domestic Magazine* 7: 341–43.

Lockyer, Sir Norman. 1870. "To Our Readers." *Nature* 2 (May 5): 1.

————. 1919. "Valedictory Memories." *Nature* 104 (November 6): 189-90.

Lodge, O. J. 1889. "Mr. Grant Allen's Notions about Force and Energy." *Nature* 39 (January 24): 289-92.

Loudon, Jane. 1840. *The Young Naturalist's Journey; or, The Travels of Agnes Merton and Her Mama*. London: William Smith.

————. 1841. *The First Book of Botany: Being a Plain and Brief Introduction to That Science, for Students and Young Persons*. London: George Bell.

————. 1842. *Botany for Ladies; or, A Popular Introduction to the Natural System of Plants, According to the Classification of De Candolle*. London: John Murray.

————. [1846]. *British Wild Flowers*, 2nd ed. London: W. S. Orr.

————. 1850. *The Entertaining Naturalist*. London: Henry G. Bohn.

————. 1857. *The Amateur Gardener's Calendar*, 2nd ed. London: Longman, Brown, Green, Longmans, and Roberts.

————. 1994. *The Mummy! A Tale of the Twenty-Second Century*. Ann Arbor: University of Michigan Press.

[Mann, R. J.]. 1886. "The Recent Progress of Astronomy." *Edinburgh Review* 163: 372-405.

[Marcet, Jane]. 1817. *Conversations on Chemistry; in which the Elements of that Science Are Familiarly Explained and Illustrated by Experiments*, 2 vols., 5th ed. London: Longman, Hurst, Rees, Orme, and Brown.

"Marlborough." [Wood]. 1884. *Marlborough Times and Wilts and Berks Country Paper* (November 8): 8.

"Miscellaneous. The Polytechnic Institution." 1861. *Chemical News* 3 (June 22): 384.

"Miss Agnes Giberne: A Pioneer of Popular Science." 1939. *Times*, August 22, 12.

"Modern Astronomy. Prof. Ball's first Lecture at the Lowell Institute." 1884. *Boston Herald*, October 15, 4.

Morris, Rev. F. O. 1853. *A History of British Butterflies*. London: Groombridge and Sons.

————. 1861. *Records of Animal Sagacity and Character*. London: Longman, Green, Longman, and Roberts.

————. 1869. *Difficulties of Darwinism. Read before the British Association at Norwich and Exeter in 1868 and 1869*. London: Longmans, Green.

————. 1872. *Dogs and Their Doings*. New York: Harper and Brothers.

Morris, Rev. M. C. F. 1897. *Francis Orpen Morris: A Memoir*. London: John C. Nimmo.

"Mr. Page's Handbook of Geological Terms." 1859. *Saturday Review* 8 (December 10): 713.

"Mrs. A. B. Fisher." 1929. *Times*, February 13, 9.

"Mrs. Leo H. Grindon, L.L.A." 1895. *Manchester Faces and Places* 7, no. 1 (October): 7-12.

"New Books and Reprints." 1890. *Saturday Review* 69: 653.

"Notices of Books. The *Boy's Playbook of Science*. Second Edition. By J. H. Pepper. *The Playbook of Metals*. Same Author." 1861. *Chemical News* 3, no. 58 (January 12): 29-31.

"Notices of Books. *The Common Objects of the Country*. By the Rev. J. G. Wood." 1858. *English Woman's Journal* (July): 347-48.

"Obituary [Becker]." 1890. *Journal of Botany* 28: 320.

"Obituary. Mrs. Brightwen." 1906. *Times*, May 7, 6.

"Obituary [Wood]." 1889a. *ARS Quatuor Coronatorum* 2: 80.

"Obituary [Wood]." 1889b. *Times*, March 3, 9.

Our Special Sightseer. 1870. "Monday Out." *Fun* (December 3): 223.

Page, David. 1854. *Introductory Text-Book of Geology*. Edinburgh and London: William Blackwood and Sons.

————. 1856. *Advanced Text-Book of Geology: Descriptive and Industrial*. Edinburgh and London: William Blackwood and Sons.

————. 1861. *The Past and Present Life of the Globe: Being a Sketch in Outline of the World's Life-System*. Edinburgh and London: William Blackwood and Sons.

————. 1868. *The Earth's Crust: A Handy Outline of Geology*, 4th ed. Edinburgh: William P. Nimmo.

————. 1869. *Chips and Chapters: A Book for Amateur and Young Geologists*. Edinburgh and London: William Blackwood and Sons.

————. 1870. *Geology for General Readers: A Series of Popular Sketches in Geology and Palaeontology*, 3rd ed. Edinburgh and London: William Blackwood and Sons.

————. 1876. *Geology and Its Influence on Modern Beliefs: Being a Popular Sketch of the Scientific Teachings and Economic Bearings*. Edinburgh and London: William Blackwood and Sons.

"Past and Present Life of the Globe." 1861. *British Quarterly Review* 34: 281.

"Patron—H. R. H. Prince Albert." 1854. *Athenaeum* 1, 396 (July 29): 945.

Pearson, Karl. 1888. "Science. *Force and Energy*: A Theory of Dynamics. By Grant Allen." *Academy* 34 (December 29): 421–22.

Pepper, John Henry. 1861a. *Playbook of Metals*. London: Routledge, Warne, and Routledge.

————. 1861b. *Scientific Amusements for Young People*. London: Routledge, Warne and Routledge.

————. 1869. *Cyclopaedic Science Simplified*. London: Frederick Warne.

————. 1890. *The True History of The Ghost and All about Metempsychosis*. London, Paris, New York, and Melbourne: Cassell and Company.

————. 2003. *The Boy's Playbook of Science*. Bristol: Thoemmes Press.

————, and Phransonbel. 1865. *The Diamond Maker; or, The Alchymist's Daughter: A Romantic Drama, in Three Acts*. London: M'Gowan and Danks.

"Playing with Lightning." 1869. *All the Year Round* (May 29): 617–20.

"Polytechnic." 1863a. *Illustrated London News* 42 (February 28): 218.

"Polytechnic." 1863b. *Times*, May 20, 9.

"Polytechnic." 1871. *Times*, April 11, 9.

"Polytechnic Institution." 1855. *Illustrated London News* 26 (May 19): 491.

"Polytechnic Institution." 1863. *Illustrated London News* 42 (January 3): 19.

"Polytechnic Institution." 1865. *Times*, October 13, 12.

"Polytechnic Institution." 1870. *Times*, February 28, 8.

"Polytechnic Institution, Regent Street." 1838. *Mirror* (September 1). University of Westminster Archives, R40.

"Polytechnic Museum." 1867. *Times*, December 23, 6.

Pratt, Anne. 1846. *Flowers and Their Associations*. London: Charles Knight.

————. 1850. *Chapters on the Common Things of the Sea-Side*. London: Society for Promoting Christian Knowledge.

————. 1853. *Our Native Songsters*. London: Society for Promoting Christian Knowledge.

————. [1873]. *The Flowering Plants, Grasses, Sedges and Ferns of Great Britain and Their Allies the Club Mosses, Pepperworts and Horsetails*, 5 vols. London: Frederick Warne.

————. n.d. *The British Grasses and Sedges*. London: Society for Promoting Christian Knowledge.

————. n.d. *The Ferns of Great Britain, and Their Allies the Club Mosses, Pepperworts, and Horsetails*. London: Society for Promoting Christian Knowledge.

————. n.d. *The Poisonous, Noxious, and Suspected Plants of Our Fields and Woods*. London: Society for Promoting Christian Knowledge.

[————]. n.d. *Wild Flowers of the Year*. London: Religious Tract Society.

Pritchard, C. 1870. "Other Worlds Than Ours." *Nature* (June 30): 161–62.

Proctor, Richard A. 1870a. *Other Worlds Than Ours*. London: Longmans, Green.

————. 1870b. "Other Worlds Than Ours." *Nature* (July 7): 190.

————. 1871. *Light Science for Leisure Hours*. London: Longmans, Green.

————. 1876. *Wages and Wants of Science-Workers*. London: Smith, Elder.

————. 1878. *The Universe of Stars*, 2nd ed. London: Longmans, Green.

————. 1881a. "Answers to Correspondents." *Knowledge* 1 (December 2): 106.

[————]. 1881b. "Our Correspondence Columns." *Knowledge* 1 (November 4): 15.

————. 1881c. "Our Correspondence Columns. Demands on Our Space." *Knowledge* 1 (December 16): 139.

————. 1881d. "Our Correspondence Columns. Plans for the New Year." *Knowledge* 1 (December 23): 160, 163-64.

————. 1881e. "Our Correspondence Columns. To Our Readers." *Knowledge* 1 (November 25): 73-74.

[————]. 1881f. "Reviews. Authors and Publishers." *Knowledge* 1 (November 25): 72.

————. 1881g. "Science and Religion." *Knowledge* 1 (November 4): 3-4.

————. 1881h. "To Our Readers." *Knowledge* 1 (November 4): 3.

[————]. 1882a. "Answers to Correspondents." *Knowledge* 1 (April 21): 539.

[————]. 1882b. "Answers to Correspondents." *Knowledge* 1 (May 12): 595.

[————]. 1882c. "Answers to Correspondents." *Knowledge* 2 (June 2): 13.

[————]. 1882d. "Answers to Correspondents." *Knowledge* 2 (September 29): 301.

[————]. 1882e. "Answers to Correspondents." *Knowledge* 2 (October 13): 332.

[————]. 1882f. "Answers to Correspondents." *Knowledge* 2 (October 27): 365.

[————]. 1882g. "The British Association." *Knowledge* 2 (September 1): 224-28.

[————]. 1882h. "The British Association." *Knowledge* 2 (September 15): 257.

[————]. 1882i. "Editorial Gossip." *Knowledge* 3 (May 19): 613.

————. 1882j. "Kew Gardens." *Knowledge* 2 (October 27): 351-52.

————. 1882k. "Letters Received." *Knowledge* 1 (February 10): 327.

————. 1882l. "Our Correspondence Columns—Our Letters, Queries and Replies." *Knowledge* 1 (February 10): 320.

[————]. 1882m. "Science and Art Gossip." *Knowledge* 2 (August 18): 191-92.

[————]. 1882n. "Science and Art Gossip." *Knowledge* 2 (October 27): 349.

[————]. 1882o. "Science and Art Gossip." *Knowledge* 2 (December 29): 489.

[————]. 1882p. "Science of the *Times*." *Knowledge* 2 (October 20): 342.

[————]. 1882q. "Special Notice." *Knowledge* 1 (February 24): 367.

[————]. 1882r. "Special Notice." *Knowledge* 1 (March 17): 434.

[————]. 1883a. "Editorial Gossip." *Knowledge* 3 (May 18): 298.

[————]. 1883b. "Editorial Gossip." *Knowledge* 3 (June 29): 391-92.

[————]. 1883c. "Editorial Gossip." *Knowledge* 4 (December 7): 349-50.

————. 1883d. "Lecture Notes." *Knowledge* 3 (January 12): 25.

————. 1883e. "Lectures and the London Press." *Knowledge* 3 (April 13), 217.

————. 1883f. "Lecturing Notes." *Knowledge* 3 (January 19): 40.

[————]. 1883g. "Letters Received." *Knowledge* 3 (April 20): 240.

[————]. 1883h. "Letters Received and Short Answers." *Knowledge* 4 (September 28): 208.

————. 1883i. "Letters Received and Short Answers." *Knowledge* 4 (November 30): 338.

[————]. 1883j. "Letters to the Editor." *Knowledge* 3 (May 4): 267.

————. 1883k. "Mathematics of the Imaginary." *Knowledge* 4 (November 9): 287-88.

————. 1883l. "Personal." *Knowledge* 3 (May 18): 287-88.

[————]. 1883m. "Science and Art Gossip." *Knowledge* 3 (April 20): 229.

————. 1883n. "Social Dynamite." *Knowledge* 3 (April 27): 244-45.

[————]. 1885a. "Editorial Gossip." *Knowledge* 7 (January 2): 12-13.

[————]. 1885b. "Editorial Gossip." *Knowledge* 7 (February 13): 133-34.

[————]. 1885c. "Editorial Gossip." *Knowledge* 7 (February 20): 155.

[————]. 1885d. "Editorial Gossip." *Knowledge* 7 (May 1): 376.

————. 1885e. "Gossip." *Knowledge* 8 (September 4): 204–5.

————. 1885f. "Gossip." *Knowledge* 8 (September 25): 273.

————. 1885g. "Gossip." *Knowledge* 9 (December 1): 67–69.

[————]. 1885h. "Letters Received and Short Answers." *Knowledge* 7 (February 20): 160.

[————]. 1885i. "Letters Received and Short Answers." *Knowledge* 7 (February 27): 182.

[————]. 1885j. "Letters Received and Short Answers." *Knowledge* 8 (July 17): 36.

[————]. 1885k. "Letters Received and Short Answers." *Knowledge* 8 (August 21): 168.

[————]. 1885l. "Letters Received and Short Answers." *Knowledge* 8 (August 28): 189.

[————]. 1885m. "Letters Received and Short Answers." *Knowledge* 8 (September 25): 278.

————. 1885n. "Letters Received and Short Answers." *Knowledge* 8 (October 9): 322.

[————]. 1885o. "Letters Received and Short Answers. Divers Correspondents." *Knowledge* 7 (May 22): 445.

[————]. 1885p. "Mr. R. A. Proctor's Lecture Tour." *Knowledge* 8 (July 31): 104.

[————]. 1885q. "To Correspondents." *Knowledge* 7 (May 15): 421.

————. 1885r. "The Unknowable; or, The Religion of Science." *Knowledge* 9 (November 1): 1–3.

————. 1885s. "The Unknowable; or, The Religion of Science." *Knowledge* 9 (December 1): 37–39.

————. 1886a. "Gossip." *Knowledge* 9 (May 1): 227–30.

————. 1886b. "Gossip." *Knowledge* 9 (August 2): 314–15.

————. 1886c. "Gossip." *Knowledge* 9 (September 1): 339–41.

————. 1886d. "The Dignity of Science." *Knowledge* 9 (January 1): 93–95.

————. 1886e. "Prize-Pig Honours for Science." *Knowledge* 9 (May 1): 215–16.

————. 1887a. "Gossip." *Knowledge* 10 (March 1): 115–17.

————. 1887b. "Gossip." *Knowledge* 10 (July 1): 209–10.

————. 1887c. *Notes on Earthquakes: With Thirteen Miscellaneous Essays.* New York: J Fitzgerald.

[————]. 1888a. "Force and Energy." *Knowledge* 11 (June 1): 171–73.

[————]. 1888b. "Mr. Lockyer on the Earth's Movements." *Knowledge* 11 (August 1): 234–35.

————. 1889a. *The Expanse of Heaven.* New York: D. Appleton.

————. 1889b. *Our Place among Infinities.* New York: D. Appleton.

————. 1895. "Autobiographical Notes." *New Science Review* 1 (April): 393–97.

————. 1902. *The Orbs around Us.* New York and Bombay: Longmans, Green.

————. 1903. *Rough Ways Made Smooth.* London, New York, and Bombay: Longmans, Green.

————. 1970. *Wages and Wants of Science-Workers.* London: Smith, Elder; 1876, repr., London: Frank Cass.

Proctor, Richard A., and the Proprietors of "Knowledge." 1881. "Our Correspondence Columns, To Our Readers." *Knowledge* 1 (December 9): 112.

"Prof. Proctor." 1880. *New York Times,* May 25, 4.

"Publications Received." 1858. *Literary Gazette, and Journal of Belles Lettres, Science, and Art* (April 17): 373.

[Ranyard, Arthur C.]. 1889. "Richard Anthony Proctor." *Royal Astronomical Society Monthly Notices* 49 (February): 164–69.

"The Rev. J. G. Wood." 1890. *Saturday Review* 69: 479.

"Reviews. Popular Zoology." 1866. *Popular Science Review* 5: 213–17.

"Richard A. Proctor Dead." 1888. *New-York Times,* September 13, 1.

Richardson, Ralph. 1871. "Obituary Notice of Dr. Page, formerly President of the Geological Society of Edinburgh." *Transactions of the Edinburgh Geological Society* 3: 220–21.

[Roberts, Mary]. 1831. *The Annals of My Village: Being a Calendar of Nature, for Every Month in the Year.* London: J. Hatchard and Son.

————. 1850. *Voices from the Woodlands, Descriptive of Forest Trees, Ferns, Mosses, and Lichens.* London: Reeve, Benham, and Reeve.

————. 1851. *A Popular History of the Mollusca; Comprising a Familiar Account of Their Classification, Instincts, and Habits, and of the Growth and Distinguishing Characters of Their Shells*. London: Reeve and Benham.

Royal Commission on Scientific Instruction and the Advancement of Science. 1872. London: George Edward Eyre and William Spottiswood, vol. 1.

"Royal Polytechnic." 1856a. *Athenaeum* (December 27): 1612.

"Royal Polytechnic." 1856b. *Illustrated London News* 29 (November 1): 453.

"Royal Polytechnic." 1858. *Illustrated London News* 32 (January 2): 11.

"Royal Polytechnic." 1864. *Illustrated London News* 45 (December 31): 666.

"Royal Polytechnic." 1866. *Illustrated London News* 48 (May 26): 511.

"Royal Polytechnic." 1866. *Times*, December 8, 1.

"Royal Polytechnic." 1867. *Illustrated London News* 50 (January 12): 30.

"Royal Polytechnic Institution." 1854. *Athenaeum* (October 28): 1306.

"Royal Polytechnic Institution." 1855. *Illustrated London News* 26 (May 12): 470.

"Royal Polytechnic Institution." 1857. *Art-Journal* 19: 35.

"Science." 1890. *Athenaeum* 3, 269 (June 21): 803-4.

"Science and the Working Classes." 1870. *Nature* 3 (November 10): 21-22.

"Science. Astronomical Literature. *The Story of the Heavens.*" 1885. *Athenaeum* 3,031 (November 28): 702-3.

"Science By-Ways." 1875. *New York Times*, December 14, 2.

"Science in the Country." 1858. *Saturday Review* 5 (April 17): 393-94.

"Science Lectures for the People." 1871. *Nature* 4 (June 1): 81.

"Sensational Science." 1875. *Saturday Review* 40 (September 11): 321-22.

"Some Astronomical Books." 1886. *Academy* 30 (September 18): 191.

Somerville, Margaret. 1874. *Personal Recollections, from Early Life to Old Age, of Mary Somerville*. Boston: Roberts Brothers.

Somerville, Mary. 1834. *The Connexion of the Physical Sciences*. London: John Murray.

Sorby, Henry Clifton. 1879-80. "The Anniversary Address of the President." *Proceedings of the Geological Society of London*, 33-92.

Spencer, Herbert. 1895-96. "Professional Institutions." *Popular Science Monthly* 47: 34-38, 164-75, 364-74, 433-45, 594-602, 739-48.

Stead, W. T. 1906. "My System." *Cassell's Magazine* (August): 293-97.

"Story of the Heavens." 1886. *British Quarterly Review* 83: 202-3.

"Story of the Heavens." 1886. *Knowledge* 9 (January 1): 97-98.

"Sudden Decease of the Rev. J. G. Wood, F.L.S." 1889. *Light* 9, no. 427 (March 9): 115.

Tait, P. G. 1873. "Letters to the Editor. Tyndall and Forbes." *Nature* 8 (September 11): 381-82.

Thiselton-Dyer, W. T. 1883. "Deductive Biology." *Nature* 27 (April 12): 554-55.

[Thompson, F.]. 1903. "The Sun & etc." *Academy and Literature* 64: 173-74.

Tuckwell, W. 1872. "Science Primers." *Nature* 6: 3-4.

Twining, Elizabeth. 1858. *Short Lectures on Plants for Schools and Adult Classes*. London: David Nutt.

————. 1866. *The Plant World*. London: T. Nelson and Sons.

————. 1868. *Illustrations of the Natural Orders of Plants with Groups and Descriptions*, 2 vols. London: Sampson Low, Son, and Marston.

W. O. "Our Book Shelf." *Nature* 24 (May 12): 27-28.

Wallace, Alfred Russel. 1905. *My Life: A Record of Events and Opinions*, 2 vols. London: Chapman and Hall.

Ward, James. 1899. *Naturalism and Agnosticism*, 2 vols. London: Adam and Charles Black.

Ward, Mary. 1859. *Telescope Teachings*. London: Groombridge and Sons.

————. 1864. *Microscope Teachings: Descriptions of Various Objects of Especial Interest and Beauty Adapted for Microscopic Observation*. London: Groombridge and Sons.

[Ward, Mary, and Lady Jane Mahon]. n.d. *Entomology in Sport, and Entomology in Earnest.* London: Paul Jerrard and Son.

Webb, Beatrice. [1950]. *My Apprenticeship.* London: Longmans, Green.

Webb, Rev. T. W. [1865]. *The Earth a Globe: the Newtonian Astronomy; Its Evidence Explained.* Cheltenham, England: Thomas Hailing.

————. 1866. "The Planet Saturn." *Intellectual Observer* 10 (October): 194-202.

————. 1867. "Gruithuisen's City in the Moon.—Jupiter's Satellites.—Occultations." *Intellectual Observer* 12 (October): 214-23.

————. 1868. *Celestial Objects for Common Telescopes,* 2nd ed. London: Longmans, Green.

————. 1871. "The Planet Jupiter." *Nature* 3 (March 20): 430-31.

————. 1880. "The Planets of the Season: Mars." *Nature* 21 (January 1): 212-13.

————. 1882. "The Great Nebula in Andromeda." *Nature* 25 (February 9): 341-45.

————. [1883]. *Optics without Mathematics.* London: Society for Promoting Christian Knowledge; New York: E. and J. B. Young.

————. 1884. "The Theory of Sunspots." *Nature* 30 (May 15): 59-60.

————. 1885a. "Saturn." *Nature* 31 (March 26): 485-86.

————. 1885b. *The Sun: A Familiar Description of His Phaenomena.* New York: Industrial Publication Company.

"Webb's *Celestial Objects.*" 1882. *Observatory* 5: 11-13.

Wells, H. G. 1894. "Popularising Science." *Nature* 50 (July 26): 300-301.

————. 1934. *Experiment in Autobiography.* Toronto: Macmillan.

————. 1937. "Introduction." In *World Natural History,* by E. G. Boulenger. London: B. T. Batsford, xv-xx.

"What Shall School Girls Read?" 1892. *Review of Reviews* 6: 159.

[Whewell, William]. 1834. "Mrs. Somerville on the Connexion of the Sciences." *Quarterly Review* 51: 54-68.

Whitehead, Alfred. 1889. "The Late Rev. J. G. Wood." *Times,* March 9, 15.

Williams, W. Mattieu. 1881. "'Knowledge' and the Scientific Societies." *Knowledge* 1 (December 16): 143-44.

Wood, Rev. J. G. 1854. "Masonic Symbols: The Hive." *Freemasons' Quarterly Magazine* n.s. 2: 45-50.

————. 1856. "Continental Freemasonry." In *The Universal Mason Library: A Republication, in Thirty Volumes, of All the Standard Publications in Masonry. Designed for the Libraries of Masonic Bodies and Individuals.* Vol. 26. New York: Juo. W. Leonard, 136-40, 218-23.

————. 1857. *Common Objects of the Sea Shore.* London: Routledge.

————. 1858. *Common Objects of the Country.* London: Routledge.

————. [1860]. *Animal Characteristics; or, Sketches and Anecdotes of Animal Life.* Second series. London: Routledge.

————. 1861a. *The Boy's Own Book of Natural History.* London: Routledge, Warne, and Routledge.

————. 1861b. *Common Objects of the Microscope.* London: Routledge, Warne, and Routledge.

————. 1870. *Homes without Hands.* New York: Harper and Brothers.

————. 1872. *Insects at Home: Being a Popular Account of British Insects, Their Structure, Habits, and Transformations.* London: Longmans, Green.

————. 1874. *Man and Beast: Here and Hereafter,* 2 vols. London: Daldy, Isbister.

————. 1877. *Nature's Teachings: Human Invention Anticipated by Nature.* London: Daldy, Isbister.

————. 187? *Common Moths of England.* London: Routledge.

————. [1882]. *Illustrated Natural History for Young People.* New York: Routledge.

————. 1883. *Insects Abroad: Being a Popular Account of Foreign Insects, Their Structure, Habits, and Transformations*. London: Longmans, Green.

————. 1887. "The Dulness of Museums." *Nineteenth Century* 21 (March): 384-96.

————. 1889. *Romance of Animal Life: Short Chapters in Nature History*. News York: Thomas Whittaker.

————. n.d. *Illustrated Natural History*, 2 vols. New York: Home Book Company.

Wood, Rev. Theodore. [1890]. *The Rev. J. G. Wood: His Life and Work*. New York: Cassell Publishing Company.

[Wright, Anne]. 1853. *The Globe Prepared for Man: A Guide to Geology*. London: W. J. Adams.

[————]. n.d. *The Observing Eye; or, Letters to Children on the Three Lowest Divisions of Animal Life*. London: Jarrold and Sons.

Youmans, Edward. 1898. "American Preface to the International Scientific Series." In *The Forms of Water: In Clouds and Rivers, Ice and Glaciers*, by John Tyndall, v-x. New York: D. Appleton.

Young, A. 1881. "The So-Called Elements." *Knowledge* 1 (December 23): 151-52.

Zornlin, Rosina. 1852. *Outlines of Geology for Families and Schools*. London: John W. Parker and Son.

————. 1855. *Physical Geography for Families and Schools*. Boston and Cambridge: James Munroe and Company.

SECONDARY REFERENCES

Alberti, Samuel J. M. M. 2001. "Amateurs and Professionals in One County: Biology and Natural History in Late Victorian Yorkshire." *Journal of the History of Biology* 34: 115-47.

————. 2007. "The Museum Affect: Visiting Collections of Anatomy and Natural History." In *Science in the Marketplace: Nineteenth-Century Sites and Experiences*, edited by Aileen Fyfe and Bernard Lightman, 371-403. Chicago: University of Chicago Press.

Allen, David Elliston. 1976. *The Naturalist in Britain: A Social History*. London: Allen Lane.

————. 2004. "Pratt, Anne." In Lightman, *Dictionary of Nineteenth-Century British Scientists*, 3: 1629-30.

Altick, Richard. 1969. "Nineteenth-Century English Best-Sellers: A Further List." *Studies in Bibliography* 22: 197-206.

————. 1978. *The Shows of London*. Cambridge, MA: Belknap Press of Harvard University Press.

————. 1983. *The English Common Reader: A Social History of the Mass Reading Public, 1800-1900*. Chicago: University of Chicago Press.

————. 1986. "Nineteenth-Century English Best-Sellers." *Studies in Bibliography* 39: 235-41.

Anderson, Katharine. 2005. *Predicting the Weather: Victorians and the Science of Meteorology*. Chicago: University of Chicago Press.

Anderson, Patricia. 1991. *The Printed Image and the Transformation of Popular Culture, 1790-1860*. Oxford: Clarendon.

————, and Jonathan Rose, eds. 1991. *British Literary Publishing Houses, 1820-1880*. Detroit: Gale Research.

Armstrong, H. E. 1973. "Our Need to Honour Huxley's Will (1933)." In *H. E. Armstrong and the Teaching of Science, 1880-1930*, edited by W. H. Brock, 55-73. Cambridge: Cambridge University Press.

Astore, William J. 2001. *Observing God: Thomas Dick, Evangelicalism, and Popular Science in Victorian Britain and America*. Aldershot, England; Burlington, VT; Singapore; Sydney: Ashgate.

Atchison, Heather. 2005. "Grant Allen, Spencer and Darwin." In *Grant Allen: Literature and Cultural Politics at the* Fin de Siècle, edited by William Greenslade and Terence Rodgers, 55-64. Aldershot, England and Burlington, VT: Ashgate.

Bahar, Saba. 2001. "Jane Marcet and the Limits to Public Science." *British Journal for the History of Science* 34: 29-49.

Ball, W. Valentine, ed. 1915. *Reminiscences and Letters of Sir Robert Ball*. London, New York, Toronto, and Melbourne: Cassell and Company.

"Ballin, Ada S. (Mrs)." 1988. *Who Was Who, 1897-1915*. London: Adam and Charles Black, 27.

Barber, Lynn. 1980. *The Heyday of Natural History, 1820-1870*. London: Jonathan Cape.

Barlow, Connie. 1998. "Evolution and the AAAS." *Science and Spirit* 9, no. 1: 12-13.

Barnes, James J., and Patience P. Barnes. 1991. "George Routledge and Sons." In Anderson and Rose, *British Literary Publishing Houses, 1820-1880*, 261-70.

Bartholomew, Michael. 1973. "Lyell and Evolution: An Account of Lyell's Response to the Prospect of an Evolutionary Ancestry for Man." *British Journal for the History of Science* 6: 261-303.

Barton, Ruth. 1981. "Scientific Opposition to Technical Education." In *Scientific and Technical Education in Early Industrial Britain*, edited by Michael D. Stephens and Gordon W. Roderick, 13-27. Nottingham: Department of Adult Education, University of Nottingham.

———. 1990. "'An Influential Set of Chaps': The X-Club and Royal Society Politics, 1864-85." *British Journal for the History of Science* 23: 53-81.

———. 1998a. "'Huxley, Lubbock, and Half a Dozen Others': Professionals and Gentlemen in the Formation of the X Club, 1851-1864." *Isis* 89: 410-44.

———. 1998b. "Just before 'Nature': The Purpose of Science and the Purpose of Popularization in Some English Popular Science Journals of the 1860s." *Annals of Science* 55: 1-33.

———. 2004. "Scientific Authority and Scientific Controversy in *Nature*: North Britain against the X Club." In *Culture and Science in the Nineteenth-Century Media*, edited by Louise Henson, Geoffrey Cantor, Gowan Dawson, Richard Noakes, Sally Shuttleworth, and Jonathan R. Topham, 223-35. Aldershot, England and Burlington, VT: Ashgate.

Baum, Richard. 2004a. "Ball, Robert Stawell." In Lightman, *Dictionary of Nineteenth-Century British Scientists*, 1: 106-7.

———. 2004b. "Webb, Thomas William." In Lightman, *Dictionary of Nineteenth-Century British Scientists*, 4: 2126-29.

Bazerman, Charles. 1988. *Shaping Written Knowledge: The Genre and Activity of the Experimental Article in Science*. Madison: University of Wisconsin Press.

Beaver, Donald de B. 1999. "Writing Natural History for Survival, 1820-1856: The Case of Sarah Bowdich, Later Sarah Lee." *Archives of Natural History* 26: 19-31.

Bell, Edward. 1924. *George Bell: Publisher*. London: Chiswick Press.

Benjamin, Marina. 1991. "Elbow Room: Women Writers on Science, 1790-1840." In *Science and Sensibility: Gender and Scientific Enquiry, 1780-1945*, edited by Marina Benjamin, 27-59. Oxford: Basil Blackwell.

Bernstein, Susan. 2004. "Becker, Lydia Ernestine." In Lightman, *Dictionary of Nineteenth-Century British Scientists*, 1: 163-68.

Blackburn, Helen. 1971. *Women's Suffrage: A Record of the Women's Suffrage Movement in the British Isles with Biographical Sketches of Miss Becker*. London: Williams and Norgate, 1902; repr.: New York: Kraus Reprint Company.

Blinderman, Charles S. 1962. "Semantic Aspects of T. H. Huxley's Literary Style." *Journal of Communication* 12: 171-78.

Block, Edwin, Jr. 1986. "T. H. Huxley's Rhetoric and the Popularization of Victorian Scientific Ideas: 1854-1874." *Victorian Studies* 29 (Spring): 363-86.

Boase, Frederick. 1965a. "Dallas, William Sweetland." In *Modern English Biography, Volume V, D to K. Supplement to Volume II*, 8. London: Frank Cass.

———. 1965b. "Houghton, William." In *Modern English Biography, Volume V, D to K. Supplement to Volume II*, 709. London: Frank Cass.

————. 1965c. "Pepper, John Henry." In *Modern English Biography, Volume VI, L to Z, Supplement to Volume III*, 386-87. London: Frank Cass.

Bonham-Carter, Victor. 1978. *Authors by Profession*, vol. 1. London: The Society of Authors.

Boulger, G. S., and Giles Hudson. 2004. "Johns, Charles Alexander." In *Oxford Dictionary of National Biography*, edited by H. C. G. Matthew and Brian Harrison. New York: Oxford University Press, 30: 223.

Bowler, Peter. 1998. "Conflict Avoidance? Anglican Modernism and Revolution in Interwar Britain." *Endeavour* 22, no. 2: 65-67.

————. 2001. *Reconciling Science and Religion: The Debate in Early-Twentieth-Century Britain*. Chicago: University of Chicago Press.

————. 2005. "From Science to the Popularisation of Science: The Career of J. Arthur Thomson." In *Science and Beliefs: From Natural Philosophy to Natural Sciences, 1700-1900*, edited by David M. Knight and Matthew D. Eddy, 231-48. Aldershot, England and Burlington, VT: Ashgate.

Breuer, Hans-Peter, ed. 1984. *The Note-Books of Samuel Butler: Volume I (1874-1883)*. Lanham, MD: University Press of America.

Brock, Claire. 2006. "The Public Worth of Mary Somerville." *British Journal for the History of Science* 39: 255-72.

Brock, W. H. 1980. "The Development of Commercial Science Journals in Victorian Britain." In *Development of Science Publishing in Europe*, edited by A. J. Meadows., 95-122. Amsterdam and New York: Elsevier Science Publishers.

————. 1996. "VII. *Glaucus*: Kingsley and the Seaside Naturalists." In *Science for All: Studies in the History of Victorian Science and Education*. Aldershot, England: Variorum.

————. 2004. "Pepper, John Henry." In Lightman, *Dictionary of Nineteenth-Century British Scientists*, 3: 1572-73.

Brockman, William S. 1991. "Grant Richards." In Anderson and Rose, *British Literary Publishing Houses, 1820-1880*, 272-79.

Broks, Peter. 1988. "Science and the Popular Press: A Cultural Analysis of British Family Magazines 1890-1914." PhD diss., University of Lancaster.

————. 1990. "Science, the Press and Empire: 'Pearson's' Publications, 1890-1914." In *Imperialism and the Natural World*, edited by John M. MacKenzie, 141-63. Manchester and New York: Manchester University Press.

————. 1993. "Science, Media, and Culture: British Magazines, 1890-1914." *Public Understanding of Science* 2: 123-39.

————. 1996. *Media Science before the Great War*. Basingstoke, England: Macmillan.

————. 2006. *Understanding Popular Science*. Maidenhead, England: Open University Press.

Brooke, John. 1974. "Natural Theology in Britain from Boyle to Paley." In *New Interactions between Theology and Natural Science*, edited by John Brooke, R. Hooykaas, and Clive Lawless. Milton Keynes, England: Open University Press, 8-54.

————. 1996. "Like Minds: The God of Hugh Miller." In *Hugh Miller and the Controversies of Victorian Science*, edited by Michael Shortland, 171-86. Oxford: Clarendon Press.

————, and Geoffrey Cantor. 1998. *Reconstructing Nature: The Engagement of Science and Religion*. Edinburgh: T. and T. Clarke.

Browne, Janet. 2004. "Twining, Elizabeth." In Lightman, *Dictionary of Nineteenth-Century British Scientists*, 4: 2047-48.

Brück, M. T. 2002. *Agnes Mary Clerke and the Rise of Astrophysics*. Cambridge: Cambridge University Press.

Burkhardt, Frederick, and Sydney Smith. 1985. *A Calendar of the Correspondence of Charles Darwin, 1832-1882*. New York and London: Garland.

————, Duncan M. Porter, Joy Harvey, and Jonathan R. Topham, eds. 1997. *The Correspondence of Charles Darwin, Volume 10, 1862*. Cambridge: Cambridge University Press.

————, Duncan M. Porter, Sheila Ann Dean, Jonathan Topham, and Sarah Wilmot, eds. 1999. *The Correspondence of Charles Darwin, Volume 11, 1863*. Cambridge: Cambridge University Press.

————, Duncan M. Porter, Sheila Ann Dean, Paul S. White, Sarah Wilmot, Samantha Evans, and Alison Pearn, eds. 2001. *The Correspondence of Charles Darwin, Volume 12, 1864*. Cambridge: Cambridge University Press.

————, Duncan M. Porter, Sheila Ann Dean, Samantha Evans, Shelley Innes, Alison M. Pearn, Andrew Sclater, Paul White, and Sarah Wilmot, eds. 2002. *The Correspondence of Charles Darwin, Volume 13, 1865*. Cambridge: Cambridge University Press.

————, Duncan M. Porter, Sheila Ann Dean, Samantha Evans, Shelley Innes, Andrew Sclater, Alison Pearn, and Paul White, eds. 2004. *The Correspondence of Charles Darwin, Volume 14, 1866*. Cambridge: Cambridge University Press.

Butterworth, Mark. 2005. "A Lantern Tour of Star-Land: The Astronomer Robert Ball and His Magic Lantern Lectures." In *Realms of Light: Uses and Perceptions of the Magic Lantern From the 17th to the 21st Century*, edited by Richard Crangle, Mervyn Heard, and Ine van Dooren. South Park, Galphay Road, Kirkby Malzeard; Ripon, North Yorkshire: The Magic Lantern Society, 162–73.

Cadbury, Deborah. 2001. *Terrible Lizard: The First Dinosaur Hunters and the Birth of a New Science*. New York: Henry Holt.

Campbell, W. W. 1903. "Reviews. *Problems in Astrophysics*." *Astrophysical Journal* 18: 156–66.

Cane, R. F. 1974–75. "John H. Pepper—Analyst and Rainmaker." *Journal of the Royal Historical Society of Queensland* 9: 116–28.

Cantor, Geoffrey, and Sally Shuttleworth, eds. 2004. *Science Serialized: Representations of the Sciences in Nineteenth-Century Periodicals*. Cambridge, MA: MIT Press.

Cantor, Geoffrey, Gowan Dawson, Graeme Gooday, Richard Noakes, Sally Shuttleworth, and Jonathan R. Topham, eds. 2004. *Science in the Nineteenth-Century Periodical: Reading the Magazine of Nature*. Cambridge: Cambridge University Press.

Carey, John, ed. 1995. *The Faber Book of Science*. London and Boston: Faber and Faber.

Chapman, Allan. 1998. *The Victorian Amateur Astronomer: Independent Astronomical Research in Britain, 1820–1920*. Chichester, NY: John Wiley.

Codell, Julie F. 1991. "T. Fisher Unwin." In Anderson and Rose, *British Literary Publishing Houses, 1820–1880*, 304–11.

Colp, Ralph, Jr. 1992. "'I Will Gladly Do My Best': How Charles Darwin Obtained a Civil List Pension for Alfred Russel Wallace." *Isis* 83: 3–26.

Collini, Stefan. 1991. *Public Moralists: Political Thought and Intellectual Life in Britain 1850–1930*. Oxford: Clarendon Press.

Colloms, Brenda. 1975. *Charles Kingsley: The Lion of Eversley*. London: Constable.

Cooter, Roger. 1984. *The Cultural Meaning of Popular Science: Phrenology and the Organization of Consent in Nineteenth-Century Britain*. Cambridge: Cambridge University Press.

———, and Stephen Pumfrey. 1994. "Separate Spheres and Public Places: Reflections of the History of Science Popularization and Science in Popular Culture." *History of Science* 32: 237–67.

Cosslett, Tess. 2003. "'Animals under Man?' Margaret Gatty's *Parables from Nature*." *Women's Writing* 10, no. 1: 137–52.

Cowie, David. 2000. "The Evolutionist at Large: Grant Allen, Scientific Naturalism, and Victorian Culture." PhD thesis, University of Kent at Canterbury.

Creese, Mary R. S. 1998. *Ladies in the Laboratory?* Lanham, MD and London: Scarecrow Press.

————. 2004a. "Bodington, Alice." In Lightman, *Dictionary of Nineteenth-Century British Scientists*, 1: 229–30.

————. 2004b. "Bowdich Lee, Sarah Eglonton." In Lightman, *Dictionary of Nineteenth-Century British Scientists*, 1: 243-44.

————. 2004c. "Brightwen, Eliza." In Lightman, *Dictionary of Nineteenth-Century British Scientists*, 1: 273-77.

————. 2004d. "Gregg, Mary." In Lightman, *Dictionary of Nineteenth-Century British Scientists*, 2: 838-40.

————. 2004e. "Ward, Mary." In Lightman, *Dictionary of Nineteenth-Century British Scientists*, 4: 2102-3.

————. 2004f. "Zornlin, Rosina Maria." In Lightman, *Dictionary of Nineteenth-Century British Scientists*, 4: 2230-31.

Cribb, Stephen. 2004. "Miller, Hugh." In Lightman, *Dictionary of Nineteenth-Century British Scientists*, 3: 1397-1402.

Cross, Nigel. 1985. *The Common Writer: Life in Nineteenth-Century Grub Street*. Cambridge: Cambridge University Press.

Crowe, Michael J. 1986. *The Extraterrestrial Life Debate, 1750-1900: The Idea of a Plurality of Worlds from Kant to Lowell*. Cambridge: Cambridge University Press.

————. 1989. "Richard Proctor and Nineteenth-Century Astronomy." History of Science Society Meeting, Gainesville, Florida.

Dalziel, George, and Edward Dalziel. 1978. *The Brothers Dalziel: A Record of Work, 1840-1890*. London: B. T. Batsford.

Darwin, Angela, and Adrian Desmond, eds. forthcoming. *The Thomas Henry Huxley Family Correspondence*. Chicago: University of Chicago Press.

Dawkins, Richard. 1998. *Unweaving the Rainbow*. Boston: Houghton Mifflin.

Dawson, Gowan. 2004. "The *Review of Reviews* and the New Journalism in Late-Victorian Britain." In *Science in the Nineteenth-Century Periodical: Reading the Magazine of Nature*, by Geoffrey Cantor, Gowan Dawson et al., 172-95. Cambridge: Cambridge University Press.

————. 2005. "Aestheticism, Immorality, and the Reception of Darwinism in Victorian Britain." In *Unmapped Countries: Biological Visions in Nineteenth-Century Literature and Culture*, edited by Anne-Julia Zwierlein, 43-54. London: Anthem Press.

————. 2007. *Darwin, Literature, and Victorian Respectability*. Cambridge: Cambridge University Press.

Dear, Peter, ed. 1991. *The Literary Structure of Scientific Argument: Historical Studies*. Philadelphia: University of Pennsylvania Press.

Desmond, Adrian. 1992. *The Politics of Evolution: Medicine, Morphology, and Reform in Radical London*. Chicago: University of Chicago Press.

————. 1997. *Huxley: From Devil's Disciple to Evolution's High Priest*. Reading, MA: Addison-Wesley.

————. 2001. "Redefining the X Axis: 'Professionals,' 'Amateurs,' and the Making of Mid-Victorian Biology—A Progress Report." *Journal of the History of Biology* 34: 3-50.

————, and James Moore. 1991. *Darwin*. London: Michael Joseph.

Dick, Steven J., and James E. Strick. 2004. *The Living Universe: NASA and the Development of Astrobiology*. New Brunswick, NJ: Rutgers University Press.

Di Gregorio, Mario A. 1984. *T. H. Huxley's Place in Natural Science*. New Haven, CT: Yale University Press.

Doyle, Sir Arthur Conan. 1974. *The Case-Book of Sherlock Holmes*. London: John Murray and Jonathan Cape.

Drain, Susan. 1994. "Marine Botany in the Nineteenth Century: Margaret Gatty, the Lady Amateurs, and the Professions." *Victorian Studies Association Newsletter* 53 (Spring): 6-11.

Early, Julie English. 1997. "The Spectacle of Science and Self." In *Natural Eloquence: Women Reinscribe Science,* edited by Barbara T. Gates and Ann B. Shteir, 215–36. Madison: University of Wisconsin Press.

Eger, Martin. 1993. "Hermeneutics and the New Epic of Science." In *The Literature of Science: Perspectives on Popular Science Writing,* edited by Murdo William McRae, 186–209. Athens: University of Georgia Press.

Eliot, Simon. 1994. *Some Patterns and Trends in British Publishing, 1800–1919.* London: The Bibliographical Society.

———. 1995. "Some Trends in British Book Production, 1800–1919." In *Literature in the Marketplace: Nineteenth-Century British Publishing and Reading Practices,* edited by John O. Jordan and Robert L. Patten, 19–43. New York: Cambridge University Press.

———. 1997. "'Patterns and Trends' and the 'NSTC': Some Initial Observations, Part One." *Publishing History* 42: 79–104.

———. 1998. "'Patterns and Trends' and the 'NSTC': Some Initial Observations, Part Two." *Publishing History* 43: 71–112.

Elliott, Brent. 2004. "Henslow, George." In Lightman, *Dictionary of Nineteenth-Century British Scientists,* 2: 933–34.

Endersby, Jim. 2004. "Kingsley, Charles." In Lightman, *Dictionary of Nineteenth-Century British Scientists,* 3: 1138–40.

England, Richard. 2003. "Introduction." In *Design after Darwin: 1860–1900,* edited by Richard England, 4 vols., 1: v–xviii. Bristol: Thoemmes.

English, Mary P. 1987. *Mordecai Cubitt Cooke: Victorian Naturalist, Mycologist, Teacher, and Eccentric.* Bristol: Biopress.

Fichman, Martin. 2004. *An Elusive Victorian: The Evolution of Alfred Russel Wallace.* Chicago: University of Chicago Press.

Fleming, Sir Ambrose. [1934]. *Memories of a Scientific Life.* London and Edinburgh: Marshall, Morgan and Scott.

Flint, Kate. 1993. *The Woman Reader, 1837–1914.* Oxford: Clarendon Press.

Foot, Michael. 1995. *H. G.: The History of Mr. Wells.* Washington, DC: Counterpoint.

Forbes, George. 1916. *David Gill: Man and Astronomer.* London: John Murray.

Ford, Katrina. 2004. "Page, David." In Lightman, *Dictionary of Nineteenth-Century British Scientists,* 3: 1521–23.

Forgan, Sophie. 2002. "'A National Treasure House of a Unique Kind' (W. L. Bragg): Some Reflections on Two Hundred Years of Institutional History." In *"The Common Purposes of Life": Science and Society at the Royal Institution of Great Britain,* edited by Frank A. J. L. James, 17–41. Aldershot, England: Ashgate.

Freeman, R. B. 1965. *The Works of Charles Darwin: An Annotated Bibliographical Handlist.* London: Dawsons of Pall Mall.

Fussell, G. E. 1955. "A Great Lady Botanist [Jane Loudon]." *Gardeners' Chronicle* 138: 192.

Fyfe, Aileen. 2002. "Publishing and the Classics: Paley's *Natural Theology* and the Nineteenth-Century Scientific Canon." *Studies in History and Philosophy of Science* 33: 729–51.

———. 2003. "Introduction to *Science for Children*." In *Science for Children,* edited by Aileen Fyfe, 7 vols., 1: xi–xxii. Bristol: Thoemmes Press and Edition Synapse.

———. 2004. *Science and Salvation: Evangelical Popular Science Publishing in Victorian Britain.* Chicago: University of Chicago Press.

———. 2005a. "Conscientious Workmen or Booksellers' Hacks? The Professional Identities of Science Writers in the Mid-Nineteenth Century." *Isis* 96: 192–223.

———. 2005b. "Expertise and Christianity: High Standards *Versus* the Free Market in Popular Publishing." In *Science and Beliefs: From Natural Philosophy to Natural Science, 1700–1900,* edited by David Knight and Matthew D. Eddy, 113-26. Aldershot, England and Burlington, VT: Ashgate.

Garwood, Christine. 2001. "Alfred Russel Wallace and the Flat Earth Controversy." *Endeavour* 25, no. 4: 139-43.

Gates, Barbara T. 1993. "Retelling the Story of Science." *Victorian Literature and Culture* 21: 289-306.

————. 1998. *Kindred Nature: Victorian and Edwardian Women Embrace the Living World*. Chicago: University of Chicago Press.

————. 2004a. "Buckley, Arabella Burton." In Lightman, *Dictionary of Nineteenth-Century British Scientists*, 1: 337-39.

————. 2004b. "Introduction." In *Wild Nature Won by Kindness* and *More about Wild Nature*. Vol. 7 of *Science Writing by Women*, edited by Bernard Lightman, v-x. Bristol: Thoemmes Continuum.

————, and Ann B. Shteir. 1997. "Introduction: Charting the Tradition." In *Natural Eloquence: Women Reinscribe Science*, edited by Barbara T. Gates and Ann B. Shteir, 3-24. Madison: University of Wisconsin Press.

Geduld, Harry M. 1987. "Introduction." In *The Definitive Time Machine: A Critical Edition of H. G. Wells's Scientific Romance*, edited by Harry M. Geduld, 1-27. Bloomington: Indiana University Press.

Gershenowitz, Harry. 1979. "George Henslow: True Darwinist." *India Journal of History of Science* 14: 25-30.

Gilbert, R. A. 2004a. "Proctor, Richard Anthony." In Lightman, *Dictionary of Nineteenth-Century British Scientists*, 3: 1641-43.

————. 2004b. "Wood, John George (1827-89)." In Lightman, *Dictionary of Nineteenth-Century British Scientists*, 4: 2193-96.

Gloag, John. 1970. *Mr. Loudon's England: The Life and Work of John Claudius Loudon, and His Influence on Architecture and Furniture Design*. Newcastle upon Tyne, England: Oriel Press.

Golden, Catherine. 1991. "Frederick Warne and Company." In Anderson and Rose, *British Literary Publishing Houses, 1820-1880*, 327-37.

Golinski, Jan. 1992. *Science as Public Culture: Chemistry and Enlightenment in Britain, 1760-1820*. Cambridge: Cambridge University Press.

Gooday, Graeme. 1990. "Precision Measurement and the Genesis of Physics Teaching Laboratories in Victorian Britain." *British Journal for the History of Science* 23: 25-51.

————. 1991. "'Nature' in the Laboratory: Domestication and Discipline with the Microscope in Victorian Life and Science." *British Journal for the History of Science* 24: 307-41.

Gould, Frederick James. 1929. *The Pioneers of Johnson's Court: A History of the Rationalist Press Association from 1899 Onwards*. London: Watts.

Gould, Paula. 1997. "Women and the Culture of University Physics in Late Nineteenth-Century Cambridge." *British Journal for the History of Science* 30: 127-49.

————. 1998. "Femininity and Physical Science in Britain, 1870-1914." PhD thesis, University of Cambridge.

Gould, Stephen Jay. 1991. *Bully for Brontosaurus*. New York: W. W. Norton.

Graham, Margaret. 1977. "A Life among the Flowers of Kent." *Country Life* 161: 1500-1501.

Grassie, William. 1998. "Science as Epic?" *Science and Spirit* 9, no. 1: 8-9, 11.

Gross, John. 1969. *The Rise and Fall of the Man of Letters: Aspects of English Literary Life since 1800*. London: Weidenfeld and Nicolson.

Hammond, John. 2001. *A Preface to H. G. Wells*. Harlow, England: Pearson Education.

"Harrison, William Jerome." 1988. *Who Was Who, 1897-1915*. London: Adam and Charles Black, 232.

Harry, Owen G. 1984a. "The Hon. Mrs. Ward (1827–1869) Artist, Naturalist, Astronomer and Ireland's First Lady of the Microscope." *The Irish Naturalists' Journal* 21: 193–200.

———. 1984b. "The Hon. Mrs. Ward and 'A Windfall for the Microscope,' of 1856 and 1864." *Annals of Science* 41: 471–82.

———. 1995. "Mary Ward at Castle Ward: The Making of a Naturalist." *Apollo* 141, no. 398 (April): 37–41.

Haught, John F. 1984. *The Cosmic Adventure: Science, Religion, and the Quest for Purpose*. Ramsey, NY: Paulist Press.

"Henslow, Rev. George." 1929. *Who Was Who, 1916–1928*. London: Adam and Charles Black, 488.

Henson, Louise, Geoffrey Cantor, Gowan Dawson, Richard Noakes, Sally Shuttleworth, and Jonathan R. Topham, eds. 2004. *Culture and Science in the Nineteenth-Century Media*. Aldershot, England and Burlington, VT: Ashgate.

Hepworth, T. C. 1978. *The Book of the Lantern*. New York: Arno Press.

Hilgartner, Stephen. 1990. "The Dominant View of Popularization: Conceptual Problems, Politics Uses." *Social Studies of Science* 20: 519–39.

Hinton, D. A. 1979. "Popular Science in England, 1830–1870." PhD thesis, University of Bath.

Hodgson, Amanda. 1999. "Defining the Species: Apes, Savages, and Humans in Scientific and Literary Writing of the 1860s." *Journal of Victorian Culture* 4: 228–51.

Holmes, Marion. 1912/13. *Lydia Becker: A Cameo Life-Sketch*, 2nd ed. London: Women's Freedom League.

Houghton, Walter E., ed. 1987. *The Wellesley Index to Victorian Periodicals, 1824–1900*, vol. 4. Toronto and Buffalo: University of Toronto Press.

Howard, Daniel F., ed. 1962. *The Correspondence of Samuel Butler with His Sister May*. Berkeley: University of California Press.

Howard, Jill. 2004. "'Physics and Fashion': John Tyndall and His Audiences in Mid-Victorian Britain." *Studies in History and Philosophy of Science* 35: 729–58.

Howsam, Leslie. 1991. "Kegan Paul, Trench, Trübner and Company Limited." In Anderson and Rose, *British Literary Publishing Houses, 1820–1880*, 238–45.

———. 2000. "An Experiment with Science for the Nineteenth-Century Book Trade: The International Scientific Series." *British Journal for the History of Science* 33: 187–207.

———. 2004. "Paul, (Charles) Kegan." *Oxford Dictionary of National Biography*. Oxford: Oxford University Press, 43: 136–38.

Hunt, Bruce. 1997. "Doing Science in a Global Empire: Cable Telegraphy and Electrical Physics in Victorian Britain." In *Victorian Science in Context*, edited by Bernard Lightman, 312–33. Chicago: University of Chicago Press.

Huyssen, Andreas. 1986. "Mass Culture as Woman: Modernism's Other." In *Studies in Entertainment: Critical Approaches to Mass Culture*, edited by Tania Modleski, 188–207. Bloomington: Indiana University Press.

"Jago, William." 1941. *Who Was Who, 1929–1940*. London: Adam and Charles Black, 702.

James, Frank A. J. L. 2002a. "Introduction." In *"The Common Purposes of Life": Science and Society at the Royal Institution of Great Britain*, edited by Frank A. J. L. James, 1–16. Aldershot, England: Ashgate.

———. 2002b. "Running the Royal Institution: Faraday as an Administrator." In *"The Common Purposes of Life": Science and Society at the Royal Institution of Great Britain*, edited by Frank A. J. L. James, 119–46. Aldershot, England: Ashgate.

Jensen, J. Vernon. 1970. "The X Club: Fraternity of Victorian Scientists." *British Journal for the History of Science* 5: 63–72.

———. 1991. *Thomas Henry Huxley: Communicating for Science*. Newark: University of Delaware Press; London and Toronto: Associated University Presses.

Jones, Greta. 2001. "Scientists against Home Rule." In *Defenders of the Union*, edited by D. George Bayes and Alan O'Day, 188-208. London and New York: Routledge.

Jones, Henry Festing. 1919. *Samuel Butler, Author of Erewhon (1835-1902): A Memoir*, 2 vols. London: Macmillan.

J. S. C. 1917. "Edwards, Amelia Ann Blanford." In *Dictionary of National Biography, Supplement*, edited by Sir Leslie Stephen and Sir Sidney Lee, 22: 601-3. London: Oxford University Press.

Katz, Wendy R. 1993. *The Emblems of Margaret Gatty: The Study of Allegory in Nineteenth-Century Children's Literature*. New York: AMS Press.

Kavanagh, Ita. 1997. "Mistress of the Microscope." In *Stars, Shells and Bluebells: Women Scientists and Pioneers*, edited by Jane Hanly, Patricia Deevy et al., 56-65. Dublin: Women in Technology and Science.

Keynes, Geoffrey, and Brian Hill, eds. 1935. *Letters between Samuel Butler and Miss E. M. A. Savage, 1871-1885*. London: Jonathan Cape.

King, Amy M. 2005. "Reorienting the Scientific Frontier: Victorian Tide Pools and Literary Realism." *Victorian Studies* 47: 153-63.

Kinraide, Rebecca. 2004. "Pinnock, William." In Lightman, *Dictionary of Nineteenth-Century British Scientists*, 3: 1602-3.

Kitteringham, Guy Stuart. 1981. "Studies in the Popularisation of Science in England, 1800-30." PhD thesis, University of Kent at Canterbury.

Knight, David. 1996. "Getting Science Across." *British Journal of the History of Science* 29: 129-38.

Kohn, David. 1997. "The Aesthetic Construction of Darwin's Theory." In *The Elusive Synthesis: Aesthetics and Science*, edited by Alfred I. Tauber, 13-48. Dordrecht, Boston, and London: Kluwer Academic Publishers.

Layton, David. 1973. *Science for the People*. London: Allen and Unwin.

———. 1977. "Founding Fathers of Science Education (4). A Victorian Showman of Science." *New Scientist* 75 (September 1): 538-39.

Lazell, David. 1972. "John G. Wood and His Wonderful Crystal Palace Lectures." *The Flower Patch* 3: 2-4, 24.

Le-May Sheffield, Suzanne. 2001. *Revealing New Worlds: Three Victorian Women Naturalists*. London and New York: Routledge.

Levin, Gerald. 1984. "Grant Allen's Scientific and Aesthetic Philosophy." *Victorians Institute Journal* 12: 77-89.

Lightman, Bernard. 1987. *The Origins of Agnosticism: Victorian Unbelief and the Limits of Knowledge*. Baltimore: Johns Hopkins University Press.

———. 1989. "Ideology, Evolution, and Late-Victorian Agnostic Popularizers." In *History, Humanity and Evolution*, edited by James R. Moore, 285-309. Cambridge: Cambridge University Press.

———. 1996. "Astronomy for the People: R. A. Proctor and the Popularization of the Victorian Universe." In *Facets of Faith and Science*, edited by Jitse M. van der Meer, 3: 31-34. Lanham, MD: The Pascal Centre for Advanced Studies in Faith and Science and University Press of America.

———. 1997a. "Constructing Victorian Heavens: Agnes Clerke and the 'New Astronomy.'" In *Natural Eloquence: Women Reinscribe Science*, edited by Barbara T. Gates and Ann B. Shteir, 61-75. Madison: University of Wisconsin Press.

———. 1997b. "'Fighting Even with Death': Balfour, Scientific Naturalism, and Thomas Henry Huxley's Final Battle." In *Thomas Henry Huxley's Place in Science and Letters: Centenary Essays*, edited by Alan Barr, 323-50. Athens: University of Georgia Press.

———. 1997c. "Introduction." In *Victorian Science in Context*, edited by Bernard Lightman, 1-12. Chicago: University of Chicago Press.

———. 1997d. "'The Voices of Nature': Popularizing Victorian Science." In *Victorian Science in Context*, edited by Bernard Lightman, 187-211. Chicago: University of Chicago Press.

———. 1999. "The Story of Nature: Victorian Popularizers and Scientific Narrative." *Victorian Review* 25, no. 2: 1-29.

———. 2000. "The Visual Theology of Victorian Popularizers of Science: From Reverent Eye to Chemical Retina." *Isis* 91: 651-80.

———. 2004a. "Brewer, Ebenezer Cobham." In Lightman, *Dictionary of Nineteenth-Century British Scientists*, 1: 266-67.

———. 2004b. "Clodd, Edward." In Lightman, *Dictionary of Nineteenth-Century British Scientists*, 1: 450-52.

———. 2004c. "Giberne, Agnes." In Lightman, *Dictionary of Nineteenth-Century British Scientists*, 2: 777-78.

———. 2004d. "Hutchinson, Henry Neville." In Lightman, *Dictionary of Nineteenth-Century British Scientists*, 2: 1040-41.

———. 2004e. "Johns, Charles Alexander." In Lightman, *Dictionary of Nineteenth-Century British Scientists*, 2: 1082-83.

———. 2004f. "Scientists as Materialists in the Periodical Press: Tyndall's Belfast Address." In *Science Serialized: Representations of the Sciences in Nineteenth-Century Periodicals*, edited by Geoffrey Cantor and Sally Shuttleworth, 199-237. Cambridge, MA: MIT Press.

———. 2006. "Depicting Nature, Defining Roles: The Gender Politics of Victorian Illustration." In *Figuring It Out: Science, Gender, and Visual Culture*, edited by Ann Shteir and Bernard Lightman. Hanover, NH and London: University Press of New England, 214-39.

———. 2007. "Lecturing in the Spatial Economy of Science." In *Science in the Marketplace: Nineteenth-Century Sites and Experiences*, edited by Aileen Fyfe and Bernard Lightman, 97-132. Chicago: University of Chicago Press.

———, ed. 2004. *Dictionary of Nineteenth-Century British Scientists*, 4 vols. Bristol: Thoemmes Continuum.

———, and Aileen Fyfe, eds. 2007. *Science in the Marketplace: Nineteenth-Century Sites and Experiences*. Chicago: University of Chicago Press.

Lindsay, Gillian. 1996. "Mary Roberts: A Neglected Naturalist." *Antiquarian Book Monthly* 23 (February): 20-22.

Linfield, Christine. 2004. "Loudon, Jane." In Lightman, *Dictionary of Nineteenth-Century British Scientists*, 3: 1263-66.

Livingstone, David N. 2003. *Putting Science in Its Place: Geographies of Scientific Knowledge*. Chicago: University of Chicago Press.

Lockyer, Mary T., and Winifred Lockyer. 1928. *Life and Work of Sir Norman Lockyer*. London: Macmillan.

Mackenzie, Norman, and Jeanne Mackenzie. 1973. *H. G. Wells: A Biography*. New York: Simon and Schuster.

MacLeod, Roy. 1968. "A Note on *Nature* and the Social Significance of Scientific Publishing, 1850-1914." *Victorian Periodicals Newsletter* 3 (November): 16-17.

———. 1969. "Science in Grub Street," "Macmillan and the Scientists," "Seeds of Competition," "Macmillan and the Young Guard," "The New Journal," "The First Issue," "Securing the Foundations," "Private Army of Contributors," "Faithful Mirror to a Profession," "Lockyer: Editor, Civil Servant and Man of Science," "Into the Twentieth Century." *Nature* 224 (November 1): 423-61.

———. 1970. "The X-Club: A Social Network of Science in Late-Victorian England." *Notes and Records of the Royal Society of London* 24: 305-22.

———. 1980. "Evolutionism, Internationalism, and Commercial Enterprise in Science: The International Scientific Series, 1871-1910." In *Development of Science Publishing in Europe*, edited by A. J. Meadows, 63-93. Amsterdam, New York, and Oxford: Elsevier Science Publishers.

———. 1996. *Public Science and Public Policy in Victorian England*. Aldershot, England: Variorum.

Mahalingam, Subbiah. 1987. "Popularizing Science: Thomas Henry Huxley's Style." PhD thesis, Oklahoma State University.

Marchant, James. 1916. *Alfred Russel Wallace: Letters and Reminiscences*. New York and London: Harper and Brothers.

Martin, Robert Bernard. 1959. *Dust of Combat: A Life of Charles Kingsley*. London: Faber and Faber.

Maxwell, Christabel. 1949. *Mrs. Gatty and Mrs. Ewing*. London: Constable Publishers.

McCabe, Joseph. 1932. *Edward Clodd: A Memoir*. London: John Lane/The Bodley Head.

McKenna-Lawlor, Susan. 1998. "The Hon. Mrs. Mary Ward (1827–1869): Astronomer, Microscopist, Artist, and Entrepreneur." In her *Whatever Shines Should Be Observed*, 29–55. Blackrock: Samton.

McLaughlin-Jenkins, Erin. 2001a. "Common Knowledge: Science and the Late Victorian Working Class Press." *History of Science* 34: 445–65.

———. 2001b. "Common Knowledge: The Victorian Working Class and the Low Road to Science, 1870–1900." PhD thesis, York University.

———. 2005. "Henry George and the Dragon: T. H. Huxley's Response to Progress and Poverty." In *Henry George's Legacy in Economic Thought.*, edited by John Laurent, 31–50. Cheltenham, UK; Northampton, MA: Edward Elgar.

McLean, Ruari, ed. 1967. *The Reminiscences of Edmund Evans*. Oxford: Clarendon Press.

McMillan, N. D., and J. Meehan. 1980. *John Tyndall: 'X'emplar of Scientific and Technological Education*. Dublin: ETA Publications.

Meadows, A. J. 1972. *Science and Controversy: A Biography of Sir Norman Lockyer*. Cambridge, MA: MIT Press.

———. 1980. "Access to the Results of Scientific Research: Developments in Victorian Britain." In *Development of Science Publishing in Europe*, edited by A. J. Meadows., 43–62. Amsterdam and New York: Elsevier Science Publishers.

Melchiori, Barbara Arnett. 2000. *Grant Allen: The Downward Path Which Leads to Fiction*. Rome: Bulzoni.

Mermin, Dorothy. 1993. *Godiva's Ride: Women of Letters in England, 1830–1880*. Bloomington: Indiana University Press.

Moore, James. 1979. *The Post-Darwinian Controversies*. Cambridge: Cambridge University Press.

———. 1986. "Geologists and Interpreters of Genesis in the Nineteenth Century." In *God and Nature: Historical Essays on the Encounter between Christianity and Science*, edited by David C. Lindberg and Ronald L. Numbers, 322–50. Berkeley: University of California Press.

———. 1987. "The Erotics of Evolution: Constance Naden and Hylo-Idealism." In *One Culture: Essays in Science and Literature*, edited by George Levine, 225–57. Madison: University of Wisconsin Press.

Morton, Peter. 2004. "Allen, Charles Grant Blairfindie." In Lightman, *Dictionary of Nineteenth-Century British Scientists*, 1: 36–39.

———. 2005. *"The Busiest Man in England": Grant Allen and the Writing Trade, 1875–1900*. New York and Houndmills, England: Palgrave Macmillan.

Morus, Iwan. 1998. *Frankenstein's Children: Electricity, Exhibition, and Experiment in Early-Nineteenth-Century London*. Princeton, NJ: Princeton University Press.

Mumby, F. A. 1934. *The House of Routledge, 1834–1934*. London: Routledge.

Mumm, S. D. 1990. "Writing for Their Lives: Women Applicants to the Royal Literary Fund, 1840–1880." *Publishing History* 27: 27–47.

———. 1991. "Jarrold and Sons." In Anderson and Rose, *British Literary Publishing Houses, 1820–1880*, 159–61.

Myers, Greg. 1989. "Science for Women and Children: The Dialogue of Popular Science in the Nineteenth Century." In *Nature Transfigured: Science and Literature, 1700–1900*, edited by John Christie and Sally Shuttleworth, 171–200. Manchester and New York: Manchester University Press.

———. 1990. *Writing Biology: Texts in the Social Construction of Scientific Knowledge*. Madison: University of Wisconsin Press.

"Naden, Constance Caroline Woodhill." 1917. *Dictionary of National Biography*, edited by Sir Leslie Stephen and Sir Sidney Lee, 14: 18–19. London: Oxford University Press.

Neeley, Kathryn A. 2001. *Mary Somerville: Science, Illumination, and the Female Mind*. Cambridge: Cambridge University Press.

Nelkin, Dorothy. 1987. *Selling Science: How the Press Covers Science and Technology*. New York: W. H. Freeman.

Noakes, Richard. 2004. "The *Boy's Own Paper* and Late-Victorian Juvenile Magazines." In *Science in the Nineteenth-Century Periodical: Reading the Magazine of Nature*, edited by Geoffrey Cantor, Gowan Dawson, Graeme Gooday, Richard Noakes, Sally Shuttleworth, and Jonathan R. Topham, 151–71. Cambridge: Cambridge University Press.

North, J. D. 1975. "Proctor, Richard Anthony." In *Dictionary of Scientific Biography*, edited by Charles Coulston Gillispie, 11: 162–63. New York: Charles Scribner's Sons.

Nottingham, Chris. 2005. "Grant Allen and the New Politics." In *Grant Allen: Literature and Cultural Politics at the Fin de Siècle*, edited by William Greenslade and Terence Rodgers, 95–110. Aldershot, England and Burlington, VT: Ashgate.

O'Connor, Ralph. 2002. "Hugh Miller and Geological Spectacle." In *Celebrating the Life and Times of Hugh Miller: Scotland in the Early 19th Century*, edited by Lester Borley, 237–58. Cromarty, UK: Cromarty Arts Trust.

———. 2003. "Thomas Hawkins and Geological Spectacle." *Proceedings of the Geologists' Association* 114: 227–41.

Ogilvie, Marilyn Bailey. 2000. "Obligatory Amateurs: Annie Maunder (1868–1947) and British Women Astronomers at the Dawn of Professional Astronomy." *British Journal for the History of Science* 33: 67–84.

Oldroyd, David R. 1996. "The Geologist from Cromarty." In *Hugh Miller and the Controversies of Victorian Science*, edited by Michael Shortland, 76–121. Oxford: Clarendon Press.

Opitz, Don. 2004a. "Aristocrats and Professionals: Country-House Science in Late-Victorian England." PhD thesis, University of Minnesota.

———. 2004b. "Introduction to *The Conchologist's Companion*." In *Science Writing by Women*, edited by Bernard Lightman, 7 vols., 3: v–ix. Bristol: Thoemmes Continuum.

———. 2004c. "Roberts, Mary." In Lightman, *Dictionary of Nineteenth-Century British Scientists*, 4: 1699–1700.

Osterbrock, Donald E. 1984. *James E. Keeler: Pioneer American Astrophysicist and the Early Development of American Astrophysics*. Cambridge: Cambridge University Press.

Otter, Sandra Den. 1996. *British Idealism and Social Explanation: A Study in Late Victorian Thought*. Oxford: Clarendon Press.

Owen, Alex. 2004. *The Place of Enchantment: British Occultism and the Culture of the Modern*. Chicago: University of Chicago Press.

Pandora, Katherine. 2001. "Knowledge Held in Common: Tales of Luther Burbank and Science in the American Vernacular." *Isis* 92, no. 3: 484–516.

Paradis, James G. 1981. "Darwin and Landscape." In *Victorian Science and Victorian Values: Literary Perspectives*, edited by James Paradis and Thomas Postlewait, 85–100. New York: New York Academy of Sciences.

———. 2004. "The Butler-Darwin Biographical Controversy in the Victorian Periodical Press." In *Science Serialized: Representations of the Sciences in Nineteenth-Century Periodicals*, edited by Geoffrey Cantor and Sally Shuttleworth, 307–29. Cambridge, MA: MIT Press.

Parker, Joan. 2001. "Lydia Becker's 'School for Science': A Challenge to Domesticity." *Women's History Review* 10, no. 4: 629–50.

Patterson, Elizabeth C. 1969. "Mary Somerville." *British Journal for the History of Science* 4: 311–39.

Paylor, Suzanne. 2004. "Scientific Authority and the Democratic Intellect: Popular Encounters with 'Darwinian' Ideas in Later Nineteenth-Century England with Special Reference to the Secularist Movement." PhD thesis, University of York.

Perkins, Maureen. 1996. *Visions of the Future: Almanacs, Time, and Cultural Change, 1775–1870.* Oxford: Clarendon Press.

Purdy, Richard Little, and Michael Millgate, eds. 1980. *The Collected Letters of Thomas Hardy*, vol. 2, *1893–1901.* Oxford: Clarendon Press.

Raby, Peter. 1997. *Bright Paradise: Victorian Scientific Travellers.* Princeton, NJ: Princeton University Press.

Rauch, Alan. 1994. "Editor's Introduction." In *The Mummy! A Tale of the Twenty-Second Century*, by Jane (Webb) Loudon, edited by Alan Rauch. Ann Arbor: University of Michigan Press, vol. 1.

———. 1997. "Parables and Parodies: Margaret Gatty's Audiences in the Parables from Nature." *Children's Literature* 25: 137–52.

———. 2001. *Useful Knowledge: The Victorians, Morality, and the March of Intellect.* Durham, NC: Duke University Press.

———. 2004. "Gatty, Margaret." In Lightman, *Dictionary of Nineteenth-Century British Scientists*, 2: 761–64.

Richards, Evelleen. 1983. "Darwin and the Descent of Woman." In *The Wider Domain of Evolutionary Thought*, edited by David Oldroyd and Ian Langham, 57–111. London: D. Reidel.

———. 1989. "Huxley and Woman's Place in Science." In *History, Humanity, and Evolution*, edited by James Moore, 253–84. Cambridge: Cambridge University Press.

———. 1997. "Redrawing the Boundaries: Darwinian Science and Victorian Women Intellectuals." In *Victorian Science in Context*, edited by Bernard Lightman, 119–42. Chicago: University of Chicago Press.

Richards, Grant. 1932. *Memories of Misspent Youth, 1872–1896.* London: William Heinemann.

Richmond, Marsha. 1997. "'A Lab of One's Own': The Balfour Biological Laboratory for Women at Cambridge University, 1884–1914." *Isis* 88: 422–55.

Riley, David. 2003. "The Manchester Science Lectures for the People, c. 1866–1879." *Bulletin of the John Rylands University Library of Manchester* 85, no. 1 (Spring): 127–45.

Ring, Katy. 1988. "The Popularisation of Elementary Science through Popular Science Books, c. 1870–c. 1939." PhD diss., University of Kent at Canterbury.

Robinson, Mark G. 2004. "Webb, Thomas William." In *Oxford Dictionary of National Biography*, edited by H. C. G. Matthew and Brian Harrison, 57: 858–59. New York: Oxford University Press.

———. 2006. "Man of the Cloth." In *The Stargazer of Hardwicke: The Life and Work of Thomas William Webb*, edited by Janet Robinson and Mark Robinson, 59–73. Leominster, England: Gracewing.

Roos, David Alan. 1981. "The 'Aims and Intentions' of *Nature*." In *Victorian Science and Victorian Values: Literary Perspectives*, edited by James Paradis and Thomas Postlewait, 159–80. New York: New York Academy of Sciences.

Roscoe, Sir Henry Enfield. 1906. *The Life and Experiences of Sir Henry Enfield Roscoe.* London and New York: Macmillan.

Ross, Robert H., ed. 1973. *In Memoriam.* Alfred, Lord Tennyson. New York: W. W. Norton.

Rozendal, Phyllis. 1988. "Grant Allen." In *British Mystery Writers, 1860–1919*, edited by Bernard Benstock and Thomas F. Staley, 3–13. Detroit: Gale Research.

Rupke, Nicolaas. 1994. *Richard Owen: Victorian Naturalist.* New Haven, CT: Yale University Press.

Russett, Cynthia Eagle. 1989. *Sexual Science: The Victorian Construction of Womanhood.* Cambridge, MA: Harvard University Press.

Sarum, Lewis O. 1999. "The Proctor Interlude in St. Joseph and in America: Astronomy, Romance, and Tragedy." *American Studies International* 37 (February): 34–54.

Savage, Gail L. 1988. "Gentleman." In *Victorian Britain: An Encyclopedia,* edited by Sally Mitchell, 325–26. New York and London: Garland.

Schaffer, Simon. 1988. "Astronomers Mark Time: Discipline and the Personal Equation." *Science in Context* 2, no. 1: 115–45.

———. 1998a. "The Leviathan of Parsonstown: Literary Technology and Scientific Representation." In *Inscribing Science: Scientific Text and the Materiality of Communication,* edited by Timothy Lenoir, 182–222. Stanford, CA: Stanford University Press.

———. 1998b. "On Astronomical Drawing." In *Picturing Science, Producing Art,* edited by Caroline A. Jones and Peter Galison, 441–74. New York: Routledge.

———. 1998c. "Physics Laboratories and the Victorian Country House." In *Making Space for Science: Territorial Themes in the Shaping of Knowledge,* edited by Crosbie Smith and Jon Agar, 149–80. New York: St. Martin's Press; Basingstoke, England: Macmillan.

Secord, Anne. 1994. "Science in the Pub: Artisan Botanists in Early Nineteenth-Century Lancashire." *History of Science* 32: 269–315.

———. 2002. "Botany on a Plate: Pleasure and the Power of Pictures in Promoting Early Nineteenth-Century Scientific Knowledge." *Isis* 93 (March): 28–57.

Secord, James A. 2000. *Victorian Sensation: The Extraordinary Publication, Deception, and Secret Authorship of* Vestiges of the Natural History of Creation. Chicago: University of Chicago Press.

———. 2002. "Quick and Magical Shaper of Science." *Science* 297 (September 6): 1648–49.

———. 2003a. "Introduction." John Henry Pepper. *The Boy's Playbook of Science.* Bristol: Thoemmes Press and Edition Synapse, v–x.

———. 2003b. "Introduction." [Samuel Clark]. *Peter Parley's Wonders of the Earth, Sea, and Sky.* Bristol: Thoemmes Press and Edition Synapse, v–x.

———. 2004a. "General Introduction." In *Collected Works of Mary Somerville,* edited by James A. Secord, 9 vols., 1: xv–xxxix. Bristol: Thoemmes Continuum.

———. 2004b. "Introduction." In *Collected Works of Mary Somerville,* edited by James A. Secord, 9 vols., 2: ix–xvi. Bristol: Thoemmes Continuum.

———. 2004c. "Introduction." In *Collected Works of Mary Somerville,* edited by James A Secord, 9 vols., 4: ix–xv. Bristol: Thoemmes Continuum.

———. 2004d. "Introduction." In *Science Writing by Women,* edited by Bernard Lightman, 7 vols., 4: v–xi. Bristol: Thoemmes Continuum.

———. 2004e. "Monsters at the Crystal Palace." In *Models: The Third Dimension of Science,* edited by Soraya de Chadarevian and Nick Hopwood, 138–69. Stanford, CA: Stanford University Press.

———. 2004f. "Page, David." In *Oxford Dictionary of National Biography,* edited by H. C. G. Matthew and Brian Harrison, 42: 322–23. Oxford: Oxford University Press.

Sheets-Pyenson, Susan. 1976. "Low Scientific Culture in London and Paris, 1820–1875." PhD thesis, University of Pennsylvania.

———. 1985. "Popular Science Periodicals in Paris and London: The Emergence of a Low Scientific Culture, 1820–1875." *Annals of Science* 42: 549–72.

Shiach, Morag. 1989. *Discourse on Popular Culture: Class, Gender, and History in Cultural Analysis, 1730 to the Present.* Cambridge: Polity Press.

Shteir, Ann B. 1996. *Cultivating Women, Cultivating Science: Flora's Daughters and Botany in England, 1760 to 1860.* Baltimore: Johns Hopkins University Press.

———. 1997a. "Elegant Recreations? Configuring Science Writing for Women." In *Victorian Science in Context,* edited by Bernard Lightman, 236–55. Chicago: University of Chicago Press.

————. 1997b. "Gender and 'Modern' Botany in Victorian England." *Osiris* 12: 29–38.

————. 2003. "Finding Phebe: A Literary History of Women's Science Writing." In *Women and Literary History: "For There She Was,"* edited by Katherine Binhammer and Jeanne Wood, 152–66. Newark: University of Delaware Press.

————. 2004a. "Lankester, Phebe." In Lightman, *Dictionary of Nineteenth-Century British Scientists*, 3: 1181–83.

————. 2004b. "'Let Us Examine the Flower': Botany in Women's Magazines, 1800–1830." In *Science Serialized: Representations of the Sciences in Nineteenth-Century Periodicals*, edited by Geoffrey Cantor and Sally Shuttleworth, 17–36. Cambridge, MA: MIT Press.

Simpson, J. A., and E. S. C. Weiner. 1989. *Oxford English Dictionary*, 2nd ed., vol. 12. Oxford: Clarendon Press.

"Slingo, Sir William." 1941. *Who Was Who, 1929–1940*. London: Adam and Charles Black, 1246–47.

Sloan, Phillip R. 2001. "'The Sense of Sublimity': Darwin on Nature and Divinity." *Osiris* 16: 251–69.

Small, Helen. 1996. "A Pulse of 124: Charles Dickens and a Pathology of the Mid-Victorian Reading Public." In *Practise and Representation of Reading in England*, edited by James Raven, Helen Small, and Naomi Tadmor, 263–90. New York: Cambridge University Press.

Smith, Crosbie. 1998. *The Science of Energy: A Cultural History of Energy Physics in Victorian Britain*. Chicago: University of Chicago Press.

Smith, David C. 1986. *H. G. Wells: Desperately Mortal*. New Haven, CT: Yale University Press.

Smith, Hobart M., Georgene E. Fawcett, James D. Fawcett, and Rogella B. Smith. 1970. "J. G. Wood and the Mexican Axolotl." *Journal of the Society for Bibliography of Natural History* 5, no. 5: 362–65.

Smith, Jonathan. 2001. "Philip Gosse and the Varieties of Natural Theology." In *Reinventing Christianity*, edited by Linda Woodhead, 251–62. Aldershot, England: Ashgate.

————. 2004. "Grant Allen, Physiological Aesthetics, and the Dissemination of Darwin's Botany." In *Science Serialized: Representations of the Sciences in Nineteenth-Century Periodicals*, edited by Geoffrey Cantor and Sally Shuttleworth, 285–305. Cambridge, MA: MIT Press.

————. 2006. *Charles Darwin and Victorian Visual Culture*. Cambridge: Cambridge University Press.

Smith, K. G. V. 2004. "Morris, Francis Orpen." In Lightman, *Dictionary of Nineteenth-Century British Scientists*, 1427–29.

Smith, Robert W. 2004. "The Story of the Heavens and Great Astronomers: Robert S. Ball and Popular Astronomy." British-North American Joint Meeting of the British Society for the History of Science, Canadian Society for the History and Philosophy of Science, and the History of Science Society, Kings College, Halifax, August.

St. Clair, William. 2004. *The Reading Nation in the Romantic Period*. Cambridge: Cambridge University Press.

Stoddart, D. R. 1975. "'That Victorian Science': Huxley's *Physiography* and Its Impact on Geography." *Transactions of the Institute of British Geographers* 66: 17–40.

Swimme, Brian, and Thomas Berry. 1992. *The Universe Story: From the Primordial Flaring Forth to the Ecozoic Era*. New York: HarperCollins.

Taylor, Geoffrey. 1951. *Some Nineteenth Century Gardeners*. London: Skeffington.

Thompson, Silvanus P. 1901. *Michael Faraday: His Life and Work*. London, Paris, New York, and Melbourne: Cassell and Company.

Thwaite, Ann. 1984. *Edmund Gosse: A Literary Landscape, 1849–1928*. London: Secker and Warburg.

————. 2002. *Glimpses of the Wonderful: The Life of Philip Henry Gosse, 1810–1888*. London: Faber and Faber.

Topham, Jonathan. 1998. "Beyond the 'Common Context': The Production and Reading of the Bridgewater Treatises." *Isis* 89: 233–62.

———. 2000. "Scientific Publishing and the Reading of Science in Nineteenth-Century Britain: A Historiographical Survey and Guide to Sources." *Studies in History and Philosophy of Science* 31: 559–612.

———. 2003. "Science, Natural Theology, and the Practice of Christian Piety in Early Nineteenth-Century Religious Magazines." In *Science Serialized: Representations of the Sciences in Nineteenth-Century Periodicals*, edited by Geoffrey Cantor and Sally Shuttleworth, 37–66. Cambridge, MA: MIT Press.

———. 2004. "A View from the Industrial Age." *Isis* 95: 431–42.

———. 2007. "Publishing 'Popular Science' in Early Nineteenth-Century Britain." In *Science in the Marketplace*, edited by Bernard Lightman and Aileen Fyfe, 135–68. Chicago: University of Chicago Press.

Tucker, Jennifer. 2005. *Nature Exposed: Photography as Eyewitness in Victorian Science*. Baltimore: Johns Hopkins University Press.

Turner, Frank. 1974. *Between Science and Religion: The Reaction to Scientific Naturalism in Late Victorian England*. New Haven, CT: Yale University Press.

———. 1993. *Contesting Cultural Authority: Essays in Victorian Intellectual Life*. Cambridge: Cambridge University Press.

Upton, John. [1910]. *Three Great Naturalists*. London: Pilgrim Press.

VanArsdel, Rosemary T. 1991. "Macmillan and Company." In Anderson and Rose, *British Literary Publishing Houses, 1820–1880*, 178–95.

Venn, J. A. 1947. "Hutchinson, Henry Neville." *Alumni Cantabrigienses, Part II from 1752 to 1900*, vol. 3. Cambridge: Cambridge University Press, 503.

Vincent, David. 1989. *Literacy and Popular Culture: England 1750–1914*. Cambridge: Cambridge University Press.

Walters, S. M., and E. A. Stow. 2001. *Darwin's Mentor: John Stevens Henslow, 1796–1861*. Cambridge: Cambridge University Press.

Wayman, Patrick A. 1986. "A Visit to Canada in 1884 by Sir Robert Ball." *Irish Astronomical Journal* 17: 185–96.

———. 1987. *Dunsink Observatory, 1785–1985: A Bicentennial History*. Dublin: Dublin Institute for Advanced Studies and Royal Dublin Society.

Weeden, Brenda. 2001. "The Rise and Fall of the Royal Polytechnic Institution." Royal Institution Centre for the History of Science and Technology, February 27.

Weedon, Alexis. 2003. *Victorian Publishing: The Economics of Book Production for a Mass Market, 1836–1916*. Aldershot, England and Burlington VT: Ashgate.

Wells, Ellen B. 1990. "J. G. Wood: Popular Natural Historian." *Book and Magazine Collector* 79 (October): 56–64.

White, Paul. 2003. *Thomas Huxley: Making the "Man of Science."* Cambridge: Cambridge University Press.

———. 2004. "Huxley, Thomas Henry." In Lightman, *Dictionary of Nineteenth-Century British Scientists*, 2: 1044–48.

Whitley, Richard. 1985. "Knowledge Producers and Knowledge Acquirers: Popularisation as a Relation between Scientific Fields and Their Publics." In *Expository Science: Forms and Functions of Popularisation*, edited by Terry Shinn and Richard Whitley, 3–28. Dordrecht: D. Reidel.

Williams, Raymond. 1984. *Keywords: A Vocabulary of Culture and Society*. London: Fontana Paperbacks.

"Williams, William Mattieu." 1921. In *Dictionary of National Biography*, edited by Sir Leslie Stephen and Sir Sidney Lee, 21: 468–69. London: Oxford University Press.

"Wilson, Andrew." 1988. *Who Was Who, 1897–1915*. London: Adam and Charles Black, 568.

Wilson, Edward O. 1978. *On Human Nature*. Cambridge, MA: Harvard University Press.

Winter, Alison. 1998. *Mesmerized: Powers of Mind in Victorian Britain*. Chicago: University of Chicago Press.

Yeo, Richard. 1984. "Science and Intellectual Authority in Mid-Nineteenth-Century Britain: Robert Chambers and *Vestiges of the Natural History of Creation*." *Victorian Studies* 28 (Autumn): 5–31.

Young, Robert. 1985. *Darwin's Metaphor: Nature's Place in Victorian Culture*. Cambridge: Cambridge University Press.

Index

Acland, Henry, 171
Adelaide Gallery, 37, 197, 199-200
Agassiz, Louis, 231
Aikin, John, 20
Airy, George, 319-20, 324
Allen, Grant: and amorality of evolution,
  275-76; and Butler, 291-93; career of,
  267-68; and Clodd, 256, 262, 266; on
  communicating with readers, 219-20; and
  contributions to *Knowledge*, 334; and
  criticisms of his work, 281-83; and
  Darwin, 279-81; and Dawkins, 499; and
  editors, 286; and the evolutionary epic,
  271-76; and Gould, 499; and Huxley,
  279, 284, 377n72; and independence from
  scientific naturalists, 222; life of, 266; and
  natural theology, 277-78; periodical
  articles of, 298; and Proctor, 303; on the
  progress of science, 4; and the publishing
  system, 284-86; and readers, 286; and
  socialism, 288-89; and specialization,
  283-84; and Spencer, 223, 268-72, 283,
  286-89, 294; and Stead, 286; and the
  sublime, 278-79; and the transformation
  of publishing, 285, 426; and Tyndall, 279;
  and writing fiction, 285
Allen, W. H., 300
almanacs, 132
Andromeda Nebula, 432-33, 480, 482
Anglican Church and natural theology, 80

Anglican clergy: and female popularizers,
  163-65; and science, 39-94, 418, 487
Anglicanism and female popularizers, 147
animal biography, 442-46
Anthropological Society, 99
Appleton and Company, 379
aquariums, 1
aristocracy, 8
Armstrong, H. E., 209
astrophotography. *See* camera in astronomy
Athenaeum, 465
audience, reading: defining of, 123-28, 227;
  mass, 99; size of, 18
authority: scientific, 155, 237, 496-97;
  submission to, 144
authorship, 15-16

Bagehot, Walter, 381
Bain, Alexander, 381
Balfour, Arthur J., 8, 428
Ball, Robert: books of, 406-17; career and life
  of, 397-406; and Clerke, 475-76; earnings
  of from lecturing and writing, 406; and
  evolutionary epic, 420; and Giberne, 430;
  and the heroes of astronomy, 415-17; and
  Hutchinson, 450-51, 455; and *Knowledge*,
  335; and lecturing, 401-8; and loss of
  faith, 400-1; and nebular hypothesis,
  400, 415; and practitioner-popularizer
  tradition, 356, 397, 418-19, 495;

535